INTRODUCTORY ALGEBRA

THIRD EDITION

**ARNOLD R. STEFFENSEN
L. MURPHY JOHNSON**
Northern Arizona University

Scott, Foresman and Company
Glenview, Illinois   London, England

To Barbara, Barbara, Becky, Cindy, and Pam

Study Aids for the Student

A *Student Solutions Manual,* by Joseph Mutter, provides complete, worked-out solutions to each practice exercise in the text, to every third exercise in set A, and to selected exercises in set C.

Tutorial software provides additional instruction and practice on selected topics that most often create difficulty. Apple and IBM versions will be available.

Cover photo © Joe Sohm/The Image Works.

Copyright © 1987, 1984, 1981 Scott, Foresman and Company.
All Rights Reserved.
Printed in the United States of America.
1 2 3 4 5 6 —WEB— 91 90 89 88 87 86
ISBN 0-673-18352-1

PREFACE

SCOPE

Introductory Algebra, Third Edition, is designed for students who have not been exposed to algebra, or who require a review before taking further mathematics or computer science courses. The primary objective is to gain familiarity with mathematical symbols and operations in order to formulate and solve first- and second-degree equations. The text begins with a thorough review of fundamental concepts that leads into a discussion of rational and real numbers. Next, linear equations and inequalities are presented with emphasis on practical applications, followed by an extensive discussion of graphing and systems. Polynomials are then introduced with careful attention given to factoring. Finally, the text concludes with chapters on rational expressions, radicals, and quadratic equations.

This book is the second in a series of worktexts including:

(1) *Fundamentals of Mathematics,*   (4) *College Algebra,*
(2) *Introductory Algebra,*   (5) *Algebra and Trigonometry,*
(3) *Intermediate Algebra,*   (6) *Trigonometry with Analytic Geometry.*

FEATURES

STUDENT GUIDEPOSTS ▷▷ specify important terms, definitions, and rules for each section. These are designed to help students locate important concepts while they study each section initially, as well as to provide direction during their review of the material for examinations.

Carefully selected **EXAMPLES** have detailed step-by-step solutions that include helpful annotations.

A **Practice Exercise** parallels each example to reinforce the concepts presented. Space for working each practice exercise is provided, with answers immediately following.

▶▶**CAUTION**◀◀ calls the attention of students to common mistakes, special problems, and exceptions to rules.

SECOND COLOR is used pedagogically to point out important steps and to emphasize methods and terminology.

Two parallel **EXERCISE SETS (A and B)** and a collection of extension exercises **(C)** follow each section and offer a wealth of practice for students and flexibility for instructors. There are over 5800 exercises including about 800 word problems.

EXERCISES A contain space for working the problems and provide answers immediately following. These exercises can be used for in-class practice or assigned as homework. Colored numbers identify selected exercises in set A that have worked-out solutions at the back of the text. Many of these solutions are to exercises with which students often experience difficulty.

EXERCISES B match set A, problem for problem, but are presented without work space or answers. These exercises can be used as homework problems, examples in class, or additional practice.

EXERCISES C are designed to give students an extra challenge. These are problems that extend the concepts of the section, are more demanding than exercises in set A or B, or require extensive use of a calculator. Answers or hints are given for selected exercises.

For Review identifies exercises that reinforce material from earlier sections of the chapter. These exercises are located at the end of most A and B exercise sets.

Each chapter concludes with a **CHAPTER SUMMARY** and a **CHAPTER TEST** The summaries contain key words and phrases for review, key concepts, and review exercises. Answers to the chapter tests are at the back of the book.

FINAL REVIEW EXERCISES keyed to each chapter, with answers supplied, are located at the back of the text.

It is assumed that most students have a hand-held calculator. However, with disagreement among mathematics educators as to the desirability of calculators in a course at this level, no specific emphasis has been made relative to their use. When working exercises that involve dividing decimals or approximating values of irrational numbers, a calculator can be a valuable tool. Exercises of this type have not been labeled as *calculator exercises,* since it is important for students to learn without specific direction when a calculator should or should not be used.

INSTRUCTIONAL FLEXIBILITY

Introductory Algebra, Third Edition, offers proven flexibility for a variety of teaching situations such as individualized instruction, lab instruction, lecture classes, or a combination of methods.

Material in each section of the book is presented in a well-paced, easy-to-follow sequence. Students in an individualized or lab instruction setting, aided by the student guideposts, can work through a section completely by reading the explanation, following the detailed steps in the examples, working the practice exercises, and then doing the exercises in set A.

The book can also serve as the basis for, or as a supplement to, classroom lectures. The straightforward presentation of material, numerous examples, practice exercises, and three sets of exercises offer the traditional lecture class an alternate approach within the convenient workbook format.

SUPPLEMENTS

▶ The *Instructor's Guide* contains: a Placement Test, Student Information Sheet, course outlines, and suggestions for using the text; eight different but equivalent tests for each chapter and two final examinations with all answers sup-

plied; answers to Exercises B and C; chapter-by-chapter instructional guidelines and other suggestions for content implementation that an instructor, tutor, or teaching assistant might find helpful.
- The *Student Solutions Manual* contains complete, worked-out solutions to each practice exercise in the text, to every third exercise in set A, and to selected exercises in set C.
- The *Computer-Assisted Testing System* (CATS) can be used with Apple and IBM computers to construct and print tests.
- *Tutorial software* provides students with additional instruction and practice on selected topics that most often create difficulty. Apple and IBM versions will be available.

ACKNOWLEDGEMENTS

We extend our gratitude to the instructors and students who used the first and second editions of this book and offered suggestions for improvement. Special thanks go to the instructors in the Individualized Instruction Program at Northern Arizona University. In particular, the assistance given over the years by Janet Condon, James Kirk, and Michael Ratliff is most appreciated. Also, we sincerely appreciate the support and encouragement of the Northern Arizona University administration, especially President Eugene M. Hughes. It is a pleasure and privilege to serve on the faculty of a university that recognizes quality teaching as its primary role.

We also express our thanks to the following reviewers for their countless beneficial suggestions at various stages of the book's development:

Thomas L. Alexander
University of Alabama, Birmingham

Judy Bergman
Louisiana State University, Baton Rouge

Douglas J. Campbell
West Valley College

Francis Champoux
Northern Essex Community College

Donley A. Chandler
El Paso Community College

James Hamilton
Vernon Regional Junior College

Donald G. Johnson
New Mexico State University

Gail Jones
Delgado Community College

Caroline Lovelace
New Mexico Military Institute

Kenneth Ohm
Sheridan College

Charles Reinauer
San Jacinto College

Chantal Shafroth
North Carolina Central University

David Steinfort
Grand Rapids Junior College

We extend special appreciation to Joseph Mutter for reviewing the entire manuscript and offering countless suggestions for improvement, for writing the supplements, and for checking the problems, and to Cindy Bown Reese for aiding in the reviewing process and typing the original manuscript. Thanks also go to Martha Hardin for typing this edition, and to Diana Denlinger Vanlandingham for typing the *Instructor's Guide*.

To everyone at Scott, Foresman and Company we are greatly indebted. Special thanks go to Jack Pritchard, Steve Quigley, Terry McGinnis, Ann Buesing, and Jacqueline Kolb.

Finally, we are deeply indebted to our families and in particular to our wives, Barbara and Barbara, who have given us unceasing support, time, and encouragement over the years as we worked on this series.

Arnold R. Steffensen
L. Murphy Johnson

CONTENTS

To the Student ix

1 Fundamentals

1.1 Fractions 1
1.2 Operations on Fractions 8
1.3 Decimals 19
1.4 Converting Fractions, Decimals, and Percents 26
1.5 Variables, Exponents, and Order of Operations 34
　　Chapter 1　Summary and Review Exercises 43
　　Chapter 1　Test 47

2 Rational and Real Numbers

2.1 Rational Numbers and the Number Line 50
2.2 Addition and Subtraction of Rational Numbers 57
2.3 Multiplication and Division of Rational Numbers 65
2.4 The Distributive Laws and Simplifying Expressions 75
2.5 Integer Exponents 82
2.6 Scientific Notation 92
2.7 Irrational and Real Numbers 96
　　Chapter 2　Summary and Review Exercises 100
　　Chapter 2　Test 105

3 Linear Equations and Inequalities

3.1 Linear Equations and the Addition-Subtraction Rule 108
3.2 The Multiplication-Division Rule 112
3.3 Solving Equations by Combining Rules 116
3.4 The Language of Problem Solving 123
3.5 Applied Problems: Number, Age, and Percent 127
3.6 Applied Problems: Geometry and Motion 136
　　Supplementary Applied Problems 144
3.7 Solving Linear Inequalities 147
　　Chapter 3　Summary and Review Exercises 154
　　Chapter 3　Test 159

4 Graphing

- 4.1 Graphing on the Number Line 162
- 4.2 The Cartesian Coordinate System 167
- 4.3 Graphing Linear Equations 174
- 4.4 Slope of a Line and Equations of a Line 186
- 4.5 Graphing Linear Inequalities 196
- Chapter 4 Summary and Review Exercises 205
- Chapter 4 Test 210

5 Systems of Linear Equations

- 5.1 Parallel, Coinciding, and Intersecting Lines 212
- 5.2 Solving Systems of Equations by Graphing 217
- 5.3 Solving Systems of Equations by Substitution 223
- 5.4 Solving Systems of Equations by Addition-Subtraction 229
- 5.5 Applications Using Systems 235
- Chapter 5 Summary and Review Exercises 243
- Chapter 5 Test 246

6 Polynomials

- 6.1 Basic Concepts of Polynomials 248
- 6.2 Addition and Subtraction of Polynomials 254
- 6.3 Multiplication of Polynomials 262
- 6.4 Special Products 269
- 6.5 Division of Polynomials 275
- Chapter 6 Summary and Review Exercises 282
- Chapter 6 Test 286

7 Factoring Polynomials

- 7.1 Common Factors and Grouping 289
- 7.2 Factoring Trinomials of the Form $x^2 + bx + c$ 296
- 7.3 Factoring Trinomials of the Form $ax^2 + bx + c$ 303
- 7.4 Factoring Perfect Square Trinomials and Difference of Squares 311
- 7.5 Factoring to Solve Equations 316
- 7.6 Applications of Factoring 321
- Chapter 7 Summary and Review Exercises 326
- Chapter 7 Test 330

8 Rational Expressions

- 8.1 Basic Concepts of Algebraic Fractions 332
- 8.2 Multiplication and Division of Fractions 339
- 8.3 Addition and Subtraction of Like Fractions 347
- 8.4 Addition and Subtraction of Unlike Fractions 352
- 8.5 Solving Fractional Equations 360
- 8.6 Ratio, Proportion, and Variation 372
- 8.7 Simplifying Complex Fractions 380
- Chapter 8 Summary and Review Exercises 386
- Chapter 8 Test 391

9 Radicals

9.1 Roots and Radicals 393
9.2 Simplifying Radicals 396
9.3 Multiplication and Division of Radicals 403
9.4 Addition and Subtraction of Radicals 409
9.5 Summary of Techniques and Rationalizing Denominators 416
9.6 Solving Radical Equations 425
9.7 Applications Using Radicals 430
Chapter 9 Summary and Review Exercises 437
Chapter 9 Test 443

10 Quadratic Equations

10.1 Factoring and Taking Roots 446
10.2 Solving by Completing the Square 454
10.3 Solving by the Quadratic Formula 459
10.4 Solving Fractional and Radical Equations 466
10.5 Applications of Quadratic Equations 472
10.6 Solving Formulas 481
10.7 Graphing Quadratic Equations 486
Chapter 10 Summary and Review Exercises 494
Chapter 10 Test 499

Final Review Exercises 502

Table of Squares and Square Roots 511

Appendix Unit Conversion and the Metric System 512

Answers to Chapter Tests 516

Solutions to Selected Exercises 518

Index 547

TO THE STUDENT

During the past few years we have taught *Introductory Algebra* to more than 1200 students and have heard the following comments more than just a few times. "I've always been afraid of math and have avoided it as much as possible. Now my major requires algebra and I'm petrified." "I don't like math, but it's required to graduate." "I can't do word problems!" If you have ever made a similar statement, now is the time to think positively, stop making negative comments, and start down the path toward success in mathematics. Don't worry about this course as a whole. The material in the text is presented in a way that enables you to take one small step at a time. Here are some general and specific guidelines that are necessary and helpful.

GENERAL GUIDELINES

1. Mastering algebra requires motivation and dedication. A successful athlete does not become a champion without commitment to his or her goal. The same is true for the successful student of algebra. Be prepared to work hard and spend time studying.

2. Algebra is not learned simply by watching, listening, or reading; *it is learned by doing*. Use your pencil and practice. When your thoughts are organized and written in a neat and orderly fashion, you have taken a giant step toward success. Be complete and write out all details. The following are samples of two students' work on a word problem. Can you tell which one was the most successful in the course?

Student A

$$\text{Let } x = \text{marked price}$$
$$0.05x = \text{tax on marked price}$$
$$x + 0.05x = \text{price plus tax}$$
$$x + 0.05x = 37.59$$
$$x(1 + 0.05) = 37.59$$
$$x(1.05) = 37.59$$
$$x = \frac{37.59}{1.05}$$
$$= \$35.80$$

Student F

$$\begin{array}{r} 37.59 \\ \underline{0.05} \\ 187.95 \end{array}$$

$$x + 0.05 = 37.59$$
$$x = 37.59 - 0.05$$

$$x = \text{tax}$$
$$37.59x = x$$

3. If the use of a calculator is permitted in your course, become familiar with its features by consulting your owner's manual. When working problems that involve computing with decimals or approximating irrational numbers, a calculator can be a time-saving device. On the other hand, you should not be so dependent on a calculator that you use it to perform simple calculations that might be done mentally. For example, it would be ridiculous to use a calculator to solve an equation such as $2x = 8$, while it would be helpful when solving $1.12x = 7343.84$. It is important that you learn when to use and when not to use your calculator.

SPECIFIC GUIDELINES

1. As you begin to study each section, glance through the material and obtain a preview of what is coming.

2. Return to the beginning of the section and start reading slowly. The STUDENT GUIDEPOSTS specify the important terms, definitions, and rules that must be understood. They also will help you locate important concepts as you progress or as you review.

3. Read through each EXAMPLE and make sure that you understand each step. The side comments in color will help you if something is not quite clear.

4. After reading each example, work the parallel PRACTICE EXERCISE. This will reinforce what you have just read and start the process of practice.

5. Periodically you will encounter a CAUTION. These warn you of common mistakes and special problems that should be avoided.

6. After you have completed the material in the body of the section, you must test your mastery of the skills and practice, practice, practice! Begin with the exercises in set A. Answers to all of the problems are placed at the end of the set for easy reference. Some of the problems, identified by colored exercise numbers, have complete solutions at the back of the book. After trying these problems, refer to the step-by-step solutions if you have difficulty. You may want to practice more by doing the exercises in set B or to challenge yourself by trying the exercises in set C. (Note: Solutions to every third problem in set A and to selected problems in set C are available in the *Student Solutions Manual*.)

7. After you have completed all of the sections in the chapter, read the CHAPTER SUMMARY that contains key words and phrases for review and key concepts. The review exercises provide additional practice before you take the CHAPTER TEST. Answers to the tests are at the back of the book.

8. To aid you in studying for your final examination, we have concluded the book with a comprehensive set of FINAL REVIEW EXERCISES

If you follow these steps and work closely with your instructor, you will greatly improve your chance of success in algebra.

Best of luck, and remember that you can do it!

1 FUNDAMENTALS

1.1 FRACTIONS

STUDENT GUIDEPOSTS

1. Set and element
2. Number and numeral
3. Natural or counting numbers
4. Whole numbers
5. Numerator and denominator of a fraction
6. Proper and improper fractions
7. Equivalent fractions
8. Fundamental principle of fractions
9. Building up and reducing fractions
10. Reducing to lowest terms
11. Prime number
12. Factoring into primes
13. Method for reducing fractions to lowest terms
14. Canceling factors

1 Two terms that are used repeatedly in algebra are *set* and *element*. A **set** is a collection of objects called **elements**. The elements of a set are often listed **2** within braces { }. Generally we are concerned with sets of numbers. A **number** is essentially a human invention, an idea or concept formed when questions about "how many" or "in what order" are asked. Names or symbols for numbers are called **numerals**. The most basic numbers are the **natural** or **counting numbers**.

3 $\{1, 2, 3, 4, \ldots\}$ Natural (counting) numbers

We use the three-dot notation to indicate that a sequence or pattern continues in the same manner. When zero is included with the natural numbers, we obtain the set of **whole numbers**.

4 $\{0, 1, 2, 3, 4, \ldots\}$ Whole numbers

2 FUNDAMENTALS

Numbers found by writing the quotient of a whole number and a natural number, such as

$$\frac{1}{2}, \frac{3}{4}, \frac{7}{3}, \frac{5}{5}, \frac{5}{1}, \frac{0}{1}, \frac{15}{1}, \frac{61}{21},$$

are called nonnegative **fractions**. The **numerator** of a fraction is the number above the fraction bar, and the **denominator** of a fraction is the number below the fraction bar. For example, 3 is the numerator and 4 is the denominator of the fraction $\frac{3}{4}$.

When the numerator of a fraction is a smaller number than the denominator, the fraction is called a **proper fraction**. Otherwise the fraction would be called an **improper fraction**. For example, $\frac{1}{2}$ and $\frac{3}{4}$ are proper fractions, while $\frac{7}{5}$, $\frac{5}{5}$, $\frac{5}{1}$, and $\frac{61}{27}$ are improper fractions.

Every whole number can be named as a fraction by letting the whole number be the numerator and 1 be the denominator. For example,

$$5 = \frac{5}{1}, \quad 2 = \frac{2}{1}, \quad 0 = \frac{0}{1}, \quad \text{and} \quad 15 = \frac{15}{1}.$$

Five could also be written as $\frac{10}{2}, \frac{15}{3}, \frac{20}{4}$, etc. Similarly, the fraction $\frac{1}{4}$ could be expressed as $\frac{1}{4}, \frac{2}{8}, \frac{3}{12}, \frac{4}{16}, \frac{5}{20}$, and so on. We can always find another name for a fraction by multiplying both the numerator and the denominator by the same nonzero number. Fractions that are names for the same number, or which have the same value, are called **equivalent fractions**. The following is an important property of fractions.

Fundamental principle of fractions

If both the numerator and denominator of a fraction are multiplied or divided by the same nonzero number, the resulting fraction is equivalent to the original fraction.

We **build up** fractions when we multiply both the numerator and denominator by the same counting number other than 1, and **reduce** fractions when we divide both by the same counting number other than 1. For example,

$$\frac{3}{4} = \frac{3 \cdot 2}{4 \cdot 2} = \frac{6}{8}, \quad \frac{3}{4} = \frac{3 \cdot 5}{4 \cdot 5} = \frac{15}{20}, \quad \frac{3}{4} = \frac{3 \cdot 10}{4 \cdot 10} = \frac{30}{40}$$

Thus, $\frac{3}{4}$ has been built up to the equivalent fractions $\frac{6}{8}, \frac{15}{20}, \frac{30}{40}$. Also,

$$\frac{12}{18} = \frac{12 \div 2}{18 \div 2} = \frac{6}{9}, \quad \frac{12}{18} = \frac{12 \div 3}{18 \div 3} = \frac{4}{6}, \quad \frac{12}{18} = \frac{12 \div 6}{18 \div 6} = \frac{2}{3}.$$

Thus, $\frac{12}{18}$ has been reduced to the equivalent fractions $\frac{6}{9}, \frac{4}{6}$, and $\frac{2}{3}$.

A fraction is **reduced to lowest terms** when 1 is the only natural number that divides both numerator and denominator.

We reduced $\frac{12}{18}$ to $\frac{6}{9}, \frac{4}{6}$, and $\frac{2}{3}$ above. Notice that $\frac{6}{9}$ and $\frac{4}{6}$ can each be reduced further (by dividing numerator and denominator by 3 and by 2, respectively), while $\frac{2}{3}$ cannot. Thus, when we reduced $\frac{12}{18}$ to $\frac{2}{3}$ we reduced it to lowest terms. The process of reducing fractions to lowest terms is easier if we understand the concept of a *prime* number. A **prime number** is a counting number greater than 1 whose only divisors are 1 and itself. For example, 2 is prime since 1 and 2 are the only divisors of 2, 5 is prime since 1 and 5 are the only divisors of 5, and 6 is not prime since 2 (also 3) is a divisor of 6.

The first few primes are

2, 3, 5, 7, 11, 13, 17, 19, 23, ...

> Every counting number greater than 1 is either prime or can be expressed as a product of primes.

12 The primes in the product are called **prime factors,** and the process of expressing a number as a product of primes is called **factoring into primes.**

EXAMPLE 1

Express as a product of primes.

(A) $15 = 3 \cdot 5$ Why not $15 = 15 \cdot 1$?

(B) $28 = 2 \cdot 2 \cdot 7$ Why not $28 = 4 \cdot 7$?

(C) $41 = 41$ 41 is a prime ◂◂

Practice Exercise 1

Express as a product of primes.

(A) 21

(B) 36

(C) 53

Answers: (A) $3 \cdot 7$ (B) $2 \cdot 2 \cdot 3 \cdot 3$
(C) 53 is a prime

13 **To reduce a fraction to lowest terms**

1. Factor the numerator and denominator into products of primes.
2. Divide all common factors from both the numerator and denominator.
3. Multiply remaining factors in the numerator and in the denominator.

14 Dividing common factors from both the numerator and denominator of a fraction is sometimes called **canceling factors.** This is shown by crossing out the common factors. For example,

$$\frac{9 \cdot \cancel{5}}{\cancel{5}} = \frac{9}{1} \quad \text{and} \quad \frac{\cancel{2} \cdot \cancel{7}}{\cancel{2} \cdot 3 \cdot \cancel{7}} = \frac{1}{3}.$$

Notice that when *all* of the factors are canceled out of a numerator or a denominator, we make the numerator or denominator 1, *not* 0.

EXAMPLE 2

Reduce the fractions to lowest terms.

(A) $\dfrac{6}{15} = \dfrac{2 \cdot \cancel{3}}{\cancel{3} \cdot 5} = \dfrac{2}{5}$ Cancel common factor 3 from both numerator and denominator

(B) $\dfrac{12}{42} = \dfrac{\cancel{2} \cdot 2 \cdot \cancel{3}}{\cancel{2} \cdot \cancel{3} \cdot 7} = \dfrac{2}{7}$ Cancel common factors 2 and 3 from both numerator and denominator

Practice Exercise 2

Reduce the fractions to lowest terms.

(A) $\dfrac{2}{4}$

(B) $\dfrac{70}{25}$

4 FUNDAMENTALS

(C) $\dfrac{14}{15} = \dfrac{2 \cdot 7}{3 \cdot 5}$ There are no common factors, so $\frac{14}{15}$ is already reduced to lowest terms. ◀◀

(C) $\dfrac{15}{77}$

Answers: (A) $\dfrac{1}{2}$ (B) $\dfrac{14}{5}$
(C) Already in lowest terms.

▶▶ **CAUTION** ◀◀ Cancel only factors (numbers that are multiplied) and never cross out parts of sums. For example,

$$\dfrac{2 \cdot 7}{2} = \dfrac{\cancel{2} \cdot 7}{\cancel{2}} = 7, \text{ but } \dfrac{2 + 7}{2} \text{ is } \textit{not} \text{ equal to } \dfrac{\cancel{2} + 7}{\cancel{2}}.$$ ◀◀

With practice, we often shorten the process of reducing fractions by dividing common factors without first factoring into primes. For example, if we note that 6 is a common factor of both 12 and 42, we could write

$$\dfrac{12}{42} = \dfrac{2 \cdot \cancel{6}}{7 \cdot \cancel{6}} = \dfrac{2}{7}.$$

Fractions are used to describe parts of quantities in many applied situations. For example, if a baseball player got three hits in five times at bat, we would say that he had a hit $\frac{3}{5}$ (three-fifths) of the time. Thus, fractions are used to indicate what part one number is of another.

EXAMPLE 3

A group of 100 students was made up of 60 girls and 40 boys. Also, exactly 75 of the students were taking math, and exactly 35 students were taking history.

(A) What fractional part of the total number of students were girls?
There were 60 girls out of the 100 students for a fractional part of $\frac{60}{100}$, or $\frac{3}{5}$, reduced to lowest terms.

(B) What fractional part of the total amount of students were taking math?
There were 75 math students out of the 100 total for a fractional part of $\frac{75}{100}$, or $\frac{3}{4}$, reduced to lowest terms.

(C) What fractional part of the total number of students were *not* taking history?
Since there were 35 students taking history, there were 65 who were not. Thus, $\frac{65}{100}$, or $\frac{13}{20}$ of the students were not taking history. ◀◀

Practice Exercise 3

In a sample of 200 pens, 120 are red and 80 are green. Also, exactly 65 of the pens have a fine point, and exactly 185 have retractable points.

(A) What fractional part of the total number of pens is green?

(B) What fractional part of the total has fine points?

(C) What fractional part of the total does not have retractable points?

Answers: (A) $\dfrac{2}{5}$ (B) $\dfrac{13}{40}$ (C) $\dfrac{3}{40}$

1.1 EXERCISES A

Answer true *or* false *in Exercises 1–10.*

1. A collection of objects is also called a set of objects.

2. The name or symbol for a number is a numeral.

3. $\{1, 2, 3, \ldots\}$ is the set of whole numbers.

4. In the fraction $\frac{5}{7}$, the number 7 is called the numerator.

5. If the numerator of a fraction is a smaller number than the denominator, the fraction is called proper.

6. Two fractions with the same value are called equivalent fractions.

7. We build up a fraction when we divide both numerator and denominator by a natural number other than 1.

8. A counting number other than 1 that is divisible only by itself and 1 is called prime.

9. The process of canceling factors from a numerator and denominator is really the process of adding common factors.

10. An idea or concept that is called to mind when considering questions such as "how many" or "in what order" is called a number.

11. Consider the fractions $\frac{1}{5}, \frac{3}{8}, \frac{9}{9}, \frac{10}{2}, \frac{6}{6}, \frac{13}{15}$, and $\frac{8}{1}$.
    (A) Which of these are proper fractions?   (B) Which of these are improper fractions?

Write three fractions that are equivalent to the following.

12. $\frac{2}{5}$
13. $\frac{5}{2}$
14. 4

15. 0
16. $\frac{105}{2}$
17. $\frac{1}{100}$

18. List all prime numbers between 1 and 20.

Express as a product of primes.

19. 70
20. 45
21. 47

22. 81
23. 180
24. 1470

Reduce to lowest terms.

25. $\frac{9}{36}$
26. $\frac{8}{28}$
27. $\frac{25}{30}$

28. $\frac{12}{35}$
29. $\frac{48}{20}$
30. $\frac{72}{16}$

31. $\frac{72}{72}$
32. $\frac{72}{114}$
33. $\frac{120}{162}$

34. Write an equivalent fraction for $\frac{2}{3}$ with 15 as the denominator.

35. Write an equivalent fraction for $\frac{7}{8}$ with 32 as the denominator.

36. Write an equivalent fraction for $\frac{5}{4}$ with 15 as the numerator.

37. Write an equivalent fraction for $\frac{6}{5}$ with 36 as the numerator.

Supply the missing numerator.

38. $\frac{3}{4} = \frac{}{12}$
39. $\frac{5}{7} = \frac{}{28}$
40. $\frac{10}{3} = \frac{}{24}$

6 FUNDAMENTALS

Supply the missing denominator.

41. $\dfrac{3}{7} = \dfrac{6}{\underline{}}$

42. $\dfrac{2}{11} = \dfrac{10}{\underline{}}$

43. $\dfrac{21}{2} = \dfrac{63}{\underline{}}$

44. A softball pitcher struck out 12 of the 21 batters she faced. What fractional part of the total number of batters did she strike out?

45. Marvin received $150 from his dad. He spent $55 for a coat and deposited the rest into his savings account. What fractional part of the $150 went for (A) the coat? (B) savings?

ANSWERS: **1.** true **2.** true **3.** false **4.** false **5.** true **6.** true **7.** false **8.** true **9.** false **10.** true
11. (A) $\dfrac{1}{5}, \dfrac{3}{8},$ and $\dfrac{13}{15}$ (B) $\dfrac{9}{9}, \dfrac{10}{2}, \dfrac{6}{6},$ and $\dfrac{8}{1}$ *Answers to 12–17 will vary; some possibilities are given.*
12. $\dfrac{4}{10}, \dfrac{6}{15}, \dfrac{8}{20}$ **13.** $\dfrac{10}{4}, \dfrac{15}{6}, \dfrac{20}{8}$ **14.** $\dfrac{4}{1}, \dfrac{8}{2}, \dfrac{12}{3}$ **15.** $\dfrac{0}{1}, \dfrac{0}{2}, \dfrac{0}{3}$ **16.** $\dfrac{210}{4}, \dfrac{315}{6}, \dfrac{420}{8}$
17. $\dfrac{2}{200}, \dfrac{3}{300}, \dfrac{4}{400}$ **18.** 2, 3, 5, 7, 11, 13, 17, 19 **19.** $2 \cdot 5 \cdot 7$ **20.** $3 \cdot 3 \cdot 5$ **21.** 47 (prime) **22.** $3 \cdot 3 \cdot 3 \cdot 3$
23. $2 \cdot 2 \cdot 3 \cdot 3 \cdot 5$ **24.** $2 \cdot 3 \cdot 5 \cdot 7 \cdot 7$ **25.** $\dfrac{1}{4}$ **26.** $\dfrac{2}{7}$ **27.** $\dfrac{5}{6}$ **28.** $\dfrac{12}{35}$ (already reduced to lowest terms)
29. $\dfrac{12}{5}$ **30.** $\dfrac{9}{2}$ **31.** 1 **32.** $\dfrac{12}{19}$ **33.** $\dfrac{20}{27}$ **34.** $\dfrac{10}{15}$ **35.** $\dfrac{28}{32}$ **36.** $\dfrac{15}{12}$ **37.** $\dfrac{36}{30}$ **38.** 9 **39.** 20 **40.** 80
41. 14 **42.** 55 **43.** 6 **44.** $\dfrac{12}{21}$, or $\dfrac{4}{7}$ when reduced to lowest terms **45.** (A) $\dfrac{55}{150}$, or $\dfrac{11}{30}$ when reduced (B) $\dfrac{95}{150}$, or $\dfrac{19}{30}$ when reduced

---

1.1 EXERCISES B

Answer true or false in Exercises 1–10.

1. The objects that belong to a set are called the elements of the set.

2. A numeral is the name or symbol for a set.

3. {0, 1, 2, 3, . . .} is the set of natural numbers.

4. The denominator of the fraction $\tfrac{5}{7}$ is 5.

5. If the numerator of a fraction is equal to or larger than the denominator, the fraction is called proper.

6. Equivalent fractions name the same number or have the same value.

7. We reduce a fraction when we divide both numerator and denominator by a natural number other than 1.

8. Prime numbers are divisible only by themselves and 1.

9. The process of dividing common factors from a numerator and denominator is sometimes called reducing common factors.

10. A number expressed as a quotient of a whole number and a natural number is called a fraction.

11. Consider the fractions $\tfrac{1}{7}, \tfrac{3}{11}, \tfrac{8}{3}, \tfrac{15}{17}, \tfrac{17}{23}, \tfrac{21}{23},$ and $\tfrac{5}{1}$.
(A) Which of these are proper fractions? (B) Which of these are improper fractions?

Write three fractions that are equivalent to the following.

12. $\dfrac{3}{5}$

13. $\dfrac{7}{2}$

14. 8

15. 2 **16.** $\dfrac{95}{3}$ **17.** $\dfrac{1}{1000}$

18. List all prime numbers between 20 and 40.

Express as a product of primes.

19. 42 **20.** 20 **21.** 53

22. 625 **23.** 1000 **24.** 2541

Reduce to lowest terms.

25. $\dfrac{7}{42}$ **26.** $\dfrac{12}{28}$ **27.** $\dfrac{20}{32}$

28. $\dfrac{15}{77}$ **29.** $\dfrac{60}{35}$ **30.** $\dfrac{88}{16}$

31. $\dfrac{54}{54}$ **32.** $\dfrac{90}{147}$ **33.** $\dfrac{168}{228}$

34. Write an equivalent fraction for $\tfrac{3}{4}$ with 20 as the denominator.

35. Write an equivalent fraction for $\tfrac{5}{8}$ with 48 as the denominator.

36. Write an equivalent fraction for $\tfrac{7}{4}$ with 21 as the numerator.

37. Write an equivalent fraction for $\tfrac{8}{5}$ with 32 as the numerator.

Supply the missing numerator.

38. $\dfrac{2}{3} = \dfrac{?}{12}$ **39.** $\dfrac{3}{7} = \dfrac{?}{35}$ **40.** $\dfrac{12}{5} = \dfrac{?}{45}$

Supply the missing denominator.

41. $\dfrac{4}{7} = \dfrac{28}{?}$ **42.** $\dfrac{8}{11} = \dfrac{24}{?}$ **43.** $\dfrac{52}{3} = \dfrac{104}{?}$

44. A basketball player hit 21 free throws in 28 attempts during a recent tournament. What fractional part of the attempts did he make?

45. Bill received $250 from his grandmother. He spent $120 to repair his car and the rest went for gasoline. What fractional part of the $250 went for **(A)** car repairs? **(B)** gasoline?

1.1 EXERCISES C

Reduce to lowest terms.

1. $\dfrac{632}{492}$ **2.** $\dfrac{246}{1230}$ **3.** $\dfrac{780}{5005}$ $\left[\text{Answer: } \dfrac{12}{77}\right]$

8 FUNDAMENTALS

1.2 OPERATIONS ON FRACTIONS

STUDENT GUIDEPOSTS

1. Division by zero
2. Multiplying fractions
3. Reciprocals
4. Dividing fractions
5. Least common denominator
6. Adding fractions
7. Mixed numbers
8. Subtracting fractions

1 The four basic operations on fractions are addition (+), subtraction (−), multiplication (·), and division (÷ or a fraction bar). We assume that these four operations applied to whole numbers are well understood. Division by zero, however, causes a problem. Why? The quotient of two numbers such as 8 and 2, written $8 \div 2$ or $\frac{8}{2}$, is the number which, when multiplied by 2, equals 8.

That is, $\quad \frac{8}{2} = 4 \quad$ because $\quad 8 = 2 \cdot 4$.

Similarly, $\quad \frac{20}{4} = 5 \quad$ because $\quad 20 = 4 \cdot 5$.

If $\quad \frac{20}{0} = $ (a number), then $\quad 20 = 0 \cdot$ (a number).

But zero times any number is zero, not 20, so there is no number equal to $\frac{20}{0}$. Since we could have used any number in our example, no number can be divided by zero (except possibly zero itself). What would $\frac{0}{0}$ equal?

$\frac{0}{0}$ could be 5 since $0 = 0 \cdot 5$

$\frac{0}{0}$ could be 259 since $0 = 0 \cdot 259$

$\frac{0}{0}$ could be (any number) since $0 \cdot$ (any number) $= 0$

Since it would be very confusing if $\frac{0}{0}$ could be any number we pleased, we agree never to divide 0 by 0. Thus, any division by 0 is said to be undefined and is excluded from mathematics. However, $0 \div 5$ or $\frac{0}{5}$ does make sense, and in fact $\frac{0}{5} = 0$ (why?).

We now review the four operations on fractions.

2
To multiply two (or more) fractions

1. Factor all numerators and denominators into products of primes.
2. Place all numerator factors over all denominator factors.
3. Divide or cancel all common factors from both numerator and denominator.
4. Multiply the remaining factors in the numerator and denominator.
5. The resulting fraction is the product (reduced to lowest terms) of the original fractions.

EXAMPLE 1

Multiply.

(A) $\dfrac{3}{8} \cdot \dfrac{20}{9} = \dfrac{3}{2 \cdot 2 \cdot 2} \cdot \dfrac{2 \cdot 2 \cdot 5}{3 \cdot 3}$ Factor all numerators and denominators

$= \dfrac{3 \cdot 2 \cdot 2 \cdot 5}{2 \cdot 2 \cdot 2 \cdot 3 \cdot 3}$ All numerator factors over all denominator factors

$= \dfrac{\cancel{3} \cdot \cancel{2} \cdot \cancel{2} \cdot 5}{\cancel{2} \cdot \cancel{2} \cdot 2 \cdot \cancel{3} \cdot 3}$ Divide or cancel common factors

$= \dfrac{5}{6}$ Multiply the remaining factors

(B) $\dfrac{3}{7} \cdot 14 = \dfrac{3}{7} \cdot \dfrac{14}{1} = \dfrac{3}{7} \cdot \dfrac{2 \cdot 7}{1} = \dfrac{3 \cdot 2 \cdot \cancel{7}}{\cancel{7} \cdot 1} = \dfrac{6}{1} = 6$ ◂◂

Practice Exercise 1

Multiply.

(A) $\dfrac{2}{15} \cdot \dfrac{5}{14}$

(B) $\dfrac{3}{8} \cdot 24$

Answers: (A) $\dfrac{1}{21}$ (B) 9

With practice we can shorten our work by not factoring completely to primes when common factors that are larger can be recognized. For example, in Example 1 (A) we might multiply as follows:

$$\dfrac{\overset{1}{\cancel{3}}}{\underset{2}{\cancel{8}}} \cdot \dfrac{\overset{5}{\cancel{20}}}{\underset{3}{\cancel{9}}} = \dfrac{5}{6}.$$ Divide out 4 and 3

3 To divide fractions, we need to understand the notion of a *reciprocal*. The **reciprocal** of a fraction is the fraction formed by interchanging its numerator and denominator. Some fractions and their reciprocals are given in the following table.

Fraction	Reciprocal
$\dfrac{2}{3}$	$\dfrac{3}{2}$
$\dfrac{7}{8}$	$\dfrac{8}{7}$
$5 \left(\text{or } \dfrac{5}{1}\right)$	$\dfrac{1}{5}$
$\dfrac{1}{3}$	$3 \left(\text{or } \dfrac{3}{1}\right)$

Any fraction multiplied by its reciprocal gives us the number 1. For example,

$$\dfrac{2}{3} \cdot \dfrac{3}{2} = 1, \quad \dfrac{7}{8} \cdot \dfrac{8}{7} = 1, \quad \dfrac{5}{1} \cdot \dfrac{1}{5} = 1, \quad \dfrac{1}{3} \cdot \dfrac{3}{1} = 1.$$

When one number is divided by a second, the first is called the **dividend**, the second is called the **divisor**, and the result is called the **quotient**. For example, in $10 \div 2 = 5$, 10 is the dividend, 2 the divisor, and 5 the quotient.

4 **To divide two fractions**

1. Replace the divisor with its reciprocal and change the division sign to multiplication.
2. Multiply the fractions.

10 FUNDAMENTALS

EXAMPLE 2

Divide.

(A) $\dfrac{13}{15} \div \dfrac{39}{5} = \dfrac{13}{15} \cdot \dfrac{5}{39}$ Replace divisor with its reciprocal and multiply

$= \dfrac{13}{3 \cdot 5} \cdot \dfrac{5}{3 \cdot 13}$ Factor

$= \dfrac{\cancel{13} \cdot \cancel{5}}{3 \cdot \cancel{5} \cdot 3 \cdot \cancel{13}}$ Indicate products and cancel

$= \dfrac{1}{9}$ Canceling is dividing so that a 1, not 0, remains in the numerator

(B) $\dfrac{4}{5} \div 44 = \dfrac{4}{5} \cdot \dfrac{1}{44} = \dfrac{2 \cdot 2}{5} \cdot \dfrac{1}{2 \cdot 2 \cdot 11} = \dfrac{\cancel{2} \cdot \cancel{2} \cdot 1}{5 \cdot \cancel{2} \cdot \cancel{2} \cdot 11} = \dfrac{1}{55}$

(C) $\dfrac{3}{7} \div \dfrac{3}{7} = \dfrac{3}{7} \cdot \dfrac{7}{3} = \dfrac{\cancel{3} \cdot \cancel{7}}{\cancel{7} \cdot \cancel{3}} = \dfrac{1}{1} = 1$ Does this answer seem reasonable? ◀◀

Practice Exercise 2

Divide.

(A) $\dfrac{11}{6} \div \dfrac{44}{21}$

(B) $\dfrac{3}{7} \div 6$

(C) $\dfrac{9}{13} \div \dfrac{9}{13}$

Answers: (A) $\dfrac{7}{8}$ (B) $\dfrac{1}{14}$ (C) 1

To add or subtract fractions with the same denominators, simply add numerators and place the result over the common denominator. For example,

$$\dfrac{3}{7} + \dfrac{5}{7} = \dfrac{3+5}{7} = \dfrac{8}{7}.$$

▶▶ **CAUTION** ◀◀ When adding fractions, we add numerators but *not* denominators. ◀◀

To add or subtract fractions with different denominators, change the fractions to equivalent fractions having a common denominator. In doing this it is a good idea to find the **least common denominator (LCD)** of the fractions, that is, the *smallest* number that has both denominators as factors. For example,

$$\dfrac{1}{2} + \dfrac{1}{3} \quad \text{can be changed to} \quad \dfrac{3}{6} + \dfrac{2}{6}.$$

Here 6 is the least common denominator of $\frac{1}{2}$ and $\frac{1}{3}$ since there is no smaller number that has both 2 and 3 as factors. Note that we could have used 12 or 18 as a common denominator, but 6 is the *least* common denominator.

To find the LCD of two (or more) fractions

1. Factor each denominator into a product of primes.
2. If there are no common factors in the denominators, the LCD is the product of *all* denominators.
3. If there are common factors in the denominators, each factor must appear in the LCD as many times as it appears in the denominator where it is found the greatest number of times.

EXAMPLE 3

Find the LCD of the fractions.

(A) $\dfrac{1}{6}$ and $\dfrac{4}{15}$

Factor the denominators: $6 = \overset{1}{2} \cdot \overset{1}{3}$ and $15 = \overset{1}{3} \cdot \overset{1}{5}$. The LCD must consist of one 2, one 3, and one 5. Thus, the LCD $= 2 \cdot 3 \cdot 5 = 30$.

(B) $\dfrac{13}{90}$ and $\dfrac{7}{24}$

Factor the denominators: $90 = 2 \cdot \overset{1}{3} \cdot \overset{2}{3} \cdot \overset{1}{5}$ and $24 = \overset{3}{2} \cdot 2 \cdot 2 \cdot \overset{1}{3}$. The LCD must consist of three 2's, two 3's, and one 5. Thus, the LCD $= 2 \cdot 2 \cdot 2 \cdot 3 \cdot 3 \cdot 5 = 360$.

(C) $\dfrac{3}{10}$ and $\dfrac{5}{21}$

Factor the denominators: $10 = \overset{1}{2} \cdot \overset{1}{5}$ and $21 = \overset{1}{3} \cdot \overset{1}{7}$. Since there are no common factors, the LCD is the product of all factors. Thus, the LCD $= 2 \cdot 3 \cdot 5 \cdot 7 = 210$.

(D) $\dfrac{1}{2}, \dfrac{5}{6},$ and $\dfrac{7}{75}$

Factor the denominators: $2 = \overset{1}{2}$, $6 = \overset{1}{2} \cdot \overset{1}{3}$, and $75 = \overset{1}{3} \cdot \overset{2}{5} \cdot 5$. Thus, the LCD $= 2 \cdot 3 \cdot 5 \cdot 5 = 150$. ◀◀

Practice Exercise 3

Find the LCD of the fractions.

(A) $\dfrac{5}{6}$ and $\dfrac{1}{21}$

(B) $\dfrac{1}{45}$ and $\dfrac{7}{20}$

(C) $\dfrac{2}{33}$ and $\dfrac{5}{14}$

(D) $\dfrac{1}{3}, \dfrac{4}{15},$ and $\dfrac{2}{75}$

Answers: **(A)** 42 **(B)** 180 **(C)** 462 **(D)** 75

To see why the rule above works, let us look again at the fractions $\frac{1}{6}$ and $\frac{4}{15}$ in Example 3(a). By definition, each denominator must be a factor of the LCD. Since 6 is a factor of the LCD, the LCD must contain the factors 2 and 3. Also, with 15 a factor of the LCD, the LCD must contain the factors 3 and 5. Since we already have 3 as a factor from the denominator 6, we only need to supply the factor 5. Thus,

$$\text{LCD} = 2 \cdot 3 \cdot 5 = 30,$$

which is what we found in Example 3(a) using the rule above.

> **To add two (or more) fractions with different denominators**
> 1. Rewrite the sum with each denominator expressed as a product of primes.
> 2. Determine the LCD.
> 3. Multiply the numerators and denominators of each fraction by all those factors present in the LCD but missing in the denominator of the particular fraction.
> 4. Place the sum of all numerators over the LCD.
> 5. Simplify the resulting numerator and reduce the fraction to lowest terms.

12 FUNDAMENTALS

EXAMPLE 4

Add.

(A) $\dfrac{5}{6} + \dfrac{7}{15} = \dfrac{5}{2 \cdot 3} + \dfrac{7}{3 \cdot 5}$ Factor denominators; the LCD is $2 \cdot 3 \cdot 5$

$= \dfrac{5 \cdot 5}{2 \cdot 3 \cdot 5} + \dfrac{7 \cdot 2}{3 \cdot 5 \cdot 2}$ Multiply numerators and denominators by missing factors

$= \dfrac{25 + 14}{2 \cdot 3 \cdot 5}$ Simplify and add over LCD

$= \dfrac{39}{2 \cdot 3 \cdot 5}$ Simplify

$= \dfrac{\cancel{3} \cdot 13}{2 \cdot \cancel{3} \cdot 5}$ Factor numerator and cancel common factors

$= \dfrac{13}{10}$ The desired sum

(B) $\dfrac{1}{2} + \dfrac{5}{6} + \dfrac{7}{25}$

$= \dfrac{1}{2} + \dfrac{5}{2 \cdot 3} + \dfrac{7}{5 \cdot 5}$ Factor: the LCD $= 2 \cdot 3 \cdot 5 \cdot 5$

$= \dfrac{1 \cdot 3 \cdot 5 \cdot 5}{2 \cdot 3 \cdot 5 \cdot 5} + \dfrac{5 \cdot 5 \cdot 5}{2 \cdot 3 \cdot 5 \cdot 5} + \dfrac{7 \cdot 2 \cdot 3}{5 \cdot 5 \cdot 2 \cdot 3}$

Multiply numerators and denominators by missing factors

$= \dfrac{75 + 125 + 42}{2 \cdot 3 \cdot 5 \cdot 5}$ Simplify and add over LCD

$= \dfrac{242}{2 \cdot 3 \cdot 5 \cdot 5}$ Simplify

$= \dfrac{\cancel{2} \cdot 11 \cdot 11}{\cancel{2} \cdot 3 \cdot 5 \cdot 5}$ Factor numerator and cancel common factors

$= \dfrac{121}{75}$ The desired sum ◀◀

Practice Exercise 4

Add.

(A) $\dfrac{7}{15} + \dfrac{8}{21}$

(B) $\dfrac{2}{3} + \dfrac{1}{5} + \dfrac{5}{6}$

Answers: (A) $\dfrac{89}{105}$ (B) $\dfrac{17}{10}$

---

7 ▶▶ It is sometimes helpful to write an improper fraction, such as $\dfrac{121}{75}$, as a *mixed number*. A **mixed number** is the sum of a whole number and a proper fraction. For example,

$$\dfrac{121}{75} \text{ can be written as } 1 + \dfrac{46}{75}.$$

We usually omit the plus sign and write this mixed number in the form

$$1\dfrac{46}{75}, \text{ read as ``one and forty-six seventy fifths.''}$$

To change an improper fraction to a mixed number, reduce the fraction to lowest terms, divide the denominator into the numerator, and write the remainder as a fraction.

EXAMPLE 5

Change each improper fraction to a mixed number.

(A) $\dfrac{13}{4}$

Divide 4 into 13. $\quad 4\overline{)13}\ \ \dfrac{3}{}$
$\underline{12}$
1

Thus, the mixed number is $3\tfrac{1}{4}$, read "three and one fourth."

(B) $\dfrac{27}{6}$

Reduce $\tfrac{27}{6}$ to lowest terms and then divide.

$$\dfrac{27}{6} = \dfrac{9 \cdot \cancel{3}}{2 \cdot \cancel{3}} = \dfrac{9}{2} \qquad 2\overline{)9}\ \ \dfrac{4}{}$$
$\underline{8}$
1

Thus, $\tfrac{27}{6} = 4\tfrac{1}{2}$. ◂◂

Practice Exercise 5

Change each improper fraction to a mixed number.

(A) $\dfrac{25}{8}$

(B) $\dfrac{40}{15}$

Answers: **(A)** $3\tfrac{1}{8}$ **(B)** $2\tfrac{2}{3}$

To change a mixed number to an improper fraction, we can think of the mixed number as a sum of two fractions. For example, the mixed number $2\tfrac{11}{18} = 2 + \tfrac{11}{18}$. Adding, we have

$$\dfrac{2}{1} + \dfrac{11}{18} = \dfrac{2}{1} + \dfrac{11}{2 \cdot 3 \cdot 3} = \dfrac{2 \cdot 2 \cdot 3 \cdot 3}{2 \cdot 3 \cdot 3} + \dfrac{11}{2 \cdot 3 \cdot 3} = \dfrac{36 + 11}{2 \cdot 3 \cdot 3} = \dfrac{47}{18}.$$

Rather than going through all of the addition steps, we often use a shortcut. Consider $2\tfrac{11}{18}$ again. Notice that if we multiply the denominator, 18, by the whole number, 2, and add the numerator, 11, we obtain

$$(18 \cdot 2) + 11 = 47.$$

When we put this result over the denominator, 18, we obtain the desired improper fraction $\tfrac{47}{18}$. We can show this as

$$\dfrac{(\textbf{denominator} \times \textbf{whole number}) + \textbf{numerator}}{\textbf{denominator}}.$$

EXAMPLE 6

Change each mixed number to an improper fraction.

(A) $4\dfrac{3}{5} = \dfrac{(5 \cdot 4) + 3}{5}$

$\phantom{4\dfrac{3}{5}} = \dfrac{20 + 3}{5} = \dfrac{23}{5}$

(B) $6\dfrac{7}{10} = \dfrac{(10 \cdot 6) + 7}{10}$

$\phantom{6\dfrac{7}{10}} = \dfrac{60 + 7}{10} = \dfrac{67}{10}$ ◂◂

Practice Exercise 6

Change each mixed number to an improper fraction.

(A) $6\dfrac{2}{7}$

(B) $3\dfrac{23}{100}$

Answers: **(A)** $\dfrac{44}{7}$ **(B)** $\dfrac{323}{100}$

To add two or more mixed numbers or mixed numbers and proper fractions, convert all mixed numbers to improper fractions and proceed as before.

EXAMPLE 7

Add.

$$3\frac{1}{4} + 6\frac{7}{10} + \frac{1}{3} = \frac{13}{4} + \frac{67}{10} + \frac{1}{3} \quad \text{Convert to improper fractions}$$

$$= \frac{13}{2 \cdot 2} + \frac{67}{2 \cdot 5} + \frac{1}{3} \quad \text{The LCD} = 2 \cdot 2 \cdot 3 \cdot 5$$

$$= \frac{13 \cdot 3 \cdot 5}{2 \cdot 2 \cdot 3 \cdot 5} + \frac{67 \cdot 2 \cdot 3}{2 \cdot 5 \cdot 2 \cdot 3} + \frac{1 \cdot 2 \cdot 2 \cdot 5}{3 \cdot 2 \cdot 2 \cdot 5}$$

$$= \frac{195 + 402 + 20}{2 \cdot 2 \cdot 3 \cdot 5}$$

$$= \frac{617}{60} \text{ or } 10\frac{17}{60} \quad \blacktriangleleft\blacktriangleleft$$

Practice Exercise 7

Add.

$$2\frac{2}{3} + 5\frac{1}{10} + \frac{1}{5}$$

Answer: $\frac{239}{30}$, or $7\frac{29}{30}$

---

8 ▶▶ To subtract two fractions

1. Rewrite the difference with each denominator expressed as a product of prime factors.
2. Determine the LCD.
3. Supply missing factors just as when adding.
4. Place the difference of the numerators over the LCD.
5. Simplify the resulting numerator and reduce the fraction to lowest terms.

EXAMPLE 8

Subtract.

(A) $\dfrac{7}{12} - \dfrac{5}{9} = \dfrac{7}{2 \cdot 2 \cdot 3} - \dfrac{5}{3 \cdot 3}$ Factor; LCD is $2 \cdot 2 \cdot 3 \cdot 3$

$$= \frac{7 \cdot 3}{2 \cdot 2 \cdot 3 \cdot 3} - \frac{5 \cdot 2 \cdot 2}{3 \cdot 3 \cdot 2 \cdot 2} \quad \text{Supply missing factors}$$

$$= \frac{21 - 20}{2 \cdot 2 \cdot 3 \cdot 3} \quad \text{Subtract and simplify}$$

$$= \frac{1}{36} \quad \text{The difference in reduced form}$$

(B) $4\dfrac{1}{2} - 3\dfrac{7}{8}$

First convert the mixed numbers to improper fractions.

$$4\frac{1}{2} - 3\frac{7}{8} = \frac{9}{2} - \frac{31}{8} = \frac{9 \cdot 4}{2 \cdot 4} - \frac{31}{8} = \frac{36 - 31}{8} = \frac{5}{8} \quad \blacktriangleleft\blacktriangleleft$$

Practice Exercise 8

Subtract.

(A) $\dfrac{8}{35} - \dfrac{1}{21}$

(B) $6\dfrac{1}{4} - 2\dfrac{6}{7}$

Answers: (A) $\dfrac{19}{105}$ (B) $\dfrac{95}{28}$, or $3\dfrac{11}{28}$

Notice in Example 8(b) that we took a shortcut and did not factor denominators into primes. It was clear that the LCD of the two fractions is 8, so we simply supplied the factor of 4 to both numerator and denominator of $\frac{9}{2}$. With practice, shortcuts such as this can save time.

The rules for adding and subtracting fractions can be extended to include combinations of these operations, as shown in the next example.

EXAMPLE 9

Perform the indicated operations.

$2\frac{1}{15} + 4\frac{1}{5} - 3\frac{1}{2}$

$= \frac{31}{15} + \frac{21}{5} - \frac{7}{2}$ Convert to improper fractions

$= \frac{31}{3 \cdot 5} + \frac{21}{5} - \frac{7}{2}$ The LCD = $2 \cdot 3 \cdot 5$

$= \frac{31 \cdot 2}{3 \cdot 5 \cdot 2} + \frac{21 \cdot 2 \cdot 3}{5 \cdot 2 \cdot 3} - \frac{7 \cdot 3 \cdot 5}{2 \cdot 3 \cdot 5}$ Supply missing factors

$= \frac{62 + 126 - 105}{2 \cdot 3 \cdot 5}$

$= \frac{83}{30}$ or $2\frac{23}{30}$ ◀◀

Practice Exercise 9

Perform the indicated operations.

$7\frac{1}{15} - 2\frac{2}{3} + 1\frac{1}{2}$

Answer: $\frac{59}{10}$, or $5\frac{9}{10}$

1.2 EXERCISES A

1. $\frac{0}{3} =$ _____

2. $\frac{3}{0} =$ _____

3. $\frac{0}{0} =$ _____

Multiply.

4. $\frac{1}{3} \cdot \frac{6}{5}$

5. $\frac{1}{2} \cdot \frac{3}{7}$

6. $\frac{6}{35} \cdot \frac{20}{12}$

7. $\frac{12}{5} \cdot 10$

*8. $4\frac{2}{5} \cdot \frac{3}{11}$

9. $2\frac{2}{3} \cdot 1\frac{1}{8} \cdot \frac{1}{6}$

Divide.

10. $\frac{3}{7} \div \frac{9}{28}$

11. $\dfrac{\frac{20}{9}}{\frac{2}{3}}$

12. $\frac{20}{27} \div \frac{35}{36}$

*A colored number indicates a complete solution in the back of the text. See *To the Student* for more details.

13. $\dfrac{8}{9} \div 4$

14. $3\dfrac{1}{5} \div \dfrac{2}{5}$

15. $3\dfrac{1}{3} \div 1\dfrac{3}{5}$

Find the LCD of the following fractions.

16. $\dfrac{2}{7}$ and $\dfrac{4}{33}$

17. $\dfrac{7}{12}$ and $\dfrac{1}{45}$

18. $\dfrac{3}{16}$ and $\dfrac{7}{20}$

19. 7 and $\dfrac{9}{2}$

20. $\dfrac{2}{3}, \dfrac{3}{4},$ and $\dfrac{1}{5}$

21. $\dfrac{1}{9}, \dfrac{5}{6},$ and $\dfrac{1}{75}$

Add.

22. $\dfrac{1}{2} + \dfrac{1}{2}$

23. $\dfrac{1}{6} + \dfrac{3}{10}$

24. $\dfrac{3}{28} + \dfrac{13}{70}$

25. $4 + \dfrac{3}{5}$

26. $2\dfrac{1}{5} + 1\dfrac{2}{3}$

27. $\dfrac{2}{3} + \dfrac{1}{6} + \dfrac{3}{4}$

Change each improper fraction to a mixed number.

28. $\dfrac{23}{8}$

29. $\dfrac{135}{4}$

30. $\dfrac{142}{11}$

Change each mixed number to an improper fraction.

31. $4\dfrac{2}{5}$

32. $66\dfrac{2}{3}$

33. $9\dfrac{7}{11}$

Subtract.

34. $\dfrac{3}{4} - \dfrac{3}{4}$

35. $\dfrac{7}{11} - \dfrac{2}{7}$

36. $\dfrac{7}{15} - \dfrac{13}{35}$

37. $\dfrac{19}{2} - 3$

38. $5\dfrac{1}{3} - 2\dfrac{3}{4}$

39. $11\dfrac{4}{5} - 5\dfrac{2}{3}$

Perform the indicated operations.

40. $\dfrac{7}{20} + \dfrac{3}{8} - \dfrac{1}{4}$

41. $\dfrac{14}{15} - \dfrac{2}{5} - \dfrac{1}{3}$

42. $\dfrac{7}{3} + \dfrac{1}{7} - 2$

Solve.

43. If $\frac{5}{6}$ of a class consists of girls and there are 72 students in the class, how many students are girls? How many are boys?

44. On a map, 1 inch represents $\frac{1}{2}$ of a mile. How many miles are represented by $4\frac{1}{4}$ inches?

45. A piece of rope 20 m long is to be cut into pieces, each of whose length is $\frac{2}{3}$ m. How many pieces can be cut?

46. A fuel tank holds $12\frac{1}{2}$ gallons when it is $\frac{3}{4}$ full. What is the capacity of the tank?

47. Alphonso hiked from Whiskey Creek to Bald Mountain, a distance of $2\frac{1}{3}$ miles. From there, he hiked to Spook Hollow, a distance of $4\frac{1}{5}$ miles. How far did he hike?

48. To obtain the right shade of paint for her living room, Tracy Bell mixed $\frac{7}{8}$ of a gallon of white paint with $\frac{2}{3}$ of a gallon of light blue paint. How much paint did she have?

49. The weight of one cubic foot of water is $62\frac{1}{2}$ pounds. How much do $3\frac{1}{5}$ cubic feet of water weigh?

50. Fahrenheit temperature can be found by multiplying Celsius temperature by $\frac{9}{5}$ and adding 32°. If Celsius temperature is 15°, what is Fahrenheit temperature?

For Review

51. Write an equivalent fraction for $\frac{7}{8}$ with
 (A) 21 for the numerator
 (B) 72 for the denominator.

52. Max received $125 for doing a particular job. If he had to pay $15 for income tax, what fractional part of the $125 went for tax?

ANSWERS: **1.** 0 **2.** undefined **3.** undefined **4.** $\frac{2}{5}$ **5.** $\frac{3}{14}$ **6.** $\frac{2}{7}$ **7.** 24 **8.** $\frac{6}{5}$ **9.** $\frac{1}{2}$ **10.** $\frac{4}{3}$ **11.** $\frac{10}{3}$ **12.** $\frac{16}{21}$ **13.** $\frac{2}{9}$ **14.** 8 **15.** $\frac{25}{12}$ **16.** 231 **17.** 180 **18.** 80 **19.** 2 **20.** 60 **21.** 450 **22.** 1 **23.** $\frac{7}{15}$ **24.** $\frac{41}{140}$ **25.** $\frac{23}{5}$ **26.** $\frac{58}{15}$ **27.** $\frac{19}{12}$ **28.** $2\frac{7}{8}$ **29.** $33\frac{3}{4}$ **30.** $12\frac{10}{11}$ **31.** $\frac{22}{5}$ **32.** $\frac{200}{3}$ **33.** $\frac{106}{11}$ **34.** 0 **35.** $\frac{27}{77}$ **36.** $\frac{2}{21}$ **37.** $\frac{13}{2}$ **38.** $2\frac{7}{12}$ **39.** $6\frac{2}{15}$ **40.** $\frac{19}{40}$ **41.** $\frac{1}{5}$ **42.** $\frac{10}{21}$ **43.** 60 girls, 12 boys **44.** $2\frac{1}{8}$ **45.** 30 **46.** $16\frac{2}{3}$ gallons **47.** $6\frac{8}{15}$ miles **48.** $1\frac{13}{24}$ gallons **49.** 200 pounds **50.** 59° **51.** (A) $\frac{21}{24}$ (B) $\frac{63}{72}$ **52.** $\frac{3}{25}$

1.2 EXERCISES B

1. $\frac{7}{0} =$ _____

2. $\frac{0}{0} =$ _____

3. $\frac{0}{7} =$ _____

18 FUNDAMENTALS

Multiply.

4. $\dfrac{2}{3} \cdot \dfrac{6}{5}$

5. $\dfrac{3}{8} \cdot \dfrac{1}{2}$

6. $\dfrac{7}{35} \cdot \dfrac{20}{14}$

7. $\dfrac{11}{4} \cdot 8$

8. $4\dfrac{4}{5} \cdot \dfrac{1}{12}$

9. $1\dfrac{1}{3} \cdot 3\dfrac{3}{8} \cdot \dfrac{2}{3}$

Divide.

10. $\dfrac{2}{7} \div \dfrac{6}{28}$

11. $\dfrac{\frac{15}{8}}{\frac{3}{4}}$

12. $\dfrac{10}{27} \div \dfrac{15}{36}$

13. $6 \div \dfrac{8}{11}$

14. $2\dfrac{3}{5} \div \dfrac{26}{5}$

15. $2\dfrac{2}{3} \div 1\dfrac{1}{5}$

Find the LCD of the following fractions.

16. $\dfrac{1}{9}$ and $\dfrac{3}{22}$

17. $\dfrac{5}{12}$ and $\dfrac{2}{27}$

18. $\dfrac{1}{18}$ and $\dfrac{3}{20}$

19. $\dfrac{2}{9}$ and 5

20. $\dfrac{1}{3}, \dfrac{3}{4},$ and $\dfrac{2}{7}$

21. $\dfrac{1}{6}, \dfrac{5}{9},$ and $\dfrac{7}{30}$

Add.

22. $\dfrac{1}{4} + \dfrac{3}{4}$

23. $\dfrac{1}{8} + \dfrac{5}{12}$

24. $\dfrac{5}{24} + \dfrac{7}{60}$

25. $\dfrac{1}{8} + 2$

26. $4\dfrac{1}{5} + 2\dfrac{1}{3}$

27. $\dfrac{1}{3} + \dfrac{5}{6} + \dfrac{1}{5}$

Change each improper fraction to a mixed number.

28. $\dfrac{81}{10}$

29. $\dfrac{245}{9}$

30. $\dfrac{167}{11}$

Change each mixed number to an improper fraction.

31. $10\dfrac{1}{5}$

32. $42\dfrac{3}{10}$

33. $8\dfrac{4}{11}$

Subtract.

34. $\dfrac{5}{8} - \dfrac{1}{4}$

35. $\dfrac{8}{15} - \dfrac{12}{35}$

36. $5 - \dfrac{1}{9}$

37. $\dfrac{9}{2} - 1\dfrac{1}{4}$

38. $6\dfrac{2}{3} - 1\dfrac{3}{4}$

39. $11\dfrac{3}{5} - 4\dfrac{1}{3}$

Perform the indicated operations.

40. $\dfrac{9}{20} + \dfrac{5}{8} - \dfrac{3}{4}$

41. $\dfrac{13}{15} - \dfrac{1}{5} - \dfrac{1}{3}$

42. $\dfrac{8}{3} + \dfrac{2}{7} - 2$

Solve.

43. Raoul took out a loan for $\dfrac{2}{3}$ of the cost of a motorbike. If the bike cost $744, how much did he borrow?

44. Sandi can hike $3\dfrac{1}{2}$ miles in one hour. How many miles can she hike in $\dfrac{3}{4}$ of an hour?

45. When Fred Vickers pounds a nail into a board, he sinks it $1\dfrac{1}{4}$ inches with each blow. How many times will he have to hit a 5-inch nail to drive it completely into the board?

46. After driving 285 miles, the Stotts had completed $\dfrac{5}{8}$ of their trip. What was the total length of the trip?

47. In July, Memphis had two rain storms, one that dropped $1\dfrac{1}{4}$ inches of rain and another that dropped $\dfrac{3}{5}$ of an inch. How much rain did Memphis receive in July?

48. It took Burford $8\dfrac{1}{2}$ hours to fix his car himself. Had he had the money, he could have hired a mechanic to do the job in $1\dfrac{1}{3}$ hours. How many hours of work could Burford have saved?

49. Cindy Bown is paid $4\dfrac{1}{2}$ dollars per hour to type a manuscript. If she types for a period of $20\dfrac{2}{3}$ hours, how much is she paid?

50. Celsius temperature can be found by subtracting 32° from Fahrenheit temperature and multiplying the result by $\dfrac{5}{9}$. If Fahrenheit temperature is 122°, what is the corresponding Celsius temperature?

For Review

51. Write an equivalent fraction for $\dfrac{8}{9}$ with **(A)** 32 as the numerator, and **(B)** 81 as the denominator.

52. A farmer collected 280 bushels of grain from his field. If he had to give 40 bushels to his landlord for rent, what fractional part of the 280 bushels went for rent?

1.2 EXERCISES C

Perform the indicated operations.

1. $22\dfrac{5}{8} + 17\dfrac{3}{10} - 8\dfrac{3}{4}$
   $\left[\text{Answer: } 31\dfrac{7}{40}\right]$

2. $42\dfrac{7}{15} - 18\dfrac{5}{6} - 11\dfrac{3}{10}$

3. $125\dfrac{16}{25} + 37\dfrac{2}{15} - 83\dfrac{7}{20}$

1.3 DECIMALS

 STUDENT GUIDEPOSTS

1 Place-value number system
2 Decimal notation
3 Adding decimals
4 Subtracting decimals
5 Multiplying decimals
6 Dividing decimals

20 FUNDAMENTALS

1 ▮▮ Recall that our number system is a **place-value system.** That is, each **digit** (0, 1, 2, 3, 4, 5, 6, 7, 8 or 9) in a numeral has a particular value determined by its location or place in the symbol. For example, the numeral

$$125.378$$

is a shorthand symbol for the **expanded numeral**

$$100 + 20 + 5 + \frac{3}{10} + \frac{7}{100} + \frac{8}{1000},$$

or $\quad 1 \cdot 100 + 2 \cdot 10 + 5 \cdot 1 + 3 \cdot \frac{1}{10} + 7 \cdot \frac{1}{100} + 8 \cdot \frac{1}{1000},$

where each of the digits 1, 2, 5, 3, 7, and 8 is multiplied by 100, 10, 1, $\frac{1}{10}$, $\frac{1}{100}$, or $\frac{1}{1000}$, depending upon the location of the digit in the symbol 125.378. We read 125.378 as "one hundred twenty-five *and* three hundred seventy-eight thousandths." This results from thinking of the number as a mixed number

$$125 + \frac{378}{1000}, \quad \text{or} \quad 125\frac{378}{1000}.$$

2 ▮▮ The numeral 125.378, representing the mixed number $125\frac{378}{1000}$, is called its **decimal notation,** and numbers expressed in this manner are referred to simply as **decimals.** Decimal notation is based on the number 10, and the period used in a decimal is called a **decimal point.** If there are no digits to the right of the decimal point we usually omit it; we write 53 rather than 53., for example. The figure below identifies the value of each position. Notice that the decimal point separates the ones digit from the tenths digit.

Hundreds	Tens	Ones	Decimal Point	Tenths	Hundredths	Thousandths
1	2	5	.	3	7	8

The following table expresses each decimal in expanded notation, as a mixed number, and as a proper or improper fraction.

Decimal	Expanded notation	Mixed number	Fraction
1.25	$1 + \frac{2}{10} + \frac{5}{100}$	$1\frac{25}{100}$	$\frac{125}{100}$
23.46	$20 + 3 + \frac{4}{10} + \frac{6}{100}$	$23\frac{46}{100}$	$\frac{2346}{100}$
43.278	$40 + 3 + \frac{2}{10} + \frac{7}{100} + \frac{8}{1000}$	$43\frac{278}{1000}$	$\frac{43278}{1000}$
562.34	$500 + 60 + 2 + \frac{3}{10} + \frac{4}{100}$	$562\frac{34}{100}$	$\frac{56234}{100}$
0.23	$\frac{2}{10} + \frac{3}{100}$	$0\frac{23}{100}$	$\frac{23}{100}$
47 or 47.0	$40 + 7$ or $40 + 7 + \frac{0}{10}$	47 or $47\frac{0}{10}$	$\frac{47}{1}$ or $\frac{470}{10}$

We have already reviewed the operations of adding, subtracting, multiplying, and dividing numbers in fractional notation. Now we review these operations when the numbers are in decimal notation.

To add two (or more) decimals

1. Arrange the numbers in a column so that the decimal points line up.
2. Add in columns from right to left.
3. Place the decimal point in the sum in line with the other decimal points.

EXAMPLE 1

Add.

(A) 21.3
    + 4.2
    ─────
     25.5

(B) 123.417
      45.
    + 1.14
    ───────
     169.557

Practice Exercise 1

Add.

(A) 43.8
    + 6.1

(B) 10.001
      0.52
    + 129.007

Answers: (A) 49.9 (B) 139.528

To subtract two decimals

1. Arrange the numbers in a column so that the decimal points line up.
2. Subtract in columns from right to left (Additional zeros may be placed to the right of the decimal points if desired).
3. Place the decimal point in the difference in line with the other decimal points.

EXAMPLE 2

Subtract.

(A) 27.5
    − 2.4
    ─────
     25.1

(B) 421.3000
    − 8.9999
    ─────────
     412.3001

(In (B), the three zeros to the right of 3 have been supplied.)

Practice Exercise 2

Subtract.

(A) 51.2
    − 9.8

(B) 654.2
    − 26.888

Answers: (A) 41.4 (B) 627.312

We will agree to say that 42.7 has *one decimal place*, 12.16 has *two decimal places*, 5.006 has *three decimal places*, and so forth.

To multiply two decimals

1. Ignore the decimal points and multiply as if the numbers were whole numbers.
2. Place the decimal point so the product has the same number of decimal places as the *sum* of the number of decimal places in the factors.

EXAMPLE 3

Multiply.

(A) 1.23   2 decimal places
   × 2.1   1 decimal place
   ─────
     123
    246
   ─────
    2.583   2 + 1 = 3 decimal places

(B)   42.103   3 decimal places
    × 2.15   2 decimal places
    ───────
    210515
    42103
    84206
    ───────
    90.52145   3 + 2 = 5 decimal places

(C) (0.02)(0.013) = 0.00026   5 decimal places ◀

Practice Exercise 3

Multiply.

(A) 41.3
   × 2.11

(B) 6.003
   × 1.09

(C) (0.004)(0.0255)

Answers: **(A)** 87.143
(B) 6.54327 **(C)** 0.000102

To divide one decimal into another

1. Move the decimal point in the divisor (the dividing number) to the right until the divisor becomes a whole number.
2. Move the decimal point in the dividend (the number divided into) to the right the same number of places adding zeros as necessary.
3. Divide as if the numbers were whole numbers.
4. Place the decimal point in the quotient directly above the decimal point in the dividend.

EXAMPLE 4

Divide.

(A) 0.12)‾14.4‾

    0.12)‾14.40‾   becomes   12)‾1440‾ = 120
       ⌣   ⌣   12
       2   2   ──
                                 24
                                 24
                                 ──
                                  0

(B) 0.003)‾42.3‾

    0.003)‾42.300‾   becomes   3)‾42,300‾ = 14,100 ◀
       ⌣   ⌣
       3   3

Practice Exercise 4

Divide.

(A) 2.1)‾28.35‾

(B) 0.0002)‾1.5448‾

Answers: **(A)** 13.5 **(B)** 7724

It is important to be aware of the procedures for operating on decimals since these same techniques are often used in the study of algebra. However, for all practical purposes, much of the tedious, time-consuming work with decimal operations has been eliminated with the availability of calculators. If you have a calculator, be sure to read the instruction manual that comes with your model. You might wish to use your calculator to check part of your work in some of the following exercises.

1.3 EXERCISES A

Express in expanded notation.

1. 3.47
2. 107.28
3. 203.005

Express in decimal notation.

4. $400 + 20 + 7 + \dfrac{3}{10}$
5. $500 + 3 + \dfrac{2}{10} + \dfrac{8}{1000}$
6. $4000 + 30 + \dfrac{2}{100} + \dfrac{5}{1000}$

Add.

7. $\begin{array}{r} 31.4 \\ +\ \ 1.07 \\ \hline \end{array}$
8. $\begin{array}{r} 403.1 \\ +\ \ 21.08 \\ \hline \end{array}$
9. $\begin{array}{r} 201.005 \\ 30.2 \\ +\ \ \ \ 1.07 \\ \hline \end{array}$

10. $23.4 + 9.75$
11. $512.3 + 2.009 + 13$
12. $15.0001 + 9.3 + 0.004$

Subtract.

13. $\begin{array}{r} 4.0003 \\ -\ 1.1 \\ \hline \end{array}$
14. $\begin{array}{r} 427.006 \\ -\ 135.04 \\ \hline \end{array}$
15. $\begin{array}{r} 36.5 \\ -\ \ 0.0007 \\ \hline \end{array}$

16. $46.01 - 13.4$
17. $923.006 - 23$
18. $47.2 - 0.0003$

Multiply.

19. $\begin{array}{r} 3.42 \\ \times\ \ 2.1 \\ \hline \end{array}$
20. $\begin{array}{r} 15.207 \\ \times\ \ \ \ 3.12 \\ \hline \end{array}$
21. $\begin{array}{r} 0.0034 \\ \times\ \ 0.209 \\ \hline \end{array}$

22. $(0.003)(0.003)$
23. $(0.02)(0.02)(0.02)$
24. $(0.0001)(0.00001)$

Divide.

25. $2.1 \overline{)70.14}$
26. $0.007 \overline{)38.717}$
27. $20.4 \overline{)0.687072}$

24 FUNDAMENTALS

28. $8\overline{)7.0}$ **29.** $2\overline{)1.0}$ **30.** $9\overline{)1.0}$

Solve.

31. Before leaving on a trip, Harley's odometer read 23,411.8. If he drove 723.9 miles on the trip, what did it read when he returned?

32. The Northern Arizona four-mile relay team members recorded times of 4.23, 4.17, 3.99, and 4.08 minutes. What was their combined time for the race?

33. Roseanne Strauch buys a pair of hose for $3.95. If she pays with a $20 bill and the sales tax amounts to $0.20, how much change should she receive?

34. When Kimmie was sick, she ran a temperature of 103.2° Fahrenheit. If normal body temperature is 98.6° Fahrenheit, how many degrees above normal was her temperature?

35. Jose is paid $7.35 per hour. If he worked a total of 35.2 hours last week, how much was he paid?

36. How many months will it take to pay off a loan of $6562.80 if the monthly payments are $182.30?

For Review

Perform the indicated operations.

37. $\dfrac{2}{9} + \dfrac{5}{12}$ **38.** $\dfrac{23}{36} - \dfrac{4}{9}$ **39.** $\dfrac{3}{4} \cdot \dfrac{16}{27}$ **40.** $1\dfrac{1}{12} \div 9\dfrac{3}{4}$

Solve.

41. Nancy Chandler plans a two-day hike from Gopher Gulch to Aspen Glen by way of North Face. It is $7\frac{3}{4}$ miles from Gopher Gulch to North Face and $9\frac{1}{8}$ miles from North Face to Aspen Glen. If she plans to travel half the total distance on each day, how far will she hike on the first day?

42. Julio has a rope $26\frac{2}{3}$ meters in length. He wishes to cut the rope into 3 pieces of equal length. What will be the length of each piece?

ANSWERS: **1.** $3 + \frac{4}{10} + \frac{7}{100}$ **2.** $100 + 7 + \frac{2}{10} + \frac{8}{100}$ **3.** $200 + 3 + \frac{5}{1000}$ **4.** 427.3 **5.** 503.208
6. 4030.025 **7.** 32.47 **8.** 424.18 **9.** 232.275 **10.** 33.15 **11.** 527.309 **12.** 24.3041 **13.** 2.9003 **14.** 291.966
15. 36.4993 **16.** 32.61 **17.** 900.006 **18.** 47.1997 **19.** 7.182 **20.** 47.44584 **21.** 0.0007106 **22.** 0.000009
23. 0.000008 **24.** 0.000000001 **25.** 33.4 **26.** 5531 **27.** 0.03368 **28.** 0.875 **29.** 0.5 **30.** 0.111...
31. 24,135.7 **32.** 16.47 minutes **33.** $15.85 **34.** 4.6° **35.** $258.72 **36.** 36 months **37.** $\frac{23}{36}$ **38.** $\frac{7}{36}$ **39.** $\frac{4}{9}$
40. $\frac{1}{9}$ **41.** $8\frac{7}{16}$ mi **42.** $8\frac{8}{9}$ meters

1.3 EXERCISES B

Express in expanded notation.

1. 35.6

2. 257.663

3. 3215.802

Express in decimal notation.

4. $700 + 30 + 8 + \frac{6}{10}$

5. $4000 + 6 + \frac{7}{100}$

6. $300 + 40 + \frac{2}{100} + \frac{7}{1000}$

Add.

7.   52.3
  + 2.09

8.   602.9
  + 43.07

9.   421.544
     2.8
  + 43.09

10. $42.6 + 8.41$

11. $631.9 + 5.007 + 21$

12. $12.0003 + 7.2 + 0.003$

Subtract.

13.   6.0004
  − 2.3

14.   637.008
  − 123.05

15.   51.6
  − 0.0004

16. $43.02 − 17.1$

17. $843.002 − 51$

18. $39.4 − 0.0006$

Multiply.

19.   2.91
  × 3.4

20.   17.603
  × 4.15

21.   0.0053
  × 0.402

22. $(0.002)(0.002)$

23. $(0.03)(0.03)(0.03)$

24. $(0.0002)(0.00002)$

Divide.

25. $4.2\overline{)94.5}$

26. $0.003\overline{)13.014}$

27. $30.1\overline{)6.52267}$

28. $8\overline{)5.0}$ **29.** $4\overline{)1.0}$ **30.** $9\overline{)4.0}$

Solve.

31. The Saturday and Sunday receipts at Waldo's Waffle Palace were $621.13 and $743.19, respectively. How much money did Waldo take in on that weekend?

32. What is the **perimeter** of (the distance around) a four-sided figure with sides of 6.031 m, 5.411 m, 4.003 m, and 5.772 m?

33. Sam Passamonte had a balance of $635.86 in his checking account when he wrote a check for $249.93. What was his new balance?

34. When George filled the gas tank on his Bronco, the odometer read 62,421.8. The next time he filled it, the odometer read 62,914.2. How far had he driven?

35. Betty Sue bought 13.4 pounds of peaches at $0.55 per pound. How much did she pay for the peaches?

36. Kate Williams paid $58.00 to rent a car for two days. If the rental fee amounts to $12.50 per day plus $0.15 per mile driven, how far did she drive the car?

For Review

Perform the indicated operations.

37. $\dfrac{12}{15} + \dfrac{8}{35}$ **38.** $\dfrac{13}{24} - \dfrac{3}{8}$ **39.** $1\dfrac{2}{3} \cdot 4\dfrac{1}{5}$ **40.** $\dfrac{6}{5} \div \dfrac{24}{45}$

Solve.

41. A wheel on a child's car is turning at a rate of $40\dfrac{1}{2}$ revolutions per minute. If the car is driven for $3\dfrac{1}{9}$ minutes, how many revolutions does the wheel make?

42. Marguerite is proofing a computer program that is $14\dfrac{1}{3}$ pages in length. If she has completed $10\dfrac{1}{5}$ pages, how much remains to be done?

1.3 EXERCISES C

Perform the indicated operation.

1.   374.26
  × 8.603

2. $0.0257\overline{)0.02584392}$ [Answer: 1.0056]

1.4 CONVERTING FRACTIONS, DECIMALS, AND PERCENTS

STUDENT GUIDEPOSTS

1 Converting fractions to decimals
2 Rational and irrational numbers
3 Converting terminating decimals to fractions
4 Percent
5 Converting percents to decimals and fractions
6 Converting decimals and fractions to percents

Every fraction can be converted to a decimal. Several problems in the previous section gave a preview of this process.

1.4 CONVERTING FRACTIONS, DECIMALS, AND PERCENTS

> **To convert a fraction to a decimal**
> 1. Divide the denominator of the fraction into the numerator.
> 2. Place the decimal point as in any decimal division problem.

EXAMPLE 1

Convert each fraction to a decimal.

(A) $\dfrac{3}{20}$

We divide 20 into 3 as follows:

$$\begin{array}{r} .15 \\ 20\overline{)3.00} \\ \underline{2\ 0} \\ 1\ 00 \\ \underline{1\ 00} \\ 0 \end{array}$$

$\dfrac{3}{20} = 0.15$

(B) $\dfrac{1}{3}$

$$\begin{array}{r} .333\ldots \\ 3\overline{)1.000} \\ \underline{9} \\ 10 \\ \underline{9} \\ 10 \\ \underline{9} \\ 1 \end{array}$$

It is clear that we will continue to obtain "3" on each division. When this happens we write $\dfrac{1}{3} = 0.333\ldots = 0.\overline{3}$, where the bar denotes the digit (sometimes digits) that continues to repeat.

(C) $\dfrac{2}{7}$

$$\begin{array}{r} .285714285714\ldots \\ 7\overline{)2.000000000000} \end{array}$$

$\dfrac{2}{7} = 0.285714285714\ldots$

$= 0.\overline{285714}$ ◀◀

Practice Exercise 1

Convert each fraction to a decimal.

(A) $\dfrac{1}{2}$

(B) $\dfrac{13}{40}$

(C) $\dfrac{2}{3}$

Answers: **(A)** 0.5 **(B)** 0.325 **(C)** $0.\overline{6}$

An important property of every fraction is that in decimal notation, the digits after the decimal either terminate, as in

$$\dfrac{1}{2} = 0.5 \qquad \dfrac{3}{20} = 0.15 \qquad \dfrac{13}{40} = 0.325,$$

or contain a block of digits that repeats without ending, as in

$$\dfrac{1}{3} = 0.\overline{3} \qquad \dfrac{2}{7} = 0.\overline{285714} \qquad \dfrac{2}{3} = 0.\overline{6}.$$

28 FUNDAMENTALS

It is also true that if a number has a decimal representation that terminates or repeats, the number can always be expressed as a fraction. This fact is often used to define the set of **rational numbers** or fractions. The rational numbers include all of the numbers we have considered thus far. Numbers that are not rational numbers, called **irrational numbers,** have decimal representations that neither terminate nor repeat. One of the most famous irrational numbers is π (the ratio of the circumference of a circle to its diameter). Other irrational numbers will be considered later.

We now turn our attention to converting a terminating decimal to a fraction. For example,

$$0.5 \text{ is } \frac{5}{10},$$

$$0.23 \text{ is } \frac{2}{10} + \frac{3}{100} = \frac{20}{100} + \frac{3}{100} = \frac{23}{100},$$

$$0.537 \text{ is } \frac{5}{10} + \frac{3}{100} + \frac{7}{1000} = \frac{500}{1000} + \frac{30}{1000} + \frac{7}{1000} = \frac{537}{1000}.$$

> **To convert a terminating decimal to a fraction**
> 1. Express the decimal in expanded notation.
> 2. Add the whole number terms and add the fractional terms to obtain a mixed number.
> 3. Convert the mixed number to a single fraction by addition.

EXAMPLE 2

Convert each decimal to a fraction.

(A) $0.47 = \frac{4}{10} + \frac{7}{100} = \frac{40}{100} + \frac{7}{100} = \frac{47}{100}$

(B) $2.35 = 2 + \frac{3}{10} + \frac{5}{100} = 2 + \frac{30}{100} + \frac{5}{100} =$

$2 + \frac{35}{100} = \frac{2 \cdot 100}{100} + \frac{35}{100} = \frac{235}{100}$

(C) $43.619 = 40 + 3 + \frac{6}{10} + \frac{1}{100} + \frac{9}{1000} = 43 + \frac{619}{1000}$

$= \frac{43619}{1000}$

Practice Exercise 2

Convert each decimal to a fraction.

(A) 0.11

(B) 3.86

(C) 51.015

Answers: (A) $\frac{11}{100}$ (B) $\frac{193}{50}$ (C) $\frac{10,203}{200}$

An important concept that is closely related to decimals and is used in a variety of applied problems is *percent*. The word **percent** literally means *per hundred*. That is, it refers to the number of parts in one hundred parts. For example, a five percent tax means five parts per one hundred parts, or a tax of

5¢ on every 100¢.

The symbol for the word *percent* is %, so that five percent would be written as 5%. Since the % symbol means "divide by 100" or "multiply by $\frac{1}{100} = 0.01$,"

$$5\% = 5(0.01) = 0.05 \quad \text{or} \quad 5\% = \frac{5}{100} = \frac{1}{20}.$$

1.4 CONVERTING FRACTIONS, DECIMALS, AND PERCENTS

> **To convert a percent to a decimal**
> Remove the % symbol and multiply by 0.01. OR, remove the % symbol and move the decimal point two places to the *left*, adding zeros as necessary.
>
> **To convert a percent to a fraction**
> Remove the % symbol and divide by 100. Reduce the resulting fraction to lowest terms.

EXAMPLE 3

Convert each percent to a decimal.

(A) $43\% = (43)(0.01) = 0.43$

$\quad\quad\quad\quad\quad\quad 43\% = 0.43$

(B) $325.5\% = (325.5)(0.01) = 3.255$

$\quad\quad\quad\quad\quad\quad 325.5\% = 3.255$

(C) $0.05\% = (0.05)(0.01) = 0.0005$

$\quad\quad\quad\quad\quad\quad 0.05\% = 0.0005$

(D) $0.2\% = (0.2)(0.01) = 0.002$

$\quad\quad\quad\quad\quad\quad 0.2\% = 0.002$ ◀◀

Practice Exercise 3

Convert each percent to a decimal.

(A) 79%

(B) 640.5%

(C) 0.09%

(D) 0.8%

Answers: (A) 0.79 (B) 6.405
(C) 0.0009 (D) 0.008

EXAMPLE 4

Convert each percent to a fraction.

(A) $28\% = \dfrac{28}{100} = \dfrac{\cancel{4}\cdot 7}{\cancel{4}\cdot 25} = \dfrac{7}{25}$

(B) $37.5\% = \dfrac{37.5}{100} = \dfrac{(37.5)(10)}{(100)(10)}$ Clear decimal by multiplying by 10

$\quad\quad\quad = \dfrac{375}{1000}$

$\quad\quad\quad = \dfrac{3\cdot\cancel{125}}{8\cdot\cancel{125}} = \dfrac{3}{8}$ Reduce fraction

(C) $66\dfrac{2}{3}\% = \left(66\dfrac{2}{3}\right)\left(\dfrac{1}{100}\right)$ Dividing by 100 is the same as multiplying by $\frac{1}{100}$

$\quad\quad\quad = \left(\dfrac{200}{3}\right)\left(\dfrac{1}{100}\right)$ $66\dfrac{2}{3} = \dfrac{(3\cdot 66)+2}{3} = \dfrac{200}{3}$

$\quad\quad\quad = \dfrac{2\cdot\cancel{100}}{3\cdot\cancel{100}} = \dfrac{2}{3}$ Reduce fraction

(D) $100\% = \dfrac{100}{100} = 1$ ◀◀

Practice Exercise 4

Convert each percent to a fraction.

(A) 65%

(B) 42.5%

(C) $16\dfrac{1}{6}\%$

(D) 500%

Answers: (A) $\dfrac{13}{20}$ (B) $\dfrac{17}{40}$
(C) $\dfrac{97}{600}$ (D) 5

> **To convert a decimal to a percent**
>
> Multiply by 100 and attach the % symbol. Or, move the decimal point two places to the *right*, adding zeros as necessary, and attach the % symbol.
>
> **To convert a fraction to a percent**
>
> Convert the fraction to a decimal by dividing the numerator by the denominator. Then change the resulting decimal to a percent.

EXAMPLE 5

Convert each decimal to a percent.

(A) $0.37 = (0.37)(100\%) = 37\%$ $\quad 0.37 = 37.\% = 37\%$

(B) $5.81 = (5.81)(100)\% = 581\%$ $\quad 5.81 = 581.\% = 581\%$

(C) $0.0004 = (0.0004)(100)\% = 0.04\%$ $\quad 0.0004 = 000.04\% = 0.04\%$

(D) $0.\overline{3} = (0.\overline{3})(100)\% = (0.333\ldots)(100)\%$

$\qquad = 33.33\ldots\%$

$\qquad = 33.\overline{3}\%$ or $33\frac{1}{3}\%$ $\quad 0.\overline{3} = \frac{1}{3}$ ◂◂

Practice Exercise 5

Convert each decimal to a percent.

(A) 0.81

(B) 9.05

(C) 0.003

(D) $0.\overline{7}$

Answers: (A) 81% (B) 905%
(C) 0.3% (D) $77.\overline{7}\%$

EXAMPLE 6

Convert each fraction to a percent.

(A) $\frac{3}{4} = 0.75$ Divide 4 into 3

$\qquad = (0.75)(100)\% = 75\%$ $\quad 0.75 = 075.\% = 75\%$

(B) $\frac{1}{8} = 0.125$ Divide 8 into 1

$\qquad = (0.125)(100)\% = 12.5\%$ $\quad 0.125 = 012.5\% = 12.5\%$

(C) $\frac{5}{3} = 1.\overline{6}$ Divide 3 into 5

$\qquad = (1.\overline{6})(100)\%$

$\qquad = (1.666\ldots)(100)\%$

$\qquad = 166.666\ldots\%$

$\qquad = 166.\overline{6}\%$ or $166\frac{2}{3}\%$ $\quad 0.\overline{6} = \frac{2}{3}$ ◂◂

Practice Exercise 6

Convert each fraction to a percent.

(A) $\frac{4}{5}$

(B) $\frac{7}{10}$

(C) $\frac{1}{9}$

Answers: (A) 80% (B) 70%
(C) $11.\overline{1}\%$

Keeping a simple example in mind will help you remember whether you should multiply by 0.01 or by 100 when making percent conversions. For example, 50% of some quantity is 0.5 or $\frac{1}{2}$ of that quantity. Thus, to convert

$\qquad$ 50% to 0.5—multiply by 0.01

and to convert

$\qquad$ 0.5 to 50%—multiply by 100.

1.4 CONVERTING FRACTIONS, DECIMALS, AND PERCENTS

Many applied problems involve finding a percent of some number. For example, we might ask

What is 5% of $40.00?

Such problems are solved by changing the percent to a decimal or fraction and then multiplying it by the given number. The word *of* means "multiply" or "times" in this setting.

$$5\% \text{ of } \$40.00$$
$$(0.05) \cdot (40.00)$$

Thus, 5% of $40.00 is (0.05)(40.00) = $2.00.

1.4 EXERCISES A

Convert each fraction to a decimal.

1. $\dfrac{1}{8}$ 2. $\dfrac{7}{20}$ 3. $\dfrac{5}{9}$

4. $\dfrac{8}{3}$ 5. $\dfrac{5}{6}$ 6. $2\dfrac{1}{7}$

Convert each decimal to a fraction.

7. 0.3   8. 0.007   9. 0.48

10. 27.95   11. 3.207   12. 25.001

Convert each percent to a decimal.

13. 65.5%   14. $\dfrac{1}{2}\%$ 15. 0.6%

16. 200%   17. $8\dfrac{1}{4}\%$ 18. $\dfrac{1}{3}\%$

Convert each percent to a fraction.

19. 57%   20. 100%   21. $\dfrac{1}{4}\%$

22. $3\dfrac{2}{3}\%$ 23. 1000%   24. $\dfrac{1}{10}\%$

Convert each decimal to a percent.

25. 2.37   26. 0.06   27. 0.005

28. $0.62\overline{3}$ **29.** $0.1\overline{6}$ **30.** 1.1

Convert each fraction to a percent.

31. $\dfrac{1}{2}$ **32.** $\dfrac{6}{5}$ **33.** $\dfrac{10}{3}$

34. $\dfrac{7}{50}$ **35.** $\dfrac{5}{6}$ **36.** $\dfrac{50}{3}$

Solve.

37. A basketball player made 15 shots in 20 attempts during a recent game. What fractional part of his shots did he make? What percent of his shots did he make?

38. Bill Schulz earned $24,000 last year and gave $3000 to charity. What fractional part of his income was given to charity? What percent of his income was given to charity?

39. There were 5000 votes cast in an election. Mr. Gomez received 55% of the votes. How many votes did he receive?

40. Snow, Montana has a normal annual snowfall of 120 inches. This year the city received 150% of normal. How many inches of snow fell this year?

For Review

41. Express 32.51 in expanded notation.

42. Add: 403.2
              1.006
            + 0.04

43. Subtract: $3.21 - 0.0005$

44. Multiply: 0.0014
                × 1.06

45. Divide: $0.004\overline{)0.086}$

46. John Hagood rented a Chevette for 3 days. If the rental fee was $8.95 per day plus $0.14 per mile, how much was he charged if he drove a total of 185.5 miles?

ANSWERS: **1.** 0.125 **2.** 0.35 **3.** $0.\overline{5}$ **4.** $2.\overline{6}$ **5.** $0.8\overline{3}$ **6.** $2.\overline{142857}$ **7.** $\dfrac{3}{10}$ **8.** $\dfrac{7}{1000}$ **9.** $\dfrac{12}{25}$ **10.** $\dfrac{559}{20}$ **11.** $\dfrac{3207}{1000}$ **12.** $\dfrac{25001}{1000}$ **13.** 0.655 **14.** 0.005 **15.** 0.006 **16.** 2 **17.** 0.0825 **18.** $0.00\overline{3}$ **19.** $\dfrac{57}{100}$ **20.** 1 **21.** $\dfrac{1}{400}$ **22.** $\dfrac{11}{300}$ **23.** 10 **24.** $\dfrac{1}{1000}$ **25.** 237% **26.** 6% **27.** 0.5% **28.** $62.\overline{3}$% or $62\dfrac{1}{3}$% **29.** $16.\overline{6}$% or $16\dfrac{2}{3}$% **30.** 110% **31.** 50% **32.** 120% **33.** $333\dfrac{1}{3}$% **34.** 14% **35.** $83\dfrac{1}{3}$% **36.** $1666\dfrac{2}{3}$% **37.** $\dfrac{3}{4}$, 75%

38. $\frac{1}{8}$, 12.5% **39.** 2750 votes **40.** 180 inches **41.** $30 + 2 + \frac{5}{10} + \frac{1}{100}$ **42.** 404.246 **43.** 3.2095
44. 0.001484 **45.** 21.5 **46.** $52.82

1.4 EXERCISES B

Convert each fraction to a decimal.

1. $\frac{5}{8}$
2. $\frac{5}{16}$
3. $\frac{11}{15}$

4. $\frac{4}{5}$
5. $\frac{1}{6}$
6. $2\frac{3}{7}$

Convert each decimal to a fraction.

7. 0.09
8. 0.002
9. 2.6

10. 31.85
11. 20.02
12. 35.001

Convert each percent to a decimal.

13. 35.5%
14. $3\frac{1}{2}$%
15. 26.54%

16. 300%
17. $7\frac{3}{4}$%
18. $\frac{2}{3}$%

Convert each percent to a fraction.

19. 90%
20. 500%
21. 275%

22. $4\frac{1}{3}$%
23. 2000%
24. $\frac{1}{100}$%

Convert each decimal to a percent.

25. 4.29
26. 0.01
27. 0.007

28. $0.57\overline{3}$
29. 7
30. 2.2

Convert each fraction to a percent.

31. $\frac{3}{5}$
32. $\frac{11}{3}$
33. $\frac{9}{50}$

34. $\dfrac{3}{10}$ **35.** $\dfrac{5}{8}$ **36.** $\dfrac{25}{3}$

Solve.

37. A football quarterback completed 21 passes in 35 attempts during a game last fall. What fractional part of his attempts did he complete? What percent of his attempts did he complete?

38. Amy received $12,000 from her grandfather. She spent $8000 on a new car. What fractional part of the money went for the car? What percent of the money went for the car?

39. There are 0.2% impurities in a laboratory solution weighing 485 grams. What is the weight of the impurities?

40. Diane received a raise giving her a salary equal to 120% of her previous salary. If she earned $16,000 last year, how much will she earn this year?

For Review

41. Express 29.65 in expanded notation.

42. Add: 209.6
           4.007
       +   0.01

43. Subtract: $4.81 - 0.0002$

44. Multiply: 0.0015
            ×   1.08

45. Divide: $0.003\overline{)0.1488}$

46. Ben Whitney rented a midsize car for two days. If the rental fee was $11.95 per day plus $0.18 per mile driven, how much was he charged if he drove a total of 230.5 miles?

1.4 EXERCISES C

Solve.

1. A retailer paid $92.00 for a dress and marked it up 30% to sell it in her store. When the dress did not sell, she sold it at a 25% discount on the marked price. What was the percent profit or loss based on the original cost?   [Answer: 2.5% loss]

2. A salesman receives a 3% commission on all sales of $50,000 or less and a 4% commission on the amount he sells over $50,000. What is his commission on sales of $85,000? What is the effective total commission rate on the $85,000 sales?

1.5 VARIABLES, EXPONENTS, AND ORDER OF OPERATIONS

STUDENT GUIDEPOSTS

1 Variable
2 Exponential notation
3 Evaluating numerical expressions
4 Symbols of grouping
5 Order of operations
6 Evaluating algebraic expressions
7 Formulas

  In algebra, we often use lower case letters such as *a, b, x,* and *y* to represent numbers. A letter that can be replaced by various numbers is called a **variable.** Lengthy verbal expressions can often be symbolized by brief algebraic expressions using variables, as in the following examples.

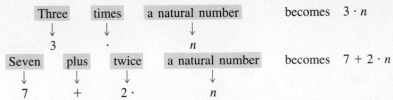

In each example, the variable *n* represents a natural number. Other notations for $3 \cdot n$ are $3(n)$, $(3)(n)$, and $3n$, with the last form preferred. The product of two natural numbers *m* and *n* would most likely be represented by *mn*, but could also be expressed by $m \cdot n$, $m(n)$, and $(m)(n)$. We usually avoid using the symbol × for multiplication since it can be confused with the letter *x* used as a variable.

When a variable or number is multiplied by itself several times, as in

$$2 \cdot 2 \cdot 2 \cdot 2 \text{ or } x \cdot x \cdot x,$$

we can use **exponential notation** to avoid writing long strings of **factors** (the individual numbers in the expressed product). For example, we write $2 \cdot 2 \cdot 2 \cdot 2$ as 2^4,

$$\underbrace{2 \cdot 2 \cdot 2 \cdot 2}_{4 \text{ factors}} = 2^{4}, \leftarrow \text{exponent} \\ \qquad\qquad\uparrow \\ \qquad\quad \text{base}$$

where 2 is called the **base**, 4 the **exponent**, and 2^4 the **exponential expression** (read 2 to the **fourth power**). Similarly, $x \cdot x \cdot x = x^3$ is called the **third power** or **cube** of *x*. The **square** or **second power** of *a* is a^2. The **first power** of *a* is a^1, which we write simply as *a*.

If *a* is any number and *n* is a natural number,

$$a^n = \underbrace{a \cdot a \cdot a \cdots a.}_{n \text{ factors}}$$

EXAMPLE 1

Write an exponential notation.

(A) $\underbrace{7 \cdot 7 \cdot 7}_{3 \text{ factors}} = 7^3$

(B) $\underbrace{a \cdot a \cdot a \cdot a \cdot a \cdot a}_{6 \text{ factors}} = a^6$

(C) $\underbrace{3 \cdot 3}_{2 \text{ factors}} \cdot \underbrace{x \cdot x \cdot x \cdot x \cdot x}_{5 \text{ factors}} = 3^2 x^5$ ◂◂

Practice Exercise 1

Write in exponential notation.

(A) $4 \cdot 4 \cdot 4 \cdot 4 \cdot 4 \cdot 4 \cdot 4$

(B) $z \cdot z \cdot z \cdot z$

(C) $6 \cdot 6 \cdot 6 \cdot y \cdot y \cdot y \cdot y$

Answers: **(A)** 4^7 **(B)** z^4 **(C)** $6^3 y^4$

When exponents or powers are used together with the operations of addition, subtraction, multiplication, and division, the resulting numerical expressions can be confusing unless we agree to an order of operations. For example,

$2 \cdot 5^2$ *could* equal 10^2 or 100,   If we first multiply then square

or $2 \cdot 5^2$ *could* equal $2 \cdot 25$ or 50.   If we first square then multiply

Similarly,

$2 \cdot 3 + 4$ *could* equal $2 \cdot 7$ or 14,   If we first add then multiply

or $2 \cdot 3 + 4$ *could* equal $6 + 4$ or 10.   If we first multiply then add

According to the following rule, we see that the second procedure in each of these examples is the correct one to use.

> **To evaluate a numerical expression**
> 1. Evaluate all powers, in any order, first.
> 2. Do all multiplications and divisions in order from left to right.
> 3. Do all additions and subtractions in order from left to right.

EXAMPLE 2

Evaluate each numerical expression.

(A) $2 + 3 \cdot 4 = 2 + 12 = 14$ Multiply first, then add

(B) $20 - 3^2 = 20 - 9 = 11$ Square first, then subtract

(C) $5 \cdot 6 - 12 \div 3 = 30 - 4 = 26$ Multiply and divide first, then subtract

(D) $25 \div 5 + 3 \cdot 2^3 = 25 \div 5 + 3 \cdot 8$ Cube first
$= 5 + 24$ Divide and multiply second
$= 29$ Add last

Practice Exercise 2

Evaluate each numerical expression.

(A) $5 \cdot 2 + 1$

(B) $6 + 2^3$

(C) $18 \div 9 + 3 \cdot 2$

(D) $16 \cdot 2 - 3 \cdot 2^2$

Answers: **(A)** 11 **(B)** 14 **(C)** 8 **(D)** 20

Suppose that we want to evaluate three times the sum of 2 and 5. If we write $3 \cdot 2 + 5$ and use the above rule, we will obtain $6 + 5$ or 11. However, it is clear from the first sentence, that we want 3 times 7 or 21 for the result. **Symbols of grouping,** such as parentheses ( ), square brackets [ ], or braces { }, can help us symbolize the problem correctly as $3 \cdot (2 + 5)$. The grouping symbols contain the expression that must be evaluated first. In this case, we must add before multiplying:

$$3 \cdot (2 + 5) = 3 \cdot (7) = 21.$$

Generally, we omit the dot and write $3(2 + 5)$ for $3 \cdot (2 + 5)$. Also, instead of writing $3 \cdot (7)$, we will remove the parentheses and write $3 \cdot 7$.

One other method of grouping is to use a fraction bar. For example, in

$$\frac{4 \cdot 5}{7 + 3},$$

the fraction bar acts like parentheses since the expression is the same as

$$(4 \cdot 5) \div (7 + 3).$$

We first multiply 4 and 5, then add 7 and 3, and finally we divide the results. Thus,

$$\frac{4 \cdot 5}{7 + 3} = \frac{20}{10} = 2.$$

> **To evaluate a numerical expression involving grouping symbols**
> Evaluate all expressions within the grouping symbols first, and begin with the innermost symbols of grouping if more than one set of symbols is present.

1.5 VARIABLES, EXPONENTS, AND ORDER OF OPERATIONS

EXAMPLE 3

Evaluate each numerical expression.

(A) $3 + (2 \cdot 5) = 3 + 10 = 13$ Evaluate inside the parentheses first

(B) $(4 + 5)2 + 3 = (9)2 + 3$ Work inside the parentheses first
$= 18 + 3$ Multiply before adding
$= 21$

(C) $[3(8 + 2) + 1]4 = [3(10) + 1]4$ Innermost grouping symbol first
$= [30 + 1]4$ Multiply before adding inside brackets
$= [31]4$ Combine numbers inside brackets before multiplying ◀◀
$= 124$

Practice Exercise 3

Evaluate each numerical expression.

(A) $3 \cdot (2 + 5)$

(B) $(6 - 1)3 - 2$

(C) $2[2(7 + 2) - 1]$

Answers: (A) 21   (B) 13   (C) 34

We now summarize the order to be followed when evaluating a numerical expression.

Order of operations

1. Evaluate within grouping symbols first, beginning with the innermost set if more than one set is used.
2. Evaluate all powers.
3. Perform all multiplications and divisions in order from left to right.
4. Perform all additions and subtractions in order from left to right.

EXAMPLE 4

Evaluate each numerical expression.

(A) $2^3 + 4^3 = 8 + 64 = 72$ Cube first, then add

(B) $(2 + 4)^3 = 6^3 = 216$ Add first inside parentheses, then cube

(C) $(3 \cdot 5)^2 = 15^2 = 225$ Multiply first, then square

(D) $3 \cdot 5^2 = 3 \cdot 25 = 75$ Square first, then multiply

(E) $5^2 - 4^2 = 25 - 16 = 9$ Square first, then subtract

(F) $(5 - 4)^2 = 1^2 = 1$ Subtract first, then square

(G) $3^2 - 15 \div 5 + 5 \cdot 2 = 9 - 15 \div 5 + 5 \cdot 2$ Square first
$= 9 - 3 + 10$ Divide and multiply in order
$= 16$ Subtract and add in order ◀◀

Practice Exercise 4

Evaluate each numerical expression.

(A) $1^2 + 6^2$

(B) $(1 + 6)^2$

(C) $(2 \cdot 7)^2$

(D) $2 \cdot 7^2$

(E) $8^2 - 6^2$

(F) $(8 - 6)^2$

(G) $2^3 - 20 \div 4 + 4 \cdot 3$

Answers: (A) 37   (B) 49
(C) 196   (D) 98   (E) 28   (F) 4
(G) 15

▶▶ **CAUTION** ◀◀   Example 4 indicates common errors to avoid when working with exponents. For example,

$$2^3 + 4^3 \neq (2 + 4)^3, \quad (3 \cdot 5)^2 \neq 3 \cdot 5^2, \quad 5^2 - 4^2 \neq (5 - 4)^2. \quad ◀◀$$

The symbol $\neq$ means "is *not* equal to." In general, if a, b, and u are numbers,
$$a^u + b^u \neq (a + b)^u, \quad (ab)^u \neq ab^u, \quad a^u - b^u \neq (a - b)^u.$$

6 An **algebraic expression** contains variables as well as numbers. We can use the order of operations for evaluating numerical expressions to evaluate algebraic expressions when specific values for the variables are given.

> **To evaluate an algebraic expression**
> 1. Replace each variable (letter) with the specified value.
> 2. Proceed as in evaluating numerical expressions.

EXAMPLE 5

Evaluate $2(a + b) - c$ when $a = 3$, $b = 4$, and $c = 5$.

$\begin{aligned} 2(a + b) - c &= 2(3 + 4) - 5 &&\text{Replace each letter with given value} \\ &= 2(7) - 5 &&\text{Evaluate inside parentheses first} \\ &= 14 - 5 = 9 \end{aligned}$

Practice Exercise 5

Evaluate $3(x - y) + z$, when $x = 4$, $y = 2$, and $z = 1$.

Answer: 7

EXAMPLE 6

Evaluate $5[12 - 3(a + 1) + b] - c$ when $a = 2$, $b = 7$, and $c = 4$.

$\begin{aligned} 5[12 - 3(a + 1) + b] - c &= 5[12 - 3(2 + 1) + 7] - 4 \\ &= 5[12 - 3(3) + 7] - 4 \\ &= 5[12 - 9 + 7] - 4 \\ &= 5[10] - 4 = 50 - 4 = 46 \end{aligned}$

Replace variables with numbers
Multiply before adding or subtracting

Practice Exercise 6

Evaluate $w - 2[3(u - 1) + v]$, when $u = 3$, $v = 1$, and $w = 14$.

Answer: 0

EXAMPLE 7

Evaluate $\dfrac{ab - 1}{c}$ for the given values.

(A) $a = 2$, $b = 3$, and $c = 0$

$$\frac{ab - 1}{c} = \frac{(2)(3) - 1}{0} \quad \text{Division by zero undefined}$$

That is, this expression is undefined when $c = 0$.

(B) $a = 3$, $b = \dfrac{1}{3}$, and $c = 5$

$$\frac{ab - 1}{c} = \frac{(3)\left(\dfrac{1}{3}\right) - 1}{5} = \frac{1 - 1}{5} = \frac{0}{5} = 0 \quad \frac{0}{5} = 0$$

Practice Exercise 7

Evaluate $\dfrac{3x}{yz - x}$ for the given values.

(A) $x = 0$, $y = 5$, $z = 8$

(B) $x = 6$, $y = 2$, $z = 3$

Answers: **(A)** 0 **(B)** undefined

EXAMPLE 8

Evaluate the following when $a = 2$, $b = 1$, $c = 3$.

(A) $3a^2 = 3(2)^2$ Not $(3 \cdot 2)^2 = 6^2 = 36$
$= 3 \cdot 4 = 12$

(B) $2ab^2c^3 = 2(2)(1)^2(3)^3$
$= 2(2)(1)(27) = 4 \cdot 27 = 108$

(C) $(2c)^2 - 2c^2 = (2 \cdot 3)^2 - 2(3)^2$ Watch the substitution
$= 6^2 - 2 \cdot 9$
$= 36 - 18 = 18$

(D) $a^a + c^c = 2^2 + 3^3$
$= 4 + 27 = 31$ ◀◀

Practice Exercise 8

Evaluate the following when $u = 5$, $v = 3$, and $w = 1$.

(A) $2u^2$

(B) $3uv^2w$

(C) $(3u)^2 - 3u^2$

(D) $v^v - w^w$

Answers: (A) 50 (B) 135
(C) 150 (D) 26

---

7 A **formula** is an algebraic statement that relates two or more quantities. For example,

$$d = rt$$

is a formula that relates the distance, d, traveled by an object moving at an average rate, r, for a period of time, t. Other familiar formulas, related to geometric figures, such as

$A = lw$ Area of a rectangle in terms of its length and width

and $P = 2l + 2w$ Perimeter of a rectangle in terms of its length and width

are summarized on the inside cover of the text. In order to use a particular formula in an applied problem, we must first understand the meaning of the variables, and next, evaluate the algebraic expression (formula) for the given values.

EXAMPLE 9

If we invest P dollars (called the principal) at an interest rate, r, compounded annually for a period of t years, it will grow to an amount, A, given by

$$A = P(1 + r)^t.$$

If a principal of $1000 is invested at 12% interest, compounded annually, how much will be in the account at the end of two years?

$A = P(1 + r)^t$ Start with the formula
$= 1000(1 + 0.12)^2$ Substitute 1000 for P, 0.12 for r, and 2 for t
$= 1000(1.12)^2$
$= 1000(1.2544)$
$= 1254.4$

Thus, there will be $1254.40 in the account. Notice that 12% was converted to a decimal, 0.12, in order to solve this problem. ◀◀

Practice Exercise 9

The perimeter of a rectangle, P, is given by $P = 2l + 2w$, where l is its length and w is its width. What is the perimeter of a rectangle having a length of 15 inches and a width of 8 inches?

Answer: 46 inches

1.5 EXERCISES A

1. What is a letter used to represent a number called?
2. Given the exponential expression x^7, (A) what is the base? (B) what is the exponent?
3. In the expression $3x$, what is the exponent on x?

Write in exponential notation.

4. $8 \cdot 8 \cdot 8 \cdot 8 \cdot 8$
5. $(2x)(2x)$
6. $2 \cdot x \cdot x$

7. $(b + c)(b + c)(b + c)$
8. $7aaaaaaa$
9. $6 \cdot 6 \cdot 6 \cdot y \cdot y \cdot z \cdot z \cdot z \cdot z$

Write without using exponents.

10. $a^2 b^3 c^4$
11. $3y^3$
12. $(3y)^3$

13. 1^{51}
14. $a^3 - c^3$
15. $\left(\dfrac{2}{3}\right)^3 x^2$

Evaluate.

16. $3 \cdot 2^2$
17. $(3 \cdot 2)^2$
18. $(3 + 2)^2$

19. $3^2 + 2^2$
20. $(5 - 2)^3$
21. $5^3 - 2^3$

22. $8 - 2^3$
23. $(8 - 2)^3$
24. $(7 - 7)^3$

25. $9 - 2 \cdot 3 + 1$
26. $(9 - 2) \cdot 3 + 4 \cdot 0$
27. $\dfrac{2(3 - 1) - 4}{5}$

28. $\dfrac{4(3 + 5) - 10}{0}$
29. $12 \div 4 + 2 \cdot 3 - 1$
30. $2[8 - 2(4 - 1) + 5]$

31. $15 - 3\{2(5 - 4) + 3\}$
32. $10 - 2[8 - 2(7 - 3)]$
33. $16 \div 4 \cdot 2 \div 4 - 2$

Evaluate when $a = 2$, $b = 3$, $c = 5$, $d = 0$, and $x = 12$.

34. $6a + b$
35. $6(a + b)$
36. $3ab \div d$

37. $2b^2$
38. $(2b)^2$
39. $(c - b)^2$

40. $c^2 - b^2$ **41.** $(a + b)^3$ **42.** $a^3 + b^3$

43. $3a^3 + 2$ **44.** $(3a)^3 + 2$ **45.** $a^a + b^b$

46. $\dfrac{2abcd - 3d}{x}$ **47.** $x - [a(b + 1) - c]$ **48.** $5a + 2[c + x \div b]$

Solve.

49. The perimeter of a rectangle, P, is given by $P = 2l + 2w$, where l is its length and w is its width. What is the perimeter of a rectangle of length 20 feet and width 13 feet?

50. The distance an automobile travels, d, at an average rate of speed, r, for a period of time, t, is given by $d = rt$. How far does a car travel in 11 hours at an average speed of 55 mph?

51. Use $A = P(1 + r)^t$ to find the amount of money in an account at the end of two years if a principal of $500 is invested at 14% interest, compounded annually.

52. The area of a square, A, is given by $A = s^2$, where s is the length of a side. What is the area of a square with side 3.5 cm?

53. The area of a triangle, A, is given by $A = \frac{1}{2} bh$, where b is the length of its base and h is its height. Find the area of a triangle with height 15 m and base 8 m.

54. The temperature measured in degrees Celsius, °C, can be obtained from degrees Fahrenheit, °F, by using $C = \frac{5}{9}(F - 32)$. Find C when F is 200°.

55. The surface area of a cube, A, is given by $A = 6e^2$ where e is the length of an edge. Find the surface area of a cube with an edge of $2\frac{1}{2}$ inches.

56. The surface area of a cylinder, A, with height, h, and base radius, r, is given by $A = 2\pi rh + 2\pi r^2$. Use 3.14 for π and find the surface area of a cylinder with a radius of 2 cm and a height of 10 cm.

For Review

57. Convert 3.7 to a fraction.

58. Convert $\frac{5}{9}$ to a decimal.

59. Convert 0.085 to a percent.

60. Convert $\frac{5}{11}$ to a percent.

61. Convert $7\frac{3}{4}\%$ to a decimal.

62. A chemist has a solution that is 0.6% salt. If the solution weighs 650 g, what is the weight of the salt?

ANSWERS: **1.** variable **2. (A)** x **(B)** 7 **3.** 1 **4.** 8^5 **5.** $(2x)^2$ **6.** $2x^2$ **7.** $(b+c)^3$, not b^3+c^3 **8.** $7a^7$ **9.** $6^3y^2z^4$ **10.** $aabbbcccc$ **11.** $3yyy$ **12.** $(3y)(3y)(3y)$ **13.** 1 **14.** $aaa - ccc$ **15.** $\left(\frac{2}{3}\right)\left(\frac{2}{3}\right)\left(\frac{2}{3}\right)xx$ or $\frac{8}{27}xx$ **16.** 12 **17.** 36 **18.** 25 **19.** 13 **20.** 27 **21.** 117 **22.** 0 **23.** 216 **24.** 0 **25.** 4 **26.** 21 **27.** 0 **28.** undefined **29.** 8 **30.** 14 **31.** 0 **32.** 10 **33.** 0 **34.** 15 **35.** 30 **36.** undefined **37.** 18 **38.** 36 **39.** 4 **40.** 16 **41.** 125 **42.** 35 **43.** 26 **44.** 218 **45.** 31 **46.** 0 **47.** 9 **48.** 28 **49.** 66 ft **50.** 605 miles **51.** $649.80 **52.** 12.25 cm² **53.** 60 m² **54.** $93.\overline{3}°$ **55.** 37.5 in² **56.** 150.72 cm² **57.** $\frac{37}{10}$ **58.** $0.\overline{5}$ **59.** 8.5% **60.** $45.\overline{45}\%$ **61.** 0.0775 **62.** 3.9 g

1.5 EXERCISES B

1. When evaluating a numerical expression involving more than one set of grouping symbols, always evaluate within which set of symbols first?

2. Given the exponential expression a^6, **(A)** what is the base? **(B)** what is the exponent?

3. In the expression $8y$, what is the exponent on y?

Write in exponential notation.

4. $c \cdot c \cdot c \cdot c \cdot c$ **5.** $3 \cdot 3 \cdot y \cdot y \cdot y \cdot y \cdot y \cdot y$ **6.** $(3a)(3a)$

7. $3 \cdot a \cdot a$ **8.** $(x-y)(x-y)(x-y)$ **9.** $5 \cdot 5 \cdot 5 \cdot a \cdot a \cdot c \cdot c \cdot c$

Write without using exponents.

10. $a^3b^4c^2$ **11.** $2z^3$ **12.** $(2z)^3$

13. $x^2 - y^2$ **14.** $a^3 + b^3$ **15.** $\left(\frac{1}{2}\right)^3 a^2$

Evaluate.

16. $2 \cdot 7^2$ **17.** $(2 \cdot 7)^2$ **18.** $(2+7)^2$

19. $2^2 + 7^2$ **20.** $(4-3)^2$ **21.** $4^2 - 3^2$

22. $4 - 2^2$ **23.** $(4-2)^2$ **24.** $(3-3)^3$

25. $14 - 5 + 2 \cdot 6$ **26.** $0 \cdot (5-2) + 0$ **27.** $\frac{2(5+1) - 3}{3}$

28. $\frac{5(2+8) - 19}{0}$ **29.** $25 \div 5 + 2 \cdot 4 - 1$ **30.** $3[10 - 2(3-1) + 6]$

31. $20 - 2[3(6-5) + 4]$ **32.** $30 - 5[7 - 2(6-3)]$ **33.** $20 \div 5 \cdot 2 \div 4 - 2$

Evaluate when $x = 3$, $y = 2$, $z = 4$, $w = 0$, and $a = 24$.

34. $5x + y$ **35.** $5(x + y)$ **36.** $4xy \div w$

37. $3y^2$ **38.** $(3y)^2$ **39.** $(z - y)^2$

40. $z^2 - y^2$
41. $(x + z)^3$
42. $x^3 + z^3$
43. $2z^3 + 1$
44. $(2z)^3 + 1$
45. $x^y + y^x$
46. $\dfrac{3wxyz - 5w}{a}$
47. $a - [x(y + 3) - z]$
48. $3x + 4[x + z \div y]$

Solve.

49. The perimeter of a square, P, with side of length s is given by $P = 4s$. What is the perimeter of a square with side of length 15 feet?

50. The area of a rectangle, A, is given by $A = lw$, where l is its length and w its width. If a rectangle is 2 cm wide and 7 cm long, what is its area?

51. Use the formula $A = P(1 + r)^t$ to find the amount of money in an account at the end of two years if a principal of $2000 is invested at 8% interest, compounded annually.

52. The area of a circle, A, with radius r, is given by $A = \pi r^2$. Use 3.14 for π and find the area of a circle with radius 20 m.

53. The volume, V, of a box with length l, width w, and height h, is given by $V = lwh$. If the box is 9 inches long, 6 inches wide, and 11 inches high, what is its volume?

54. The temperature measured in degrees Fahrenheit, °F, can be obtained from degrees Celsius, °C, by using $F = \frac{9}{5}C + 32$. Find F when C is 20°.

55. The surface area of a sphere, A, with radius r, is given by $A = 4\pi r^2$. Use 3.14 for π and find the surface area of a sphere with radius 50 cm.

56. The surface area of a cylinder, A, is given by $A = 2\pi rh + 2\pi r^2$, where r is the radius of its base and h is its height. Find A if $r = 10$ inches, $h = 8$ inches, and $\pi = 3.14$.

For Review

57. Convert 8.3 to a fraction.

58. Convert $\frac{6}{11}$ to a decimal.

59. Convert 0.017 to a percent.

60. Convert $\frac{8}{9}$ to a percent.

61. Convert $10\frac{1}{4}\%$ to a decimal.

62. It has been estimated that 32% of the population of Northern, Minnesota is of Danish ancestry. If 19,270 people live in Northern, approximately how many of them are of Danish background?

1.5 EXERCISES C

Evaluate.

1. $2(5 - 3)^2 - [7 - (8 - 6)^2]$
2. $13 - [11 - (7 - 4)^2]^3$
3. $[(6 \div 3)^2 + 9 \cdot (3 - 1)^2]^2$
   [Answer: 1600]

CHAPTER 1 SUMMARY

Key Words and Phrases for Review

1.1   set
       element
       numeral
       natural (counting) numbers
       whole numbers
       proper fraction
       improper fraction
       equivalent fractions

       built up fractions
       reduced to lowest terms
       prime number
       canceling

1.2   reciprocal
       least common denominator (LCD)
       mixed number

1.3
place-value system
digit
expanded numeral
decimal

1.4
rational number
irrational number
percent

1.5
variable
exponent
power (first, square, cube, etc.)
base
symbols of grouping
algebraic expression

Key Concepts

1.1
1. When reducing fractions by canceling common factors, do not make the numerator 0 when it is actually 1. For example,

$$\frac{5}{5 \cdot 3} = \frac{\cancel{5}}{\cancel{5} \cdot 3} = \frac{1}{3} \text{ NOT } \frac{0}{3}$$

2. Factors common to the numerator and denominator of a fraction may be divided or canceled. However, do not cancel numbers that are not factors. For example,

$$\frac{2 \cdot 3}{2} = \frac{\cancel{2} \cdot 3}{\cancel{2}} = 3,$$

but $\frac{2 + 3}{2}$ is *not* equal to $\frac{\cancel{2} + 3}{\cancel{2}}$

1.2
1. To add or subtract fractions with different denominators, find the LCD. However, *do not* find the LCD to multiply or divide fractions.

2. For any nonzero number a, $\frac{0}{a} = 0$, but $\frac{a}{0}$ is undefined.

1.4
1. To convert **a percent to a decimal,** remove the % symbol and multiply by 0.01 (move the decimal point two places to the *left*).

For example,
$$48\% = 48\,(0.01) = 0.48$$

2. To convert **a decimal to a percent,** multiply by 100 (move the decimal point two places to the *right*) and attach the % symbol. For example,

$$0.172 = 0.172\,(100)\% = 17.2\%.$$

1.5
1. In an expression such as $2y^3$, only y is cubed—not $2y$. That is, $2y^3$ is not the same as $(2y)^3$.

2. When simplifying a numerical expression,
   - First: Evaluate within grouping symbols.
   - Second: Evaluate all powers.
   - Third: Perform all multiplications and divisions in order from left to right.
   - Fourth: Perform all additions and subtractions in order from left to right.

Review Exercises

1.1
1. Express 490 as a product of primes.

2. Reduce $\frac{110}{385}$ to lowest terms.

3. Write an equivalent fraction for $\frac{3}{11}$ having 44 for the denominator.

4. Supply the missing numerator. $\frac{9}{13} = \frac{}{39}$

5. Kelly received 250 of the 600 votes cast for the two candidates in an election. What fractional part of the total number of votes did she receive?

CHAPTER 1 SUMMARY 45

1.2 *Perform the indicated operations.*

6. $\dfrac{3}{14} \cdot \dfrac{21}{9}$

7. $3\dfrac{1}{3} \div \dfrac{20}{9}$

8. $0 \div \dfrac{5}{4}$

9. $\dfrac{5}{9} + \dfrac{7}{12}$

10. $\dfrac{17}{18} - \dfrac{2}{3}$

11. $1\dfrac{1}{5} + \dfrac{9}{10} - \dfrac{1}{3}$

12. Change $\dfrac{219}{4}$ to a mixed number.

13. Change $5\dfrac{2}{7}$ to an improper fraction.

14. On a map, one inch represents $\dfrac{3}{4}$ of a mile. How many miles are represented by $2\dfrac{1}{3}$ inches?

15. Julie Simpson bought two pieces of fabric, one $3\dfrac{1}{8}$ yards long and the other $4\dfrac{3}{4}$ yards long. How many yards of fabric did she purchase?

1.3 16. Express 29.34 in expanded notation.

17. Express $2000 + 4 + \dfrac{4}{10} + \dfrac{2}{1000}$ in decimal notation.

18. Add: 14.5
          0.007
       + 325.16

19. Subtract: 307.2
             − 12.009

20. Subtract: $4.7 - 0.001$

21. Multiply: 1.31
            × 0.02

22. Multiply: (1.04)(1.04)(1.04)

23. Divide: $0.005\overline{)36.15}$

24. Graydon Bell works in Perko's Delicatessen as a registered cheese cutter. If he is paid $9.85 per hour and worked a total of 35.6 hours last week, what did he earn?

1.4 25. Convert $\dfrac{3}{8}$ to a decimal.

26. Convert $\dfrac{3}{11}$ to a decimal.

27. Convert 27.235 to a fraction.

28. Convert 35.2% to a decimal.

29. Convert 0.5% to a decimal.

30. Convert $\dfrac{4}{3}$ to a percent.

31. Convert 0.03 to a percent.

32. Convert 25 to a percent.

33. Sherman's Shoes has an inventory consisting of 1980 pairs of men's and women's shoes. If 55% of the inventory consists of women's shoes, how many pairs of women's shoes are in stock?

1.5

Write in exponential notation.

34. $aaaaa$

35. $(3z)(3z)(3z)$

36. $(a + w)(a + w)$

Write without using exponents.

37. b^7

38. $2x^3$

39. $(2x)^3$

Evaluate.

40. $2 \cdot 5 - 9 \div 3 + 1$

41. $2 \cdot 3^3 - 3 \cdot 2^3$

Evaluate when $a = 2$, $b = 5$, and $x = 4$.

42. $3a^2$

43. $(3a)^2$

44. $3^2 a$

45. $(b + x)^2$

46. $b^2 + x^2$

47. $abx - 10x$

48. $2 + 5[x + (b - a)]$

49. $2a + 3[x \div a \cdot b - 4]$

50. Use $A = P(1 + r)^t$ to find the amount of money in an account at the end of two years if a principal of $5000 is invested at 15% interest, compounded annually.

ANSWERS: **1.** $2 \cdot 5 \cdot 7 \cdot 7$ **2.** $\frac{2}{7}$ **3.** $\frac{12}{44}$ **4.** 27 **5.** $\frac{5}{12}$ **6.** $\frac{1}{2}$ **7.** $\frac{3}{2}$ **8.** 0 **9.** $\frac{41}{36}$ **10.** $\frac{5}{18}$ **11.** $1\frac{23}{30}$ **12.** $54\frac{3}{4}$ **13.** $\frac{37}{7}$ **14.** $1\frac{3}{4}$ miles **15.** $7\frac{7}{8}$ yards **16.** $20 + 9 + \frac{3}{10} + \frac{4}{100}$ **17.** 2004.402 **18.** 339.667 **19.** 295.191 **20.** 4.699 **21.** 0.0262 **22.** 1.124864 **23.** 7230 **24.** $350.66 **25.** 0.375 **26.** $0.\overline{27}$ **27.** $\frac{5447}{200}$ **28.** 0.352 **29.** 0.005 **30.** $133\frac{1}{3}$% **31.** 3% **32.** 2500% **33.** 1089 **34.** a^5 **35.** $(3z)^3$ **36.** $(a + w)^2$ **37.** $bbbbbbb$ **38.** $2xxx$ **39.** $(2x)(2x)(2x)$ **40.** 8 **41.** 30 **42.** 12 **43.** 36 **44.** 18 **45.** 81 **46.** 41 **47.** 0 **48.** 37 **49.** 22 **50.** $6612.50

CHAPTER 1 TEST

1. Reduce $\frac{322}{350}$ to lowest terms.

   1. _____

2. Write an equivalent fraction for $\frac{2}{13}$ with 52 as the denominator.

   2. _____

3. Supply the missing term. $\frac{5}{6} = \frac{35}{}$

   3. _____

4. In a test of 560 video games, 420 were working. What fractional part were working? Reduce the fraction.

   4. _____

Perform the indicated operation.

5. $\frac{2}{5} \cdot \frac{15}{22}$

   5. _____

6. $\frac{9}{14} \div \frac{2}{7}$

   6. _____

7. $\frac{5}{6} + \frac{7}{9}$

   7. _____

8. $\frac{17}{24} - \frac{2}{3}$

   8. _____

9. Change $\frac{31}{9}$ to a mixed number.

   9. _____

CHAPTER 1 TEST Continued

10. Change $5\frac{23}{27}$ to an improper fraction.

10. _____

11. There were 942 books in a shipment. If $\frac{1}{6}$ of these were science books, how many science books were in the shipment?

11. _____

12. A bottle contains $7\frac{2}{5}$ liters of solution. If $2\frac{7}{10}$ liters are used, how many liters are left in the bottle?

12. _____

Perform the indicated operation.

13. $14.06 + 3.057$

13. _____

14. $26.3 - 7.441$

14. _____

15. $(0.42)(5.06)$

15. _____

16. $46.2 \div 1.05$

16. _____

17. Convert $\frac{1}{5}$ to a percent.

17. _____

18. Convert 1.85 to a fraction.

18. _____

19. The tax rate on purchases is 4%. What is the tax on a $27 purchase?

19. _____

20. Write $a \cdot a \cdot a$ in exponential notation.

20. _____

21. Evaluate. $42 \div 6 - (4 - 2) \cdot 3$

21. _____

Evaluate when $x = 3$, $y = 2$, and $z = 5$.

22. $3z^2$

22. _____

23. $(3z)^2$

23. _____

24. $(x + y)^3$

24. _____

25. $x^3 + y^3$

25. _____

26. $z + 2[(x + y) - z]$

26. _____

27. Use $A = P(1 + r)^t$ to find the amount of money in an account at the end of two years if a principal of $500 is invested at 9% interest, compounded annually.

27. _____

2 RATIONAL AND REAL NUMBERS

2.1 RATIONAL NUMBERS AND THE NUMBER LINE

STUDENT GUIDEPOSTS

1. Number lines
2. Less than and greater than
3. Integers
4. Equal (unequal) numbers
5. Plotting numbers
6. Rational numbers
7. Cross products
8. Absolute values

 Many ideas in algebra can be better understood if we "picture" them in some way. Numbers are often pictured using a *number line*. Consider the set of whole numbers. We draw a line like the one in Figure 2.1, and select some unit of length. Then, starting at an arbitrary point that we label 0, we mark off unit lengths to the right, labeling the points 1, 2, 3, ... The result is a **number line** displaying the whole numbers, shown in Figure 2.1. Every whole number is paired with a point on this line. The arrowhead points to the direction in which the whole numbers continue to be identified.

Figure 2.1

EXAMPLE 1

What whole numbers are paired with a, b, and c on the number line in Figure 2.2?

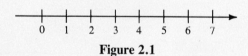

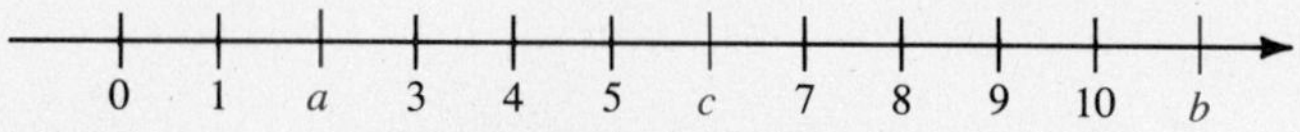

Figure 2.2

Point a is paired with 2, point b with 11, and point c with 6. ◀◀

Practice Exercise 1

What whole numbers are paired with x, y, and z on the following number line?

Answer: x paired with 4, y paired with 3, z paired with 1

2.1 RATIONAL NUMBERS AND THE NUMBER LINE

The whole numbers occur in a natural order. For example, we know that 3 is less than (has a smaller value than) 8, and that 8 is greater than (has a larger value than) 3. Note on the number line in Figure 2.2, that 3 is to the left of 8, while 8 is to the right of 3. The symbol $<$ means **"is less than."** We write "3 is less than 8" as

$$3 < 8.$$

Similarly, the symbol $>$ means **"is greater than."** We write "8 is greater than 3" as

$$8 > 3.$$

Notice that $3 < 8$ and $8 > 3$ have the same meaning even though they are read differently.

Suppose a and b are any two whole numbers.

1. If a is to the left of b on a number line, then $a < b$.
2. If a is to the right of b on a number line, then $a > b$.

EXAMPLE 2

Place the correct symbol ($<$ or $>$) between the numbers in the given pairs. Use part of a number line if necessary.

(A) 2   7

$2 < 7$. Notice that 2 is to the left of 7 on the number line in Figure 2.3.

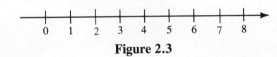

Figure 2.3

(B) 12   9

$12 > 9$. Notice that 12 is to the right of 9 on the number line in Figure 2.4.

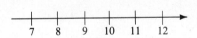

Figure 2.4

Practice Exercise 2

Place the correct symbol ($<$ or $>$) between the given pairs of numbers.

(A) 3   11

(B) 15   7

Answers: **(A)** $3 < 11$ **(B)** $15 > 7$

Notice in Example 2 that when $<$ and $>$ are used, *the symbol always points to the smaller of the two numbers*.

The number lines we have considered thus far display no numbers to the left of 0. What meaning could we give to a number to the left of 0? One example of such a number is a temperature of 5° below zero on a cold day, which is sometimes represented as $-5°$ (read "negative 5°"). Others are shown in the following table.

Measurement	Number
5° above zero	5° (or +5°)
5° below zero	$-5°$
100 ft above sea level	100 ft (or +100 ft)
100 ft below sea level	-100 ft
$16 deposit into an account	$16 (or +$16)
$16 check written on an account	$-$16

52 RATIONAL AND REAL NUMBERS

3 When we put a minus sign in front of each of the counting numbers, we obtain the **negative integers.** The collection of negative integers together with the counting numbers (sometimes called **positive integers**) and zero is called the set of **integers.**

$$\{\ldots, -3, -2, -1, 0, 1, 2, 3, \ldots\} \quad \text{Integers}$$

The whole numbers are often called **nonnegative integers.** The number line in Figure 2.5 shows negative integers, positive integers, and zero. Note that we can write 1 as +1, 2 as +2, and so on. Also, we read −1 as "negative 1," for example.

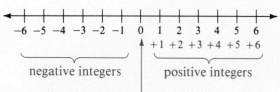

Figure 2.5

4 When using the number line, we say that two numbers are **equal** (for example, 3 + 2 = 5) if they correspond to the same point on a number line. If two numbers correspond to different points on a number line, they are **unequal.** We use $\neq$ to represent "is unequal to." We can expand our definitions of *less than* and *greater than* for whole numbers to include the integers. That is, one integer is **less than (greater than)** a second if it is to the left (right) of the second on a number line. If one number is **less than or equal to** another number, we use the symbol $\leq$ (similarly $\geq$ represents **greater than or equal to**). For example,

$$-4 \leq 4, \quad -2 \leq 0, \quad 0 \leq 3, \quad 2 \leq 2, \quad \text{and} \quad -1 \leq 1.$$

Also, all negative integers are less than zero (left of zero), all positive integers are greater than zero (right of zero), and any positive integer is greater than any negative integer.

EXAMPLE 3

Some relationships between numbers on the number line in Figure 2.6 follow.

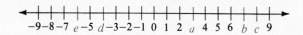

Figure 2.6

	In Words	*In Symbols*
(A)	1 is less than a	$1 < a$
(B)	c is greater than d	$c > d$
(C)	e is unequal to -5	$e \neq -5$
(D)	d is equal to -4	$d = -4$
(E)	b is less than or equal to b	$b \leq b$
(F)	a is greater than or equal to 2	$a \geq 2$

Practice Exercise 3

Use Figure 2.6 and write the following in symbols.

(A) d is less than 0

(B) a is greater than or equal to -1

(C) b is equal to 7

(D) e is less than or equal to e

(E) a is greater than d

(F) c is unequal to -8

Answers: **(A)** $d < 0$ **(B)** $a \geq -1$ **(C)** $b = 7$ **(D)** $e \leq e$ **(E)** $a > d$ **(F)** $c \neq -8$

5 In addition to the integers, the fractions we studied in Chapter 1, together with their negatives, can be shown or **plotted** on a number line. All of the fractions to the right of zero are called **positive rational numbers.** Those to the left of zero are called **negative rational numbers,** and together, along with zero, they form the set of **rational numbers.** We can think of the rational numbers as

{all numbers that can be written as a quotient of two integers}.

Remember from Chapter 1 that any integer can be written as a quotient of two integers, for example $-2 = -\frac{2}{1}$ and $1 = \frac{1}{1}$, so all integers are also rational numbers. Some examples of rational numbers are plotted in Figure 2.7.

Figure 2.7

Just as with integers, we can define the order relationships of *less than* and *greater than* for rational numbers. For example, we see that

$$-\frac{1}{2} < \frac{1}{2}, \quad \frac{3}{2} < \frac{9}{4}, \quad 2\frac{7}{8} > -\frac{3}{2}, \quad \text{and} \quad -\frac{9}{4} > -3\frac{1}{3}$$

by considering the number line in Figure 2.7.

A very important property of our number system states that if two numbers are unequal, then one of them must be less than the other. Given two integers, it is easy to determine if they are equal and, if not, which is less than (or which is greater than) the other. However, this is not quite so easy when considering two rational numbers.

Given two positive rational numbers $\frac{a}{b}$ and $\frac{c}{d}$,

1. $\frac{a}{b} = \frac{c}{d}$ whenever $ad = bc$.

2. $\frac{a}{b} > \frac{c}{d}$ whenever $ad > bc$.

3. $\frac{a}{b} < \frac{c}{d}$ whenever $ad < bc$.

7 ▶ The product ad is called the **first cross product,** and the product bc is called the **second cross product.** Thus, two positive fractions are equal if the cross products are equal, and two unequal fractions are in the same order as the first and second cross products. Similar results can be stated for negative rational numbers.

EXAMPLE 4

Place the appropriate symbol $=$, $>$, or $<$ between the given pairs of fractions.

(A) $\frac{4}{6} \quad \frac{28}{42}$

Since $4 \cdot 42 = 168$ and $6 \cdot 28 = 168$, the fractions are equal.

$$\frac{4}{6} = \frac{28}{42}.$$

Practice Exercise 4

Place the appropriate symbol, $=$, $>$, or $<$, between the given pairs of fractions.

(A) $\frac{14}{3} \quad \frac{23}{5}$

54 RATIONAL AND REAL NUMBERS

(B) $\dfrac{24}{7}$ $\dfrac{39}{11}$

Since $24 \cdot 11 = 264$ and $7 \cdot 39 = 273$, and $264 < 273$,

$$\dfrac{24}{7} < \dfrac{39}{11}.$$

(C) $\dfrac{15}{9}$ $\dfrac{31}{20}$

Since $15 \cdot 20 = 300$ and $9 \cdot 31 = 279$, and $300 > 279$,

$$\dfrac{15}{9} > \dfrac{31}{20}.$$ ◀◀

(B) $\dfrac{21}{13}$ $\dfrac{18}{11}$

(C) $\dfrac{9}{28}$ $\dfrac{36}{112}$

Answers: (A) > (B) < (C) =

In the previous chapter we reviewed the four basic operations of addition, subtraction, multiplication, and division on the nonnegative rational numbers. When operating on *all* rational numbers we use the notion of absolute value.

8 ▶▶▶   The **absolute value** of a number x is the distance from zero to x on a number line. We denote the absolute value of x by $|x|$, and read the symbol as "the absolute value of x."

EXAMPLE 5

(A) As shown in Figure 2.8, 3 is 3 units from 0. Thus, $|3| = 3$.

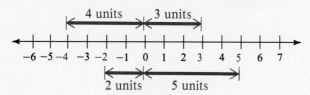

Figure 2.8

(B) 5 is 5 units from 0. Thus, $|5| = 5$.
(C) 0 is 0 units from 0. Thus, $|0| = 0$.
(D) -2 is 2 units from 0. Thus, $|-2| = 2$.
(E) -4 is 4 units from 0. Thus, $|-4| = 4$. ◀◀

Practice Exercise 5

Find the following absolute values.

(A) $|-8|$

(B) $|8|$

(C) $|0.05|$

(D) $|-0.05|$

(E) $\left|-\dfrac{3}{8}\right|$

Answers: (A) 8 (B) 8 (C) 0.05
(D) 0.05 (E) $\dfrac{3}{8}$

When finding the absolute value of a number, remember that the absolute value of a positive number or zero is the number itself. The absolute value of a negative number is the positive number formed by removing the minus sign. Thus, *the absolute value of a number is always greater than or equal to zero, never negative.*

2.1 EXERCISES A

Answer true *or* false.

1. $\{0, 1, 2, 3, \ldots\}$ is called the set of nonnegative integers.

2. If a number x is located to the left of a number y on a number line, then $x > y$.

2.1 RATIONAL NUMBERS AND THE NUMBER LINE

3. The numeral -7 is read "negative seven."

4. A temperature of 32° below zero could be written as $+32°$.

5. The distance from zero to a number y on a number line is called the absolute value of y.

6. An elevation of 4500 feet above sea level would be denoted by -4500.

7. A withdrawal of $200 from a savings account could be denoted by $-\$200$.

Place the correct symbol =, >, or < between the given pairs of numbers.

8. 7   1
9. 0   3
10. -2 0
11. 2   9
12. -3 2
13. $\frac{1}{2}$ $-\frac{1}{2}$
14. $-\frac{3}{4}$ $\frac{1}{4}$
15. -5 -5

Is the given statement true or false?

16. $-12 \leq 0$
17. $6 \leq 6$
18. $0 \leq -3$
19. $-5 \leq -5$
20. $-2 \leq -7$
21. $\frac{2}{3} \leq -\frac{2}{3}$
22. $4\frac{1}{2} \geq 4.5$
23. $-0.5 \leq -\frac{1}{2}$

24. Plot the numbers $\frac{2}{3}, -\frac{7}{8}, \frac{5}{4}, -\frac{7}{4}, \frac{5}{2}$, and $-\frac{10}{3}$ on the given number line.

25. Insert the appropriate symbol =, >, or < between the two positive fractions.

   (A) If $ad = bc$, then $\frac{a}{b}$ $\frac{c}{d}$.

   (B) If $ad < bc$, then $\frac{a}{b}$ $\frac{c}{d}$.

   (C) If $ad > bc$, then $\frac{a}{b}$ $\frac{c}{d}$.

   (D) The products ad and bc are called _____.

Place the appropriate symbol =, >, or < between the given pairs of fractions.

26. $\frac{17}{43}$ $\frac{28}{79}$
27. $\frac{35}{11}$ $\frac{41}{13}$
28. $\frac{4}{11}$ $\frac{7}{19}$
29. $\frac{13}{50}$ $\frac{39}{150}$
30. $\frac{5}{7}$ $\frac{60}{83}$
31. $\frac{2}{9}$ $\frac{21}{91}$

Evaluate the absolute values.

32. $|17|$
33. $|-1|$
34. $|0|$
35. $\left|-\frac{3}{4}\right|$
36. $|3.1|$
37. $\left|-\frac{5}{2}\right|$
38. $|45|$
39. $|-0.8|$

56 RATIONAL AND REAL NUMBERS

ANSWERS: 1. true 2. false 3. true 4. false 5. true 6. false 7. true 8. > 9. < 10. < 11. <
12. < 13. > 14. < 15. = 16. true 17. true 18. false 19. true 20. false 21. false 22. true 23. true
24. [number line with points at $-\frac{10}{3}$, $-\frac{7}{4}$, $-\frac{7}{8}$, $\frac{2}{3}$, $\frac{5}{4}$, $\frac{5}{2}$] 25. (A) = (B) < (C) > (D) cross products 26. > 27. > 28. <
29. = 30. < 31. < 32. 17 33. 1 34. 0 35. $\frac{3}{4}$ 36. 3.1 37. $\frac{5}{2}$ 38. 45 39. 0.8

2.1 EXERCISES B

Answer true or false.

1. $\{\ldots, -3, -2, -1\}$ is called the set of negative integers.
2. If a number w is located to the right of a number v on a number line, then $w < v$.
3. The numeral -11 is read "negative eleven."
4. The low temperature last winter in Twin Falls, Idaho of 38° below zero could be written $-38°$.
5. The absolute value of z, denoted by $|z|$, is the distance from zero to z on a number line.
6. An airplane flying at an altitude of 25,000 feet could be described as flying at an altitude of $-25,000$ feet.
7. A check in the amount of $35.00 written on an account could be denoted by $-$35.00.

Place the correct symbol =, >, or < between the given pairs of numbers.

8. 4   9
9. 5   0
10. 0 -7
11. 3   12
12. -5 1
13. $\frac{2}{3}$ $-\frac{2}{3}$
14. -0.2 0.2
15. -6 -6

Is the given statement true or false?

16. $-3 \geq -4$
17. $-1 \geq -1$
18. $0 \leq -1$
19. $-8 \leq -8$
20. $-5 \leq -40$
21. $\frac{1}{4} \geq -\frac{1}{4}$
22. $3\frac{1}{5} \leq 3.2$
23. $-0.1 \leq -\frac{1}{10}$

24. Plot the numbers $\frac{1}{3}$, $-\frac{3}{4}$, $\frac{7}{4}$, $-\frac{9}{4}$, $\frac{7}{2}$, and $-\frac{11}{3}$ on a number line.

25. Insert the appropriate symbol =, >, or < between the two positive fractions.

   (A) If $xw < yv$ then $\frac{x}{y}$ $\frac{v}{w}$.

   (B) If $xw = yv$ then $\frac{x}{y}$ $\frac{v}{w}$.

   (C) If $xw > yv$ then $\frac{x}{y}$ $\frac{v}{w}$.

   (D) The cross products used to determine order on the two rational numbers $\frac{x}{y}$ and $\frac{v}{w}$ are _____ and _____.

Place the appropriate symbol =, >, or < between the given pairs of fractions.

26. $\frac{3}{5}$ $\frac{7}{10}$
27. $\frac{36}{11}$ $\frac{77}{20}$
28. $\frac{19}{43}$ $\frac{57}{129}$

29. $\frac{6}{7}$ $\frac{47}{55}$
30. $\frac{121}{9}$ $\frac{55}{3}$
31. $\frac{27}{201}$ $\frac{13}{121}$

Evaluate the absolute values.

32. $|25|$ 33. $|-2|$ 34. $\left|-\dfrac{1}{2}\right|$ 35. $|-0|$

36. $|8.7|$ 37. $\left|-\dfrac{9}{4}\right|$ 38. $|112|$ 39. $|-0.03|$

2.1 EXERCISES C

Place the appropriate symbol =, >, or < between the pairs of fractions.

1. $-\dfrac{14}{19}$ $-\dfrac{2}{3}$ 2. $-\dfrac{21}{8}$ $-\dfrac{31}{11}$ 3. $-\dfrac{77}{14}$ $-\dfrac{11}{2}$

2.2 ADDITION AND SUBTRACTION OF RATIONAL NUMBERS

▶▶ STUDENT GUIDEPOSTS

1 Adding numbers with the same signs
2 Adding numbers with different signs
3 Additive inverse (negative)
4 Additive identity
5 Commutative law of addition
6 Associative law of addition
7 Adding more than two numbers
8 Subtracting numbers

In Chapter 1 we reviewed the four basic operations on the nonnegative rational numbers. Now we extend our treatment to addition and subtraction of all rational numbers. Consider the addition problem $3 + 2$ on the number line in Figure 2.9.

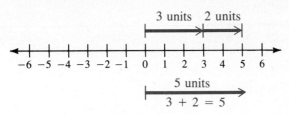

Figure 2.9

To find $3 + 2$, we start at 0, move 3 units to the *right*, and then move 2 more units to the *right*. We are then 5 units to the *right* of zero. Thus, $3 + 2 = 5$.

We can also find $3 + (-2)$ on a number line. We first move 3 units to the *right* and then 2 units to the *left*. This puts us 1 unit to the *right* of 0, as shown in Figure 2.10. Thus, $3 + (-2) = 1$.

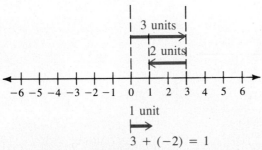

Figure 2.10

58 RATIONAL AND REAL NUMBERS

Intuitively, if we think of numbers as temperatures, an increase of 3° (+3) followed by a second increase of 2° (+2) results in a total increase of 5° (+5). Similarly, an increase of 3° (+3) followed by a decrease of 2° (−2) results in a net increase of 1° (+1).

Notice that when adding numbers using a number line, we move to the *right when a number is positive* and to the *left when it is negative*.

EXAMPLE 1

Find $(-3) + (-2)$.

Draw a number line. Move 3 units to the left for −3 and then 2 more units to the left for −2. As seen in Figure 2.11, we end up 5 units to the left of 0, at −5. Thus, $(-3) + (-2) = -5$. ◀◀

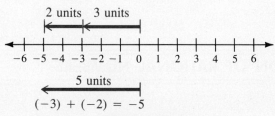

Figure 2.11

Practice Exercise 1

Find $(-11) + (-8)$.

Answer: −19

EXAMPLE 2

Find $(-3) + 2$.

Draw a number line. Move 3 units to the left and then 2 units to the right. As Figure 2.12 shows, we are then at −1. Thus, $(-3) + 2 = -1$. ◀◀

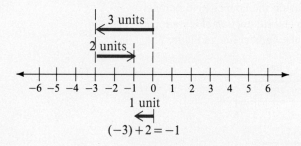

Figure 2.12

Practice Exercise 2

Find $(-11) + 8$.

Answer: −3

EXAMPLE 3

Find $2 + (-3)$.

Draw a number line. Move 2 units to the right and then 3 units to the left. (See Figure 2.13.) This puts us at −1. Thus, $2 + (-3) = -1$. ◀◀

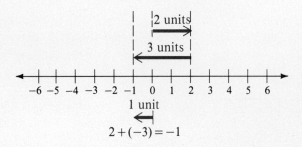

Figure 2.13

Practice Exercise 3

Find $8 + (-11)$.

Answer: −3

2.2 ADDITION AND SUBTRACTION OF RATIONAL NUMBERS

The number-line method of adding numbers shows us what addition means, but it takes too much time. For example, we found $(-3) + (-2)$ to be -5 using a number line (Figure 2.11). However, we could have added $3 + 2$ and attached a minus sign:

$$(-3) + (-2) = -(3 + 2) = -5.$$

This is an example of the following rule.

> **1** **Adding numbers having the same sign**
>
> To add two numbers having the same sign, add their absolute values. The sum has the same sign as the numbers being added.

When the signs of the numbers being added are not the same, we need another rule. For $3 + (-2) = 1$ (Figure 2.10), we could have found the difference between 3 and 2 and attached a plus sign:

$$3 + (-2) = +(3 - 2) = +1 = 1.$$

Also, for $(-3) + 2 = 2 + (-3) = -1$ (Figures 2.12 and 2.13), we could have subtracted 2 from 3 and attached a minus sign:

$$(-3) + 2 = 2 + (-3) = -(3 - 2) = -1.$$

These are examples of the following rule.

> **2** **Adding numbers having different signs**
>
> To add two numbers having different signs, subtract the smaller absolute value from the larger absolute value. The result has the same sign as the number with the larger absolute value. If the absolute values are the same, the sum is zero.

Remember that if a number (other than zero) has no sign, this is the same as having a plus sign attached. For example, 5 and $+5$ are the same.

EXAMPLE 4

Add using the preceding rules.

(A) $6 + 7 = 13$

(B) $(-6) + (-7) = -(6 + 7) = -13$

(C) $6 + (-7) = -(7 - 6) = -1 \qquad |-7| = 7 > 6 = |6|$

(D) $(-6) + 7 = +(7 - 6) = +1 = 1 \qquad |-6| = 6 < 7 = |7|$

(E) $12 + (-12) = 0 \qquad\qquad |12| = 12 = |-12|$ ◀◀

Practice Exercise 4

Add.

(A) $5 + 14$

(B) $(-5) + (-14)$

(C) $5 + (-14)$

(D) $(-5) + 14$

(E) $(-14) + 14$

Answers: **(A)** 19 **(B)** -19 **(C)** -9 **(D)** 9 **(E)** 0

3 Example 4(e) illustrates an important property of our number system. If a is any number, we call $-a$ the **additive inverse**, or **negative** of a, which means

$$a + (-a) = (-a) + a = 0.$$

4 The number 0 also plays a special role in the system since for any number a,

$$a + 0 = 0 + a = a.$$

60 RATIONAL AND REAL NUMBERS

Intuitively, any number a remains identically the same when added to 0. For this reason, 0 is called the **additive identity.**

EXAMPLE 5
Add.

(A) $0 + 7 = 7 + 0 = 7$ 0 is the additive identity

(B) $0 + (-7) = (-7) + 0 = -7$ 0 is the additive identity

(C) $(-2.3) + (-4.1) = -(2.3 + 4.1) = -6.4$

(D) $(-3.5) + 1.7 = -(3.5 - 1.7) = -1.8$

(E) $\frac{1}{2} + \left(-\frac{3}{4}\right) = -\left(\frac{3}{4} - \frac{1}{2}\right) = -\left(\frac{3}{4} - \frac{2}{4}\right) = -\frac{1}{4}$ ◀◀

Practice Exercise 5
Add.

(A) $13 + 0$

(B) $(-13) + 0$

(C) $(-4.02) + (-1.35)$

(D) $(-6.35) + 2.11$

(E) $\left(-\frac{2}{9}\right) + \frac{4}{11}$

Answers: (A) 13   (B) -13
(C) -5.37 (D) -4.24 (E) $\frac{14}{99}$

Two additional properties of addition involve changing the order of addition, and rearranging grouping symbols in sums of three (or more) numbers. If a and b are two numbers, the **commutative law of addition** states that

5 $$a + b = b + a.$$

The order of addition can be changed by the commutative law. For example, $16 + 5 = 5 + 16$. Given another number c, the **associative law of addition** states that

6 $$(a + b) + c = a + (b + c).$$

The grouping symbols can be rearranged by the associative law. For example, $(5 + 2) + 3 = 5 + (2 + 3)$.

EXAMPLE 6

(A) Use the commutative law to complete the equation $5 + \frac{1}{4} =$ _____. The commutative law says the order of addition can be changed, so $5 + \frac{1}{4} = \frac{1}{4} + 5$.

(B) Use the associative law to complete the equation $(7 + \frac{1}{2}) + (-2) =$ _____. The associative law says the grouping symbols can be rearranged, so $(7 + \frac{1}{2}) + (-2) = 7 + [\frac{1}{2} + (-2)]$. ◀◀

Practice Exercise 6

(A) Complete the following using the commutative law.
$(-\frac{1}{2}) + 9 =$ _____

(B) Complete the following using the associative law.
$(7 + 0.5) + 2 =$ _____

Answers: (A) $9 + (-\frac{1}{2})$
(B) $7 + (0.5 + 2)$

Taken together, the commutative and associative laws of addition allow us to add several signed numbers by reordering and regrouping the numbers to obtain the sum in the most convenient manner.

7 **Adding more than two numbers**
1. Add all positive numbers.
2. Add all negative numbers.
3. Add the resulting pair of numbers (one positive and one negative).

2.2 ADDITION AND SUBTRACTION OF RATIONAL NUMBERS

EXAMPLE 7
Add.

$(-2) + (-3) + 2 + 7 + (-5) + 6 + 9 + (-1)$
$= [2 + 7 + 6 + 9] + [(-2) + (-3) + (-5) + (-1)]$
$= [2 + 7 + 6 + 9] + [-(2 + 3 + 5 + 1)]$
$= [24] + [-11] = +[24 - 11] = 13$ ◀◀

Practice Exercise 7
Add.

$5 + (-2) + (-1) + 7 + (-9) + 6$

Answer: 6

With practice we should be able to skip many of the intermediate steps in the preceding examples and make computations mentally.

Subtraction of numbers can be defined as addition of an additive inverse or a negative. For example, we can think of $5 - 3 = 2$ as $5 + (-3) = 2$.

8 ▶▶ Subtracting numbers
To subtract one number from another, change the sign of the number being subtracted and add.

EXAMPLE 8
Subtract.

(A) $5 - (+3) = 5 + (-3)$ Change sign and add
    $= 5 - 3 = 2$
(B) $5 - (-3) = 5 + 3 = 8$ Change sign and add
(C) $(-5) - 3 = (-5) + (-3)$ Change sign and add
    $= -(5 + 3) = -8$
(D) $(-5) - (-3) = (-5) + 3$ Change sign and add
    $= -(5 - 3) = -2$
(E) $(-5) - 0 = (-5) + (-0)$ Change sign and add
    $= (-5) + 0$ The negative of zero is zero:
    $= -5$ ◀◀ $-0 = 0$

Practice Exercise 8
Subtract.

(A) $17 - (+2)$

(B) $17 - (-2)$

(C) $(-17) - 2$

(D) $(-17) - (-2)$

(E) $0 - (-17)$

Answers: (A) 15 (B) 19 (C) −19
(D) −15 (E) 17

EXAMPLE 9
Subtract.

(A) $5 - (+5) = 5 + (-5) = 0$

(B) $5 - (-5) = 5 + 5 = 10$

(C) $(-5) - 5 = (-5) + (-5) = -(5 + 5) = -10$

(D) $(-5) - (-5) = (-5) + 5 = 0$

(E) $(-5.8) - (-3.2) = (-5.8) + 3.2 = -(5.8 - 3.2) = -2.6$

(F) $\left(-\dfrac{2}{3}\right) - \dfrac{1}{6} = \left(-\dfrac{2}{3}\right) + \left(-\dfrac{1}{6}\right) = \left(-\dfrac{4}{6}\right) + \left(-\dfrac{1}{6}\right) = -\dfrac{5}{6}$ ◀◀

Practice Exercise 9
Subtract.

(A) $12 - (+12)$

(B) $12 - (-12)$

(C) $(-12) - 12$

(D) $(-12) - (-12)$

(E) $(-1.2) - (-5.7)$

(F) $\left(-\dfrac{1}{8}\right) - \dfrac{3}{5}$

Answers: (A) 0 (B) 24 (C) −24
(D) 0 (E) 4.5 (F) $-\dfrac{29}{40}$

62 RATIONAL AND REAL NUMBERS

▶▶ **CAUTION** ◀◀ Although addition satisfies both the commutative and associative laws, subtraction does not. For example,

$$3 = 5 - 2 \neq 2 - 5 = -3.$$ Subtraction is *not* commutative

Likewise,

$$(8 - 4) - 3 = 4 - 3 = 1$$

but Subtraction is *not* associative

$$8 - (4 - 3) = 8 - 1 = 7,$$ ◀◀

EXAMPLE 10

John had $932 in his bank account on May 1. He deposited $326 and $791 during the month and wrote checks for $816, $315, and $940 during the month. What was his balance at the end of May?

We can consider the deposits as positive numbers and the checks written as negative numbers. Thus, we must add 932, 326, 791, −816, −315, and −940.

```
   932              − 816
   326              − 315           − 2071
 + 791              − 940             2049
  ────                ────            ────
  2049  Sum of deposits   − 2071  Sum of checks   − 22  Final sum
```

Thus, John is overdrawn (in the red) $22 on his account. ◀◀

Practice Exercise 10

Marlo had a temperature of 100.1° at noon. It rose 2.3° before dropping 3.5° by 3:00 P.M. What was her temperature at 3:00 P.M.?

Answer: 98.9°

2.2 EXERCISES A

Perform the indicated operation.

1. $4 + 3$
2. $(-4) + (-3)$
3. $4 + (-3)$
4. $(-4) + 3$
5. $4 + 0$
6. $0 + (-4)$
7. $7 + (-7)$
8. $(-7) + 7$
9. $(-7) + (-7)$
10. $7 + 7$
11. $(-8) + (-5)$
12. $9 + (-1)$
13. $19 + (-25)$
14. $(-16) + 14$
15. $4 - 3$
16. $(-4) - (-3)$
17. $4 - (-3)$
18. $(-4) - 3$
19. $0 - 4$
20. $(-4) - 0$
21. $7 - (-7)$
22. $(-7) - 7$
23. $(-7) - (-7)$
24. $7 - 7$
25. $(-8) - (-5)$
26. $9 - (-1)$
27. $(-1) - 9$
28. $14 - 21$
29. $19 - (-25)$
30. $(-16) - 14$
31. $1.3 + (-2.5)$
32. $(-1.3) + 2.5$
33. $(-1.3) + (-2.5)$
34. $1.3 - (-2.5)$
35. $(-1.3) - 2.5$
36. $(-1.3) - (-2.5)$
37. $\frac{1}{2} + \left(-\frac{2}{5}\right)$
38. $\left(-\frac{1}{2}\right) + \frac{2}{5}$
39. $\left(-\frac{1}{2}\right) + \left(-\frac{2}{5}\right)$
40. $\frac{1}{2} - \left(-\frac{2}{5}\right)$
41. $\left(-\frac{1}{2}\right) - \frac{2}{5}$
42. $\left(-\frac{1}{2}\right) - \left(-\frac{2}{5}\right)$

2.2 ADDITION AND SUBTRACTION OF RATIONAL NUMBERS

43. $(-3) + 2 + (-7) + (-8) + 9$

44. $5 + 8 + 9 + (-3) + (-7)$

45. $(-8) + (-6) + 9 + (-3) + 5 + (-6)$

46. $(-4) + (-2) + 6 + 7 + (-1)$

47. Show by example that subtraction does not satisfy the commutative law. Can you find numbers a and b for which $a - b$ does equal $b - a$?

Add.

48. 23
    -16
    -36
    5
    81

49. -16
    -81
    -14
    -70
    -93

50. -64
    25
    38
    -17
    -83

51. When a cold front came through Cut Bank, Montana, the temperature dropped from 35° above zero to 13° below zero ($-13°$). What was the total change in temperature?

52. On March 1, Allen had $64 in his bank account. During the month, he wrote checks for $81, $108, and $192 and made deposits of $123 and $140. What was the value of his account at the end of the month?

53. A helicopter is 600 ft above sea level and a submarine directly below it is 250 ft below sea level (-250 ft). How far apart are they?

54. At 5:30 A.M., it was $-42°$F. By 2:00 P.M. the same day, the temperature had risen 57° to the recorded high temperature. Shortly thereafter, a cold front passed through dropping the temperature 41° by 5:00 P.M. What was the temperature at 5:00 P.M.?

For Review

Place the correct symbol =, <, or > between the given pairs of numbers.

55. -5 -4

56. $\dfrac{2}{3}$ $\dfrac{77}{110}$

57. $\dfrac{5}{12}$ $\dfrac{30}{72}$

58. -4.7 -1.2

Evaluate the absolute values.

59. $|-81|$

60. $\left|-\dfrac{5}{4}\right|$

61. $|17|$

62. $|-7.25|$

ANSWERS: **1.** 7 **2.** -7 **3.** 1 **4.** -1 **5.** 4 **6.** -4 **7.** 0 **8.** 0 **9.** -14 **10.** 14 **11.** -13 **12.** 8
13. -6 **14.** -2 **15.** 1 **16.** -1 **17.** 7 **18.** -7 **19.** -4 **20.** -4 **21.** 14 **22.** -14 **23.** 0 **24.** 0
25. -3 **26.** 10 **27.** -10 **28.** -7 **29.** 44 **30.** -30 **31.** -1.2 **32.** 1.2 **33.** -3.8 **34.** 3.8 **35.** -3.8
36. 1.2 **37.** $\dfrac{1}{10}$ **38.** $-\dfrac{1}{10}$ **39.** $-\dfrac{9}{10}$ **40.** $\dfrac{9}{10}$ **41.** $\dfrac{9}{10}$ **42.** $-\dfrac{1}{10}$ **43.** -7 **44.** 12 **45.** -9 **46.** 6
47. $2 - 1 = 1 \neq -1 = 1 - 2$; if $a = b$ then $a - b = 0 = b - a$ **48.** 57 **49.** -274 **50.** -101 **51.** 48° **52.** $-\$54$
53. 850 ft **54.** $-26°$ **55.** < **56.** < **57.** = **58.** < **59.** 81 **60.** $\dfrac{5}{4}$ **61.** 17 **62.** 7.25

2.2 EXERCISES B

Perform the indicated operations.

1. $5 + 2$

2. $(-5) + (-2)$

3. $5 + (-2)$

4. $(-5) + 2$

5. $0 + 5$

6. $(-5) + 0$

7. $5 + (-5)$
8. $(-5) + 5$
9. $(-5) + (-5)$
10. $5 + 5$
11. $(-9) + (-3)$
12. $(-2) + 11$
13. $18 + (-23)$
14. $(-15) + 19$
15. $8 - 7$
16. $(-8) - (-7)$
17. $8 - (-7)$
18. $(-8) - 7$
19. $10 - 0$
20. $0 - (-10)$
21. $6 - (-6)$
22. $(-6) - 6$
23. $(-6) - (-6)$
24. $6 - 6$
25. $(-11) - (-4)$
26. $(-13) - (-1)$
27. $(-1) - 7$
28. $(-12) - 24$
29. $18 - (-27)$
30. $(-17) - 11$
31. $1.5 + (-3.8)$
32. $(-1.5) + 3.8$
33. $(-1.5) + (-3.8)$
34. $1.5 - (-3.8)$
35. $(-1.5) - 3.8$
36. $(-1.5) - (-3.8)$
37. $\frac{1}{4} + \left(-\frac{3}{8}\right)$
38. $\left(-\frac{1}{4}\right) + \frac{3}{8}$
39. $\left(-\frac{1}{4}\right) + \left(-\frac{3}{8}\right)$
40. $\frac{1}{4} - \left(-\frac{3}{8}\right)$
41. $\left(-\frac{1}{4}\right) - \frac{3}{8}$
42. $\left(-\frac{1}{4}\right) - \left(-\frac{3}{8}\right)$
43. $(-5) + 4 + (-9) + (-10) + 6$
44. $5 + 9 + (-3) + 1 + (-10)$
45. $(-5) + (-7) + 10 + (-3) + 4 + (-7)$
46. $(-6) + (-3) + 6 + 10 + (-2)$

47. Show by example that subtraction does not satisfy the associative law. Can you find numbers, a, b, and c for which $(a - b) - c$ does equal $a - (b - c)$?

Add.

48.
26
−14
−51
3
92

49.
−17
−20
−43
−5
−110

50.
−78
32
51
−60
−92

51. In a five-hour period of time, the temperature in Empire, N.Y. dropped from 41° to 12° below zero. What was the total temperature change?

52. On July 4, Uncle Sam had $97 in his checking account. During the next week he wrote checks for $15, $102, and $312, and made deposits of $145 and $170. What was the value of his account at the end of the week?

53. A balloon is 1200 ft above sea level, and a diving bell directly below it is 340 ft beneath the surface of the water. How far apart are the two?

54. At 8:00 A.M., Leah Johnson had a temperature of 99.8°. It rose another 3.1° before falling 4.3° by noon. What was her temperature at noon?

For Review

Place the correct symbol =, <, or > between the given pairs of numbers.

55. $-6 \quad -7$
56. $\frac{3}{4} \quad \frac{96}{120}$
57. $\frac{7}{8} \quad \frac{77}{88}$
58. $0 \quad -11.5$

Evaluate the absolute values.

59. $\left|-\frac{9}{10}\right|$
60. $|-210|$
61. $|42|$
62. $|-0.01|$

2.2 EXERCISES C

Perform the indicated operations.

1. $(-962) - (-508)$
2. $-48.62 - 40.002$
3. $64\dfrac{7}{24} - 98\dfrac{13}{16}$

   $\left[\text{Answer: } -34\dfrac{25}{48}\right]$

2.3 MULTIPLICATION AND DIVISION OF RATIONAL NUMBERS

▶▶ STUDENT GUIDEPOSTS

1 Multiplying numbers
2 Multiplicative inverse (reciprocal)
3 Multiplicative identity
4 Commutative law of multiplication
5 Associative law of multiplication
6 Dividing numbers
7 Double negative property
8 Double sign properties

In multiplying and dividing numbers, we need to decide what sign to give the product or quotient. To help us, we look for patterns in several products.

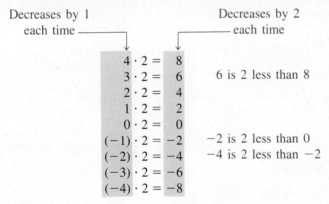

We know the products $4 \cdot 2$, $3 \cdot 2$, $2 \cdot 2$, $1 \cdot 2$, and $0 \cdot 2$, and we can see that these products decrease by 2 each time. For this pattern to continue, a negative number times a positive number must be negative. For example,

$$(-1) \cdot 2 = -2, \quad (-2) \cdot 2 = -4, \quad (-3) \cdot 2 = -6, \quad (-4) \cdot 2 = -8.$$

Similarly, any positive number times any negative number must be negative. We use this to find a pattern in the following products.

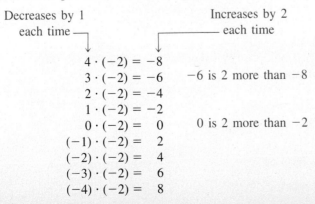

66 RATIONAL AND REAL NUMBERS

For this pattern to continue, a negative number times a negative number must be a positive number. For example,

$$(-1) \cdot (-2) = 2, \quad (-2) \cdot (-2) = 4, \quad (-3) \cdot (-2) = 6, \quad (-4) \cdot (-2) = 8.$$

Our results are summarized in the following rule.

1
> **Multiplying numbers**
> 1. Multiply the absolute values of the numbers.
> 2. If both signs are positive or both are negative, the product is positive.
> 3. If one sign is positive and one is negative, the product is negative.
> 4. If one number (or both numbers) is zero, the product is zero.

When multiplying numbers, remember that *like signs have a positive product and unlike signs have a negative product.*

EXAMPLE 1

Multiply.

(A) $5 \cdot 2 = 10$ Like signs

(B) $(-5)(-2) = (5 \cdot 2) = 10$ Like signs

(C) $5 \cdot (-2) = -(5 \cdot 2) = -10$ Unlike signs

(D) $(-5) \cdot 2 = -(5 \cdot 2) = -10$ Unlike signs

(E) $0 \cdot (-8) = 0$

(F) $0 \cdot 0 = 0$

(G) $(3)\left(\dfrac{1}{3}\right) = 1$ ◀◀

Practice Exercise 1

Multiply.

(A) $4 \cdot 11$

(B) $(-4)(-11)$

(C) $4 \cdot (-11)$

(D) $(-4) \cdot 11$

(E) $(12) \cdot 0$

(F) $0 \cdot (-12)$

(G) $\left(\dfrac{3}{7}\right)\left(\dfrac{7}{3}\right)$

Answers: **(A)** 44 **(B)** 44 **(C)** −44 **(D)** −44 **(E)** 0 **(F)** 0 **(G)** 1

Example 1(g) illustrates a property of reciprocals that we have seen before. If a is any number except 0, we call $\dfrac{1}{a}$ the **multiplicative inverse** (or **reciprocal**) of a, which means that

2

$$a \cdot \dfrac{1}{a} = \dfrac{1}{a} \cdot a = 1.$$

The number 1 plays a role in multiplication similar to the one 0 plays in addition. Since for any number a,

3

$$a \cdot 1 = 1 \cdot a = a,$$

we see that a number remains identically the same when multiplied by 1. For this reason, 1 is sometimes called the **multiplicative identity.**

EXAMPLE 2

Multiply.

(A) $(-5)\left(-\dfrac{1}{5}\right) = 1$ -5 and $-\dfrac{1}{5}$ are reciprocals or multiplicative inverses

Practice Exercise 2

Multiply.

(A) $\left(-\dfrac{1}{9}\right)(-9)$

(B) $1 \cdot (-7) = (-7) \cdot 1 = -7$ 1 is the multiplicative identity

(C) $(-2.1) \cdot 3.5 = -(2.1)(3.5) = -7.35$

(D) $(-2.1) \cdot (-3.5) = (2.1)(3.5) = 7.35$

(E) $\left(\dfrac{1}{4}\right)\left(-\dfrac{3}{5}\right) = -\left(\dfrac{1}{4}\right)\left(\dfrac{3}{5}\right) = -\dfrac{3}{20}$

(F) $\left(-\dfrac{1}{4}\right)\left(-\dfrac{3}{5}\right) = \left(\dfrac{1}{4}\right)\left(\dfrac{3}{5}\right) = \dfrac{3}{20}$ ◀◀

(B) $1 \cdot (-13)$

(C) $(-4.1) \cdot (-2.9)$

(D) $(-4.1) \cdot (2.9)$

(E) $\left(\dfrac{1}{7}\right)\left(-\dfrac{3}{8}\right)$

(F) $\left(-\dfrac{1}{7}\right)\left(-\dfrac{3}{8}\right)$

Answers: **(A)** 1 **(B)** -13 **(C)** 11.89 **(D)** -11.89 **(E)** $-\dfrac{3}{56}$ **(F)** $\dfrac{3}{56}$

Two more properties of multiplication, similar to the ones for addition, involve changing the order of multiplication and rearranging grouping symbols in products of three (or more) numbers. If a and b are two numbers, the **commutative law of multiplication** states that

$$a \cdot b = b \cdot a.$$

The order of multiplication can be changed by the commutative law. For example, $4 \cdot 7 = 7 \cdot 4$. Given another number c, the **associative law of multiplication** states that

$$(a \cdot b) \cdot c = a \cdot (b \cdot c).$$

The grouping symbols can be changed by the associative law. For example, $(2 \cdot 5) \cdot 3 = 2 \cdot (5 \cdot 3)$.

EXAMPLE 3

(A) Use the commutative law to complete the equation $7 \cdot \dfrac{3}{4} = $ _____.
The commutative law says the order of multiplication can be changed, so $7 \cdot \dfrac{3}{4} = \dfrac{3}{4} \cdot 7$.

(B) Use the associative law to complete the equation $(3 \cdot \dfrac{1}{5}) \cdot (-5) = $ _____. The associative law says the grouping symbols can be rearranged, so $(3 \cdot \dfrac{1}{5}) \cdot (-5) = 3 \cdot (\dfrac{1}{5} \cdot (-5))$. ◀◀

Practice Exercise 3

(A) Use the commutative law to complete the following.
$(0.09) \cdot (1000) = $ _____

(B) Use the associative law to complete the following.
$(\dfrac{1}{2} \cdot \dfrac{3}{10}) \cdot (-6) = $ _____

Answers: **(A)** $(1000) \cdot (0.09)$

(B) $\left(\dfrac{1}{2}\right) \cdot \left(\dfrac{3}{10} \cdot (-6)\right)$

Often we will need to multiply more than two numbers. To do this, multiply in pairs and keep track of the sign. Notice in Example 4 that *products involving an odd number of minus signs are negative while those with an even number are positive.*

EXAMPLE 4

Multiply.

(A) $(3)(-2)(-4) = (-6)(-4) = 24$ Multiply 3 times -2 first

(B) $(-1)(-1)(-2)(-3)(-4) = (1)(-2)(-3)(-4)$
$= (-2)(-3)(-4)$ $(-1)(-1) = 1$
$= (6)(-4)$ $(1)(-2) = -2$
$= -24$ $(-2)(-3) = 6$

Practice Exercise 4

Multiply.

(A) $(2)(-1)(-4)(3)$

(B) $(-6)(-1)(-5)(-2)(-1)$

(C) $(-3)^4$

68 RATIONAL AND REAL NUMBERS

(C) $(-2)^3 = (-2)(-2)(-2) = (4)(-2)$　　　$(-2)(-2) = 4$
　　　　　　　　　　　　$= -8$ ◀◀

Answers: (A) 24　(B) -60
(C) 81

We have seen that the only difference between multiplying positive numbers and multiplying rational numbers is finding the sign of the product. Recall that division can be thought of as multiplication by a reciprocal. For example,

$$15 \div 3 \text{ or } \frac{15}{3} \quad \text{is the same as} \quad 15 \cdot \frac{1}{3}.$$

Thus, we can use the same rules for signs in division that we used for multiplication.

6 ▶▐▪▐▪ **Dividing numbers**

1. Divide the absolute values of the numbers.
2. If both signs are positive or both are negative, the quotient is positive.
3. If one sign is positive and one sign is negative, the quotient is negative.
4. Zero divided by any number (except zero) is zero, and division of any number by 0 is undefined.

Remember that the rule of signs for division is the same as for multiplication: *like signs have a positive quotient and unlike signs have a negative quotient*.

EXAMPLE 5

Divide.

(A) $6 \div 3 = 2$　　　　　　　　　　Like signs

(B) $(-6) \div (-3) = 6 \div 3 = 2$　　　Like signs

(C) $6 \div (-3) = -(6 \div 3) = -2$　　Unlike signs

(D) $(-6) \div 3 = -(6 \div 3) = -2$　　Unlike signs

(E) $0 \div (-3) = 0$

(F) $(-3) \div 0$ is undefined

(G) $\dfrac{1}{2} \div \left(-\dfrac{3}{5}\right) = \dfrac{1}{2} \cdot \left(-\dfrac{5}{3}\right) = -\left(\dfrac{1}{2} \cdot \dfrac{5}{3}\right) = -\dfrac{5}{6}$ ◀◀

Practice Exercise 5

Divide.

(A) $12 \div 4$

(B) $(-12) \div (-4)$

(C) $12 \div (-4)$

(D) $(-12) \div 4$

(E) $(-0.3) \div 0$

(F) $0 \div (-0.3)$

(G) $\left(-\dfrac{3}{5}\right) \div \dfrac{9}{10}$

Answers: (A) 3　(B) 3　(C) -3
(D) -3　(E) undefined　(F) 0
(G) $-\dfrac{2}{3}$

▶▶ **CAUTION** ◀◀　Although multiplication satisfies both the commutative and associative laws, division does not. For example,

$$3 = 6 \div 2 \neq 2 \div 6 = \frac{2}{6} = \frac{1}{3}.$$ Division is not commutative

Likewise,

$$(8 \div 4) \div 2 = 2 \div 2 = 1$$

but   Division is not associative

$$8 \div (4 \div 2) = 8 \div 2 = 4.\ \blacktriangleleft$$

EXAMPLE 6

Willie wrote 3 checks for $75 each and 4 checks for $105 each. By how much did this change his bank balance?

The 3 checks for $75 could be calculated as follows:

$$(3)(-75) = -225.$$

There are 4 checks for $105, so

$$(4)(-105) = -420.$$

Thus, his account is changed by

$$-225 + (-420) = -225 - 420 = -645 \text{ dollars.}$$

His balance is $645 less. $\blacktriangleleft$

Practice Exercise 6

A diver descended below the surface of the ocean by diving 25 ft each minute for 11 minutes. How far below sea level was he at this time?

Answer: 275 ft (-275)

EXAMPLE 7

Mary, Sue, and Maria share an apartment with total rent of $720. By how much is each woman's account changed if they each write a check for an equal share of the rent?

Since the rent is a decrease in their accounts, we use -720. To find the shares, we divide -720 by 3:

$$-720 \div 3 = -240.$$

Thus, each woman's account is decreased by $240. $\blacktriangleleft$

Practice Exercise 7

During a 5-hour period, the temperature decreased 42°. What was the average decrease per hour?

Answer: 8.4° ($-8.4°$)

When evaluating numerical expressions involving rational numbers, follow the order of operations and rules of grouping given in Section 1.5.

EXAMPLE 8

Evaluate each numerical expression.

(A) $(-3) + 5 \cdot (-1) - (-4)$
$= (-3) + (-5) - (-4)$ Multiply first
$= (-8) - (-4)$ Add first from left
$= (-8) + 4$ Then subtract
$= -4$

(B) $(-2)^3 + 1 = (-2)(-2)(-2) + 1$ $a^3 = aaa$
$= (4)(-2) + 1$ $(-2)(-2) = 4$
$= (-8) + 1$ Multiply first
$= -7$ Then add

Practice Exercise 8

Evaluate each numerical expression.

(A) $5 + (-3) \cdot (-2) - 6$

(B) $(-4)^2 + (-1)^3$

70 RATIONAL AND REAL NUMBERS

(C) $\frac{1}{2} - \left\{\frac{1}{4} - \left[\left(-\frac{3}{4}\right) - \left(-\frac{1}{4}\right)\right]\right\}$

$= \frac{1}{2} - \left\{\frac{1}{4} - \left[\left(-\frac{3}{4}\right) + \frac{1}{4}\right]\right\}$

$= \frac{1}{2} - \left\{\frac{1}{4} - \left[-\frac{2}{4}\right]\right\}$ Innermost parentheses first

$= \frac{1}{2} - \left\{\frac{1}{4} + \frac{2}{4}\right\}$

$= \frac{1}{2} - \left(\frac{3}{4}\right)$

$= \frac{2}{4} - \frac{3}{4} = -\frac{1}{4}$ ◀◀

(C) $-0.15 + [3.2 - (5.65 - 1.6)]$

Answers: (A) 5   (B) 15   (C) -1

Notice in Example 8 that we use parentheses to avoid writing plus and minus signs right next to each other. The parentheses may be removed and the two signs replaced by one according to the following rule, part 4 of which is called the **double negative property.**

Double sign properties

If a is any number,

1. $+(+a) = +a$ 2. $+(-a) = -a$
3. $-(+a) = -a$ 4. $-(-a) = +a$

Notice that when a plus sign precedes parentheses, we *do not* change the sign inside when the parentheses and plus sign are removed. However, when a minus sign precedes parentheses, we *do* change the sign inside when the parentheses and minus sign are removed.

EXAMPLE 9

Write without parentheses.

(A) $+(+6) = +6 = 6$

(B) $+(-8) = -8$

(C) $-(+7) = -7$

(D) $-(-5) = +5 = 5$

(E) $-[-(-2)] = -[+2] = -2$ Remove innermost parentheses first ◀◀

Practice Exercise 9

Write without parentheses.

(A) $+(+11)$

(B) $+(-19)$

(C) $-(+17)$

(D) $-(-15)$

(E) $-[+(-12)]$

Answers: (A) 11   (B) -19 (C) -17 (D) 15   (E) 12

EXAMPLE 10

Evaluate when $a = -3$ and $b = -2$.

(A) $a - b = (-3) - (-2) = (-3) + 2 = -1$ Use parentheses when substituting

(B) $b - [-a] = (-2) - [-(-3)] = (-2) - [3]$
$= -2 - 3 = -5$

Practice Exercise 10

Evaluate when $x = -5$ and $y = -1$.

(A) $x - y$

(B) $y - (-x)$

(C) $2x - 6y$

(C) $2a - 3b = 2(-3) - 3(-2) = -6 - (-6) = -6 + 6 = 0$
(D) $|a - 2b| = |(-3) - 2(-2)| = |(-3) - (-4)|$
$= |(-3) + 4| = |1| = 1$ ◀◀

(D) $|y - 3x|$

Answers: (A) -4 (B) -6
(C) -4 (D) 14

EXAMPLE 11

Evaluate $3[2(a + 2) - 3(b + c)]$ when $a = -2$, $b = 4$, and $c = -8$.

$3[2(a + 2) - 3(b + c)] = 3[2(-2 + 2) - 3(4 + (-8))]$ Replace variables with numbers

$= 3[2(0) - 3(-4)]$ Add inside parentheses first
$= 3[0 + 12]$
$= 3[12]$
$= 36$ ◀◀

Practice Exercise 11

Evaluate $-4[(u - v) - (w - u)]$ when $u = -3$, $v = 5$, and $w = -1$.

Answer: 40

2.3 EXERCISES A

Perform the indicated operations.

1. $(2)(6)$
2. $(-2)(-6)$
3. $(2)(-6)$

4. $(-2)(6)$
5. $0 \cdot (-6)$
6. $(-5) \cdot 7$

7. $(-3)(-9)$
8. $(-10)(8)$
9. $(9)(-12)$

10. $(12)(-12)$
11. $(-12)(-12)$
12. $(-15) \cdot 20$

13. $6 \div 2$
14. $(-6) \div (-2)$
15. $6 \div (-2)$

16. $(-6) \div 2$
17. $0 \div (-6)$
18. $(-6) \div 0$

19. $\dfrac{99}{-11}$
20. $\dfrac{-36}{-4}$
21. $\dfrac{-12}{12}$

22. $\dfrac{0}{-38}$
23. $\dfrac{-48}{-16}$
24. $\dfrac{52}{-13}$

25. $(1.2)(-2.3)$
26. $(-1.2)(-2.3)$
27. $(-1.2)(2.3)$

28. $\left(\dfrac{2}{5}\right)\left(-\dfrac{15}{4}\right)$
29. $\left(-\dfrac{2}{5}\right)\left(-\dfrac{15}{4}\right)$
30. $\left(-\dfrac{2}{5}\right)\left(\dfrac{15}{4}\right)$

31. $6.4 \div (-0.8)$

32. $(-6.4) \div (-0.8)$

33. $\dfrac{-6.4}{0.8}$

34. $\dfrac{3}{5} \div \left(-\dfrac{9}{15}\right)$

35. $\left(-\dfrac{3}{5}\right) \div \left(-\dfrac{9}{15}\right)$

36. $\dfrac{-\dfrac{3}{5}}{\dfrac{9}{15}}$

37. $(-1)(-1)(-1)$

38. $(-1)(2)(-3)$

39. $(-1)(-2)(-3)(-4)$

40. $(-1)(-2)(-3)(4)(-5)(6)(-1)$

41. $(-1)(-1)(-1)(-1)(-1)(-1)(-1)$

42. Show by example that division does not satisfy the associative law. Can you find numbers a, b, and c for which $(a \div b) \div c$ does equal $a \div (b \div c)$?

43. Tony had to write four $12 checks during the month of June. By how much did this change his bank balance?

44. Martha had 423 points in a contest, but then she received 20 penalty points 3 times. What was her total after the 3 penalties?

Evaluate each numerical expression.

45. $(-3)^3 + 27$

46. $(-1)^5 + (-1)^3$

47. $2 - [3 - (4 - 1)]$

48. $4 + 2[(-1) - (-5)]$

49. $(3 - 8) \cdot 4 + 1$

50. $3 - \dfrac{2 + (-4)}{1 - 5}$

Write without parentheses using only one sign (or no sign).

51. $-(-3)$

52. $-(+4)$

53. $+(-x)$

54. $-(-x)$

55. $-[+(-2)]$

56. $-[-(+3)]$

57. $-[-(-4)]$

58. $+[-(-10)]$

Evaluate the following expressions when $a = -1$, $b = 2$, and $c = -4$.

59. $a^2 - 1$

60. $abc + b^2$

61. $3(a + b) + c$

62. $2(b - a) - c$

63. $2[3(a + 1) + 2(4 + c)]$

64. $a^2 + b^2 + c$

For Review

Perform the following operations.

65. $(-12) + 9$

66. $(-12) + (-9)$

67. $15 + (-8)$

68. $(-12) - 9$

69. $(-12) - (-9)$

70. $15 - (-8)$

Add.

71. -13
       12
       26
       -18

72. -28
       -29
       -42
       -36

73. -92
       81
       27
       -61

74. -102
       321
       -421
       -60

75. The Brisebois family owed the credit union $3000. They repaid $1200 and later borrowed $650. What number represents the status of their account?

ANSWERS: **1.** 12 **2.** 12 **3.** -12 **4.** -12 **5.** 0 **6.** -35 **7.** 27 **8.** -80 **9.** -108 **10.** -144 **11.** 144 **12.** -300 **13.** 3 **14.** 3 **15.** -3 **16.** -3 **17.** 0 **18.** undefined **19.** -9 **20.** 9 **21.** -1 **22.** 0 **23.** 3 **24.** -4 **25.** -2.76 **26.** 2.76 **27.** -2.76 **28.** $-\frac{3}{2}$ **29.** $\frac{3}{2}$ **30.** $-\frac{3}{2}$ **31.** -8 **32.** 8 **33.** -8 **34.** -1 **35.** 1 **36.** -1 **37.** -1 **38.** 6 **39.** 24 **40.** -720 **41.** -1 **42.** $(4 \div 2) \div 2 = 2 \div 2 = 1 \neq 4 = 4 \div 1 = 4 \div (2 \div 2)$, if $a = b = c = 1$, then $(1 \div 1) \div 1 = 1 \div 1 = 1 = 1 \div 1 = 1 \div (1 \div 1)$ **43.** $-\$48$ **44.** 363 points **45.** 0 **46.** -2 **47.** 2 **48.** 12 **49.** -19 **50.** $\frac{5}{2}$ **51.** 3 **52.** -4 **53.** $-x$ **54.** x **55.** 2 **56.** 3 **57.** -4 **58.** 10 **59.** 0 **60.** 12 **61.** -1 **62.** 10 **63.** 0 **64.** 1 **65.** -3 **66.** -21 **67.** 7 **68.** -21 **69.** -3 **70.** 23 **71.** 7 **72.** -135 **73.** -45 **74.** -262 **75.** $-\$2450$

2.3 EXERCISES B

Perform the indicated operations.

1. $(5)(9)$

2. $(-5)(-9)$

3. $(5)(-9)$

4. $(-5)(9)$

5. $(-7)(0)$

6. $(-2)(-7)$

7. $(12)(-2)$

8. $(-12)(2)$

9. $(-6)(-13)$

10. $(-6)(13)$

11. $(-11)(-11)$

12. $(-15)(30)$

13. $9 \div 3$

14. $(-9) \div (-3)$

15. $9 \div (-3)$

16. $(-9) \div 3$

17. $0 \div (-5)$

18. $(-5) \div 0$

19. $\dfrac{88}{-11}$

20. $\dfrac{-24}{-8}$

21. $\dfrac{-15}{-15}$

22. $\dfrac{0}{-25}$

23. $\dfrac{-100}{10}$

24. $\dfrac{52}{-26}$

25. $(1.5)(-2.2)$

26. $(-1.5)(-2.2)$

27. $(-1.5)(2.2)$

28. $\left(\dfrac{3}{5}\right)\left(-\dfrac{10}{9}\right)$

29. $\left(-\dfrac{3}{5}\right)\left(-\dfrac{10}{9}\right)$

30. $\left(-\dfrac{3}{5}\right)\left(\dfrac{10}{9}\right)$

31. $4.8 \div (-0.6)$ **32.** $(-4.8) \div (-0.6)$ **33.** $\dfrac{-4.8}{-0.6}$

34. $\dfrac{2}{5} \div \left(-\dfrac{6}{15}\right)$ **35.** $\left(-\dfrac{2}{5}\right) \div \left(-\dfrac{6}{15}\right)$ **36.** $\dfrac{-\dfrac{2}{5}}{\dfrac{6}{15}}$

37. $(-1)(-1)(-1)(-1)$ **38.** $(-1)(-2)(-3)$ **39.** $(-1)(2)(-3)(4)$

40. $(-1)(-2)(3)(-1)(5)(-4)$ **41.** $(-1)(-1)(-1)(-1)(-1)(-1)(-1)(-1)(-1)$

42. Show by example that division does not satisfy the commutative law. Can you find numbers a and b for which $a \div b$ does equal $b \div a$?

43. During each of 5 consecutive hours, the temperature dropped 7°. By how much did this change the temperature?

44. Hank Anderson received 330 points for a performance, but then was penalized 25 points each for four rule infractions. What was his point total after the penalties were imposed?

Evaluate each numerical expression.

45. $(-3) + 2 \cdot (-4)$ **46.** $(-2)^2 + 3$ **47.** $4 - [2 - (5 - 3)]$

48. $6 + 3[(-2) - (-4)]$ **49.** $(4 - 7) \cdot 5 + 2$ **50.** $2 - \dfrac{4 + (-5)}{3 - 5}$

Write without parentheses using only one sign (or no sign).

51. $-(-5)$ **52.** $+(-2)$ **53.** $-(+a)$ **54.** $-(-a)$

55. $-[+(-3)]$ **56.** $-[-(+2)]$ **57.** $-[-(-7)]$ **58.** $+[-(-8)]$

Evaluate the following expressions when $x = -2$, $y = 1$, and $z = -3$.

59. $3xy + z$ **60.** $xyz + z^2$ **61.** $3(y - x) - z$

62. $x + \dfrac{y}{z}$ **63.** $2[4(x + 2) + 7(z + 3)]$ **64.** $x^2 + y^2 + z$

For Review

Perform the following operations.

65. $(-17) + 6$ **66.** $(-17) + (-6)$ **67.** $17 + (-6)$

68. $(-17) - 6$ **69.** $(-17) - (-6)$ **70.** $17 - (-6)$

Add.

71. -16 **72.** -40 **73.** -97 **74.** -205
      15 -27 27 420
      3 -19 13 -318
      -27 -52 -47 -65

75. A football team lost 6 yards on first down, gained 11 yards on second down, and lost 4 yards on third down. If the series started on their 25 yard line, where was the ball placed on fourth down?

2.3 EXERCISES C

Evaluate each numerical expression.

1. $-2 - [3 - (4 - 2)^2]^2$
2. $[17 - (-5 + 8)^2 - 6 \div 2]^2$
3. $-[2^3 - (2 - 4)]^2 - \dfrac{16 - (-6)}{15 - 4}$

[Answer: -102]

2.4 THE DISTRIBUTIVE LAWS AND SIMPLIFYING EXPRESSIONS

STUDENT GUIDEPOSTS

1. Distributive laws
2. Terms, factors, and coefficients
3. Similar or like terms
4. Factoring
5. Collecting like terms
6. Simplified expressions
7. Removing (clearing) parentheses

1 The operations of addition (or subtraction) and multiplication are related by two very important properties called the **distributive laws.** Consider the numerical expression $2(3 + 5)$. Since the order of operations tells us to work within the parentheses first, we evaluate it as follows:

$$2(3 + 5) = 2(8) = 16.$$

However, in this case if we were to "distribute" the product of 2 over the sum of $3 + 5$,

$$2(3 + 5) = 2 \cdot 3 + 2 \cdot 5 = 6 + 10 = 16,$$

we obtain the same result. Similarly,

$$4(8 - 3) = 4(5) = 20,$$

and

$$4(8 - 3) = 4 \cdot 8 - 4 \cdot 3 = 32 - 12 = 20.$$

Thus multiplication can be "distributed over" addition and subtraction. These examples illustrate the following laws.

Distributive laws of multiplication over addition and subtraction

If a, b and c are numbers,

1. $a(b + c) = ab + ac$
2. $a(b - c) = ab - ac.$

Since multiplication is commutative, products are not affected by changing the order of multiplication. Thus we also have

$$(b + c)a = ba + ca \quad \text{and} \quad (b - c)a = ba - ca.$$

Also, multiplication distributes over sums with more than two terms. For example,

$$a(b + c + d) = ab + ac + ad.$$

EXAMPLE 1

Evaluate each numerical expression in two ways.

(A) $7(3 + 2) = 7 \cdot 3 + 7 \cdot 2 = 21 + 14 = 35$. Also,
$7(3 + 2) = 7(5) = 35$.

(B) $5(8 - 3) = 5 \cdot 8 - 5 \cdot 3 = 40 - 15 = 25$. Also,
$5(8 - 3) = 5(5) = 25$.

(C) $6(7 + 3 - 5) = 6 \cdot 7 + 6 \cdot 3 - 6 \cdot 5 = 42 + 18 - 30 = 30$. Also,
$6(7 + 3 - 5) = 6(5) = 30$.

(D) $-7(3 + 2) = (-7)(3) + (-7)(2) = (-21) + (-14) = -35$. Also,
$-7(3 + 2) = (-7)(5) = -35$.

(E) $-2(-3 - 6) = (-2)(-3) - (-2)(6) = 6 - (-12) = 6 + 12 = 18$. Also,
$-2(-3 - 6) = -2(-9) = (2)(9) = 18$.

Practice Exercise 1

Evaluate each numerical expression in two ways.

(A) $4(1 + 9)$

(B) $3(2 - 7)$

(C) $5(4 - 6 + 3)$

(D) $-3(5 - 1)$

(E) $-9(-2 - 8)$

Answers: (A) 40 (B) −15 (C) 5 (D) −12 (E) 90

The distributive laws are useful in computation, but they are more useful when we work with algebraic expressions. In Section 1.5, we introduced algebraic expressions involving sums, differences, products, or quotients of numbers and variables.

A part of an expression that is a product of numbers and variables and that is separated from the rest of the expression by plus (or minus) signs is called a **term.** The numbers and letters that are multiplied in a term are called **factors** of the term. The numerical factor is called the **(numerical) coefficient** of the term. For example,

$$3x, \quad 4a + 7, \quad 3x + 7y - z, \quad 4x + 5a + 3 - 8x$$

are algebraic expressions with one, two, three, and four terms, respectively. In $4a + 7$, the term $4a$ has factors 4 and a, and 4 is the coefficient of the term.

Two terms are **similar** or **like terms** if they contain the same variables. In $4x + 5a + 3 - 8x$, the terms $4x$ and $-8x$ are like terms. Note that the minus sign goes with the term and thus the coefficient of $-8x$ is -8.

If the terms of an expression have a common factor, the distributive laws can be used in reverse to **remove the common factor** by a process called **factoring.**

EXAMPLE 2

Use the distributive laws to factor.

(A) $3x + 3y = 3(x + y)$ The distributive law in reverse order

(B) $4a - 4b = 4(a - b)$ 4 and $(a - b)$ are factors of $4a - 4b$

(C) $5u + 5v - 5w = 5(u + v - w)$

(D) $3x + 6 = 3 \cdot x + 3 \cdot 2$ 6 is $3 \cdot 2$
$= 3(x + 2)$ Factor out 3

Practice Exercise 2

Use the distributive laws to factor.

(A) $7a + 7b$

(B) $3u - 3w$

(C) $11c - 11d + 11v$

(D) $6u + 6$

(E) $3w - 15$

(F) $-12a - 12b$

2.4 THE DISTRIBUTIVE LAWS AND SIMPLIFYING EXPRESSIONS

(E) $8a - 8 = 8 \cdot a - 8 \cdot 1$ Express 8 as $8 \cdot 1$
        $= 8(a - 1)$ Factor out 8

(F) $-3x - 3y = (-3)x + (-3)y$ Factor out -3
          $= (-3)(x + y)$ ◀◀

Answers: (A) $7(a + b)$
(B) $3(u - w)$ (C) $11(c - d + v)$
(D) $6(u + 1)$ (E) $3(w - 5)$
(F) $-12(a + b)$

In any factoring problem, we check by multiplying. For example, since

$$3(x + y) = 3x + 3y \quad \text{and} \quad 4(a - b) = 4a - 4b,$$

our factoring in the first two parts of Example 2 is correct.

5   When an expression contains like terms, it can be simplified by **collecting like terms.** This process is illustrated in the next example.

EXAMPLE 3

Use the distributive laws to collect like terms.

(A) $5x + 7x = (5 + 7)x = 12x$ Factor out x

(B) $-8x + 2x = (-8 + 2)x = -6x$ Factor out x

(C) $-2a + 7a - 9a = (-2 + 7 - 9)a = -4a$ Factor out a

(D) $6y + 2y - y + 5 = 6 \cdot y + 2 \cdot y - 1 \cdot y + 5$ $-y = -1 \cdot y$
               $= (6 + 2 - 1)y + 5$ Factor y out of first three terms
               $= 7y + 5$

The terms $7y$ and 5 *cannot* be collected since they are not like terms. Thus, $7y + 5$ *is not* $12y$. You can see this more easily if you replace y by some number. For example, when $y = 2$,

$$7y + 5 = 7(2) + 5 = 14 + 5 = 19, \quad \text{but} \quad 12y = 12(2) = 24.$$

(E) $4a + 7b - a + 6b = 4a - a + 7b + 6b$ Commutative law
                   $= 4 \cdot a - 1 \cdot a + 7 \cdot b + 6 \cdot b$ $-a = -1 \cdot a$
                   $= (4 - 1)a + (7 + 6)b$ Distributive law
                   $= 3a + 13b$

With practice some steps can be left out.

(F) $0.07x + x = (0.07)x + 1 \cdot x$ $x = 1 \cdot x$
             $= (0.07 + 1)x$ Distributive law
             $= 1.07x$ ◀◀

Practice Exercise 3

Use the distributive laws to collect like terms.

(A) $3u + 10u$

(B) $-9v + 5v$

(C) $-w + 2w - 5w$

(D) $3z - z + 2z + 13$

(E) $6x + 6y - x - 7y$

(F) $p + (0.14)p$

Answers: (A) $13u$ (B) $-4v$
(C) $-4w$ (D) $4z + 13$ (E) $5x - y$
(F) $(1.14)p$

6   When all like terms of an expression have been collected, we say that the expression has been **simplified.**
     Using the distributive law and the fact that $-x = (-1) \cdot x$, we get

$$-(a + b) = (-1)(a + b) = (-1)(a) + (-1)(b) = -a - b,$$
$$-(a - b) = (-1)(a - b) = (-1)(a) - (-1)(b) = -a + b,$$
$$-(-a - b) = (-1)(-a - b) = (-1)(-a) - (-1)(b) = a + b.$$

These observations lead us to the next rule.

7 To simplify an expression by removing parentheses

1. When a minus sign precedes parentheses, remove the parentheses by changing the sign of every term within the parentheses.
2. When a plus sign precedes parentheses, remove the parentheses without changing any of the signs of the terms.

EXAMPLE 4

Simplify by removing parentheses.

(A) $-(x + 1) = -x - 1$ Change all signs

(B) $-(x - 1) = -x + 1$ Change all signs

(C) $-(-x + y + 5) = +x - y - 5 = x - y - 5$ Change all signs

(D) $+(-x + y + 5) = -x + y + 5$ Change no signs

(E) $x - (y - 3) = x - y + 3$ Change all signs within parentheses ◀

Practice Exercise 4

Simplify by removing parentheses.

(A) $-(2 + y)$

(B) $-(-2 - y)$

(C) $-(5 - u - w)$

(D) $+(5 - u - w)$

(E) $a - (1 - b)$

Answers: (A) $-2 - y$ (B) $2 + y$ (C) $-5 + u + w$ (D) $5 - u - w$ (E) $a - 1 + b$

The process of removing parentheses is sometimes called **clearing parentheses.**

EXAMPLE 5

Clear parentheses and collect like terms.

(A) $3x + (2x - 7) = 3x + 2x - 7$ Do not change signs
                 $= (3 + 2)x - 7$ Collect like terms
                 $= 5x - 7$

(B) $y - (4y - 4) = y - 4y + 4$ Change all signs within parentheses
                $= (1 - 4)y + 4$ Collect like terms
                $= -3y + 4$

(C) $3 - (5a + 2) + 7a = 3 - 5a - 2 + 7a$ Change all signs within parentheses
                         $= 7a - 5a + 3 - 2$ Commutative law
                         $= 2a + 1$ Collect like terms

(D) $3x - (-2x - 7) + 5 = 3x + 2x + 7 + 5$ Change all signs within parentheses
                           $= (3 + 2)x + 7 + 5$
                           $= 5x + 12$ ◀

Practice Exercise 5

Clear parentheses and collect like terms.

(A) $a + (2 - 3a)$

(B) $b - (1 - b)$

(C) $2x - (3x + 4) - x$

(D) $5 - (-8 - 2w) + 8$

Answers: (A) $-2a + 2$ (B) $2b - 1$ (C) $-2x - 4$ (D) $2w + 21$

▶ **CAUTION** ◀   Change *all* signs within the parentheses.

$$-(-2x - 7) = 2x + 7, \quad \text{not} \quad 2x - 7. \blacktriangleleft$$

EXAMPLE 6

Evaluate the expressions when $a = -2$ and $b = -1$.

(A) $3a + 7 = 3(-2) + 7 = -6 + 7 = 1.$

Note how using parentheses at the substitution step helps us to avoid ambiguous statements. For example, without parentheses above we would have $3 \cdot -2 + 7$, which is confusing.

(B) $4a^2 = 4(-2)^2 = 4 \cdot 4 = 16$ Only -2 is squared. Compare this example with the next one

(C) $(4a)^2 = (4 \cdot (-2))^2 = (-8)^2 = 64$ From (b) and (c) we see that $4a^2 \neq (4a)^2$

(D) $-a^2 = -(-2)^2 = -(4) = -4$ Compare this with the next example

(E) $(-a)^2 = (-(-2))^2 = (2)^2 = 4$ From (d) and (e) we see that $-a^2 \neq (-a)^2$

(F) $-a - b = -(-2) - (-1) = 2 + 1 = 3$

(G) $-a^3 = -(-2)^3 = -(-8) = +8 = 8$

Practice Exercise 6

Evaluate the expressions when $u = -3$ and $w = -1$.

(A) $2u - 3$

(B) $3w^2$

(C) $(3w)^2$

(D) $-u^3$

(E) $(-u)^3$

(F) $-u - w$

(G) $-(-u + w^2)$

Answers: (A) -9 (B) 3 (C) 9 (D) 27 (E) 27 (F) 4 (G) -4

▶▶ **CAUTION** ◀◀ Two common mistakes to avoid are: (1) forgetting to change *all* signs when a minus sign appears in front of a set of parentheses (for example, $-(x - 2)$ is not $-x - 2$), and (2) making sign errors when evaluating expressions such as $-a^2$ (for example, $-a^2$ is not the same as $(-a)^2$). ◀◀

2.4 EXERCISES A

1. (A) Compute $-2(3 - 8)$ (B) Compute $(-2)(3) - (-2)(8)$ (C) Why are these two equal?

2. (A) Compute $-3(4 + 2)$ (B) Compute $(-3)(4) + (-3)(2)$ (C) Why are these two equal?

3. In $2x - 3y + 7 - 9x$ the terms $2x$ and $-9x$ are called similar or _____ terms.

How many terms does each expression have?

4. $2a + b - 3$ 5. $-2x - 3y + 4 - z$ 6. $3w$

Multiply.

7. $4(x + y)$ 8. $5(x - y)$ 9. $10(x + 2y)$

10. $10(x + 2y + 3z)$ 11. $2(2a + 1 + 6b)$ 12. $-5(x + y)$

Use the distributive laws to factor.

13. $4x + 4y$ 14. $5x - 5y$ 15. $10x + 20y$

16. $10x + 20y + 30z$

17. $4a + 2 + 12b$

18. $-5x - 5y$

Simplify by removing parentheses.

19. $-(y + 2)$

20. $-(-a - b)$

21. $-(-x + 2)$

22. $-(2y - 2)$

23. $x - (3 - b)$

24. $-(x - y - 4)$

25. $+(-x - y - z)$

26. $-(1 + x) + (a - b)$

27. $+(u - v) - (w - 3)$

Use the distributive laws and collect like terms.

28. $3x - 8x$

29. $-4z - 9z$

30. $1 - 2x + x$

31. $y - 3 - 5y$

32. $2x - y - 5x + y$

33. $a - b + a - b$

34. $1 - x + 2y - 3x - y$

35. $-a + 3 + b - 2a + b$

36. $u - 3 + 3v - 2u + 1$

37. $\frac{1}{2}x - \frac{2}{3}y - \frac{5}{2}x - \frac{1}{3}y$

38. $-\frac{3}{4}a + \frac{1}{4} + \frac{3}{4}a - \frac{1}{4}b$

39. $2.1u - 3 - 5.8u$

40. $x - 3(x + 1)$

41. $2a - (1 - 3a)$

42. $2(1 - 2u) + u$

43. $2x - (-x + 1) + 3$

44. $a + (2 - 5a) - 3$

45. $7u - (-3u - 1) + 5$

46. $-2[x - 3(x + 1)]$

47. $a - [3a - (1 - 2a)]$

48. $2 - [3u - (-2 + 3u)]$

49. $3(x - 2) - 2[5y - (2x - y)]$

50. $2(3a - 4b) - (-4a + b) - (-4b - 5a)$

51. $-[2u - (u - v)] - 3[(u - 2v) - 3v]$

Evaluate the following expressions when $x = -3$, $y = -1$ and $z = 2$.

52. $-2x - y$

53. $2x^2$

54. $(2x)^2$

55. $-2x^2$

56. $(-2x)^2$

57. $-x - y + z$

58. $x^2 - y^2 - z^2$

59. $-3y^2 - (x + z)$

60. $z^3 - x^3$

61. $x^2 - 4yz$

62. $(x - y)^3$

63. $x^3 - y^3$

64. $x^2y^2z^2$

65. $2x + 3z + y + 1$

66. $-(-x)$

67. $|x + y|$ **68.** $|x - y|$ **69.** $|x^2 + y^2|$

For Review

Perform the indicated operations.

70. $(3)(-11)$ **71.** $(-3)(-11)$ **72.** $(-12) \div 6$

73. $(-12) \div (-6)$ **74.** $\left(-\dfrac{1}{3}\right)\left(-\dfrac{3}{4}\right)$ **75.** $(-6.3) \div (-0.9)$

ANSWERS: **1.** (A) 10 (B) 10 (C) distributive law   **2.** (A) -18 (B) -18 (C) distributive law   **3.** like   **4.** 3
5. 4   **6.** 1   **7–12.** Answers given in exercises 13–18   **13–18.** Answers given in exercises 7–12   **19.** $-y - 2$
20. $a + b$ **21.** $x - 2$ **22.** $-2y + 2$ **23.** $x - 3 + b$ **24.** $-x + y + 4$ **25.** $-x - y - z$ **26.** $-1 - x + a - b$
27. $u - v - w + 3$ **28.** $-5x$ **29.** $-13z$ **30.** $1 - x$ **31.** $-4y - 3$ **32.** $-3x$ **33.** $2a - 2b$ **34.** $1 - 4x + y$
35. $-3a + 3 + 2b$ **36.** $-u - 2 + 3v$ **37.** $-2x - y$ **38.** $\dfrac{1}{4} - \dfrac{1}{4}b$ **39.** $-3 - 3.7u$ **40.** $-2x - 3$ **41.** $5a - 1$
42. $2 - 3u$ **43.** $3x + 2$ **44.** $-4a - 1$ **45.** $10u + 6$ **46.** $4x + 6$ **47.** $-4a + 1$ **48.** 0   **49.** $7x - 12y - 6$
50. $15a - 5b$ **51.** $-4u + 14v$ **52.** 7   **53.** 18   **54.** 36   **55.** -18 **56.** 36   **57.** 6   **58.** 4   **59.** -2 **60.** 35
61. 17   **62.** -8 **63.** -26 **64.** 36   **65.** 0   **66.** -3 **67.** 4   **68.** 2   **69.** 10   **70.** -33 **71.** 33   **72.** -2
73. 2   **74.** $\dfrac{1}{4}$ **75.** 7

2.4 EXERCISES B

1. (A) Compute $-3(5 - 2)$.   (B) Compute $(-3)(5) - (-3)(2)$.   (C) Why are these two equal?

2. (A) Compute $-4(6 + 2)$.   (B) Compute $(-4)(6) + (-4)(2)$.   (C) Why are these two equal?

3. In $2a + b + 8 - 10a$ the terms $2a$ and $-10a$ are called like or _____ terms.

How many terms does each expression have?

4. $2a + 7 + 6w - 3$ **5.** $4a + 8b$ **6.** $2w - 3 + 7b + a + c$

Multiply.

7. $4(x - y)$ **8.** $a(x + y)$ **9.** $10(a + 4b)$

10. $11(x + 2y + 3z)$ **11.** $2(3x + 1 + 4y)$ **12.** $-6(a + w)$

Use the distributive laws to factor.

13. $4x - 4y$ **14.** $ax + ay$ **15.** $10a + 40b$

16. $11x + 22y + 33z$ **17.** $6x + 2 + 8y$ **18.** $-6a - 6w$

Simplify by removing parentheses.

19. $-(x + 3)$ **20.** $-(-x - y)$ **21.** $-(-w + 5)$

22. $-(4a - 4)$ **23.** $a - (2 - y)$ **24.** $-(a - b - 9)$

25. $+(-a - b - w)$ **26.** $-(1 + a) + (w - c)$ **27.** $+(x - y) - (a - 8)$

Use the distributive laws and collect like terms.

28. $-3x + 8x$ **29.** $-3w - 10w$ **30.** $1 - 3x + 2x$

31. $5y + 2 - 3y$ **32.** $x + y - x - y$ **33.** $2y - u + y - 3u$

82 RATIONAL AND REAL NUMBERS

34. $1 - 2a + b - a - 3b$

35. $w + 3 - 2w + b - 4$

36. $z - 4 + 3z - a + 2$

37. $\frac{1}{4}a - \frac{1}{5}w + \frac{3}{4}a + \frac{2}{5}w$

38. $-\frac{1}{3}x + \frac{1}{2} - \frac{2}{3}x + \frac{1}{3}y$

39. $2.5w - 3 + 4.1w$

40. $a - 2(a + 5)$

41. $2w - (4 - 5w)$

42. $3(1 - 5x) + x$

43. $2b - (-b + 2) + 5$

44. $y + (3 - 2y) - 5$

45. $2y - (-5y - 2) + 7$

46. $-3[a - 2(a + 2)]$

47. $x - [4x - (3 - 4x)]$

48. $4 - [2w - (-3 - 2w)]$

49. $-6(x + 2y) - 5[8x - (-2x + y)]$

50. $-3(-a + 3b) + 5(-2a - 2b) - 2(3a - b)$

51. $-[(6u - v) - 5v] - 4[-u - (2u - v)]$

Evaluate the following expressions when $a = -2$, $b = -1$, and $c = 3$.

52. $-3a - b$

53. $3a^2$

54. $(3a)^2$

55. $-3a^2$

56. $(-3a)^2$

57. $-a - b + c$

58. $a^2 - b^2 - c^2$

59. $-2b^2 - (a + c)$

60. $c^3 - a^3$

61. $a^2 - 3bc$

62. $(a - b)^3$

63. $a^3 - b^3$

64. $a^2b^2c^2$

65. $2a + b - 3c + 5$

66. $-(-a)$

67. $|a + c|$

68. $|a - c|$

69. $|b^2 + c^2|$

For Review

Perform the indicated operations.

70. $(4)(-10)$

71. $(-4)(-10)$

72. $(-18) \div 9$

73. $(-18) \div (-9)$

74. $\left(-\frac{1}{5}\right)\left(-\frac{5}{6}\right)$

75. $(-5.4) \div (-0.9)$

2.4 EXERCISES C

Use the distributive laws and collect like terms.

1. $3\{x - 2[x - (y - x)] - (2x - y)\}$

2. $4\{[5(a - 2b) - 3b] - 2[6a - (3b - a)]\}$
   [Answer: $-36a - 28b$]

3. $7x - \{[6x - 2y - 3(2x - y)] - 3[-x - (3y - x)]\}$

4. $-5\{[-4(a + 5b) - 2b] - 7[a - (-a - b)]\}$
   [Answer: $90a + 145b$]

2.5 INTEGER EXPONENTS

STUDENT GUIDEPOSTS

1 Review of exponential notation
2 Product rule
3 Quotient rule
4 Power rule
5 Powers of products and quotients
6 Zero as an exponent
7 Negative exponents
8 Summary of rules for exponents

In Section 1.5 we introduced exponents and exponential notation.

1 ▶||▶||▶ If a is any number and n is a positive integer,

$$a^n = \underbrace{a \cdot a \cdot a \cdots a}_{n \text{ factors}},$$

where a is the base, n the exponent, and a^n the exponential expression.

EXAMPLE 1

Write without using exponents.

(A) $4^5 = \underbrace{4 \cdot 4 \cdot 4 \cdot 4 \cdot 4}_{5 \text{ factors}}$ The product is 1024

(B) $3y^2 = 3\underbrace{yy}_{2 \text{ }ys \text{ as factors}}$ 3 *is not* squared

(C) $(3y)^2 = \underbrace{(3y)(3y)}_{2 \text{ }(3y)s}$ 3 *is* squared

(D) $2^2 + 3^2 = \underbrace{2 \cdot 2}_{2 \text{ }2s} + \underbrace{3 \cdot 3}_{2 \text{ }3s}$ This simplifies to 13, *not* $(2 + 3)^2 = 5^2 = 25$

(E) $1^{32} = \underbrace{1 \cdot 1 \cdot 1 \cdots 1}_{32 \text{ factors}} = 1$ 1 to any power is always 1

(F) $(-1)^3 = (-1)(-1)(-1)$
$= (+1)(-1) = -1$ -1 to an *odd* power is -1

(G) $(-1)^4 = (-1)(-1)(-1)(-1)$
$= (-1)(-1) = 1$ -1 to an *even* power is $+1$ or 1 ◀◀

Practice Exercise 1

Write without using exponents.

(A) 9^7

(B) $6w^3$

(C) $(6w)^3$

(D) $4^2 + 8^2$

(E) 1^{27}

(F) $(-1)^5$

(G) $(-1)^{14}$

Answers: (A) $9 \cdot 9 \cdot 9 \cdot 9 \cdot 9 \cdot 9 \cdot 9$
(B) $6 \cdot w \cdot w \cdot w$ (C) $(6w)(6w)(6w)$
(D) $4 \cdot 4 + 8 \cdot 8$ (E) 1   (F) -1
(G) 1

▶▶ **CAUTION** ◀◀ Two of the most common errors made when working with exponents have been shown in the above examples:

$$3y^2 \neq (3y)^2 \quad \text{and} \quad 2^2 + 3^2 \neq (2 + 3)^2. \text{ ◀◀}$$

Parts (F) and (G) of Example 1 illustrate the following general rule: An odd power of a negative number is negative, and an even power of a negative number is positive.

When we combine terms containing exponential expressions by multiplication, division, or taking powers, our work can be simplified by using the basic properties of exponents. For example,

$$a^3 \cdot a^2 = \underbrace{(a \cdot a \cdot a)}_{3 \text{ factors}}\underbrace{(a \cdot a)}_{2 \text{ factors}} = \underbrace{a \cdot a \cdot a \cdot a \cdot a}_{5 \text{ factors}} = a^5.$$

When two exponential expressions *with the same base* are multiplied, the product is that base raised to the sum of the exponents on the original expressions.

 Product rule for exponents

If a is any number, and m and n are positive integers,

$$a^m \cdot a^n = a^{m+n}.$$

(To multiply powers with the same base, add exponents.)

84 RATIONAL AND REAL NUMBERS

EXAMPLE 2
Find the product.

(A) $a^3 \cdot a^4 = a^{3+4} = a^7$

(B) $5^3 \cdot 5^7 = 5^{3+7} = 5^{10}$ Not 25^{10}

(C) $2^3 \cdot 2^2 \cdot 2^6 = 2^{3+2+6} = 2^{11}$ The rule applies to more than two factors

(D) $3x^3 \cdot x^2 = 3x^{3+2} = 3x^5$ The 3 is not raised to the powers ◀◀

Practice Exercise 2
Find the product.

(A) $w^5 \cdot w^8$

(B) $8^2 \cdot 8^8$

(C) $3^2 \cdot 3^3 \cdot 3^5 \cdot 3^{11}$

(D) $7a^5 \cdot a^{12}$

Answers: (A) w^{13} (B) 8^{10} (C) 3^{21} (D) $7a^{17}$

When two powers with the same base are divided, for example,

$$\frac{a^5}{a^2} = \frac{\overbrace{a \cdot a \cdot a \cdot a \cdot a}^{5 \text{ factors}}}{\underbrace{a \cdot a}_{2 \text{ factors}}} = \frac{a \cdot a \cdot a \cdot \cancel{a} \cdot \cancel{a}}{\cancel{a} \cdot \cancel{a}} = \underbrace{a \cdot a \cdot a}_{3 \text{ factors}} = a^3,$$

the quotient can be found by raising the base to the difference of the exponents $(5 - 2 = 3)$.

3 ▶▶ Quotient rule for exponents

If a is any number except zero and m, n, and $m - n$ are positive integers, then

$$\frac{a^m}{a^n} = a^{m-n}.$$

(To divide powers with the same base, subtract exponents.)

EXAMPLE 3
Find the quotient.

(A) $\dfrac{a^7}{a^3} = a^{7-3} = a^4$

(B) $\dfrac{5^8}{5^5} = 5^{8-5} = 5^3$

(C) $\dfrac{2^3}{3^4}$ Cannot be simplified using the rule of exponents since the bases are different

(D) $\dfrac{3x^3}{x^2} = 3x^{3-2} = 3x^1 = 3x$ $x^1 = x$

(E) $\dfrac{2yy^3}{y^2} = \dfrac{2y^1 y^3}{y^2} = \dfrac{2y^{1+3}}{y^2}$

$= \dfrac{2y^4}{y^2} = 2y^{4-2} = 2y^2$ ◀◀

Practice Exercise 3
Find the quotient.

(A) $\dfrac{x^{10}}{x^4}$

(B) $\dfrac{7^{12}}{7^9}$

(C) $\dfrac{8^3}{5^2}$

(D) $\dfrac{2a^7}{a}$

(E) $\dfrac{7z^2 z^8}{z^7}$

Answers: (A) x^6 (B) 7^3 (C) cannot be simplified using the quotient rule (D) $2a^6$ (E) $7z^3$

When raising a power to a power, for example,

$$(a^2)^3 = \underbrace{(a^2) \cdot (a^2) \cdot (a^2)}_{3 \text{ factors}} = (a \cdot a) \cdot (a \cdot a) \cdot (a \cdot a)$$
$$= \underbrace{a \cdot a \cdot a \cdot a \cdot a \cdot a}_{6 \text{ factors}} = a^6,$$

the resulting exponential expression can be found by raising the base to the product of the exponents ($2 \cdot 3 = 6$).

Power rule

If a is any number, and m and n are positive integers,

$$(a^m)^n = a^{mn}.$$

(To raise a power to a power, multiply exponents.)

▶▶ **CAUTION** ◀◀ Do not confuse the power rule with the product rule. For example,

$$(a^5)^2 = a^{5 \cdot 2} = a^{10},$$

but

$$a^5 a^2 = a^{5+2} = a^7. \blacktriangleleft$$

EXAMPLE 4
Find the powers.

(A) $(a^3)^8 = a^{3 \cdot 8} = a^{24}$

(B) $2(x^3)^2 = 2x^{3 \cdot 2} = 2x^6$ ◀◀

Practice Exercise 4
Find the powers.

(A) $(w^5)^7$

(B) $8(a^4)^3$

Answers: **(A)** w^{35} **(B)** $8a^{12}$

A product or quotient of expressions is often raised to a power. For example,

$$(3x^2)^4 = \underbrace{(3x^2) \cdot (3x^2) \cdot (3x^2) \cdot (3x^2)}_{4 \text{ factors}}$$
$$= \underbrace{3 \cdot 3 \cdot 3 \cdot 3}_{4 \text{ factors}} \cdot \underbrace{x^2 \cdot x^2 \cdot x^2 \cdot x^2}_{4 \text{ factors}} = 3^4(x^2)^4,$$

and

$$\left(\frac{2}{y^2}\right)^3 = \underbrace{\frac{2}{y^2} \cdot \frac{2}{y^2} \cdot \frac{2}{y^2}}_{3 \text{ factors}} = \frac{\overbrace{2 \cdot 2 \cdot 2}^{3 \text{ factors}}}{\underbrace{y^2 \cdot y^2 \cdot y^2}_{3 \text{ factors}}} = \frac{2^3}{(y^2)^3}.$$

These illustrate the next rule.

If a and b are any numbers, and n is a positive integer, then

$$(a \cdot b)^n = a^n \cdot b^n \quad \text{and} \quad \left(\frac{a}{b}\right)^n = \frac{a^n}{b^n} \quad (b \text{ not zero}).$$

EXAMPLE 5

Simplify.

(A) $(2y)^5 = 2^5 \cdot y^5 = 2^5 y^5 = 32y^5$

(B) $(3a^2b^3)^4 = 3^4 \cdot (a^2)^4 \cdot (b^3)^4$ Raise each factor to the fourth power

$\qquad = 3^4 a^8 b^{12}$

$\qquad = 81 a^8 b^{12}$

(C) $\left(\dfrac{2}{x^3}\right)^4 = \dfrac{2^4}{(x^3)^4}$ $\left(\dfrac{a}{b}\right)^n = \dfrac{a^n}{b^n}$

$\qquad = \dfrac{16}{x^{3 \cdot 4}}$ $(a^m)^n = a^{mn}$

$\qquad = \dfrac{16}{x^{12}}$

(D) $\left(\dfrac{3a^2}{b}\right)^3 = \dfrac{(3a^2)^3}{b^3}$ $\left(\dfrac{a}{b}\right)^n = \dfrac{a^n}{b^n}$

$\qquad = \dfrac{3^3(a^2)^3}{b^3}$ $(a \cdot b)^n = a^n \cdot b^n$

$\qquad = \dfrac{27a^6}{b^3}$ ◀

Practice Exercise 5

Simplify.

(A) $(5a)^3$

(B) $(2u^3 w^5)^2$

(C) $\left(\dfrac{3}{z^2}\right)^3$

(D) $\left(\dfrac{4x}{y^3}\right)^2$

Answers: (A) $125a^3$ (B) $4u^6 w^{10}$ (C) $\dfrac{27}{z^6}$ (D) $\dfrac{16x^2}{y^6}$

▶▶ **CAUTION** ◀◀   Rules similar to the product and quotient rules above do not exist for sums and differences. For example,

$$(2^2 + 3^2)^3 \quad \text{is not} \quad (2^2)^3 + (3^2)^3,$$

and

$$(1^3 - 4^3)^2 \quad \text{is not} \quad (1^3)^2 - (4^3)^2.$$

Also, exponential expressions with different bases cannot be combined by adding exponents. For example, in general,

$$a^2 \cdot b^5 \quad \text{is not} \quad (ab)^7. \quad \blacktriangleleft$$

We know that if a is not zero,

$$\dfrac{a^m}{a^n} = a^{m-n}.$$

Suppose we let $m = n$. Then

$$\dfrac{a^m}{a^m} = a^{m-m} = a^0 \quad \text{and also} \quad \dfrac{a^m}{a^m} = 1.$$

(Any number divided by itself is 1.) This suggests the following definition.

6 ▶▶▶   If a is any number except zero,

$$a^0 = 1.$$

Remember that 0^0 is not defined.

2.5 INTEGER EXPONENTS

EXAMPLE 6

Simplify.

(A) $5^0 = 1$

(B) $(2a^2b^3)^0 = 1$ (assuming $a \neq 0$ and $b \neq 0$) ◀◀

Practice Exercise 6

Simplify.

(A) $21^0 = $ _____ .

(B) If $x \neq 0$ and $y \neq 0$, $(7xy^5)^0 = $ _____ .

Answers: (A) 1 (B) 1

Again, consider

$$\frac{a^m}{a^n} = a^{m-n} \quad (a \neq 0).$$

What happens when $n > m$? For example, if we let $n = 5$ and $m = 2$ and we extend the quotient rule to include $m - n < 0$, we have

$$\frac{a^m}{a^n} = \frac{a^2}{a^5} = a^{2-5} = a^{-3}.$$

If we look at the problem another way, we have

$$\frac{a^2}{a^5} = \frac{\cancel{a} \cdot \cancel{a}}{\cancel{a} \cdot \cancel{a} \cdot a \cdot a \cdot a} = \frac{1}{a \cdot a \cdot a} = \frac{1}{a^3}.$$

Thus, we conclude that $a^{-3} = \frac{1}{a^3}$. This suggests a way to define exponential expressions with negative integer exponents.

7 ▶▶ If $a \neq 0$ and n is a positive integer ($-n$ is a negative integer), then

$$a^{-n} = \frac{1}{a^n}.$$

EXAMPLE 7

Simplify and write without negative exponents.

(A) $5^{-3} = \frac{1}{5^3} = \frac{1}{125}$ 5^{-3} is not -5^3 nor $(-3)(5)$

(B) $4^{-2} = \frac{1}{4^2} = \frac{1}{16}$

(C) $\frac{1}{3^{-2}} = \frac{1}{\frac{1}{3^2}} = \frac{1}{\frac{1}{9}} = 1 \cdot \frac{9}{1} = 9 = 3^2$

(D) $(-2)^{-5} = \frac{1}{(-2)^5} = \frac{1}{-32} = -\frac{1}{32}$ ◀◀

Practice Exercise 7

Simplify and write without negative exponents.

(A) 7^{-2}

(B) 2^{-5}

(C) $\frac{1}{5^{-1}}$

(D) $(-6)^{-2}$

Answers: (A) $\frac{1}{49}$ (B) $\frac{1}{32}$

(C) 5 (D) $\frac{1}{36}$

Example 7 shows that we can "remove" negative exponents simply by moving an exponential expression with a negative exponent from numerator to denominator (or denominator to numerator) and changing the sign of the exponent.

88 RATIONAL AND REAL NUMBERS

▶▶ **CAUTION** ◀◀ a^{-n} is not equal to $-a^n$ nor to $(-n)a$. As shown in Example 7 (a), 5^{-3} is $\frac{1}{125}$ and not -5^3, which is -125, nor $(-3)(5)$, which is -15. ◀◀

All the rules of exponents stated for positive integer exponents apply to all integer exponents: positive, negative, and zero. These are summarized in the next set of rules.

Rules for exponents

Let a and b be any two numbers, m and n any two integers.

1. $a^m \cdot a^n = a^{m+n}$
2. $\frac{a^m}{a^n} = a^{m-n}$ $(a \neq 0)$
3. $(a^m)^n = a^{mn}$
4. $(a \cdot b)^n = a^n b^n$
5. $\left(\frac{a}{b}\right)^n = \frac{a^n}{b^n}$ $(b \neq 0)$
6. $a^0 = 1$ $(a \neq 0)$
7. $a^{-n} = \frac{1}{a^n}$ $(a \neq 0)$
8. $\frac{1}{a^{-n}} = a^n$ $(a \neq 0)$

EXAMPLE 8

Simplify and write without negative exponents.

(A) $(2a)^{-1} = \frac{1}{(2a)^1} = \frac{1}{2a}$ $(2a)^{-1}$ is not $-2a$

(B) $2a^{-1} = 2\frac{1}{a} = \frac{2}{a}$ The exponent -1 is only on a, not on 2

(C) $y^3 y^{-5} = y^{3+(-5)} = y^{-2} = \frac{1}{y^2}$

(D) $\frac{a^2 b^{-3}}{a^{-1} b^5} = a^{2-(-1)} b^{-3-5} = a^3 b^{-8} = a^3 \cdot \frac{1}{b^8} = \frac{a^3}{b^8}$

(E) $\frac{1}{x^{-3}} = \frac{1}{\frac{1}{x^3}} = 1 \cdot \frac{x^3}{1} = x^3$

(F) $(2a^{-2}b)^{-3} = 2^{-3}(a^{-2})^{-3} b^{-3} = \frac{1}{2^3} a^{(-2)(-3)} \frac{1}{b^3}$

$= \frac{1}{8} \cdot a^6 \frac{1}{b^3} = \frac{a^6}{8b^3}$

(G) $\left(\frac{a^2}{2y^{-3}}\right)^{-2} = \frac{(a^2)^{-2}}{2^{-2}(y^{-3})^{-2}} = \frac{a^{-4}}{\frac{1}{2^2} y^6} = \frac{\frac{1}{a^4}}{\frac{y^6}{2^2}}$

$= \frac{1}{a^4} \cdot \frac{2^2}{y^6} = \frac{4}{a^4 y^6}$ ◀◀

EXAMPLE 9

Evaluate the following when $a = -2$ and $b = 3$.

(A) $a^{-1} = (-2)^{-1} = \frac{1}{-2} = -\frac{1}{2}$ $(-2)^{-1}$ is not $+2$

Practice Exercise 8

Simplify and write without negative exponents.

(A) $(5w)^{-2}$

(B) $5w^{-2}$

(C) $u^{-7} u^5$

(D) $\frac{x^4 y^{-3}}{x^{-1} y^2}$

(E) $\frac{1}{m^{-7}}$

(F) $(6y^{-3} z^{-1})^{-2}$

(G) $\left(\frac{3w^{-1}}{x^2}\right)^{-3}$

Answers: (A) $\frac{1}{25w^2}$ (B) $\frac{5}{w^2}$
(C) $\frac{1}{u^2}$ (D) $\frac{x^5}{y^5}$ (E) m^7
(F) $\frac{y^6 z^2}{36}$ (G) $\frac{w^3 x^6}{27}$

Practice Exercise 9

Evaluate the following when $x = -3$ and $y = 5$.

(A) x^{-1}

(B) $\dfrac{a^{-2}}{b} = \dfrac{(-2)^{-2}}{3} = \dfrac{\frac{1}{(-2)^2}}{3} = \dfrac{\frac{1}{4}}{3} = \dfrac{1}{4} \cdot \dfrac{1}{3} = \dfrac{1}{12}$

(C) $(a+b)^{-1} = (-2+3)^{-1} = 1^{-1} = \dfrac{1}{1} = 1$

(D) $a^{-1} + b^{-1} = (-2)^{-1} + (3)^{-1}$

$= \dfrac{1}{-2} + \dfrac{1}{3} = -\dfrac{3}{6} + \dfrac{2}{6} = -\dfrac{1}{6}$ ◀◀

(B) $\dfrac{x^2}{y^{-1}}$

(C) $(y - x)^{-1}$

(D) $y^{-1} - x^{-1}$

Answers: (A) $\dfrac{1}{-3}$ (B) 45   (C) $\dfrac{1}{8}$ (D) $\dfrac{8}{15}$

2.5 EXERCISES A

Write in exponential notation.

1. $8 \cdot 8 \cdot 8 \cdot 8$
2. $2 \cdot 2 \cdot y \cdot y \cdot y$
3. $(2x)(2x)(2x)(2x)$

Write without using exponents.

4. $2y^4$
5. $(2y)^4$
6. $a^2 + b^2$

Simplify and write without negative exponents.

7. $x^2 \cdot x^5$
8. $a^3 \cdot a^2 \cdot a^4$
9. $2y^2 \cdot y^8$

10. $\dfrac{a^4}{a^3}$
11. $\dfrac{2y^5}{y^3}$
12. $(a^3)^4$

13. $(2x^3)^4$
14. $2(x^3)^4$
15. $\left(\dfrac{2}{x^3}\right)^4$

16. $\dfrac{a^3}{b^5}$
17. 5^0
18. 0^0

19. $(2a)^0$ $(a \neq 0)$
20. $(2x)^{-1}$
21. $2x^{-1}$

22. $\dfrac{2x^7}{x^9}$
23. $3y^4 y^{-7}$
24. $(3y)^{-2}$

25. $3y^{-2}$
26. $(3y^{-2})^3$
27. $\left(\dfrac{2y}{x^3}\right)^{-2}$

28. $\dfrac{b^{-2}}{a^{-4}}$
29. $\dfrac{a^{-4}b^2}{b^{-2}}$
30. $\dfrac{a^{-2}b^2}{a^4 b^{-3}}$

31. $\dfrac{x^3 y^{-5}}{x^4 y^{-6}}$

32. $\dfrac{3^0 a^3 b^{-8}}{ab^4}$

33. $\dfrac{3^{-1} x^{-1} y^{-5}}{x^{-6} y^2}$

34. $(x^{-2} y^{-1})^{-2}$

35. $(x^2 y^{-3})^{-4}$

36. $(3x^{-1} y)^{-2}$

37. $\left(\dfrac{a^{-5}}{b^{-1}}\right)^{-1}$

38. $\left(\dfrac{2a^{-3}}{b^3}\right)^{-2}$

39. $\left(\dfrac{2a^3 b^{-2}}{a^{-5} b}\right)^{-3}$

Evaluate when $a = -2$ and $b = 3$.

40. $3a^2$

41. $(3a)^2$

42. $-3a^2$

43. $(-3a)^2$

44. $-b^2$

45. $(-b)^2$

46. $a^2 - b^2$

47. $(a - b)^2$

48. a^{-2}

49. $-2a$

50. $-a^2$

51. $a^{-2} + b^{-2}$

52. $(a + b)^{-2}$

53. $\dfrac{a^{-1}}{b^{-2}}$

54. a^{-3}

55. $(-a)^{-3}$

56. $a^{-1} b^{-1}$

57. $(ab)^{-1}$

For Review

Remove parentheses and collect like terms.

58. $2a - (4a - 1)$

59. $-(1 - a) + (a + 1)$

60. $-[-(1 - a)]$

61. $2[y - (3y - 2)]$

62. $-3[-y + (1 - y)]$

63. $-[-(-y)] + y$

ANSWERS: **1.** 8^4 **2.** $2^2 y^3$ **3.** $(2x)^4$ **4.** $2 \cdot y \cdot y \cdot y \cdot y$ **5.** $(2y)(2y)(2y)(2y)$ **6.** $a \cdot a + b \cdot b$ **7.** x^7 **8.** a^9 **9.** $2y^{10}$ **10.** $a^1 = a$ **11.** $2y^2$ **12.** a^{12} **13.** $16x^{12}$ **14.** $2x^{12}$ **15.** $\dfrac{16}{x^{12}}$ **16.** cannot be simplified further **17.** 1 **18.** undefined **19.** 1 **20.** $\dfrac{1}{2x}$ **21.** $\dfrac{2}{x}$ **22.** $\dfrac{2}{x^2}$ **23.** $\dfrac{3}{y^3}$ **24.** $\dfrac{1}{9y^2}$ **25.** $\dfrac{3}{y^2}$ **26.** $\dfrac{27}{y^6}$ **27.** $\dfrac{x^6}{4y^2}$ **28.** $\dfrac{a^4}{b^2}$ **29.** $\dfrac{b^4}{a^4}$ **30.** $\dfrac{b^5}{a^6}$ **31.** $\dfrac{y}{x}$ **32.** $\dfrac{a^2}{b^{12}}$ **33.** $\dfrac{x^5}{3y^7}$ **34.** $x^4 y^2$ **35.** $\dfrac{y^{12}}{x^8}$ **36.** $\dfrac{x^2}{9y^2}$ **37.** $\dfrac{a^5}{b}$ **38.** $\dfrac{a^6 b^6}{4}$ **39.** $\dfrac{b^9}{8a^{24}}$ **40.** 12 **41.** 36 **42.** -12 **43.** 36 **44.** -9 **45.** 9 **46.** -5 **47.** 25 **48.** $\dfrac{1}{4}$ **49.** 4 **50.** -4 **51.** $\dfrac{13}{36}$ **52.** 1 **53.** $-\dfrac{9}{2}$ **54.** $-\dfrac{1}{8}$ **55.** $\dfrac{1}{8}$ **56.** $-\dfrac{1}{6}$ **57.** $-\dfrac{1}{6}$ **58.** $-2a + 1$ **59.** $2a$ **60.** $1 - a$ **61.** $-4y + 4$ **62.** $6y - 3$ **63.** 0

2.5 EXERCISES B

Write in exponential notation.

1. $4 \cdot 4 \cdot z \cdot z \cdot z$
2. $(3w)(3w)(3w)$
3. $3www$

Write without using exponents.

4. $4y^3$
5. $(4y)^3$
6. $x^2 + y^2$

Simplify and write without negative exponents.

7. $y^2 \cdot y^7$
8. $x^3 \cdot x^2 \cdot x^6$
9. $2z^2 \cdot z^5$
10. $\dfrac{b^5}{b^2}$
11. $\dfrac{2y^7}{y^4}$
12. $(w^3)^5$
13. $(2c^2)^4$
14. $2(y^3)^5$
15. $\left(\dfrac{2}{a^2}\right)^3$
16. $\dfrac{x^3}{y^4}$
17. 7^0
18. -0^0
19. $(4x)^0 (x \neq 0)$
20. $(5y)^{-1}$
21. $5y^{-1}$
22. $\dfrac{3a^3}{a^7}$
23. $4z^4 z^{-9}$
24. $(5w)^{-2}$
25. $5w^{-2}$
26. $(2a^{-2})^3$
27. $\left(\dfrac{2y}{x^3}\right)^2$
28. $\dfrac{b^{-5}}{a^{-3}}$
29. $\dfrac{a^{-3}b^{-3}}{b^{-4}}$
30. $\dfrac{x^{-2}y^4}{x^3 y^{-3}}$
31. $\dfrac{a^4 b^{-3}}{a^{-2} b^2}$
32. $\dfrac{5^0 a^{-5} b^{-2}}{a^4 b^{-1}}$
33. $\dfrac{2^{-2} x^3 y^{-4}}{x^{-2} y^{-2}}$
34. $(a^{-1} b^{-5})^{-1}$
35. $(a^{-6} b^3)^{-3}$
36. $(4a^{-1} b^{-1})^{-3}$
37. $\left(\dfrac{x^{-4}}{y^{-2}}\right)^{-2}$
38. $\left(\dfrac{3x^4}{y^{-3}}\right)^{-1}$
39. $\left(\dfrac{3x^{-2} y^{-1}}{x^3 y^{-5}}\right)^{-4}$

Evaluate when $x = -3$ and $y = 2$.

40. $4x^2$
41. $(4x)^2$
42. $-4x^2$
43. $(-4x)^2$
44. $-x^2$
45. $(-x)^2$
46. $y^2 - x^2$
47. $(y - x)^2$
48. x^{-2}
49. $-3y$
50. $-y^3$
51. $x^{-2} + y^{-2}$
52. $(x + y)^{-2}$
53. $\dfrac{x^{-1}}{y^{-2}}$
54. x^{-3}
55. $(-x)^{-3}$
56. $x^{-1} y^{-1}$
57. $(xy)^{-1}$

For Review

Remove parentheses and collect like terms.

58. $4y - (6y - 3)$
59. $-(2 - y) + (2 + y)$
60. $-[-(3 - y)]$
61. $3[w - (2w - 4)]$
62. $-4[-w + (2 - w)]$
63. $-[-(-w)] - w$

2.5 EXERCISES C

Simplify and write without negative exponents.

1. $\dfrac{a^{-2}b^3c^2}{a^{-3}b^{-2}c^{-2}}$

2. $\dfrac{3^0 x^{-6}(y^{-1})^{-2}}{x^{-2}y^{-3}}$

3. $\dfrac{2^{-2}(x^2)^{-3}y^3 z^{-2}}{3^{-1}x^{-1}(yz)^{-1}}$
   $\left[\text{Answer: } \dfrac{3y^4}{4x^5 z}\right]$

4. $\left(\dfrac{5^0 a^{-6}b^2 c^3}{a^2 b^{-1} c^{-2}}\right)^{-1}$

5. $\left(\dfrac{2^{-1}x^{-5}y^{-8}}{4^{-1}x^3 y^{-4}}\right)^{-2}$

6. $\left[\left(\dfrac{a^2 b^{-2}}{a^{-5}b^{-1}}\right)^{-1}\right]^{-2}$
   $\left[\text{Answer: } \dfrac{a^{14}}{b^2}\right]$

2.6 SCIENTIFIC NOTATION

▶▶ STUDENT GUIDEPOSTS

1 Scientific notation
2 Calculations using scientific notation

When computing with very large or very small numbers using a calculator, problems can arise due to limitations on the display (usually 8 digits). For example, if we use a calculator to multiply

$$(290{,}000)(15{,}000),$$

the product might be given as shown in Figure 2.14.

Figure 2.14

The number 4.35×10^9 is a shortened notation for 4,350,000,000, which has ten digits, too many for the display. Knowledge of integer exponents is extremely important to the understanding of this type of notation known as *scientific notation*.

For example, a scientist might use the number

$$235{,}000{,}000{,}000{,}000{,}000{,}000$$

but, instead of writing out all the zeros, he or she would write

$$2.35 \times 10^{20}.$$

This short form is easier to use in computations. Likewise, the number

$$0.000000000057$$

could be written as

$$5.7 \times 10^{-11}.$$

1 ▶▶ A number is written in **scientific notation** if it is the product of a number between 1 and 10 and a power of 10.

2.6 SCIENTIFIC NOTATION

> **To write a number in scientific notation**
> 1. Move the decimal point to the position immediately to the right of the first nonzero digit.
> 2. Multiply by the power of ten that is equal in absolute value to the number of decimal places moved. The exponent on 10 is positive if the original number is greater than 10 and negative if the number is less than 1.

EXAMPLE 1

Write in scientific notation.

(A) $2{,}500{,}000 = 2.5 \times 10^6$ Count 6 decimal places
(first nonzero digit; 6 places)

(B) $0.0000025 = 2.5 \times 10^{-6}$ Count 6 decimal places
(first nonzero digit; 6 places)

(C) $4{,}321{,}000{,}000 = 4.321 \times 10^9$
(9 places)

(D) $0.00000000001 = 1 \times 10^{-11}$
(11 places)

(E) $0.1 = 1 \times 10^{-1}$
(1 place)

(F) $4.8 = 4.8 \times 10^0$ ◀◀

Practice Exercise 1

Write in scientific notation.

(A) 18,300,000

(B) 0.000087

(C) 65,240,000,000,000

(D) 0.0000000001

(E) 0.5

(F) 9.7

Answers: (A) 1.83×10^7
(B) 8.7×10^{-5} (C) 6.524×10^{13}
(D) 1×10^{-10} (E) 5×10^{-1}
(F) 9.7×10^0

EXAMPLE 2

Write in standard notation.

(A) $5.4 \times 10^5 = 540{,}000$ Count 5 decimal places

(B) $5.4 \times 10^{-5} = 0.000054$ Count 5 decimal places

(C) $8.94 \times 10^{13} = 89{,}400{,}000{,}000{,}000$

(D) $2.113 \times 10^{-8} = 0.00000002113$ ◀◀

Practice Exercise 2

Write in standard notation.

(A) 6.1×10^7

(B) 6.1×10^{-7}

(C) 5.35×10^{12}

(D) 9.03×10^{-6}

Answers: (A) 61,000,000
(B) 0.00000061
(C) 5,350,000,000,000
(D) 0.00000903

2 Scientific notation not only shortens the notation for many numbers, but also helps in calculations.

EXAMPLE 3

Perform the indicated operations using scientific notation.

(A) $(30{,}000)(2{,}000{,}000) = (3 \times 10^4)(2 \times 10^6)$
$\qquad = (3 \cdot 2) \times (10^4 \times 10^6)$ Change order
$\qquad = 6 \times 10^{10}$ Add exponents

Practice Exercise 3

Perform the indicated operations using scientific notation.

(A) $(5{,}000{,}000)(200{,}000)$

(B) $(6.4 \times 10^{-10})(0.8 \times 10^9)$

94 RATIONAL AND REAL NUMBERS

(B) $(2.4 \times 10^{-12})(4.0 \times 10^{11}) = (2.4)(4.0) \times (10^{-12} \times 10^{11})$
$= 9.6 \times 10^{-1}$

(C) $\dfrac{3.2 \times 10^{-1}}{1.6 \times 10^5} = \dfrac{3.2}{1.6} \times \dfrac{10^{-1}}{10^5} = 2 \times 10^{-6}$ ◂◂

(C) $\dfrac{8.1 \times 10^{-2}}{2.7 \times 10^{-6}}$

Answers: **(A)** 1×10^{12}
(B) 5.12×10^{-1} **(C)** 3×10^4

2.6 EXERCISES A

Write in scientific notation.

1. 370
2. 0.0037
3. 98,000

4. 0.00012
5. 8360
6. 0.00279

7. 0.0000000000756
8. 2,650,000,000
9. 0.01

Write in standard notation.

10. 2.3×10^2
11. 2.3×10^{-2}
12. 8.7×10^{-5}

13. 4.58×10^6
14. 7.51×10^{-8}
15. 6.64×10^{10}

Perform the indicated operations using scientific notation.

16. $(4 \times 10^5)(1 \times 10^6)$
17. $(40,000,000)(20,000)$

18. $(1 \times 10^{-10})(5 \times 10^7)$
19. $(0.0000022)(300)$

20. $\dfrac{3.3 \times 10^{12}}{1.1 \times 10^{-2}}$
21. $\dfrac{0.0000006}{0.03}$

22. The measure of one calorie is equal to 0.000000278 kilowatt-hours. Write this number in scientific notation.

23. The distance that light will travel in 1 year is approximately 5,870,000,000,000 miles. Write the number in scientific notation.

For Review

Evaluate when $a = -2$ and $b = 5$.

24. a^{-1}
25. a^{-2}
26. $-2a$

27. b^{-2}
28. $-2b$
29. $a^{-1} + b^{-1}$

30. $(a + b)^{-1}$ **31.** $(a - b)^{-2}$ **32.** $a^{-2} - b^{-2}$

ANSWERS: **1.** 3.7×10^2 **2.** 3.7×10^{-3} **3.** 9.8×10^4 **4.** 1.2×10^{-4} **5.** 8.36×10^3 **6.** 2.79×10^{-3}
7. 7.56×10^{-11} **8.** 2.65×10^9 **9.** 1×10^{-2} **10.** 230 **11.** 0.023 **12.** 0.000087 **13.** 4,580,000
14. 0.0000000751 **15.** 66,400,000,000 **16.** 4×10^{11} **17.** 8×10^{11} **18.** 5×10^{-3} **19.** 6.6×10^{-4} **20.** 3×10^{14}
21. 2×10^{-5} **22.** 2.78×10^{-7} **23.** 5.87×10^{12} **24.** $-\dfrac{1}{2}$ **25.** $\dfrac{1}{4}$ **26.** 4 **27.** $\dfrac{1}{25}$ **28.** -10 **29.** $-\dfrac{3}{10}$
30. $\dfrac{1}{3}$ **31.** $\dfrac{1}{49}$ **32.** $\dfrac{21}{100}$

2.6 EXERCISES B

Write in scientific notation.

1. 5400

2. 0.0054

3. 386

4. 0.000028

5. 33,000

6. 0.1

7. 0.000000000254

8. 4,620,000,000

9. 0.001

Write in standard notation.

10. 8.4×10^3

11. 8.4×10^{-3}

12. 5.42×10^{-7}

13. 7.25×10^7

14. 2.06×10^{-6}

15. 4.99×10^{12}

Perform the indicated operations using scientific notation.

16. $(5 \times 10^3)(1 \times 10^4)$

17. $(60{,}000{,}000)(50{,}000)$

18. $(1 \times 10^{-12})(3 \times 10^8)$

19. $(0.00000044)(500)$

20. $\dfrac{4.4 \times 10^{15}}{1.1 \times 10^{-3}}$

21. $\dfrac{0.00000008}{0.04}$

22. The earth is approximately 93,000,000 miles from the sun. Write this number in scientific notation.

23. An important number used by chemists is 602,000,000,000,000,000,000,000, known as Avogadro's number. Write this number in scientific notation.

For Review

Evaluate when $x = -3$ and $y = 2$.

24. x^{-1}

25. x^{-2}

26. $-2x$

27. y^{-2}

28. $-2y$

29. $x^{-1} + y^{-1}$

30. $(x + y)^{-1}$

31. $(x - y)^{-2}$

32. $x^{-2} - y^{-2}$

2.6 EXERCISES C

Perform the indicated operations using scientific notation.

1. $\dfrac{(2.5 \times 10^{-3})(4.2 \times 10^{-8})}{(5.0 \times 10^7)(8.4 \times 10^{-10})}$

2. $\dfrac{(0.000036)(0.0001)^{-1}}{(1{,}200{,}000)(1 \times 10^7)^{-2}}$ [Answer: 3×10^7]

2.7 IRRATIONAL AND REAL NUMBERS

STUDENT GUIDEPOSTS

1 Irrational numbers
2 Real numbers
3 Perfect squares and square roots
4 Principal square roots and radicals
5 Radicands

Thus far we have concentrated on working with the rational numbers. Remember that a **ratio**nal number is the quotient (**ratio**) of two integers. Equivalently a rational number is a number with a decimal form that either terminates or repeats a block of digits.

There are many numbers that cannot be expressed in this way. One of the most familiar of these is π, the number equal to the ratio of the circumference of any circle to its diameter. Such numbers are called *irrational numbers*. In decimal notation **irrational numbers** do not terminate nor do they have a repeating block of digits. The **real numbers** are found by taking the rational numbers together with the irrational numbers.

Before identifying some of the more familiar irrational numbers, we look at the notion of perfect squares and their *square roots*.

> When an integer is squared, the result is called a **perfect square**. Either of the identical factors of a perfect square number is called a **square root** of the number.

EXAMPLE 1

(A) 4 is a perfect square since $4 = 2^2$. We call 2 a square root of 4. Since $4 = (-2)^2$, -2 is also a square root of 4.

(B) 25 is a perfect square since $25 = 5^2$. The two square roots of 25 are 5 and -5.

(C) 0 is a perfect square since $0 = 0^2$. Unlike other perfect squares, 0 has only one square root, namely itself, 0. ◀◀

Practice Exercise 1

(A) Is 1 a perfect square?

(B) Is 81 a perfect square?

(C) Is 40 a perfect square?

Answers: (A) Yes; $1 = 1^2$
(B) Yes; $81 = 9^2$ (C) No; no integer times itself is equal to 40.

Do not confuse the terms *square* and *square root*!

$$5^2 = 5 \cdot 5 = 25$$

25 is the **square** of 5
5 is a **square** root of 25

All perfect squares other than 0 have two square roots, one positive and one negative. When we refer to the square root of a number, do we mean the positive root or the negative root? To avoid confusion, we refer to the positive root as the **principal square root** and we use the symbol $\sqrt{}$, called a **radical**, to represent it. For example, the principal or positive square root of 4 is $\sqrt{4}$, or 2. The number under the radical (4 in this case) is called the **radicand**. Since the radical symbol only designates the principal or positive (possibly zero) square root of a number, we designate the negative square root by $-\sqrt{}$. For example, $-\sqrt{4} = -2$.

The first sixteen perfect squares and their square roots are listed in the following table.

Perfect square N	Positive square root of N $\sqrt{N}$	Negative square root of N $-\sqrt{N}$
0	0	$-0 = 0$
1	1	-1
4	2	-2
9	3	-3
16	4	-4
25	5	-5
36	6	-6
49	7	-7
64	8	-8
81	9	-9
100	10	-10
121	11	-11
144	12	-12
169	13	-13
196	14	-14
225	15	-15

In addition to whole number perfect squares, there are also fractional perfect squares. A fraction that can be factored into the product of two identical fractional factors is called a **perfect square,** and each factor is called a **square root** of the fraction.

EXAMPLE 2

(A) $\frac{4}{9}$ is a perfect square, with $\sqrt{\frac{4}{9}} = \frac{2}{3}$ and $-\sqrt{\frac{4}{9}} = -\frac{2}{3}$.

(B) $\frac{25}{81}$ is a perfect square, with $\sqrt{\frac{25}{81}} = \frac{5}{9}$ and $-\sqrt{\frac{25}{81}} = -\frac{5}{9}$.

Practice Exercise 2

(A) Is $\frac{36}{121}$ a perfect square?

(B) Is $\frac{50}{200}$ a perfect square?

Answers: (A) Yes; $\frac{36}{121} = \left(\frac{6}{11}\right)^2$

(B) Yes; first reduce the fraction to $\frac{1}{4}$, and $\frac{1}{4} = \left(\frac{1}{2}\right)^2$.

Example 2 shows that a fraction is a perfect square if its numerator and denominator are whole number perfect squares. However, a fraction can be a perfect square without this feature. For example, if we reduce $\frac{8}{18}$ to lowest terms, we change it to a ratio of whole number perfect squares:

$$\sqrt{\frac{8}{18}} = \sqrt{\frac{2 \cdot 4}{2 \cdot 9}} = \sqrt{\frac{4}{9}} = \frac{2}{3}.$$

Thus far we have only discussed square roots of perfect squares. What about square roots of other numbers? This problem can be divided into two parts, square roots of positive numbers and square roots of negative numbers. Square

98 RATIONAL AND REAL NUMBERS

roots of negative numbers are not real numbers and will not be discussed in this text. However, square roots of positive numbers that are not perfect squares supply us with many examples of irrational numbers. These include $\sqrt{2}$, $\sqrt{3}$, $\sqrt{5}$, $\sqrt{6}$, and $\sqrt{7}$ among many, many more.

Suppose we consider $\sqrt{2}$, the number which when squared is 2. We know that

$$\sqrt{1} = 1 \text{ and } \sqrt{4} = 2.$$

Since $1 < 2 < 4$, we would expect $\sqrt{1} < \sqrt{2} < \sqrt{4}$ so that

$$1 < \sqrt{2} < 2.$$

Since there is no integer between 1 and 2, $\sqrt{2}$ cannot be an integer. In more advanced work it can be shown that $\sqrt{2}$ is not a rational number either.

Most of our work in this text will be with rational numbers until Chapter 9 when we consider operations with radicals.

2.7 EXERCISES A

Evaluate the square roots.

1. $\sqrt{36}$
2. $-\sqrt{49}$
3. $\sqrt{225}$
4. $-\sqrt{196}$

5. $\sqrt{0}$
6. $-\sqrt{1}$
7. $-\sqrt{169}$
8. $\sqrt{81}$

9. $\sqrt{\dfrac{9}{25}}$
10. $\sqrt{\dfrac{64}{196}}$
11. $-\sqrt{\dfrac{49}{81}}$
12. $-\sqrt{\dfrac{36}{225}}$

13. $\sqrt{\dfrac{1}{144}}$
14. $\sqrt{\dfrac{4}{169}}$
15. $\sqrt{\dfrac{49}{196}}$
16. $\sqrt{\dfrac{9}{81}}$

17. $\sqrt{\dfrac{8}{18}}$
18. $-\sqrt{\dfrac{2}{50}}$
19. $\sqrt{\dfrac{12}{48}}$
20. $\sqrt{\dfrac{75}{363}}$

21. (A) Compute $(16)^2$ (B) $\sqrt{256} =$ 22. (A) Compute $(17)^2$ (B) $\sqrt{289} =$

23. (A) Compute $(18)^2$ (B) $\sqrt{324} =$ 24. (A) Compute $(19)^2$ (B) $\sqrt{361} =$

25. If x is any whole number, what is $\sqrt{x^2}$?   26. If x is any natural number, what is $-\sqrt{x^2}$?

27. **(A)** What is the square of 9? **(B)** What is the principal square root of 9?

28. **(A)** What is the square of $\frac{1}{4}$? **(B)** What is the principal square root of $\frac{1}{4}$?

29. Between what two positive integers is $\sqrt{15}$ located?

30. Between what two positive integers is $\sqrt{150}$ located?

For Review

Write in scientific notation.

31. 0.000000785

32. 438,000,000,000

Write without using scientific notation.

33. 4.02×10^{10}

34. 3.5×10^{-6}

35. The shutter speed on a Panomate camera in 0.026 seconds. Write this number in scientific notation.

ANSWERS: **1.** 6 **2.** −7 **3.** 15 **4.** −14 **5.** 0 **6.** −1 **7.** −13 **8.** 9 **9.** $\frac{3}{5}$ **10.** $\frac{4}{7}$ **11.** $-\frac{7}{9}$ **12.** $-\frac{2}{5}$ **13.** $\frac{1}{12}$ **14.** $\frac{2}{13}$ **15.** $\frac{1}{2}$ **16.** $\frac{1}{3}$ **17.** $\frac{2}{3}$ **18.** $-\frac{1}{5}$ **19.** $\frac{1}{2}$ **20.** $\frac{5}{11}$ **21.** (A) 256 (B) 16 **22.** (A) 289 (B) 17 **23.** (A) 324 (B) 18 **24.** (A) 361 (B) 19 **25.** x **26.** $-x$ **27.** (A) 81 (B) 3 **28.** (A) $\frac{1}{16}$ (B) $\frac{1}{2}$ **29.** 3 and 4 **30.** 12 and 13 **31.** 7.85×10^{-7} **32.** 4.38×10^{11} **33.** 40,200,000,000 **34.** 0.0000035 **35.** 2.6×10^{-2}

2.7 EXERCISES B

Evaluate the square roots.

1. $\sqrt{196}$
2. $-\sqrt{36}$
3. $\sqrt{49}$
4. $-\sqrt{121}$
5. $\sqrt{144}$
6. $-\sqrt{64}$
7. $-\sqrt{100}$
8. $\sqrt{169}$
9. $\sqrt{\frac{4}{9}}$
10. $\sqrt{\frac{25}{196}}$
11. $-\sqrt{\frac{81}{121}}$
12. $-\sqrt{\frac{36}{49}}$
13. $\sqrt{\frac{1}{225}}$
14. $\sqrt{\frac{4}{36}}$
15. $\sqrt{\frac{49}{121}}$
16. $\sqrt{\frac{4}{16}}$
17. $\sqrt{\frac{8}{32}}$
18. $-\sqrt{\frac{50}{98}}$
19. $\sqrt{\frac{12}{75}}$
20. $\sqrt{\frac{147}{363}}$

21. **(A)** Compute $(20)^2$ **(B)** $\sqrt{400} =$

22. **(A)** Compute $(21)^2$ **(B)** $\sqrt{441} =$

100 RATIONAL AND REAL NUMBERS

23. (A) Compute $(24)^2$ (B) $\sqrt{576} =$
24. (A) Compute $(25)^2$ (B) $\sqrt{625} =$
25. If a is any whole number, what is $\sqrt{a^2}$?
26. If a is any natural number what is $-\sqrt{a^2}$?
27. (A) What is the square of 16? (B) What is the principal square root of 16?
28. (A) What is the square root of $\frac{1}{9}$? (B) What is the principal square root of $\frac{1}{9}$?
29. Between what two positive integers is $\sqrt{30}$ located?
30. Between what two positive integers is $\sqrt{200}$ located?

For Review

Write in scientific notation.

31. 0.000000000225
32. 14,900,000,000

Write without using scientific notation.

33. 6.05×10^{11}
34. 8.4×10^{-7}

35. The thickness of a new type of experimental fabric is 0.00065 cm. Write this number in scientific notation.

2.7 EXERCISES C

Evaluate the square roots where a is a natural number.

1. $\sqrt{25a^2}$
2. $\sqrt{\dfrac{16}{a^4}}$
3. $-\sqrt{\dfrac{625}{576}}$
4. $-\sqrt{\dfrac{847}{1008}}$ $\left[\text{Answer: } -\dfrac{11}{12}\right]$

CHAPTER 2 SUMMARY

Key Words and Phrases for Review

2.1
- number line
- integers
- equal (=)
- unequal ($\neq$)
- less than (<)
- greater than (>)
- less than or equal ($\leq$)
- greater than or equal ($\geq$)
- plotted
- rational numbers
- cross products
- absolute value

2.2
- additive inverse (negative)
- additive identity
- commutative law of addition
- associative law of addition

2.3
- multiplicative inverse (reciprocal)
- multiplicative identity
- commutative law of multiplication
- associative law of multiplication
- double negative property

2.4
- distributive laws
- term
- factors
- (numerical) coefficient
- removing common factor
- collecting like terms
- clearing parentheses

2.6
- scientific notation

2.7
- irrational numbers
- real numbers
- perfect square
- square root
- principal square root
- radical
- radicand

Key Concepts

2.1
1. If a and b are any numbers such that a is to the left of b on a number line, then $a < b$ or $b > a$.
2. The absolute value of a number x, denoted by $|x|$, is the distance from x to zero on a number line and as a result, is never negative.

2.2 If a, b, and c are any numbers,

(A) $a + (-a) = (-a) + a = 0$ Additive inverses (negatives)

(B) $a + 0 = 0 + a = a$ Additive identity

(C) $a + b = b + a$ Commutative law of addition

(D) $(a + b) + c = a + (b + c)$ Associative law of addition

2.3 If a, b, and c are any numbers,

(A) $a \cdot \dfrac{1}{a} = \dfrac{1}{a} \cdot a = 1$ $(a \neq 0)$ Multiplicative inverses (reciprocals)

(B) $a \cdot 1 = 1 \cdot a = a$ Multiplicative identity

(C) $ab = ba$ Commutative law of multiplication

(D) $(ab)c = a(bc)$ Associative law of multiplication

2.4
1. The distributive laws,
$$a(b + c) = ab + ac$$
and
$$a(b - c) = ab - ac,$$
are used to factor and to collect like terms in algebraic expressions.

2. Only like terms can be combined. For example, $4y + 5 \neq 9y$ since $4y$ and 5 are *not* like terms.

3. A minus sign before a set of parentheses changes the sign of *every* term inside when the parentheses are removed. For example, $-(2 - a) = -2 + a$, *not* $-2 - a$.

2.5
1. If a and b are any numbers, and m and n are natural numbers, the following rules for exponents hold.

(A) $a^m \cdot a^n = a^{m+n}$

(B) $\dfrac{a^m}{a^n} = a^{m-n}$, if $a \neq 0$

(C) $(a^m)^n = a^{mn}$

(D) $(a \cdot b)^n = a^n \cdot b^n$

(E) $\left(\dfrac{a}{b}\right)^n = \dfrac{a^n}{b^n}$, if $b \neq 0$

(F) $a^0 = 1$, if $a \neq 0$ and 0^0 is undefined.

2. In an expression such as $2y^3$, only y is cubed, not $2y$. That is, $2y^3$ is not the same as $(2y)^3$, which is equal to $8y^3$.

2.6 The radical symbol by itself only designates the principal (positive or zero) square root of a number. For example, $\sqrt{9} = 3$ not -3. We must write $-\sqrt{9}$ to represent -3.

Review Exercises

2.1 *Find the absolute values.*

1. $|12|$ 2. $|-7|$ 3. $|-2.51|$

Are the following expressions true or false?

4. $-11 < -1$ **5.** $0 \leq -4$ **6.** $5 \leq 5$

Place the appropriate symbol, =, >, or <, between the pairs of fractions.

7. $\dfrac{2}{17}$ $\quad$ $\dfrac{10}{85}$ **8.** $\dfrac{13}{5}$ $\quad$ $\dfrac{28}{11}$ **9.** $\dfrac{6}{7}$ $\quad$ $\dfrac{11}{12}$

10. The lowest temperature ever recorded in Hawley Lake, Arizona was 41° below zero. How could this be written?

2.2

Perform the indicated operations.

11. $(-2) + (-5)$ **12.** $2 + (-5)$ **13.** $(-2) - (-5)$

14. $(-2) - 5$ **15.** $2 - (-5)$ **16.** $0 - (-2)$

17. $(-2.3) + (-1.5)$ **18.** $\left(\dfrac{1}{5}\right) - \left(-\dfrac{3}{10}\right)$ **19.** $\left(-\dfrac{1}{5}\right) - \left(-\dfrac{3}{10}\right)$

20. $(-5) + (-3) + 5 + (-8) + (-4) + (-6)$

21. On first down the LSU football team made 7 yards. It lost 9 yards on second down, then picked up 11 yards on third down. If the series started at the LSU 30 yard line, where was the ball spotted on fourth down?

2.3

Perform the indicated operations.

22. $(-4)(2)$ **23.** $(-14) \div 7$ **24.** $(-14) \div (-7)$

25. $\dfrac{0}{-12}$ **26.** $\left(-\dfrac{1}{3}\right)\left(\dfrac{15}{8}\right)$ **27.** $(-2.5)(-1.6)$

28. $(-2)(-2)(-1)(2)(-3)(-1)(-1)(-1)$

29. When Bertha was on a crash diet, she lost 7 pounds in each of 6 consecutive weeks. Use a number to express her total weight loss.

Evaluate the following:

30. $(-2) + (-8) \div (-4)$ **31.** $(-1)^3 + 5$

32. $3 - 2[1 - (4 - 3)]$

33. $(6 - 8) \cdot 2 - (-4)$

Write without parentheses using only one sign (or no sign).

34. $-(-9)$ **35.** $-(-y)$ **36.** $-[-(-8)]$ **37.** $-[+(-x)]$

2.4 *Factor.*

38. $-4x + 12$ **39.** $-2 - 6x - 10y$ **40.** $-5x - 25y$

Collect like terms.

41. $3a - 2 - 7a + 4 - a$

42. $-x + 2 - 3y + 4x + y$

Clear parentheses and collect like terms.

43. $y - (2y - 3) + y - (-y + 2)$

44. $-3[2a - (4 - a)] - (-a - 3)$

Evaluate when $b = -2$ and $c = 4$.

45. $7b^2$ **46.** $-7b^2$ **47.** $(-7b)^2$

48. $(c - b)^2$ **49.** $-b^3$ **50.** $|b^2 - c|$

2.5 *Write in exponential notation.*

51. $aaaaa$ **52.** $(3z)(3z)(3z)$ **53.** $(a + b)(a + b)$

Write without using exponents.

54. b^7 **55.** $-2x^3$ **56.** $(-2x)^3$

Simplify and write without negative exponents.

57. $2y^7 y^2$ **58.** $(3x^2)^3$ **59.** $(x^2 y^3)^5$

60. $2a^0 \quad a \neq 0$ **61.** $(2a)^0 \quad a \neq 0$ **62.** $\dfrac{2b^2 b^{-5}}{b^7}$

104 RATIONAL AND REAL NUMBERS

63. $5a^{-1}$ 64. $(5a)^{-1}$ 65. $\dfrac{a^{-1}}{b^{-1}}$

66. $(a^2b^{-3})^{-2}$ 67. $\dfrac{3^0 a^{-5} b^7}{a^4 b^{-1}}$ 68. $\left(\dfrac{2^{-1} x^2 y^{-4}}{x^{-2} y^{-2}}\right)^{-1}$

2.6 *Write in scientific notation.*

69. 0.000000411   70. 549,000,000,000

Write without using scientific notation.

71. 6.15×10^{12} 72. 4×10^{-8}

2.7 *Evaluate the square roots.*

73. $-\sqrt{169}$ 74. $\sqrt{\dfrac{48}{75}}$ 75. $-\sqrt{\dfrac{27}{3}}$

76. **(A)** What is the square of 25? **(B)** What is the principal square root of 25?

ANSWERS: **1.** 12 **2.** 7 **3.** 2.51 **4.** true **5.** false **6.** true **7.** = **8.** > **9.** < **10.** −41° **11.** −7
12. −3 **13.** 3 **14.** −7 **15.** 7 **16.** 2 **17.** −3.8 **18.** $\dfrac{1}{2}$ **19.** $\dfrac{1}{10}$ **20.** −21 **21.** 39 yard line **22.** −8
23. −2 **24.** 2 **25.** 0 **26.** $-\dfrac{5}{8}$ **27.** 4 **28.** −24 **29.** −42 pounds **30.** 0 **31.** 4 **32.** 3 **33.** 0 **34.** 9
35. y **36.** −8 **37.** x **38.** $-4(x-3)$ **39.** $-2(1+3x+5y)$ **40.** $-5(x+5y)$ **41.** $-5a+2$ **42.** $3x-2y+2$
43. $y+1$ **44.** $-8a+15$ **45.** 28 **46.** −28 **47.** 196 **48.** 36 **49.** 8 **50.** 0 **51.** a^5 **52.** $(3z)^3$ **53.** $(a+b)^2$
54. $bbbbbbb$ **55.** $-2xxx$ **56.** $(-2x)(-2x)(-2x)$ **57.** $2y^9$ **58.** $27x^6$ **59.** $x^{10}y^{15}$ **60.** 2 **61.** 1 **62.** $\dfrac{2}{b^{10}}$ **63.** $\dfrac{5}{a}$
64. $\dfrac{1}{5a}$ **65.** $\dfrac{b}{a}$ **66.** $\dfrac{b^6}{a^4}$ **67.** $\dfrac{b^8}{a^9}$ **68.** $\dfrac{2y^2}{x^4}$ **69.** 4.11×10^{-7} **70.** 5.49×10^{11} **71.** 6,150,000,000,000
72. 0.00000004 **73.** −13 **74.** $\dfrac{4}{5}$ **75.** −3 **76. (A)** 625 **(B)** 5

CHAPTER 2 TEST

1. True or false: the number $\frac{2}{7}$ is called the negative of $\frac{7}{2}$.

    1. _____

2. Find the absolute value. $|-13|$

    2. _____

3. Is the following true or false? $-15 < -5$

    3. _____

4. Place the appropriate symbol, $=$, $<$, or $>$, between the pair of fractions. $\frac{6}{13}$ $\frac{3}{5}$

    4. _____

5. The lowest temperature recorded in Polar, Idaho was 49° below zero. If the highest temperature was 92°, what is the difference between these extremes?

    5. _____

Perform the indicated operations.

6. $(-8) + (-14)$

    6. _____

7. $5 - (-3)$

    7. _____

8. $\left(\frac{1}{3}\right)\left(-\frac{3}{8}\right)$

    8. _____

9. $(-4.5) \div (-1.5)$

    9. _____

10. $(-3) + 8 - (-7) - 9 - (-2)$

    10. _____

11. $(-2)(-1)(-1)(-1)(3)(-3)$

    11. _____

12. Evaluate.
    $(2 - 7) \cdot 5 - (-6)$

    12. _____

13. Write $-[-(-6)]$ without parentheses using only one sign (or no sign).

    13. _____

CHAPTER 2 TEST Continued

14. Factor.
$-3y + 15$

14. _____

15. Collect like terms.
$3y - 2b + 5 + b - 5y$

15. _____

16. Clear parentheses and collect like terms.
$y - (2y - 3) + y$

16. _____

17. Evaluate $x^2 - y^2$ when $x = 4$ and $y = 1$.

17. _____

18. Multiply. $-3(4y - 2)$

18. _____

19. Write in exponential notation. $5 \cdot 5 \cdot a \cdot a \cdot a$

19. _____

Simplify and write without negative exponents.

20. $2y^5 y^2$

20. _____

21. $\dfrac{a^7}{a^3}$

21. _____

22. $(-2x^2)^3$

22. _____

23. z^{-3}

23. _____

24. $\dfrac{x^{-6}y^3}{x^2y^{-4}}$

24. _____

25. Evaluate y^{-3} when $y = -4$.

25. _____

26. Evaluate. $\dfrac{3.9 \times 10^{-4}}{1.3 \times 10^{-2}}$

26. _____

27. Evaluate.
$\sqrt{121}$

27. _____

28. Evaluate.
$-\sqrt{\dfrac{50}{32}}$

28. _____

3

LINEAR EQUATIONS AND INEQUALITIES

3.1 LINEAR EQUATIONS AND THE ADDITION-SUBTRACTION RULE

STUDENT GUIDEPOSTS

1 Equation
2 Solution of an equation
3 Conditional equation
4 Identity
5 Contradiction
6 Linear (first-degree) equation
7 Addition-subtraction rule

1 An **equation** is a statement that two quantities or expressions are equal. The two quantities are written with an equal sign ($=$) between them. The expression to the left of the equal sign is called the **left side** of the equation, and the expression to the right of the equal sign is called the **right side** of the equation. Some equations are true, some are false, and for some the truth value cannot be determined. For example,

$1 + 2 = 3$ is true

$2 + 5 = 3 - 7$ is false

$x + 2 = 9$ is neither true nor false since the value of x is not known.

In algebra we use letters called **variables** to represent numbers. Equations that contain a variable are important. When the variable in an equation can be replaced by a number that makes the resulting equation true, that number is called a **2** **solution** of the equation. The process of finding all solutions is called **solving the equation.** The equation

$$x + 2 = 9$$

has 7 as a solution because when x is replaced by 7,

$7 + 2 = 9$ is true.

Note that 3 is not a solution since

$3 + 2 = 9$ is false.

22. $\dfrac{1}{3} + x = 5$

23. $\dfrac{1}{2} + y = -3$

24. $-\dfrac{1}{4} + z = \dfrac{3}{4}$

25. $x - 2.1 = 3.4$

26. $-4.2 = y + 1.7$

27. $-2.6 = z - 3.9$

28. $x + 2\dfrac{1}{2} = 5$

29. $y - 3\dfrac{2}{3} = 4$

30. $1\dfrac{3}{8} + z = \dfrac{1}{4}$

Answer Exercises 31–34 as true or false.

31. -2 is a solution to $x + 2 = 0$. true

32. $y + 3 = 5$ is a linear equation. true

33. Zero is a solution of $x = 2x$. false

34. $x + 2 = x$ is an identity. false

ANSWERS 1–3. Answers will vary 4. 2 5. 4 6. 0 7. Any number is a solution 8. No solution 9. 0 10. y
11. $y - 1$ 12. no 13. 4 14. 13 15. -16 16. 3 17. -1 18. 2 19. 3 20. -4 21. -5 22. $\dfrac{14}{3}$
23. $-\dfrac{7}{2}$ 24. 1 25. 5.5 26. -5.9 27. 1.3 28. $2\dfrac{1}{2}$ 29. $7\dfrac{2}{3}$ 30. $-1\dfrac{1}{8}$ 31. true 32. true 33. true
34. false

3.1 EXERCISES B

Give an example of each of the following.

1. A false equation.

2. An identity.

3. An equation that is neither true nor false.

Solve the following equations by direct observation or inspection.

4. $x + 5 = 12$

5. $3y = 15$

6. $4 + z = 4 - z$

7. $1 + x = 1 + x$

8. $y + 2 = y + 8$

9. $8z = 3z$

Use the equation $x + 3 = 19$ to answer Exercises 10–12.

10. What is the right side of the equation?

11. What is the solution?

12. Is this a contradiction?

Solve and check.

13. $x + 2 = 12$

14. $y - 3 = 10$

15. $z + 13 = -2$

16. $x - 11 = -9$

17. $y + 17 = 18$

18. $z + 9 = 18$

19. $5 = x + 7$

20. $-3 = y - 3$

21. $-4 = z + 3$

22. $\dfrac{1}{3} + x = 8$

23. $\dfrac{1}{3} + y = \dfrac{7}{3}$

24. $-\dfrac{1}{4} + z = 1$

112 LINEAR EQUATIONS AND INEQUALITIES

25. $x - 2.3 = 5.7$
26. $-5.7 = y - 2.3$
27. $-3.8 = z - 1.2$
28. $x + 1\frac{3}{4} = 3$
29. $y - 2\frac{1}{5} = 7$
30. $3\frac{2}{3} + z = \frac{1}{9}$

Answer Exercises 31–34 as true or false.

31. -4 is a solution to $4 + x = 0$.
32. $z + 2 = 8$ is a linear equation.
33. A solution of $3y = y$ is 0.
34. $y + 1 = 1 - y$ is a contradiction.

3.1 EXERCISES C

Solve.

1. $3.75 - x = 7\frac{2}{5}$
2. $0.00125 + y = \frac{1}{1000}$
3. $125\frac{7}{8} = 95.2 - z$
   [Answer: -30.675]

3.2 THE MULTIPLICATION-DIVISION RULE

STUDENT GUIDEPOSTS

1 Equivalent equations
2 Multiplication-division rule

1 When two equations have the same solution, we say that the equations are **equivalent**. For example,

$$x - 2 = 7 \quad \text{and} \quad x = 9$$

are equivalent equations since both have 9 as their solution. As we have seen, adding or subtracting the same number on both sides of an equation gives an equivalent equation. Thus, solving an equation is a process of changing a given equation into an equivalent equation for which the solution or solutions will be clearly recognized. Another rule used in this process is the multiplication-division rule.

2 **Multiplication-division rule**

If each side of an equation is multiplied or divided by the same nonzero number or expression, another equation with the same solution is obtained.

EXAMPLE 1

Solve. $2x = 10$

$\frac{1}{2} \cdot 2x = \frac{1}{2} \cdot 10$ Multiply both sides by the reciprocal of 2, which is $\frac{1}{2}$

$1 \cdot x = 5$
$x = 5$

Practice Exercise 1

Solve. $9y = -27$

The solution is 5.

Check: $2 \cdot 5 \stackrel{?}{=} 10$ Replace x with 5 in the original equation
          $10 = 10$ ◀◀

Answer: -3

Equivalently, in Example 1 we could divide both sides of the equation by 2.

$$2x = 10$$

$$\frac{2x}{2} = \frac{10}{2} \qquad \text{Divide both sides by 2}$$

$$\frac{2}{2} \cdot x = 5$$

$$1 \cdot x = 5$$

$$x = 5$$

EXAMPLE 2

Solve. $\frac{1}{3}x = 18$

$3 \cdot \frac{1}{3}x = 3 \cdot 18$ Multiply both sides by the reciprocal of $\frac{1}{3}$, which is 3

$1 \cdot x = 54$

$x = 54$

The solution is 54. Check. ◀◀

Practice Exercise 2

Solve. $\frac{1}{8}y = -1$

Answer: -8

Notice in Examples 1 and 2 that we used the multiplication-division rule to make the coefficient of the variable equal to 1. This is the important final step when solving an equation. Also, with practice, many of the steps may be performed mentally.

EXAMPLE 3

Solve. $\dfrac{x}{\frac{4}{5}} = 20$

Practice Exercise 3

Solve. $\dfrac{z}{\frac{1}{7}} = -35$

In an equation of this type, simplify the left side first.

$$\frac{x}{\frac{4}{5}} = \frac{\frac{x}{1}}{\frac{4}{5}} = \frac{x}{1} \div \frac{4}{5} = \frac{x}{1} \cdot \frac{5}{4} = \frac{5}{4} \cdot x$$

Thus we are actually solving

$$\frac{5}{4} \cdot x = 20$$

$$\frac{4}{5} \cdot \frac{5}{4} \cdot x = \frac{4}{5} \cdot 20 \qquad \text{Multiply both sides by } \frac{4}{5}$$

$$1 \cdot x = 16 \qquad \frac{4}{5} \cdot 20 = \frac{4 \cdot 20}{5} = \frac{4 \cdot 4}{1} = 16$$

$$x = 16$$

The solution is 16.

Check: $\dfrac{16}{\frac{4}{5}} \stackrel{?}{=} 20.$

$\dfrac{5}{4} \cdot 16 \stackrel{?}{=} 20,$

$20 = 20$ ◀◀

Answer: -5

EXAMPLE 4

Solve.
$$-2.7x = 54$$
$$\dfrac{-2.7x}{-2.7} = \dfrac{54}{-2.7} \quad \text{Divide both sides by } -2.7$$
$$\dfrac{-2.7}{-2.7} \cdot x = \dfrac{54}{-2.7}$$
$$1 \cdot x = -20$$
$$x = -20$$

The solution is -20.

Check: $-2.7(-20) \stackrel{?}{=} 54$

$54 = 54$ ◀◀

Practice Exercise 4

Solve. $\quad 0.6y = -7.2$

Answer: -12

Intuitively, we remember the multiplication-division rule with the phrase "multiply or divide both sides of the equation by the same thing."

3.2 EXERCISES A

Use the multiplication-division rule to solve. Check all solutions.

1. $5x = 25$
2. $56 = 8y$
3. $-11z = 77$

4. $12x = -84$
5. $-72 = -y$
6. $-2z = 2$

7. $-5x = -50$
8. $-60y = 360$
9. $\dfrac{z}{1} = 10$

10. $\dfrac{2}{3}x = 10$
11. $\dfrac{y}{4} = -3$
12. $-\dfrac{2}{5}z = 8$

13. $\dfrac{x}{\frac{1}{3}} = 27$
14. $\dfrac{y}{\frac{7}{2}} = -6$
15. $\dfrac{z}{-\frac{1}{8}} = 16$

16. $\dfrac{2x}{3} = 4$
17. $\dfrac{3y}{2} = -6$
18. $\dfrac{4z}{-3} = -8$

19. $4x = -24.8$
20. $1.2y = 7.2$
21. $-0.8z = -6.4$

22. $2\dfrac{1}{2}x = 10$
23. $-3\dfrac{1}{3}y = 20$
24. $-4\dfrac{1}{5}z = -63$

For Review

Solve.

25. $-8 + x = -12$
26. $-\dfrac{3}{4} + y = -\dfrac{7}{4}$
27. $7.4 + z = -3.2$

28. Is $2 + x = x + 2$ an identity?
29. Is 0 a solution of $x - 3 = -3 - x$?

ANSWERS: **1.** 5 **2.** 7 **3.** −7 **4.** −7 **5.** 72 **6.** −1 **7.** 10 **8.** −6 **9.** 10 **10.** 15 **11.** −12 **12.** −20 **13.** 9 **14.** −21 **15.** −2 **16.** 6 **17.** −4 **18.** 6 **19.** −6.2 **20.** 6 **21.** 8 **22.** 4 **23.** −6 **24.** 15 **25.** −4 **26.** −1 **27.** −10.6 **28.** yes **29.** yes

3.2 EXERCISES B

Use the multiplication-division rule to solve. Check.

1. $7x = 49$
2. $72 = 6y$
3. $-11z = 44$
4. $12x = -96$
5. $-63 = -y$
6. $-3z = 48$
7. $-20x = -300$
8. $30y = -240$
9. $\dfrac{z}{1} = -30$
10. $\dfrac{2}{3}x = 14$
11. $\dfrac{y}{5} = -3$
12. $-\dfrac{2}{7}z = 8$
13. $\dfrac{x}{\tfrac{1}{2}} = 10$
14. $\dfrac{y}{\tfrac{1}{9}} = 27$
15. $\dfrac{z}{-\tfrac{2}{3}} = 12$
16. $\dfrac{3x}{5} = 6$
17. $\dfrac{4y}{3} = -2$
18. $\dfrac{8z}{-5} = -16$
19. $3x = -25.5$
20. $1.2y = 8.4$
21. $-1.2z = -4.8$
22. $3\dfrac{2}{3}x = 22$
23. $-4\dfrac{1}{4}y = 17$
24. $-1\dfrac{3}{5}z = -32$

116 LINEAR EQUATIONS AND INEQUALITIES

For Review

Solve.

25. $-21 + x = -25$

26. $-\dfrac{5}{6} + y = \dfrac{7}{12}$

27. $1.9 + z = -6.7$

28. Is $x + 1 = x - 1$ a contradiction?

29. Is -6 a solution of $x - 6 = 0$?

3.2 EXERCISES C

Solve.

1. $0.0568x = 0.006816$

2. $9\dfrac{7}{12}y = -7\dfrac{3}{16}$

3. $-15\dfrac{3}{4}z = 0.525$
[Answer: $-0.0\overline{3}$]

3.3 SOLVING EQUATIONS BY COMBINING RULES

STUDENT GUIDEPOSTS

1. Equations with similar terms
2. Equations with a variable on both sides
3. Equations with parentheses
4. Summary of the method for solving a linear equation

When we solve an equation, we isolate the variable on one side so we can find the solution by inspection. Sometimes we need to use both the addition-subtraction rule and the multiplication-division rule to do this. Generally, we use the addition-subtraction rule before using the multiplication-division rule, as in the following example.

EXAMPLE 1

Solve. $3x + 5 = 11$

$3x + 5 - 5 = 11 - 5$ Subtract 5 from both sides

$3x + 0 = 6$

$3x = 6$

$\dfrac{1}{3} \cdot 3x = \dfrac{1}{3} \cdot 6$ Multiply both sides by $\dfrac{1}{3}$

$1 \cdot x = 2$

$x = 2$

The solution is 2.

Check: $3 \cdot 2 + 5 \stackrel{?}{=} 11$

$6 + 5 \stackrel{?}{=} 11$

$11 = 11$ ◀◀

Practice Exercise 1

Solve. $4y + 10 = 2$

Answer: -2

3.3 SOLVING EQUATIONS BY COMBINING RULES

1 Sometimes the sides of an equation have similar terms. These terms should be combined before attempting to apply the rules.

EXAMPLE 2
Solve.
$$15 = 4y + 3y$$
$$15 = 7y \quad \text{Combine similar terms}$$
$$\frac{1}{7} \cdot 15 = \frac{1}{7} \cdot 7y \quad \text{Multiply both sides by } \frac{1}{7}$$
$$\frac{15}{7} = 1 \cdot y$$
$$\frac{15}{7} = y$$

The solution is $\frac{15}{7}$. Check by substituting in the original equation. ◀◀

Practice Exercise 2
Solve. $9z - 3z = 2$

Answer: $\frac{1}{3}$

2 Sometimes an equation has terms involving the variable on both sides. We use the addition-subtraction rule to get them together on the same side so that they can then be combined.

EXAMPLE 3
Solve.
$$7x - 2 = 2x + 3$$
$$7x - 2 + 2 = 2x + 3 + 2 \quad \text{Add 2 to both sides}$$
$$7x + 0 = 2x + 5$$
$$7x = 2x + 5$$
$$7x - 2x = 2x - 2x + 5 \quad \text{Subtract 2x from both sides}$$
$$5x = 0 + 5$$
$$5x = 5$$
$$\frac{1}{5} \cdot 5x = \frac{1}{5} \cdot 5 \quad \text{Multiply both sides by } \frac{1}{5}$$
$$1 \cdot x = 1$$
$$x = 1$$

The solution is 1. Check. ◀◀

Practice Exercise 3
Solve. $4 + 3y = -8 - y$

Answer: -3

To solve an equation, use the addition-subtraction rule to isolate all terms involving the variable on one side of the equation and then combine them. Next apply, if necessary, the multiplication-division rule. Always remember to combine similar terms before applying the rules.

EXAMPLE 4
Solve.
$$2 - x + 10 = 3x + 2x + 24$$
$$-x + 12 = 5x + 24 \quad \text{Combine similar terms}$$
$$-x + 12 - 12 = 5x + 24 - 12 \quad \text{Subtract 12 from both sides}$$
$$-x = 5x + 12$$
$$-x - 5x = 5x - 5x + 12 \quad \text{Subtract 5x from both sides}$$
$$-6x = 12$$

Practice Exercise 4
Solve. $4y + 2 - y = 3 + 5y - 9$

118 LINEAR EQUATIONS AND INEQUALITIES

$$\left(-\frac{1}{6}\right)(-6x) = \left(-\frac{1}{6}\right)12 \qquad \text{Multiply both sides by } -\frac{1}{6}$$

$$1 \cdot x = -2$$

$$x = -2$$

The solution is -2.

Check: $2 - (-2) + 10 \stackrel{?}{=} 3(-2) + 2(-2) + 24$

$2 + 2 + 10 \stackrel{?}{=} -6 - 4 + 24$

$14 = 14$ ◀◀

Answer: 4

---

3 ▮▮▮ When solving an equation containing parentheses, remove the parentheses using the distributive property. The resulting equation can then be solved using previous methods.

EXAMPLE 5

Solve. $3(x + 6) = 21$

$3 \cdot x + 3 \cdot 6 = 21$ Remove parentheses by multiplying both 6 and x by 3

$3x + 18 = 21$

$3x + 18 - 18 = 21 - 18$ Subtract 18

$3x = 3$

$\frac{1}{3} \cdot 3x = \frac{1}{3} \cdot 3$ Multiply by $\frac{1}{3}$

$1 \cdot x = 1$

$x = 1$

The solution is 1. Check. ◀◀

Practice Exercise 5

Solve. $5(z + 2) = 10$

Answer: 0

EXAMPLE 6

Solve. $6z - (3z - 4) = 14$

$6z - 3z + 4 = 14$ Remove parentheses by changing signs

$3z + 4 = 14$

$3z + 4 - 4 = 14 - 4$ Subtract 4

$3z = 10$

$z = \frac{10}{3}$

The solution is $\frac{10}{3}$.

Check: $6\left(\frac{10}{3}\right) - \left(3\left(\frac{10}{3}\right) - 4\right) \stackrel{?}{=} 14$

$20 - (10 - 4) \stackrel{?}{=} 14$

$20 - 6 \stackrel{?}{=} 14$

$14 = 14$ ◀◀

Practice Exercise 6

Solve. $4y - (y - 8) = -1$

Answer: -3

EXAMPLE 7

Solve. $2x - 4(2x - 4) = 22 - 3(x - 5)$

$$2x - 8x + 16 = 22 - 3x + 15 \quad \text{Remove parentheses}$$
$$-6x + 16 = 37 - 3x \quad \text{Combine like terms}$$
$$-6x + 16 - 16 = 37 - 3x - 16 \quad \text{Subtract 16}$$
$$-6x = 21 - 3x$$
$$-6x + 3x = 21 - 3x + 3x \quad \text{Add } 3x$$
$$-3x = 21$$
$$\left(-\frac{1}{3}\right)(-3x) = \left(-\frac{1}{3}\right)21 \quad \text{Multiply by } -\frac{1}{3}$$
$$1 \cdot x = -7$$
$$x = -7$$

The solution is -7. Check. ◀◀

Practice Exercise 7

Solve. $3z - 2(1 - z) = 7 - 5(z + 3)$

Answer: $-\frac{3}{5}$

We conclude this section by summarizing the method to solve linear equations. If when solving an equation an identity results, every real number is a solution. If a contradiction results, there is no solution.

To solve a linear equation

1. Simplify both sides by removing parentheses and collecting like terms.
2. Use the addition-subtraction rule to isolate all variable terms on one side and all constant terms on the other side. Collect like terms when possible.
3. Divide both sides by the coefficient of the variable.
4. Check the solution by substituting in the original equation.

3.3 EXERCISES A

Solve each equation. Check all solutions.

1. $2x + 1 = 5$

2. $3y - 1 = 8$

3. $-4z + 3 = 5$

4. $\frac{3}{4}x + \frac{1}{4} = 1$

5. $\frac{2}{5} - y = \frac{3}{10}$

6. $12 = -1.2z + 24$

7. $2x + 3x = 10$

8. $3y - y = 7$

9. $\frac{1}{4}z + \frac{1}{2}z = -9$

10. $9.2x - 3.1x = 12.2$

11. $-\dfrac{1}{7}y + y = 18$

12. $3z = 5 - 7z$

13. $32 - 8x = 3 - 9x$

14. $-7 + 21y + 23 = 16$

15. $z + 3z = 8 - 2z + 10$

16. $x + 9 + 6x = 2 + x + 1$

17. $6y - 4y + 1 = 12 + 2y - 11$

18. $2.1x - 3.2 = -8.4x - 45.2$

19. $3y + \dfrac{5}{2}y + \dfrac{3}{2} = \dfrac{1}{2}y + \dfrac{5}{2}y$

Remove parentheses and solve.

20. $3(2x + 1) = 21$

21. $10 = 5(y - 20)$

22. $2(5z + 1) = 42$

23. $3x - 4(x + 2) = 5$

24. $2y - (13 - 2y) = 59$

25. $3z - (z + 4) = 0$

26. $5(x + 4) - 4(x + 3) = 0$

27. $4(y - 3) - 6(y + 1) = 0$

28. $9z - (3z - 18) = 36$

29. $5(2 - 3x) = 15 - (x + 7)$

30. $8y - (2y + 5) = 0$

31. $9z - (5z - 2) = 8$

32. $7(x - 5) = 10 - (x + 1)$ **33.** $(y - 9) - (y + 6) = 4y$ **34.** $8(2z + 1) = 4(7z + 7)$

35. $5 + 3(x + 2) = 3 - (x + 2)$ **36.** $\frac{1}{3}(6x - 9) = \frac{1}{2}(8x - 4)$ **37.** $5(x + 1) - 4x = x - 5$

38. $3(2 - 4x) = 4(2x - 1) - 2(1 + x)$ **39.** $2 + y = 3 - 2[1 - 2(y + 1)]$

40. $-2(z - 1) - 3(2z + 1) = -9$ **41.** $-4(-x + 1) + 3(2x + 3) - 7x = -10$

For Review

Solve.

42. $\dfrac{x}{-\frac{1}{5}} = 20$ **43.** $\dfrac{y}{-5} = -2$ **44.** $\dfrac{3z}{7} = -6$

45. $-2\dfrac{3}{8}x = -19$ **46.** $y - \dfrac{2}{3} = \dfrac{1}{9}$ **47.** $3.6 - z = 7.2$

ANSWERS: 1. 2 2. 3 3. $-\frac{1}{2}$ 4. 1 5. $\frac{1}{10}$ 6. 10 7. 2 8. $\frac{7}{2}$ 9. -12 10. 2 11. 21 12. $\frac{1}{2}$
13. -29 14. 0 15. 3 16. -1 17. every real number (identity) 18. -4 19. $-\frac{3}{5}$ 20. 3 21. 22 22. 4
23. -13 24. 18 25. 2 26. -8 27. -9 28. 3 29. $\frac{1}{7}$ 30. $\frac{5}{6}$ 31. $\frac{3}{2}$ 32. $\frac{11}{2}$ 33. $-\frac{15}{4}$ 34. $-\frac{5}{3}$
35. $-\frac{5}{2}$ 36. $-\frac{1}{2}$ 37. no solution 38. $\frac{2}{3}$ 39. -1 40. 1 41. -5 42. -4 43. 10 44. -14 45. 8
46. $\frac{7}{9}$ 47. -3.6

3.3 EXERCISES B

Solve and check.

1. $3x + 1 = 7$
2. $4y - 2 = 10$
3. $-2z + 5 = 9$
4. $\frac{2}{3}x + \frac{1}{3} = 1$
5. $\frac{3}{5} - y = \frac{7}{10}$
6. $10 = -0.5z + 5$
7. $3x + 7x = 40$
8. $4y - y = 27$
9. $\frac{1}{8}z + \frac{5}{8}z = -6$
10. $1.3x + 2.7x = 4.4$
11. $-\frac{1}{5}y + y = 8$
12. $5z = 2 - 9z$
13. $25 - 7x = 4 - 10x$
14. $-2 + 8y + 27 = 1$
15. $z + 2z = 3 - 2z + 17$
16. $x + 3 + 4x = 2 + x + 1$
17. $3y - y + 10 = 14 + 2y - 4$
18. $6.4 - 4.2x = 16.8x + 90.4$
19. $2x - \frac{1}{3} + \frac{2}{3}x = \frac{4}{9} + \frac{1}{3}x$

Clear parentheses and solve.

20. $2(3x + 1) = 14$
21. $9 = 3(4y - 1)$
22. $7(2z - 1) = -35$
23. $4x - 2(x + 1) = -1$
24. $7y - (4 - 3y) = 11$
25. $2z - (z + 1) = 0$
26. $6(x + 2) - 2(2x + 1) = 0$
27. $4(2y - 1) - 7(y + 2) = 0$
28. $6z - (2z - 5) = 0$
29. $3(2 - 4x) = 13 - (x + 1)$
30. $7y - (2y + 15) = 0$
31. $8z - (3z - 1) = 0$
32. $4(x - 2) = 12 - (x + 3)$
33. $(y - 2) - (y + 2) = 4y$
34. $9(2z - 1) = 3(z + 2)$
35. $7 + 3(x + 1) = 5 - (x + 1)$
36. $\frac{1}{2}(2y - 4) = \frac{2}{3}(9y + 3)$
37. $6(z - 1) - 3z = 3z + 8$
38. $2(1 - 3x) = 5(2x + 1) - 3(1 + x)$
39. $4 + y = 8 - 3[2 - (y + 2)]$
40. $-3(z - 4) - 2(3z + 1) = -8$
41. $-7(-x + 2) + 5(2x + 1) - 11x = 3$

For Review

Solve.

42. $-\frac{3}{5}x = 6$
43. $\dfrac{y}{-\frac{1}{7}} = 14$
44. $\frac{2z}{9} = -12$

45. $-3\frac{1}{7}x = -88$ **46.** $\frac{3}{4} - y = \frac{1}{2}$ **47.** $8.4 + z = -9.2$

3.3 EXERCISES C

Solve and check.

1. $0.02x - (0.56x - 7.33) = 2.15 - (0.7x - 0.11)$

2. $3\frac{2}{3}x + 7\frac{1}{9} - \left(\frac{x}{6} - 4\frac{1}{9}\right) = -\left(2\frac{1}{6} - 6\frac{2}{9}x\right)$

$\left[\text{Answer: } \frac{241}{49}\right]$

3. $x(x - 3) - 3x[1 - (x - 1] = 4x(x - 5) + 22$

4. $-3[1 - (x^2 - 2)] - 2x(1 - x) = -5[4 - x(x - 2)]$

$\left[\text{Answer: } -\frac{11}{8}\right]$

3.4 THE LANGUAGE OF PROBLEM SOLVING

To solve a word problem using algebra, we must do two things. First the *words* of the problem must be translated into an algebraic equation, and second, the equation must be solved. We have learned how to solve many simple equations; now we will concentrate on translating problems into equations. Some common terms and their symbolic translations are given below.

Symbol	Stands for
$+$	and, sum, sum of, added to, increased by, more than
$-$	minus, less, subtracted from, less than, diminished by, difference between, difference, decreased by
$\cdot$	times, product, product of, multiplied by, of
$\div$	divided by, quotient of, ratio
$=$	equals, is equal to, is as much as, is, is the same as

Any letter (we often use x) can be used to stand for the unknown or desired quantity. Some examples of translation follow.

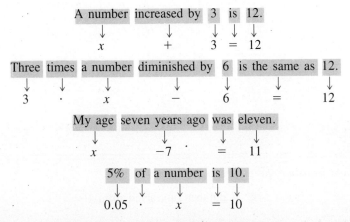

Although we often use x as the variable in a translation, it is sometimes helpful to use another letter more indicative of the quantity it represents. For example, we might use t for time, w for wages, A for area, or M for Mary's age.

124 LINEAR EQUATIONS AND INEQUALITIES

EXAMPLE 1

Select a variable to represent each quantity and translate the phrase into symbols.

Word phrase	Symbolic translation
(A) The time in 4 hours	$t + 4$
(B) My wages less $100 for taxes	$w - 100$
(C) One half of the area	$\frac{1}{2}A$
(D) Twice Mary's age in 3 years	$2(M + 3)$
(E) $2000 less the cost	$2000 - c$ ◀◀

Practice Exercise 1

Select a variable and translate each phrase into symbols.

(A) My age in 6 years
(B) The price plus $10 in tax
(C) Twice the area
(D) Five miles less than the distance
(E) Five miles less the distance

Answers: (A) $a + 6$ (B) $p + 10$
(C) $2A$ (D) $d - 5$ (E) $5 - d$

EXAMPLE 2

Use x for the variable and translate each word sentence into symbols.

Word expression	Symbolic translation
(A) Five times a number is 20.	$5x = 20$
(B) The product of a number and 7 is 35.	$7x = 35$
(C) Twice a number, increased by 3, is 11.	$2x + 3 = 11$
(D) Six is 4 less than twice a number.	$6 = 2x - 4$
(E) Six is 4 less twice a number.	$6 = 4 - 2x$
(F) One tenth of a number is 13.	$\frac{1}{10}x = 13$
(G) Five times a number, minus twice the number, equals 10.	$5x - 2x = 10$
(H) A number subtracted from 5 is 12.	$5 - x = 12$
(I) 5 subtracted from a number is 12.	$x - 5 = 12$
(J) A number divided by 9 is equivalent to $\frac{2}{3}$.	$\frac{x}{9} = \frac{2}{3}$ ◀◀

Practice Exercise 2

Use x for the variable and translate into symbols.

(A) Twice a number is 50.
(B) The product of a number and 8 is 64.
(C) Three times a number, increased by 2, is 25.
(D) Ten is 2 less a number.
(E) Ten is 2 less than a number.
(F) Three fifths of a number is 9.
(G) Twice a number, decreased by four times the number, is the same as the number.
(H) Her age diminished by 3 is 1.
(I) Three diminished by her age is 1.
(J) His weight divided by 3 is the same as 24.

Answers: (A) $2x = 50$ (B) $8x = 64$
(C) $3x + 2 = 25$ (D) $10 = 2 - x$
(E) $10 = x - 2$ (F) $\frac{3}{5}x = 9$
(G) $2x - 4x = x$ (H) $x - 3 = 1$
(I) $3 - x = 1$ (J) $\frac{x}{3} = 24$

3.4 EXERCISES A

Select a variable to represent each quantity and translate the phrase into symbols.

1. The sum of a number and 7
2. A number subtracted from 10
3. The product of a number and 3
4. A number divided by 13
5. The time in 8 hours
6. The time 6 hours ago

7. His salary plus $200

8. My wages less $50

9. Twice the volume

10. $3000 less the cost

11. A number added to its reciprocal

12. Her score less 20 points

13. 300 more than twice the number of votes

14. 300 more than the number of votes, doubled

15. 4% of the selling price

16. Cost plus 8% of the cost

Using x for the variable, translate the following into symbols.

17. 7 times a number is 25.

18. A number increased by 8 equals 12.

19. A number decreased by 15 is equal to 35.

20. The product of a number and 6 is 42.

21. Twice a number, increased by 3, is 27.

22. Three times a number, diminished by 8, equals 24.

23. A number less 5 is 12.

24. 8 is 4 less than three times a number.

25. A number diminished by 3 is 9.

26. Seven times a number, less 2, is 31.

27. The product of a number and 3, decreased by 5, equals 34.

28. When 10 is added to three times a number the result is the same as twice the number, plus 12.

29. 3% of a number is 12.

30. The sum of two numbers is 24 and one of them is three times the other.

31. Seven times 2 less than a number is 31. (Compare with Exercise 26.)

32. Twice the sum of a number and 5 is three times the number.

33. Seven more than a number is 13.

34. Four times my age in 3 years is 48.

35. When 4 is divided by some number the quotient is the same as $\frac{1}{2}$.

36. One half a number, less twice the reciprocal of the number is $\frac{3}{2}$.

For Review

Solve and check.

37. $-\frac{1}{4}y + y = 27$

38. $15 - 3z = 5 + 7z$

39. $2(y - 3) = 2(4y + 1) - 3(y + 1)$

40. $2 + x = 4 - 2[1 - (x - 3)]$

ANSWERS: **1.** $n + 7$ **2.** $10 - n$ **3.** $3n$ **4.** $n/13$ **5.** $t + 8$ **6.** $t - 6$ **7.** $s + 200$ **8.** $w - 50$ **9.** $2V$
10. $3000 - c$ **11.** $n + \frac{1}{n}$ **12.** $s - 20$ **13.** $2v + 300$ **14.** $2(v + 300)$ **15.** $0.04p$ **16.** $c + 0.08c$ **17.** $7x = 25$
18. $x + 8 = 12$ **19.** $x - 15 = 35$ **20.** $6x = 42$ **21.** $2x + 3 = 27$ **22.** $3x - 8 = 24$ **23.** $x - 5 = 12$
24. $8 = 3x - 4$ **25.** $x - 3 = 9$ **26.** $7x - 2 = 31$ **27.** $3x - 5 = 34$ **28.** $10 + 3x = 2x + 12$ **29.** $0.03x = 12$
30. $x + 3x = 24$ **31.** $7(x - 2) = 31$ **32.** $2(x + 5) = 3x$ **33.** $7 + x = 13$ **34.** $4(x + 3) = 48$ **35.** $\frac{4}{x} = \frac{1}{2}$
36. $\frac{1}{2}n - 2\frac{1}{n} = \frac{3}{2}$ or $\frac{n}{2} - \frac{2}{n} = \frac{3}{2}$ **37.** 36 **38.** 1 **39.** $-\frac{5}{3}$ **40.** 6

3.4 EXERCISES B

Select a variable to represent each quantity and translate the phrase into symbols.

1. The sum of a number and 12
2. A number subtracted from 33
3. The product of a number and 5
4. A number divided by 7
5. The time in 3 hours
6. The time 14 hours ago
7. Her salary plus $500
8. My wages less $75
9. Three times the volume
10. $5000 less the cost
11. A number less its reciprocal
12. His score less 28 points
13. 100 more than triple the number of votes
14. 100 more than the number of votes, tripled
15. 24% of a number
16. Cost plus 10% of the cost

Using x for the variable, translate the following into symbols.

17. 4 times a number is 32.
18. A number increased by 3 equals 14.
19. A number decreased by 13 is equal to 49.
20. The product of a number and 8 is 96.
21. Twice a number, increased by 4, is 86.
22. Three times a number, diminished by 17, equals 42.
23. A number less 4 is 21.
24. 9 is 3 less than five times a number.
25. A number diminished by 2 is 18.
26. Twelve times a number, less 17, is 39.
27. The product of a number and 7, decreased by 8, equals 29.
28. When 15 is added to three times a number the result is the same as twice the number, plus 40.
29. 9% of a number is 15.
30. The sum of two numbers is 86, and one of them is three times the other.
31. Eleven times 4 less than a number is 25.
32. Twice the sum of a number and 8 is four times the number.
33. Six more than a number is 29.
34. Five times my age in 2 years is 200.
35. When 7 is divided by some number the quotient is the same as $\frac{4}{9}$.
36. One third a number, less three times the reciprocal of the number is $\frac{4}{3}$.

For Review

Solve and check.

37. $-\frac{1}{6}y + y = 75$
38. $12 - 4z = 2z + 12$

39. $3(y - 4) = 2(3y + 1) - (y - 3)$

40. $1 + x = 3 - 2[2 - (x - 5)]$

3.4 EXERCISES C

Using x for the variable, translate the following into symbols.

1. If two thirds of eight more than a number is increased by five times the reciprocal of the number, the result is twice the number, decreased by 8.

2. A number divided by six less than the number is the same as the product of the number and three more than the number.

3.5 APPLIED PROBLEMS: NUMBER, AGE, AND PERCENT

 STUDENT GUIDEPOSTS

    1 Method for solving applied problems   4 Commission (rate)
    2 Consecutive integers   5 Interest applications
    3 Percent translations

Success with applied problems (or word problems) comes with practice. Problem solving will be easier if you follow these steps.

To solve an applied problem

1. Read the problem (perhaps several times) and determine what quantity (or quantities) you are asked to find.
2. Represent the unknown quantity (or quantities) using a letter.
3. Determine which expressions are equal and write an equation.
4. Solve the equation and state the answer to the problem.
5. Check to see if the answer (or answers) satisfy the conditions of the problem.

EXAMPLE 1

Twice a number, increased by 13, is 71. What is the number?

Let x = the desired number. The problem may be symbolized as follows.

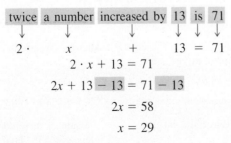

$$2 \cdot x + 13 = 71$$
$$2x + 13 - 13 = 71 - 13$$
$$2x = 58$$
$$x = 29$$

The desired number is 29.

Check: Twice 29 plus 13 is indeed 71.

Practice Exercise 1

Three times a number, decreased by 24, is 60. What is the number?

Answer: 28

128 LINEAR EQUATIONS AND INEQUALITIES

▶▶ **CAUTION** ◀◀ Always be neat and complete when working word problems. Time-consuming errors result when you attempt to take shortcuts. Writing out what the variable means helps in solving the problem. ◀◀

EXAMPLE 2

Bob is three times as old as Dick, and the sum of their ages is 48 years. How old is each?

Let x = Dick's age,

$3x$ = Bob's age.

$$x + 3x = 48$$
$$4x = 48$$
$$x = 12$$
$$\text{and} \quad 3x = 36$$

Dick is 12 years old and Bob is 36 years old.

Check: $36 = 3 \cdot 12$ and $12 + 36 = 48$. ◀◀

Practice Exercise 2

The sum of two numbers is 77. If one number is three fourths of the other, find the two numbers.

Answer: 33 and 44

2 ▶▶▶▶ **Consecutive integers** are integers which immediately follow each other in regular counting order. For example, 13, 14, and 15 are consecutive integers. In general, if x is an integer, the next consecutive integer is $x + 1$, and the next consecutive integer after it is $x + 2$. Consecutive even or odd integers such as 2, 4, 6 or 3, 5, 7 can also be considered. In these cases, if x is an integer, the next consecutive even (or odd) integer is $x + 2$, and the next consecutive even (or odd) integer after it is $x + 4$.

EXAMPLE 3

The sum of three consecutive integers is 96. Find the numbers.

Let x = first integer,

$x + 1$ = next consecutive integer (Why?),

$x + 2$ = third consecutive integer (Why?).

$$x + (x + 1) + (x + 2) = 96$$
$$3x + 3 = 96 \quad \text{Combine like terms}$$
$$3x = 93 \quad \text{Subtract 3}$$
$$x = 31$$
$$\text{and} \quad x + 1 = 32$$
$$\text{and} \quad x + 2 = 33$$

The numbers are 31, 32, and 33.

Check: $31 + 32 + 33 = 96$ and the three integers are consecutive. ◀◀

Practice Exercise 3

Wendy and Diane were born in consecutive years. If the sum of their ages is 43 and Diane is the older, how old is each?

Answer: Wendy is 21 and Diane is 22

EXAMPLE 4

The sum of two consecutive even integers is 2 less than twice the larger. Find the numbers.

Let x = the first even integer,

$x + 2$ = the second (larger) even integer.

Practice Exercise 4

The sum of three consecutive odd integers is 135. Find the numbers.

3.5 APPLIED PROBLEMS: NUMBER, AGE, AND PERCENT

$$x + (x + 2) = 2(x + 2) - 2$$
$$2x + 2 = 2x + 4 - 2 \quad \text{Clear parentheses}$$
$$2x + 2 = 2x + 2 \quad \text{Combine like terms}$$
$$2 = 2 \quad \text{Subtract } 2x$$

Since we obtained an identity, any value of x will make the equation true. Therefore, *any* two consecutive even integers solve the problem. ◀◀

Answer: 43, 45, 47

   Many applied problems involve percent and can be restated into one of three basic forms. The following examples illustrate these three types.

What is 25% of 40?
What percent of 60 is 12?
8 is 40% of what number?

Letting x represent the unknown number in each statement and recalling that the word *of* translates to times, these three statements can be translated into equations. The first becomes

What is 25% of 40?
$\quad\downarrow\quad\downarrow\quad\downarrow\quad\downarrow\quad\downarrow$
$\quad x \;\; = \;\; 0.25 \;\; \cdot \;\; 40.$

Recall from Chapter 1 that 25% = 0.25

Solving, we have

$$x = (0.25)(40) = 10.$$

The second statement translates to

What percent of 60 is 12?
$\quad\downarrow\qquad\downarrow\quad\downarrow\quad\downarrow$
$\quad x \qquad \cdot \;\; 60 \;\; = \;\; 12.$

Solving,

$$x = \frac{12}{60} = 0.2.$$

In percent notation, x becomes 20%.

Finally, the third statement becomes

8 is 40% of what number?
$\downarrow\;\downarrow\;\downarrow\;\downarrow\qquad\downarrow$
$8 \;=\; 0.40 \;\cdot\; x.$

Thus,

$$x = \frac{8}{0.40} = 20.$$

EXAMPLE 5

(A) 75% of 128 is what?

The equation we must solve is $(0.75)(128) = x$.
$$96 = x$$

Practice Exercise 5

(A) 36% of 225 is what?

(B) 18 is what percent of 48?

The equation we must solve is $18 = x \cdot 48$.

$$\frac{18}{48} = x \qquad \text{Divide by 48}$$

$$0.375 = x$$

In percent notation, $x = 37.5\%$. ◀◀

(B) 102 is what percent of 68?

Answers: **(A)** 81 **(B)** 150%

EXAMPLE 6

The player had a batting percentage of 40% in a series. If he had 12 hits, how many times was he at bat?

In effect, we are asked: 12 is 40% of what?

$$12 = (0.40)x$$

$$\frac{12}{0.40} = x \qquad \text{Divide by 0.40}$$

$$30 = x$$

Thus, he went to bat 30 times in the series. ◀◀

Practice Exercise 6

For the season, basketball player Andy Hurd shot 90.5% from the free throw line. If he made 181 free throws, how many times did he go to the free throw line?

Answer: 200

4   Some salespersons receive all or part of their income as a percent of their sales. This is called a **commission.** The **commission rate** is the percent of the sales that the person receives. That is,

(commission) = (commission rate) · (total sales).

EXAMPLE 7

Paul received a commission of $720 on the sale of a machine. If his commission rate is 8% of the selling price, what was the selling price?

The problem can be written: 8% of what is $720?

$$(0.08)x = 720$$

$$x = \frac{720}{0.08}$$

$$x = 9000$$

Thus, the machine sold for $9000. ◀◀

Practice Exercise 7

Kerry received a commission of $462 on the sale of a new car. If her commission rate is 3% of the selling price, what was the selling price?

Answer: $15,400

EXAMPLE 8

Due to inflation, the price of an item rose 25% over its price the year before. What was the price last year if the price this year is 60¢?

Problems such as this can be solved by using the equation

(previous price) + (increase in price) = (new price).

Let x be last year's price.

$$x + (0.25) \cdot x = 60$$

$$1.25x = 60 \qquad x + (0.25) \cdot x = 1 \cdot x + (0.25) \cdot x$$

$$= (1 + 0.25) \cdot x = 1.25x$$

Practice Exercise 8

Due to overstocking, a dealer is forced to sell a typewriter at a 20% discount. What was the original price of the typewriter if he sells it for $288?

$$x = \frac{60}{1.25} = 48$$

Last year's price was 48¢. ◀◀

Answer: $360

   Money that we pay to borrow money or the money earned on savings is called **interest**. The money borrowed or saved is called the **principal,** and the interest is often calculated as a percent of the principal. This percent is called the **rate of interest.** For example, suppose you borrow $500 for one year and the interest rate is 12% per year, the interest charged would be

$$(0.12)(\$500) = \$60.$$

Had you needed the money for two years, the interest rate would be

$$2(0.12) = 0.24 = 24\%,$$

with the total interest for two years amounting to

$$(0.24)(\$500) = \$120.$$

This type of interest is called **simple interest** since interest is charged only on the principal and not on the interest itself.

EXAMPLE 9

What sum of money invested at 4% simple interest will amount to $988 in one year?

Interest problems like this can be solved by using the equation

(amount invested) + (interest) = (total amount).

Let x be the amount invested.

$$x + (0.04) \cdot x = 988 \quad \text{Interest} = (0.04)x$$
$$(1.04) \cdot x = 988$$
$$x = \frac{988}{1.04} = 950$$

The amount invested is $950. ◀◀

Practice Exercise 9

Missie must have $1100 available at the end of one year. What amount must she invest now at 10% simple interest to have the desired amount at the end of the year?

Answer: $1000

3.5 EXERCISES A

Solve. (Some problems have been started.)

1. If 14 is added to four times a number, the result is 38. Find the number.

   Let x = the desired number,
   $4x$ = four times the number.

2. The sum of the two numbers is 180. The larger number is five times the smaller number. Find the two numbers.

   Let x = the smaller number.

132 LINEAR EQUATIONS AND INEQUALITIES

3. The sum of two numbers is 98. If one is six times the other, find the two numbers.

4. Two thirds of a number is 124; what is the number?

    Let x = the desired number,
    $\frac{2}{3}x$ = two thirds the desired number.

5. Two thirds of the human body is water. If your body contains 124 pounds of water, how much do you weigh? [*Hint:* Compare with Exercise 4.]

6. The first of two numbers is four times the second. If the first is 30,168, what is the second?

7. The area of Lake Superior is four times the area of Lake Ontario. If the area of Lake Superior is 30,168 mi^2, what is the area of Lake Ontario? [*Hint:* Compare with Exercise 6.]

8. The sum of three consecutive integers is 78. Find the three integers.

    Let x = the first integer,
    $x + 1$ = the second integer,
    _____ = the third integer.

9. The sum of two consecutive odd integers is 96. Find the two integers.

    Let x = the first odd integer,
    $x + 2$ = the next odd integer.

10. The sum of two consecutive even integers is 162. Find the two integers.

11. Maria is 4 years older than Juan. The sum of their ages is 32 years. How old is each?

    Let x = Juan's age,
    $x + 4$ = Maria's age.

12. George is 5 years older than Sue. If the sum of their ages is 47, how old is each?

13. How old is Marvin if you get 100 years when you multiply his age by 6 and add 16 years?

14. If three fourths of a number is 27, what is the number?

15. The sum of two numbers is 9, and one is twice the other. What are the numbers?

    Let x = the first number,
    $2x$ = the second number.

16. A board is 9 ft long. It is to be cut into two pieces in such a way that one piece is twice as long as the other. How long is each piece? [*Hint:* Compare with Exercise 15.]

17. A board is 17 ft long. It is to be cut into two pieces in such a way that one piece is seven ft more than the other. How long is each piece?

18. Tony is four times as old as Angela and half as old as Theresa. The sum of the three ages is 39 years. How old is each?

19. 45 is 15% of what?
$45 = 0.15 \cdot x$

20. 105 is 20% of what?

21. 25% of 164 is what?
$0.25 \cdot 164 = x$

22. 18% of 90 is what?

23. What is 30% of 420?

24. What percent of 90 is 15?
$x \cdot 90 = 15$

25. What percent of 320 is 20?

26. 40 is what percent of 200?

27. 25 is what percent of 500?

28. 25 is 4% of what?

29. In a 150 ml acid solution there are 45 ml of acid. What is the percent of acid in the solution?

30. A basketball player hit 120 free throws in 150 attempts. What percent of her shots did she make?

31. Juan received a commission of $104.50 on the sale of a typewriter. If his commission rate is 11% of the selling price, what was the selling price?

Let x = selling price,
$0.11x$ = his commission on sales.

32. Percy received $31.50 interest on his savings last year. If he is paid 6% simple interest, what amount did he have invested?

33. A baseball player got 42 hits one season. If his batting average was 28%, how many times was he up to bat?

34. A dress is discounted 30% and the sale price is $51.45. What was the original price?

35. If the sales-tax rate is 5% and the marked price plus tax of a mixer is $37.59, what is the marked price?

   Let x = marked price,
   $0.05x$ = tax on the marked price,
   $x + 0.05x$ = price plus tax.

36. The price of a package of gum rose 20% last year. If the present price is 42¢, what was the price last year?

37. If the present retail price of an item is 65¢ and due to increased overhead the price will have to be raised 20%, what will be the new price?

38. What sum invested at $4\frac{1}{2}$% simple interest will amount to $1254 in one year?

39. A baseball player got 21 hits in 70 times at bat. What was his batting average?

40. A family spent $200 a month for food. This was 16% of their monthly income. What was their monthly income?

41. The sales-tax rate in Murphyville is 4%. How much tax would be charged on a purchase of $12.50?

42. What sum at 4% simple interest will amount to $1508 in 1 year?

43. The area of Greenland is 25% of the area of the United States. What is the area of Greenland if the area of the U.S. is 3,615,000 mi^2?

44. The human brain is $2\frac{1}{2}$% of the total body weight. If Nick's brain weighs 4 lb, how much does Nick weigh?

For Review

Select a variable to represent each quantity and translate the phrase into symbols.

45. Price plus 10% of the price

46. The reciprocal of a number less twice the number

ANSWERS: **1.** 6 **2.** 150, 30 **3.** 14, 84 **4.** 186 **5.** 186 lb **6.** 7542 **7.** 7542 mi² **8.** 25, 26, 27 **9.** 47, 49 **10.** 80, 82 **11.** Juan is 14, Maria is 18 **12.** George is 26 and Sue is 21 **13.** 14 **14.** 36 **15.** 3, 6 **16.** 3 ft and 6 ft **17.** 5 ft and 12 ft **18.** Angela is 3, Tony is 12, Theresa is 24 **19.** 300 **20.** 525 **21.** 41 **22.** 16.2 **23.** 126 **24.** $16\frac{2}{3}\%$ **25.** $6\frac{1}{4}\%$ **26.** 20% **27.** 5% **28.** 625 **29.** 30% **30.** 80% **31.** $950 **32.** $525 **33.** 150 **34.** $73.50 **35.** $35.80 **36.** 35¢ **37.** 78¢ **38.** $1200 **39.** 30% **40.** $1250 **41.** 50¢ **42.** $1450 **43.** 903,750 mi² **44.** 160 lb **45.** $p + 0.10p$ **46.** $\frac{1}{n} - 2n$

3.5 EXERCISES B

1. If 9 is added to six times a number, the result is 39. Find the number.

2. The sum of two numbers is 72. The larger number is five times the smaller number. Find the two numbers.

3. The sum of two numbers is 48. If one is five times the other, find the two numbers.

4. Three fourths of a number is 288; what is the number?

5. Three fourths of a filled container is composed of water. If the container holds 288 gallons of water, how much fluid is in it?

6. The first of two numbers is five times the second. If the first is 1250, what is the second?

7. The number of acres in the Henderson farm is five times the number of acres in the Carlson farm. If the Henderson farm has 1250 acres, how many acres are there in the Carlson farm?

8. The sum of three consecutive odd integers is 195. Find the integers.

9. The sum of two consecutive even integers is 90. What are the integers?

10. The sum of two consecutive integers is 203. Find the integers.

11. Barb is 9 years older than Larry. Twice the sum of their ages is 126. How old is each?

12. Hector is 7 years younger than Pedro. If the sum of their ages is 39, how old is each?

13. How old is Burford if you get 99 years when you multiply his age by 10 and subtract 31 years?

14. Two thirds of a number is 62; what is the number?

15. The sum of two numbers is 168, and one is three times the other. What are the numbers?

16. A rope is 168 feet long. It is to be cut into two pieces in such a way that one piece is three times as long as the other. How long is each piece?

17. A steel rod is 22 m long. It is cut into two pieces in such a way that one piece is 6 m longer than the other. How long is each piece?

18. Hortense is five times as old as Wally and half as old as Lew. The sum of their ages is 80 years. How old is each?

19. 48 is 12% of what?

20. 18 is 4% of what?

21. 35% of 180 is what?

22. 50% of 280 is what?

23. What is 3% of 1250?

24. What percent of 85 is 17?

25. What percent of 60 is 210?

26. 15 is what percent of 300?

27. 2.5 is what percent of 7.5?

28. 1230 is 60% of what?

29. In a 250 ml salt solution there are 15 ml of salt. What is the percent of salt in the solution?

30. A basketball player hit 12 shots in 20 attempts during one game. What was his shooting percent?

31. Blue Carpet Realty received a commission of $4950 on the sale of a house. If the commission rate is 6%, what was the selling price of the house?

32. Terry McGinnis received $845 interest on her savings last year. If she is paid 13% simple interest, what amount did she have invested?

33. A quarterback completed 21 passes one game for a completion percentage of 60%. How many passes did he attempt?

34. A sport coat was discounted 20% and the sale price was $68. What was the original price?

35. Fred now makes $18,920 per year. What was his salary before he received a 10% raise?

36. The price of a pet mouse rose 30% last year. If the present price is $1.56, what was the price one year ago?

37. Due to inflation, the cost of an item rose $7.20. This was a 12% increase. What was the former price? The new price?

38. What sum invested at $8\frac{1}{2}$% simple interest will amount to $1410.50 in one year?

39. A prize fighter won 38 of his 40 professional fights. What was his percent of wins?

40. The Perez family spent $125 last week for food. This was 20% of their weekly income. What was their weekly income?

41. The sales-tax rate in Canyon City is 4%. How much tax would be charged on a purchase of $327?

42. What sum at 11% simple interest will amount to $7215 in 1 year?

43. The area of the Wyler Ranch is 35% of the area of the Rodriguez Ranch. What is the area of the Wyler Ranch if the area of the Rodriguez Ranch is 1200 mi^2?

44. One year the population of the United States was 5% of the total world population. If the U.S. population was 200,000,000, what was the world population?

For Review

Select a variable to represent each quantity and translate the phrase into symbols.

45. Cost increased by 15% of the cost

46. A number less twice its reciprocal

3.5 EXERCISES C

Solve.

1. A rope 27 m long is to be cut into three pieces in such a way that the second piece is one fourth the first and the third is 3 m longer than the second. Find the length of each piece.

2. A retailer uses a 30% markup to make a profit on suits. If Chris Katsaropoulos bought a suit for $124.02 and that included a 6% sales tax, what was the original cost of the suit to the dealer? [Answer: $90.00]

3.6 APPLIED PROBLEMS: GEOMETRY AND MOTION

 **STUDENT GUIDEPOSTS**

1 Geometry problems
2 Approximately equal (≈)
3 Distance is the product of rate and time
4 Motion problems

1 Many geometry problems require knowledge of basic formulas that are summarized inside the front cover. As you read through the problem, try to determine which formula is appropriate. It is always helpful to make a sketch.

EXAMPLE 1

Find the width of a rectangle if its area is 192 ft² and its length is 16 ft.

In order to work this problem, we must know that the area of a rectangle is given by $A = lw$. Make a sketch as in Figure 3.1.

Let w = width of rectangle.

$A = lw$
$192 = 16w$
$12 = w.$

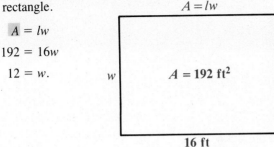

Figure 3.1

The width is 12 ft. ◀◀

Practice Exercise 1

The area of a rectangular garden is 19.8 m² and its width is 3.6 m. Find the length of the garden.

Answer: 5.5 m

EXAMPLE 2

If the perimeter of a rectangle is 32 ft and its length is three times its width, find its dimensions. Make a sketch as in Figure 3.2.

The perimeter of a rectangle is given by $P = 2l + 2w$.

Let w = width of rectangle

$3w$ = length of rectangle.

$P = 2l + 2w$
$32 = 2(3w) + 2(w)$
$32 = 6w + 2w$
$32 = 8w$
$4 = w$

and $12 = 3w$

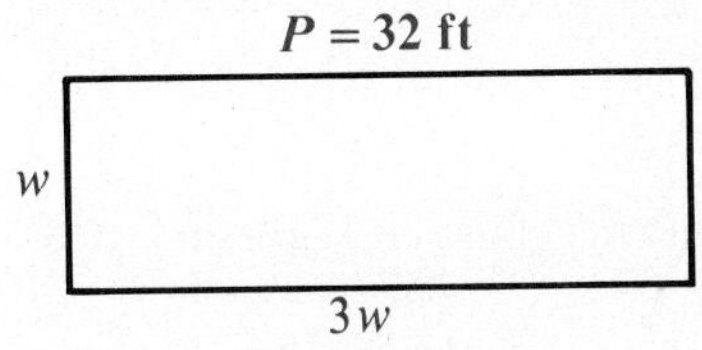

Figure 3.2

The length is 12 ft and the width is 4 ft. ◀◀

Practice Exercise 2

A farmer encloses a pasture with 20 miles of fence. If the width of the pasture is one fourth the length, find the dimensions of the pasture.

Answer: 2 miles by 8 miles

EXAMPLE 3

The first angle of a triangle is three times as large as the second. The third angle is 40° larger than the first angle. What is the measure of each angle?

The sum of the measures of the angles of any triangle is 180°. Make a sketch as in Figure 3.3.

Let $3x$ = measure of the first angle,
x = measure of the second angle,
$3x + 40$ = measure of the third angle.

$x + 3x + 3x + 40 = 180$
$7x + 40 = 180$
$7x = 140$
$x = 20$

and $3x = 60$

and $3x + 40 = 100$

The measures are 20°, 60°, and 100°. ◀◀

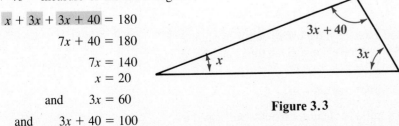

Figure 3.3

Practice Exercise 3

The second angle of a triangle is one half as large as the first, and the third is 45° less than three times the first. Find the measure of each.

Answer: 50°, 25°, 105°

EXAMPLE 4

The circumference of a circular flower bed is 182 ft. How many feet of pipe are required to reach from the edge of the bed to a fountain in the center of the bed? The circumference of a circle is given by $C = 2\pi r$. [Use 3.14 for π]. Make a sketch as in Figure 3.4.

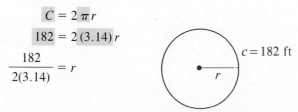

$$C = 2\pi r$$
$$182 = 2(3.14)r$$
$$\frac{182}{2(3.14)} = r$$

Figure 3.4

$$28.98 \approx r \quad \text{To the nearest hundredth}$$

Thus, we need about 29 ft of pipe.

Practice Exercise 4

The circumference of a circle is 46 in. Use 3.14 for π and find the approximate length (to the nearest tenth) of the diameter.

Answer: 14.6 in

In Example 4 we used the symbol $\approx$ to represent the phrase "is *approximately equal* to." It is important to use $\approx$ whenever an approximate or rounded number is indicated.

EXAMPLE 5

A cylindrical storage bin with radius 10 ft is to be built. How high will it need to be to store 10,200 ft^3 of grain?

The volume of a cylinder is given by $V = \pi r^2 h$. [Use 3.14 for π.]

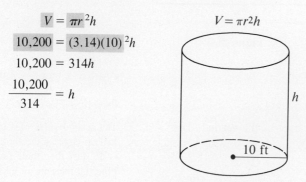

$$V = \pi r^2 h$$
$$10,200 = (3.14)(10)^2 h$$
$$10,200 = 314h$$
$$\frac{10,200}{314} = h$$

Figure 3.5

$$32.48 \approx h \quad \text{To the nearest hundredth}$$

Thus, the bin must be about 32.5 ft high.

Practice Exercise 5

A cylindrical tank with radius 8 m has surface area 965 m^2. Use 3.14 for π and find the approximate height of the tank.
Let h = height of tank
The formula for surface area is
$A = 2\pi rh + 2\pi r^2$.

Answer: 11.2 m

Problems that involve distances and rates of travel often result in simple linear equations. The distance d that an object travels in a given time t at a fixed rate r is given by the formula

$$(\text{distance}) = (\text{rate}) \cdot (\text{time}) \quad \text{or} \quad d = rt.$$

For example, a boy walking at a rate of 4 miles per hour for 3 hours will travel a distance of

$$d = rt = 4 \cdot 3 = 12 \text{ miles.}$$

3.6 APPLIED PROBLEMS: GEOMETRY AND MOTION

4 ▶ Most types of motion problems depend in some way on the formula $d = rt$.

EXAMPLE 6

An automobile is driven 336 miles in 7 hours. How fast (at what rate) was the car driven?

Let r = average rate of travel,

336 = distance d driven,

7 = time t of travel.

We must solve the following equation:

$$336 = r(7)$$
$$\frac{336}{7} = r$$
$$48 = r$$

Thus, the average rate of travel was 48 mph. ◀◀

Practice Exercise 6

Kevin rides his bike 162 km at an average rate of 36 km/hr. How long did he ride?

Answer: 4.5 hr

EXAMPLE 7

Two hikers leave the same camp, one traveling east and the other traveling west. The one going east is hiking at a rate of 1 mph faster than the other, and after 5 hours they are 35 miles apart. How fast is each hiking?

Make a sketch like the one in Figure 3.6.

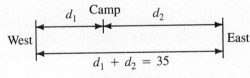

Figure 3.6

Let r = rate of hiker going west,

$r + 1$ = rate of hiker going east,

$d_1 = 5r$ = distance the westbound hiker travels,

$d_2 = 5(r + 1)$ = distance the eastbound hiker travels.

The equation we must solve is

$$d_1 + d_2 = 5r + 5(r + 1) = 35.$$
$$5r + 5r + 5 = 35$$
$$10r + 5 = 35$$
$$10r = 30$$
$$r = 3.$$

Thus, the westbound hiker is traveling 3 mph and the eastbound hiker is traveling 4 mph. ◀◀

Practice Exercise 7

Two boats leave an island at the same time, one heading north and the other south. If one travels 5 mph faster than the other, and they are 140 miles apart in 4 hours, at what rate is each sailing?

Answer: 15 mph, 20 mph

3.6 EXERCISES A

Solve. If necessary, check the inside front cover for a list of formulas. (Some problems have been started.)

1. The perimeter of a rectangle is 60 in. If the length is 4 in more than the width, find the dimension.

   Let w = width of rectangle,
   $w + 4$ = length of rectangle.

   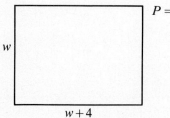

2. If one angle of a triangle is 60° more than the smallest angle and the third angle is six times the smallest angle, determine the measure of each angle.

   Let x = smallest angle,
   $x + 60$ = another angle,
   $6x$ = third angle.

3. Find the length of a rectangle if its area is 104 in² and its width is 8 in.

4. Find the altitude of a triangle whose base is 12 cm and whose area is 84 cm².

5. A circular patio has a radius of 14 ft. How many feet of edging material will be needed to enclose the patio? [Use $\pi \approx 3.14$.]

6. In a triangle, the longest side is twice the shortest side, and the third side is 5 ft shorter than the longest side. Find the three sides if the perimeter is 45 ft.

7. Find the altitude of a parallelogram whose base is 19 in and whose area is 133 in².

   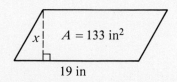

8. If two angles of a triangle are equal and the third angle is equal to the sum of the first two, find the measure of each.

9. An isosceles triangle has two equal sides called legs and a third side called the base. If each leg of an isosceles triangle is four times the base and the perimeter is 108 m, find the length of the legs and base.

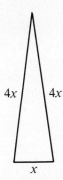

10. The perimeter of a rectangle is 630 in. Find the dimensions if the length is 25 in more than the width.

11. Find the altitude of a trapezoid if the area is 247 m^2 and the bases are 16 m and 22 m.

12. Find the base of a parallelogram if the perimeter is 88 cm and a side is 20 cm.

13. Find the area of a circular garden having a diameter of 96 yd.

14. The volume of a grain storage silo in the shape of a rectangular solid is 4725 m^3. If both the width and the height are 15 m, what is the length?

15. Find the height of a cylindrical tank with volume 3200 m^3 and radius 8 m.

16. The perimeter of a square is 48 ft. What is the length of a side?

17. A cube with edge 10 ft is submerged in a rectangular tank containing water. If the tank is 40 ft by 50 ft, how much does the level of water in the tank rise?

18. Find the volume, rounded to the nearest tenth, of a sphere with radius 2 m.

19. The sphere in Exercise 18 is dropped into a rectangular tank that is 10 m long and 8 m wide. To the nearest tenth, how much does this raise the water level?

20. A rancher wishes to enclose a pasture that is 3 mi long and 2 mi wide with a fence selling for 35¢ per linear foot. How much will the project cost?

21. Mary Lou Mercer walks for 7 hours at a rate of 5 km/hr. How far does she walk?

22. Bob Packard runs a distance of 20 miles at a rate of 8 mph. How many minutes does he run? [*Hint:* Note the units used.]

23. A hiker crossed a valley by walking 5 mph the first 2 hours and 3 mph the next 4 hours. What was the total distance that she hiked?

24. Two trains leave Omaha, one traveling east and the other traveling west. If one is traveling 20 mph faster than the other, and if after 4 hours they are 520 miles apart, how fast is each traveling?

25. Two ships sail north from Bermuda. One is traveling at 22 knots and the other at 17 knots. How many nautical miles will they be from each other after 8 hours?

26. Two cars that are 550 mi apart and whose speeds differ by 10 mph are traveling towards each other. What is the speed of each if they meet in 5 hours?

For Review

27. The sum of three consecutive odd integers is 135. What are the integers?

28. Twice a number, less 7, is the same as three times the number, less 30. What is the number?

29. Gus is 7 years older than Jan. If twice the sum of their ages is 66, how old is each?

30. A wire is 42 in long. It is to be cut into three pieces in such a way that the second is twice the first and the third is equal to the sum of the lengths of the first two. How long is each piece?

31. 54 is 120% of what?

32. 30 is what percent of 150?

33. If the cost of an item rose 8¢, a 16% increase, what was the old price? What is the new price?

34. If the sales-tax rate is 3%, how much tax would be charged on a purchase of $18.00?

35. If a student answered 68 questions correctly on an 80 question exam, what was her percent score?

36. Betsy earned $11,130 one year after she received a 6% raise. What was her former salary?

ANSWERS: Some answers are rounded. **1.** 17 in, 13 in **2.** 15°, 75°, 90° **3.** 13 in **4.** 14 cm **5.** 87.9 ft **6.** 10 ft, 15 ft, 20 ft **7.** 7 in **8.** 45°, 45°, 90° **9.** 48 m, 48 m, 12 m **10.** 145 in, 170 in **11.** 13 m **12.** 24 cm **13.** 7234.6 yd^2 **14.** 21 m **15.** 15.9 m **16.** 12 ft **17.** 0.5 ft **18.** 33.5 m^3 **19.** 0.4 m **20.** $18,480.00 **21.** 35 km **22.** 150 minutes **23.** 22 mi **24.** 75 mph, 55 mph **25.** 40 nautical miles **26.** 50 mph, 60 mph **27.** 43, 45, 47 **28.** 23 **29.** Gus is 20, Jan is 13 **30.** 7 in, 14 in, 21 in **31.** 45 **32.** 20% **33.** 50¢, 58¢ **34.** 54¢ **35.** 85% **36.** $10,500

3.6 EXERCISES B

Solve.

1. The perimeter of a rectangle is 52 ft. If the length is 8 ft more than the width, find the dimensions.

2. If one angle of a triangle is 72° more than the smallest angle, and the third angle is seven times the smallest angle, determine the measure of each angle.

3. The area of a rectangle is 32.5 ft^2 and the width is 5 ft. What is the length?

4. Find the altitude of a triangle whose area is 78 cm^2 and whose base is 12 cm.

5. A circular garden is to be enclosed with edging material. How many meters of edging will be needed if the radius of the garden is 5 m? [Use $\pi \approx 3.14$.]

6. In a triangle, the longest side is three times as long as the shortest side, and the third side is 2 ft shorter than the longest side. Find the three sides if the perimeter is 26 ft.

7. Find the altitude of a parallelogram with a base of 17 cm and an area of 93.5 cm^2.

8. If two angles of a triangle are equal and the third angle is equal to twice their sum, find the measure of each.

9. An isosceles triangle with a base of 13 in has a perimeter of 53 in. Find the length of its legs.

10. The perimeter of a rectangular table is 90 in. Find its dimensions if the length is 15 in more than the width.

11. What is the altitude of a trapezoid with area 24 yd^2 and bases of length 4.5 yd and 3.5 yd?

12. Find the base of a parallelogram if the perimeter is 100 in and a side is 15 in.

13. What is the area of a circular rug having diameter 82 inches?

14. The volume of a rectangular solid is 26,250 cm^3. If both the width and the height are 25 cm, what is the length?

15. Find the height of a cylindrical storage tank with volume 1105.3 ft^3 and radius 4 ft.

16. The perimeter of a square is 72 yd. What is the length of a side.

17. A cube with edge 8 in is submerged in a rectangular tank containing water. If the tank is 10 in by 20 in, how much does the level of water in the tank rise?

18. Find the volume of a sphere with radius 8 cm.

19. If the sphere in Exercise 18 is submerged in a rectangular tank 20 cm wide by 30 cm long, how much will the level of the water increase?

20. How much will it cost to enclose a garden with a picket fence if the garden is 12 yd long, 8 yd wide, and fencing costs 85¢ per linear foot?

21. A ship travels for 15 hours at 20 knots, where a knot is 1 nautical mile per hour. How many nautical miles has it traveled?

22. Jessie Meyer drove her boat a distance of 45 miles at a rate of 30 mph. How many minutes did she drive?

23. Barry Silverstein traveled by car for 6 hours at a rate of 55 mph and then by boat for 2 hours at 24 mph. What was the total distance traveled?

24. Two campers leave Rocky Mountain National Park, one traveling north and the other traveling south. After 3 hours they are 255 miles apart, and one has been traveling 5 mph faster than the other. How fast is each traveling?

25. Two truckers leave Atlanta at the same time heading west. How far apart will they be in 13 hours if one is traveling at a rate of 60 mph and the other at 56 mph?

26. Two families leave home at 8:00 A.M., planning to meet for a picnic at a point between them. One travels at a rate of 45 mph and the other travels at 55 mph. If they live 500 miles apart, at what time do they meet?

For Review

27. The sum of three consecutive even integers is 222. What are the integers?

28. Five times a number, decreased by 8, is the same as four times 1 more than the number. Find the number.

29. Lucy is seven years older than her sister. In 3 years she will be twice as old as her sister is now. How old is each now?

30. A 36 m steel rod is cut into three pieces. The second piece is twice as long as the first and the third is three times as long as the second. How long is each piece?

31. 540 is 150% of what?

32. 98 is what percent of 140?

33. If the cost of a calculator rose $1.92 and this was a 12% increase, what was the old price? What is the new price?

34. If the sales-tax rate is 6%, how much tax will be charged on the purchase of a wallet costing $14.50?

35. During July a car dealer sold 45 cars out of his total inventory of 54 cars. What percent of his inventory was sold?

36. After an 8% salary increase, Henry makes $24,300. What was his salary before the raise?

3.6 EXERCISES C

Solve.

1. The area shown required 124 m of fencing to enclose. Determine the dimensions by finding the value of x.

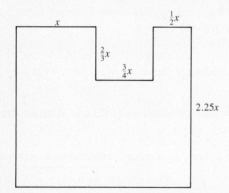

2. Pam drove her car at an average speed of 55 mph for one half the distance of a trip and then traveled by boat at 20 mph the second half. How many miles did she travel if the total trip required 9 hours? [Answer: 264 mi]

SUPPLEMENTARY APPLIED PROBLEMS

The following exercises use the techniques presented in the previous three sections.

1. Seven more than twice Herman's age is equal to 73. How old is Herman?

2. The perimeter of Ivan Danielson's yard is 528 yd. If the yard is rectangular in shape with length 24 yd more than the width, what are the dimensions?

3. On a recent field trip, Cindy noticed that the bus was completely full and that there were 14 more girls than boys on the bus. If the bus has a capacity of 54 children, how many boys were on the bus?

4. Bob, Mary, and Sharon were born in consecutive years. If the sum of their ages is 36, how old is each?

5. A pole is standing in a pond. If one fourth of the height of the pole is in the sand at the bottom of the pond, 9 ft are in the water, and three eighths of the total height is in the air above the water, what is the length of the pole?

6. In an election with only two candidates, the winner received 50 votes more than twice the total for the loser. If a total of 1310 votes were cast, how many did each candidate receive?

7. Four fifths of the operating budget of a community college goes for faculty salaries. If the total of all faculty salaries is $4,000,000, what is the total operating budget?

8. If $16.80 is charged as tax on the sale of a washing machine priced at $420, what is the tax rate?

9. A fuel tank contained 840 gallons of fuel. An additional quantity of fuel was added making a total of 1092 gallons. What was the percent increase?

10. The measure of the second angle of a triangle is one fourth the measure of the third and 30° more than the first. Find the measure of each.

11. International Car Rental will rent a compact car for $12.50 per day plus 10¢ per mile driven. If Charlene rented a compact for two days and was billed a total of $73.00, how many miles did she drive?

12. Tim Chandler bought a pair of running shoes on sale for $33.60. If the sale price was discounted 30% from the original price, what was the original price?

13. A rectangular patio is twice as long as it is wide. If its length were decreased 8 ft and the width were increased 8 ft, it would be square. What are its dimensions?

14. The relationship between °F and °C is given by the formula $F = \frac{9}{5}C + 32$. Determine °C for a temperature of 50°F.

15. How much simple interest will be earned on a deposit of $800 for one year at an interest rate of 12% per year?

16. Barbara purchased a Ford Bronco for $16,500. If the options totaled $5800, what percent of the purchase price was the base price of the vehicle?

17. Leah was able to do 40 more situps than Lynn on a recent fitness test. If the total for both girls was 290, how many situps did Lynn do?

18. The perimeter of a square is eleven cm less than five times the length of a side. Find the length of a side.

19. The population of Deserted, Utah was 740 in 1980. This was a 20% decrease in the population in 1970. What was the population in 1970?

20. Gene won $10,000 in the Illinois Lottery. He invested part in a savings account that earned 14% simple interest, and the rest in a fund that paid 11% simple interest. If at the end of one year he earned an income of $1310 from the two, how much was invested at each rate?

21. The Arnotes plan to sell their home. They must receive $82,000 after deducting the sales commission of 6% on the selling price. Rounded to the nearest dollar, at what selling price should the house be listed?

22. Dr. Cotera wishes to enclose a rectangular plot of land with 140 yards of fencing in such a way that the length is four times the width. What will be the dimensions of the plot?

23. On a recent vacation, Leona traveled one third the total distance by car, 400 miles by boat, and one half the total distance by air. What was the total distance traveled on the vacation?

24. Becky must have an average of 90 on four tests in geology to get an A in the course. What is the lowest score she can make on the fourth test if her first three grades are 96, 78, and 91?

ANSWERS: **1.** 33 yr **2.** 120 yd, 144 yd **3.** 20 **4.** Bob is 11, Mary is 12, Sharon is 13 **5.** 24 ft **6.** 420, 890 **7.** $5,000,000 **8.** 4% **9.** 30% **10.** 5°, 35°, 140° **11.** 480 mi **12.** $48.00 **13.** 32 ft, 16 ft **14.** 10°C **15.** $96 **16.** 64.8% **17.** 125 **18.** 11 cm **19.** 925 **20.** $7000 in savings, $3000 in the fund **21.** $87,234 **22.** 14 yd, 56 yd **23.** 2400 mi **24.** 95

3.7 SOLVING LINEAR INEQUALITIES

STUDENT GUIDEPOSTS

1 Linear inequalities
2 Addition-subtraction rule
3 Multiplication-division rule
4 Solving by a combination of rules
5 Solving inequalities involving parentheses
6 Summary of the method for solving linear inequalities
7 Applications involving inequalities

1 Statements such as

$$x + 1 < 7 \qquad 2x - 1 \geq 3 \qquad 3x + 1 \leq 4x - 5$$

are called **linear inequalities**. In each case, if the inequality symbols, $<$, $\geq$, and $\leq$ were replaced with the equality symbol $=$, we would know how to solve the resulting linear equations. Fortunately, solving an inequality is much like solving an equation since most of the same basic rules apply. There is one exception that will be discussed shortly. If we start with the true inequality

$$7 < 15$$

and add 4 to both sides

$$7 + 4 < 15 + 4$$
$$11 < 19$$

we obtain another true inequality. Likewise if we subtract 3 from both sides

$$7 - 3 < 15 - 3$$
$$4 < 12$$

we again obtain a true inequality. These observations lead to the addition-subtraction rule for inequalities.

2 **Addition-subtraction rule**

If the same number or expression is added to or subtracted from both sides of a true inequality, another true inequality with the same solution is obtained.

148 LINEAR EQUATIONS AND INEQUALITIES

We use this rule to solve inequalities in very much the same way that we used the similar addition-subtraction rule to solve equations. The key is to isolate the variable on one side of the inequality.

EXAMPLE 1
Solve.

(A)
$$x + 1 < 7$$
$$x + 1 - 1 < 7 - 1 \quad \text{Subtract 1 from both sides}$$
$$x + 0 < 6$$
$$x < 6$$

All numbers that are less than 6 are solutions.

(B)
$$x - 3 \geq 5$$
$$x - 3 + 3 \geq 5 + 3 \quad \text{Add 3 to both sides}$$
$$x + 0 \geq 8$$
$$x \geq 8$$

The solutions are all numbers that are greater than or equal to 8. Generally, we will not make this statement and simply indicate the solution by writing $x \geq 8$. ◀◀

Practice Exercise 1
Solve.

(A) $y + 11 > 3$

(B) $y - 15 \leq -12$

Answers: (A) $y > -8$ (B) $y \leq 3$

If we start with the true inequality

$$6 < 14$$

and multiply both sides by 3, we obtain the true inequality

$$18 < 42.$$

However, if we multiply both sides by -2, we obtain the false inequality

$$-12 < -28.$$

In order to obtain a true inequality, we must reverse the symbol of inequality, that is, change from $<$ to $>$. Thus, we have the multiplication-division rule for inequalities.

3 ▶▶

Multiplication-division rule

If both sides of a true inequality are multiplied or divided by the same *positive* number, a true inequality results. If both sides of a true inequality are multiplied or divided by the same *negative* number, a true inequality results when the symbol of inequality is *reversed*.

For example starting with

$$4 < 10$$

dividing both sides by -2 and reversing the inequality from $<$ to $>$, we obtain the true inequality

$$-2 > -5.$$

Likewise, starting with
$$-8 \le 4,$$
multiplying both sides by -3 and reversing we obtain
$$24 \ge -12.$$

▶▶ **CAUTION** ◀◀ Always remember to reverse the symbol of inequality when multiplying or dividing by a negative number. This is the only substantial difference between solving an equation and solving an inequality. ◀◀

EXAMPLE 2
Solve.

(A) $\qquad 4x > 16$

$\quad \dfrac{1}{4} \cdot 4x > \dfrac{1}{4} \cdot 16 \qquad \dfrac{1}{4}$ is positive so inequality remains the same

$\qquad 1 \cdot x > 4$

$\qquad x > 4$

The solution is $x > 4$.

(B)
$$-\dfrac{1}{2}x \le 7$$

$$(-2)\left(-\dfrac{1}{2}x\right) \ge (-2)(7) \qquad \text{Multiply both sides by } -2 \text{ and reverse the inequality}$$

$$1 \cdot x \ge -14$$

$$x \ge -14$$

The solution is $x \ge -14$. ◀◀

Practice Exercise 2
Solve.

(A) $\dfrac{1}{2}y < 3$

(B) $-7y \ge -35$

Answers: (A) $y < 6$ (B) $y \le 5$

4 ▶▶▶ We now solve inequalities that require both the addition-subtraction rule and the multiplication-division rule. As is the case with solving equalities, the basic idea is to isolate the variable on one side of the inequality.

EXAMPLE 3
Solve. $\quad 2x + 5 < 7$

$\qquad 2x + 5 - 5 < 7 - 5 \qquad$ Subtract 5

$\qquad 2x < 2$

$\qquad \dfrac{1}{2} \cdot 2x < \dfrac{1}{2} \cdot 2 \qquad$ Multiply by $\dfrac{1}{2}$ or divide by 2

$\qquad 1 \cdot x < 1$

$\qquad x < 1$

The solution is $x < 1$. ◀◀

Practice Exercise 3
Solve. $\quad 3y - 7 \ge 5$

Answer: $y \ge 4$

EXAMPLE 4
Solve. $\quad 6 - 4x > 2 - 3x$

$\quad 6 - 6 - 4x > 2 - 6 - 3x \qquad$ Subtract 6

$\qquad -4x > -4 - 3x$

Practice Exercise 4
Solve. $\quad 9 - 8y < 6 - 5y$

150 LINEAR EQUATIONS AND INEQUALITIES

$$-4x + 3x > -4 - 3x + 3x \quad \text{Add } 3x$$
$$-x > -4$$
$$(-1)(-x) < (-1)(-4) \quad \text{Multiply by } -1 \text{ and reverse inequality}$$
$$x < 4$$

The solution is $x < 4$. ◀◀

Answer: $y > 1$

5 Often an inequality involves parentheses. When this occurs, first remove all parentheses and then proceed as in previous cases.

EXAMPLE 5

Solve. $-6(2 + x) \geq 2(4x - 2)$
$$-12 - 6x \geq 8x - 4 \quad \text{Remove parentheses}$$
$$-12 + 12 - 6x \geq 8x - 4 + 12 \quad \text{Add 12}$$
$$-6x \geq 8x + 8$$
$$-6x - 8x \geq 8x - 8x + 8 \quad \text{Subtract } 8x$$
$$-14x \geq 8$$
$$\left(-\frac{1}{14}\right)(-14x) \leq \left(-\frac{1}{14}\right) \cdot 8 \quad \text{Multiply by } -\frac{1}{14} \text{ and reverse inequality}$$
$$x \leq -\frac{8}{14}$$
$$x \leq -\frac{4}{7}$$

The solution is $x \leq -\frac{4}{7}$. ◀◀

Practice Exercise 5

Solve. $5(z - 3) > -3(7 - z)$

Answer: $z > -3$

6 The method to be used when solving a linear inequality is summarized below.

> **To solve a linear inequality**
> 1. Simplify both sides of the inequality by removing parentheses and collecting like terms.
> 2. Use the addition-subtraction rule to isolate all variable terms on one side and all constant terms on the other side. Collect like terms when possible.
> 3. Use the multiplication-division rule to obtain a variable with coefficient of 1. Be sure to reverse the symbol of inequality whenever multiplying or dividing both sides by a negative number.

7 Some applied problems result in inequalities.

EXAMPLE 6

In order for Beth to get an A in her Spanish course, she must earn a total of 360 points on four tests each worth 100 points. If she got scores of 87,

Practice Exercise 6

If Angie's age is tripled then increased by 5, the result is less than

96, and 91 on the first three tests, determine (using an inequality) those scores she could make on the fourth test to get an A.

Let s = the score Beth must get on the fourth test. Since the total of her four test scores must be greater than or equal to 360, we must solve

$$87 + 96 + 91 + s \geq 360$$
$$271 + s \geq 360$$
$$s \geq 86. \quad \text{Subtract 274 from both sides}$$

Thus, Beth must make a score of 86 or better on the fourth test in order to get an A in the course. ◀◀

or equal to 25 plus her age. Angie is at most how old?

Answer: 10 yr ($a \leq 10$)

3.7 EXERCISES A

Solve.

1. $x + 9 > 12$

2. $x - 3 < 14$

3. $x + 1 \geq -5$

4. $z + 2.3 > 4.7$

5. $3.9 > z - 5.2$

6. $z - \dfrac{3}{4} \leq \dfrac{2}{3}$

7. $2a + 1 > a - 3$

8. $a - 4 \leq 2a + 5$

9. $3a - 11 \geq 2a + 9$

10. $3(b + 1) > 2b - 5$

11. $3x > 12$

12. $\dfrac{1}{3}x < -4$

13. $-\dfrac{1}{4}y \geq 2$

14. $-3y < 7$

15. $2.1y > 4.2$

16. $-2 < -5z$

17. $-a \leq -7$

18. $-3.1z \leq 9.3$

19. $-\dfrac{2}{3}z > \dfrac{4}{3}$

20. $3b \geq -\dfrac{1}{2}$

21. $-3b \geq \dfrac{1}{2}$

22. $1 - 3x \geq -8$

23. $-8y - 7 \geq 21 - 15y$

24. $4(z - 3) \leq 3(2z + 4)$

25. $3(2x + 3) - (3x + 2) < 12$

26. $5(x + 3) + 4 \geq x - 1$

27. $y + 3 - 4y > 2(y + 12)$

28. $3 - 2z < 5(z - 7)$

29. $0.05 + 3(z - 1.2) > 2z$

30. $\dfrac{3}{4}z - \dfrac{3}{8} < \dfrac{3}{2} + \dfrac{1}{8}z$

Are the following statements true or false?

31. If $x < 9$, then $-2x < -18$.

32. If $x > 4$, then $3x > 12$.

33. If $x \leq 7$, then $x + 1 \leq 8$.

34. If $x < -1$, then $-x > 1$.

35. If $x < 3$, then $x - 7 > -4$.

36. If $-x < -10$, then $1 - x < -9$.

Solve.

37. The product of a number and 3 is greater than or equal to the number less 8. Find the numbers that satisfy this.

38. If twice my age, increased by 7, is greater than or equal to 31 less my age, I am at least how old?

39. Professor Packard will give an F to any student with a point total less than 180 in a course having three 100 point exams. Burford made 38 points on the first exam and 54 on the second. Determine the minimum score he could make on the third exam in order to avoid failing the course.

40. In order for Darrell Fosberg to win a trip to Bermuda, his new car sales must average 50 over the three-month period June, July, and August. If he sold 47 cars in June and 62 cars in July, how many cars must he sell in August to qualify for the trip?

For Review

41. The length of a rectangular room is 5 yd more than its width. If its perimeter is 54 yd, find its dimensions.

42. Find the volume, rounded to the nearest tenth, of a sphere with radius 3 cm. [Use 3.14 for π.]

43. Roy drove 378 miles in 7 hours. What was his average speed?

44. Two families leave their homes at the same time planning to meet at a point between them. If one travels 55 mph, the other travels 50 mph, and they live 273 miles apart, how long will it take for them to meet?

ANSWERS: **1.** $x > 3$ **2.** $x < 17$ **3.** $x \geq -6$ **4.** $z > 2.4$ **5.** $9.1 < z$ ($z > 9.1$) **6.** $z \leq \frac{17}{12}$ **7.** $a > -4$
8. $-9 \leq a$ ($a \geq -9$) **9.** $a \geq 20$ **10.** $b > -8$ **11.** $x > 4$ **12.** $x < -12$ **13.** $y \leq -8$ **14.** $y > -\frac{7}{3}$ **15.** $y > 2$
16. $\frac{2}{5} > z$ ($z < \frac{2}{5}$) **17.** $a \geq 7$ **18.** $z \geq -3$ **19.** $z < -2$ **20.** $b \geq -\frac{1}{6}$ **21.** $b \leq -\frac{1}{6}$ **22.** $x \leq 3$ **23.** $y \geq 4$
24. $z \geq -12$ **25.** $x < \frac{5}{3}$ **26.** $x \geq -5$ **27.** $y < -\frac{21}{5}$ **28.** $z > \frac{38}{7}$ **29.** $z > 3.55$ **30.** $z < 3$ **31.** false
32. true **33.** true **34.** true **35.** false **36.** true **37.** $x \geq -4$ **38.** 8 yr ($a \geq 8$) **39.** 88 ($s \geq 88$)
40. At least 41 ($n \geq 41$) **41.** 11 yd, 16 yd **42.** 113.0 cm³ **43.** 54 mph **44.** 2.6 hr

3.7 EXERCISES B

Solve.

1. $x + 3 > 2$

2. $x - 4 \leq 7$

3. $x + 3 \geq -4$

4. $z + 1.9 < 3.9$

5. $4.2 < z - 8.1$

6. $z - \frac{2}{3} \leq \frac{3}{4}$

7. $a - 5 \leq 2a + 3$

8. $2a + 11 > a - 3$

9. $4a - 12 \geq 3a - 1$

10. $2(b + 5) \geq b + 7$

11. $2x > 14$

12. $\frac{1}{3}x < -8$

13. $-\frac{1}{3}y < 2$

14. $-5y < 11$

15. $2.1y \leq 6.3$

16. $-5a > -30$

17. $-a > -10$

18. $-4.1z > 8.2$

19. $-\frac{1}{3}z \leq \frac{2}{3}$

20. $4b \geq -\frac{1}{5}$

21. $-4b \geq \frac{1}{5}$

22. $1 - 4x \geq -7$

23. $-5y - 2 < 32 + 12y$

24. $5(y - 2) > 4(2y + 1)$

25. $2(3x + 1) - (5x + 4) \geq 2$

26. $3x - 1 \leq x - (3x - 14)$

27. $x - 2 + 2x \leq 2(x + 5)$

28. $3 - 2(z + 1) \geq 5 - z$

29. $0.01 + 2(z + 1.5) > 3z$

30. $\dfrac{1}{4}z + \dfrac{3}{8} \geq \dfrac{1}{2} - \dfrac{3}{8}z$

Are the following statements true or false?

31. If $x > 7$, then $-2x < -14$.

32. If $x < 5$, then $3x < 15$.

33. If $x \leq -1$, then $x + 1 \leq 0$.

34. If $-x \geq 1$, then $x \geq 1$.

35. If $x < 5$, then $x - 4 < 1$.

36. If $-x \leq -5$, then $2 - x \leq -3$.

Solve.

37. The product of a number and 5 is less than or equal to that number increased by 8. Find the numbers that satisfy this.

38. If Jeff's age is doubled, then diminished by 4, the result is greater than or equal to 5 plus his age. Jeff is at least how old?

39. In order to get an A in chemistry, Paula must have at least 270 points out of a total of 300 in the course. She has made 87 points on the first test and 98 on the second. At least how many points must she get on the third and final exam to receive an A?

40. Each team of four members in a tug-of-war must have a total weight of at most 600 pounds in order to qualify for the finals. Ron's team has three members weighing 120 lb, 140 lb, and 128 lb. What is the greatest amount Ron can weigh in order for his team to qualify?

For Review

41. The width of a rectangular table is 22 in less than its length. If its perimeter is 216 in, find its dimensions.

42. A circular cylinder has volume 117.75 cm³ and radius 5 cm. What is the height of the cylinder? [Use 3.14 for π.]

43. Martha hiked 25 miles at an average rate of 3 mph. What length of time did she hike?

44. Two planes leave Atlanta at the same time, one heading east at 400 mph and the other heading west at 460 mph. How long will it take for them to be 3010 miles apart?

3.7 EXERCISES C

Solve.

1. $\dfrac{x - 5}{3} - \dfrac{2x + 1}{2} \leq x - (2x - 1)$

2. The temperature of a hot-spring bath in Mexico is advertised to stay between 35°C and 40°C. What is this range of temperature in degrees Fahrenheit? [Answer: $95° \leq F \leq 104°$]

CHAPTER 3 SUMMARY

Key Words and Phrases for Review

3.1	equation	3.2	equivalent equations	3.6	approximately equal ($\approx$)
	variable		multiplication-division rule	3.7	linear inequalities
	solution	3.5	consecutive integers		
	conditional		commission (rate)		
	identity		principal		
	contradiction		rate of interest		
	addition-subtraction rule		simple interest		

Key Concepts

3.1 When using the addition-subtraction rule to solve an equation, add or subtract the same expression on both sides.

3.2 When using the multiplication-division rule to solve an equation, multiply by a number that makes the coefficient of x equal to 1. For example, to solve

$$\frac{x}{\frac{4}{5}} = 20$$

multiply by $\frac{4}{5}$, *not* by $\frac{5}{4}$.

3.3
1. Use the addition-subtraction rule before the multiplication-division rule when a combination is necessary to solve an equation.
2. When solving an equation involving parentheses, first use the distributive law to clear all parentheses.

3.5
1. Write out complete descriptions of the variables and terms when solving applied problems.
2. Two consecutive even (or odd) integers can be named x and $x + 2$. Two consecutive integers can be named x and $x + 1$.

3. When solving percent-increase problems add the increase to the original amount. For example, an amount of money x plus 5% interest can be represented as $x + 0.05x = (1 + 0.05)x = (1.05)x$.

3.6
1. An accurate sketch is often helpful when solving a geometry problem.
2. The formula $d = rt$, or (distance) = (rate) · (time), is used in many motion problems.

3.7
1. The same expression can be added or subtracted on both sides of an inequality.
2. When you multiply or divide both sides of an inequality by a negative number, *reverse* the symbol of inequality. For example,

$$-x < 5$$
$$(-1)(-x) > (-1)(5)$$
$$x > -5.$$

Review Exercises

3.1
1. Use the equation $2(x + 1) = 2x + 2$ to answer Exercises (A)–(F).
   - (A) What is the variable? _____
   - (B) What is the left side? _____
   - (C) What is the right side? _____
   - (D) What is the solution? _____
   - (E) Is this an identity? _____
   - (F) Is this a contradiction? _____

Solve for x.

2. $x + 9 = 13$

3. $\frac{2}{3} - x = \frac{4}{9}$

3.2 4. $-2.3x = 4.6$

5. $\dfrac{x}{\frac{1}{3}} = 15$

3.3 **6.** $2x + 3 = x + 3 - 4x$ **7.** $x - \dfrac{1}{3} = -8x$

8. $3(x - 4) + 5 = 2(x + 1) - 3$ **9.** $5(2x - 2) - 3(x - 4) = 0$

10. $x - (x + 3) = x - 6$ **11.** $20 - 3(x + 5) = 0$

3.4 *Letting x represent the unknown number, translate the following into symbols.*

12. A number increased by 5 is 7.   **13.** Four times a number is 23.

14. Twice a number, increased by 7, is -12.   **15.** When 5 is added to six times a number, the result is the same as 3 less than the number.

3.5 **16.** Fred is 12 years older than Bertha. Twice the sum of their ages is 100. How old is each?   **17.** The sum of three consecutive even integers is 138. Find the integers.

18. If the sales-tax rate is 4%, how much tax would be charged on a purchase of $420?   **19.** 30% of what is 198?

20. 14% of 50 is what?

21. What percent of 900 is 585?

22. The price of an item rose 15% last year. If the present price is $41.40, what was the price last year?

23. What sum of money invested at 12% simple interest will amount to $716.80 in 1 year?

3.6 24. The diameter of a circle is 7.56 cm. What is the radius?

25. The area of a rectangle is 25.2 ft² and the length is 6 ft. What is the width?

26. The perimeter of a rectangle is 80 ft. If the length is 10 ft more than the width, what are the dimensions?

27. Two trains leave the same city, one heading north and the other south. If one train is moving 5 mph faster than the other, and if after 4 hours they are 556 miles apart, how fast is each traveling?

3.7 *Solve.*

28. $x - 3 > 4$

29. $y + 2 \leq -5$

30. In solving an inequality, when is the symbol of inequality reversed?

Solve.

31. $\frac{1}{5}x \leq 35$

32. $-4y > 24$

33. $6 - 2x > -4 + 3x$

34. $5 + x > 3 - (x + 10)$

35. $3(x + 2) \leq 5x - 4$

36. $2 - (x - 2) < 7(x - 3)$

ANSWERS: **1.** (A) x (B) $2(x + 1)$ (C) $2x + 2$ (D) every real number (E) yes (F) no   **2.** 4   **3.** $\frac{2}{9}$ **4.** -2 **5.** 5   **6.** 0   **7.** $\frac{1}{27}$ **8.** 6   **9.** $-\frac{2}{7}$ **10.** 3   **11.** $\frac{5}{3}$ **12.** $x + 5 = 7$ **13.** $4x = 23$ **14.** $2x + 7 = -12$ **15.** $6x + 5 = x - 3$ **16.** Fred is 31, Bertha is 19   **17.** 44, 46, 48   **18.** $16.80   **19.** 660   **20.** 7   **21.** 65%   **22.** $36   **23.** $640   **24.** 3.78 cm   **25.** 4.2 ft   **26.** 15 ft, 25 ft   **27.** 72 mph, 67 mph   **28.** $x > 7$ **29.** $y \leq -7$ **30.** When both sides are multiplied or divided by a negative number   **31.** $x \leq 175$ **32.** $y < -6$ **33.** $x < 2$ **34.** $x > -6$ **35.** $x \geq 5$ **36.** $x > \frac{25}{8}$

CHAPTER 3 TEST

Solve.

1. $x - 5 = 10$

1. _____

2. $4x = 32$

2. _____

3. $x + \dfrac{3}{4} = \dfrac{5}{4}$

3. _____

4. $5.1x = -10.2$

4. _____

5. $\dfrac{1}{4}x = 9$

5. _____

6. $\dfrac{x}{\frac{1}{5}} = 20$

6. _____

CHAPTER 3 TEST Continued

7. $3x - 5 = 8x + 10$ 7. _____

8. $3(x + 2) - 5(x - 4) = 0$ 8. _____

9. $4x - (x + 6) = 3$ 9. _____

10. The sum of two consecutive odd integers is 168. What are the integers?   10. _____

11. The price of a dress was $54 but the price was increased by 15%. What is the new price?   11. _____

12. The area of a triangle is 240 cm² and its height is 12 cm. What is the length of the base?   12. _____

160

13. Two trains that are 840 miles apart and whose speeds differ by 9 mph are traveling towards each other. If they will meet in 8 hours, at what speed is each traveling?

13. _____

Solve the inequalities.

14. $3x - 6 \leq 9x + 12$

14. _____

15. $3 - (2x - 5) > 6x + 4$

15. _____

4 GRAPHING

4.1 GRAPHING ON THE NUMBER LINE

STUDENT GUIDEPOSTS

1. Number line
2. Origin
3. Graphing solutions to equations
4. Plotting points
5. Graphing solutions to inequalities

When working with abstract ideas such as numbers and equations it is often helpful to picture these ideas. This was done in Chapter 2 with a number line. **1** Recall that a **number line** is a line which has been marked off in unit lengths (any unit will do) with each point on the line associated with a number, as in Figure 4.1.

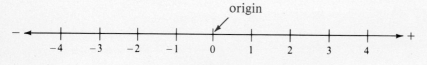

Figure 4.1

2 The **origin** is identified with the number zero and the positive numbers correspond to points to the right of zero while the negative numbers correspond to points to the left of zero. We think of this line as extending infinitely far in both directions and indicate this by arrows. Thus, we can associate any number we want to consider with a point on this line, and given any point on the line a number can be identified with it.

For example, on the number line in Figure 4.2, point A corresponds to the number 1, B to the number 3, C to the number $-\frac{1}{2}$ (C is halfway between -1

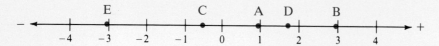

Figure 4.2

and 0), D to the number $\frac{7}{4}$, and E to the number -3. If we wanted to find the point associated with the number 10, we would have to extend the line to the right and continue marking unit lengths until we reached the desired position.

In Chapter 3 we learned how to solve linear equations. All of these equations had only one variable. If we were to graph these equations, we would first solve the equation, then find the point (or points) on a number line which corresponds to the solution (or solutions) of the equation. We call this **plotting points.**

EXAMPLE 1

Graph $2x + 1 = 3$.

First solve the equation.

$$2x + 1 - 1 = 3 - 1 \quad \text{Subtract 1}$$
$$2x = 2$$
$$x = 1$$

The solution is $x = 1$. Plot the point corresponding to 1 on a number line, as in Figure 4.3.

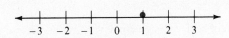

Figure 4.3

Practice Exercise 1

Graph $3x - 2 = 4$.

Answer: Plot the solution $x = 2$.

EXAMPLE 2

Graph $x + 1 = x$.

If we attempt to solve the equation, we obtain the following result:

$$x - x + 1 = x - x \quad \text{Subtract } x$$
$$1 = 0.$$

This equation is a contradiction because $1 \neq 0$ (no number plus 1 can equal the number itself). There are no solutions to the equation so we have no points to plot. The graph is the plain number line shown in Figure 4.4.

Figure 4.4

Practice Exercise 2

Graph $5x + 3 = 5x$.

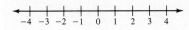

Answer: Equation is a contradiction. Plot no points.

EXAMPLE 3

Graph $2x + 1 = 1 + 2x$.

Attempt to solve the equation.

$$2x + 1 - 1 = 1 - 1 + 2x \quad \text{Subtract 1}$$
$$2x = 2x$$
$$2x - 2x = 2x - 2x \quad \text{Subtract } 2x$$
$$0 = 0$$

Practice Exercise 3

Graph $4(x - 1) = 4x - 4$.

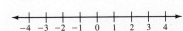

Since $0 = 0$ is always true, this equation is an identity and every number is a solution. Thus every number must be plotted, so that the graph of the solution is the entire number line shown in Figure 4.5. ◀◀

Figure 4.5

Answer: Equation is an identity. Every number is plotted.

Graphing equations is a relatively simple procedure of plotting points. The procedure for graphing linear inequalities is illustrated in the following figures.

The graph of $x > 3$ is shown in Figure 4.6. The open circle at 3 indicates that the number 3 is not included among the solutions while the colored line shows that all points to the right of 3 are included.

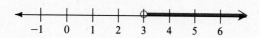

Figure 4.6

The graph of $x < 0$ is shown in Figure 4.7, while the graph of $x \geq -1$ is given in Figure 4.8. Notice that the circle at the point -1 is solid indicating that -1 is included in the graph of $x \geq -1$.

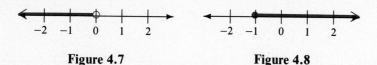

Figure 4.7   **Figure 4.8**

Finally, the graph of $x \leq 1$ is given in Figure 4.9.

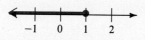

Figure 4.9

We can graph linear inequalities we studied in Chapter 3 by using the techniques we have just described.

EXAMPLE 4

Graph $2(x - 1) < 3x - 3$.

$$2(x - 1) < 3x - 3$$
$$2x - 2 < 3x - 3 \quad \text{Remove parentheses}$$
$$2x - 2 + 2 < 3x - 3 + 2 \quad \text{Add 2}$$
$$2x < 3x - 1$$
$$2x - 3x < 3x - 3x - 1 \quad \text{Subtract } 3x$$
$$-x < -1$$
$$x > 1 \quad \text{Multiply by } -1 \text{ and reverse the inequality}$$

Practice Exercise 4

Graph $2x - 6 \geq 5(x - 3)$

The solution, $x > 1$, is graphed in Figure 4.10.

Figure 4.10

Answer: Graph $x \leq 3$.

4.1 EXERCISES A

Graph the equation on a number line.

1. $2x - 3 = 1$

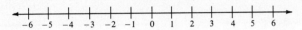

2. $x + 5 = x$

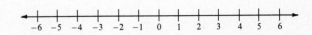

3. $x - 1 = x - 1$

4. $\dfrac{y}{\frac{1}{4}} = 20$

5. $-\dfrac{1}{10} - \dfrac{3}{5}y = \dfrac{1}{5}$

6. $a - 3 + 4a = 2 + a - 5$

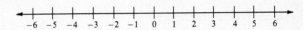

7. $3(z - 1) - 2(z + 3) = -7$

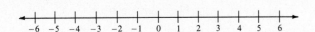

8. $12 - x = 3 + 2[4 - (x - 1)]$

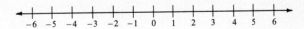

Graph the inequality on a number line.

9. $x > 2$

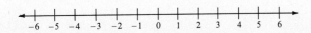

10. $x \leq 2$

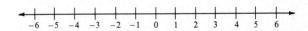

11. $y \leq -1$

12. $y > -1$

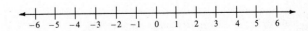

13. $2a > 10$

14. $-2a \leq -8$

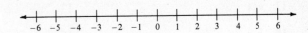

15. $1 - 3x < 8$

16. $2x + 3 > 5x - 3$

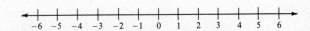

17. $-8y - 21 \geq 7 - 15y$

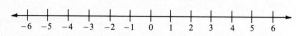

18. $25 \geq 5(3 - z) - 10$

19. $4(a - 1) < 2(3a + 2)$

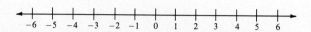

20. $-2a + 3 + a < 2(a + 3)$

ANSWERS: 1.  (solution is $x = 2$) 2.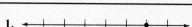
(no solution—equation is a contradiction) 3. 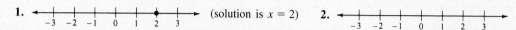 (every number is a solution—equation is an identity) 4. 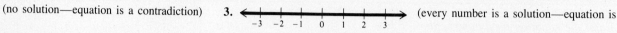 5. $\left(y = -\frac{1}{2}\right)$

6. 7. 8.
9. 10. 11.
12. 13. 14.
15. 16. 17.
18. 19. 20.

4.1 EXERCISES B

Graph the equation on a number line.

1. $2x - 5 = 1$
2. $4x + 3 = 4x - 3$
3. $x - 5 = x - 5$
4. $\dfrac{y}{\frac{1}{3}} = 9$
5. $-\dfrac{1}{4} - \dfrac{3}{8}y = \dfrac{1}{2}$
6. $a - 2 + 3a = 4 + a - 5$
7. $2(z - 2) - 3(z + 1) = -4$
8. $7 - 2x = 3 + [2 - (x - 3)]$

Graph the inequality on a number line.

9. $x > 1$
10. $x \leq 1$
11. $y \leq -3$
12. $y > -3$
13. $3a < 12$
14. $-3a > -9$
15. $1 - 2x < 4$
16. $3x + 1 < x - 5$
17. $-4y - 4 \geq 3 - 11y$
18. $3(z + 7) - 18 \leq 2z$
19. $5(a - 2) \geq 2(3a - 4)$
20. $-3a - 4 + 2a \geq 3(a + 2)$

4.1 EXERCISES C

Graph the inequality on a number line.

1. $-2 \leq x < 3$
2. $x < -2$ or $x \geq 3$

4.2 THE CARTESIAN COORDINATE SYSTEM

STUDENT GUIDEPOSTS

1. Graphing terminology
2. Plotting points
3. Solutions as ordered pairs

In Section 4.1 we graphed linear equations and inequalities in one variable by plotting points on a number line. Now we develop a system in which equations and inequalities in two variables can be graphed. Consider a horizontal number line and a vertical number line as shown in Figure 4.11.

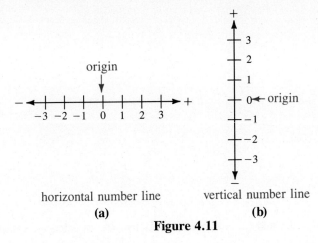

horizontal number line vertical number line
(a) (b)
Figure 4.11

1 When the horizontal number line and the vertical number line are placed together so that the two origins coincide and the lines are perpendicular, as in Figure 4.12, the result is called a **rectangular** or **Cartesian coordinate system** (named after French mathematician René Descartes), or a **coordinate plane** (a plane is a flat surface that extends infinitely far in all directions).

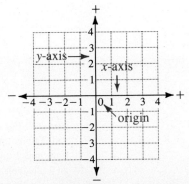

Rectangular or Cartesian Coordinate System
Figure 4.12

168 GRAPHING

The horizontal number line is called the **horizontal axis** or the ***x*-axis,** and the vertical number line is called the **vertical axis** or ***y*-axis.** Collectively the number lines are referred to as **axes** and the point of intersection of the lines is the **origin.**

Just as there is one and only one point on a number line associated with each number, there is one and only one point in a plane associated with each **ordered pair** of numbers. For example, the ordered pair (3, 2) is identified with a point in a coordinate plane as follows:

The first number, 3, called the **first coordinate** or ***x*-coordinate** of the point, is associated with a point on the *x*-axis.

The second number, 2, called the **second coordinate** or ***y*-coordinate** of the point, is associated with a point on the *y*-axis.

The ordered pair, (3, 2), is identified with the point where the vertical line through 3 on the *x*-axis intersects the horizontal line through 2 on the *y*-axis. See Figure 4.13.

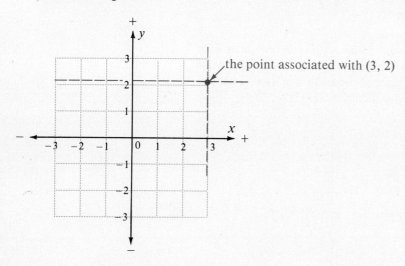

Figure 4.13

EXAMPLE 1

The points associated with (1, 3), (−2, 1), (−3, −2), and (2, −2) are given in the coordinate plane in Figure 4.14.

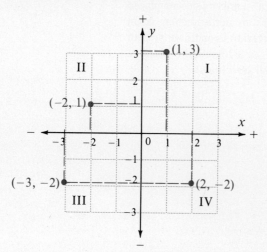

Figure 4.14

Practice Exercise 1

Indicate the points associated with (−1, 2), (3, −2), (−1, −1), and (1, 2) in the given coordinate plane.

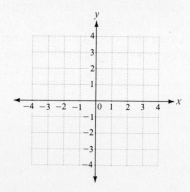

Answer: For (−1, 2) go 1 unit left and 2 units up; for (3, −2) go 3 units right and 2 units down; for (−1, −1) go 1 unit left and 1 unit down; for (1, 2) go 1 unit right and 2 units up.

4.2 THE CARTESIAN COORDINATE SYSTEM

The axes in a coordinate system divide the plane into four sections called **quadrants.** The first, second, third, and fourth quadrants are identified by the Roman numerals I, II, III, and IV, respectively, in Figure 4.14. The *x*-coordinate (first) and the *y*-coordinate (second) have the following signs in each quadrant:

$$\text{I: } (+, +), \quad \text{II:} (-, +), \quad \text{III:} (-, -), \quad \text{IV:} (+, -).$$

We often use (x, y) to refer to a general ordered pair of numbers. The point in the plane associated with the pair (x, y) has x as its *x*-coordinate and y as its *y*-coordinate. We **plot** a point when we identify it with a given pair of numbers in a plane.

EXAMPLE 2

The points A, B, C, D, E, F, G, and H in Figure 4.15 have coordinates $A(4, 1), B(0, 2), C(0, 0), D(-2, 1), E(-4, -2), F(0, -3), G(1, -2)$, and $H(3, 0)$.

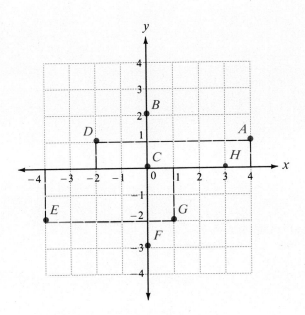

Figure 4.15

Practice Exercise 2

Give the coordinates of each point.

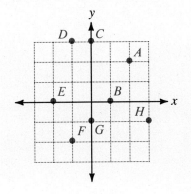

Answer: $A(2, 2), B(1, 0), C(0, 3), D(-1, 3), E(-2, 0), F(-1, -2), G(0, -1), H(3, -1)$

Consider the equation $x + 3y = 10$, in the two variables x and y. A **solution** to an equation in two variables x and y is an ordered pair of numbers which when substituted for the variables results in a true equation. In an ordered pair, the *x*-value is always written first and the *y*-value is always written second. For example, $(1, 3)$ is a solution to $x + 3y = 10$ since when x is replaced by 1 and y is replaced by 3, the resulting equation is true.

$$\begin{aligned} x + 3y &= 10 \\ 1 + 3(3) &= 10 \quad x = 1 \text{ and } y = 3 \\ 1 + 9 &= 10 \\ 10 &= 10 \end{aligned}$$

You can verify that $(10, 0), (4, 2)$, and $(13, -1)$ are also solutions.

EXAMPLE 3

Given the equation $3x + 2y = 6$, complete the ordered pairs so that they are solutions to the equation.

$$(0, \ ), \quad (\ ,0), \quad (1, \ ), \quad (\ ,1), \quad (-2, \ )$$

To complete the ordered pair $(0, \ )$, substitute 0 for x in $3x + 2y = 6$ and solve for y.

$$3(0) + 2y = 6$$
$$2y = 6$$
$$y = 3$$

Thus, the completed ordered pair is $(0, 3)$.

To complete the ordered pair $(\ , 0)$, substitute 0 for y in $3x + 2y = 6$ and solve for x.

$$3x + 2(0) = 6$$
$$3x = 6$$
$$x = 2$$

Thus, the completed ordered pair is $(2, 0)$.
To complete the ordered pair $(1, \ )$, substitute 1 for x and solve for y.

$$3(1) + 2y = 6$$
$$3 + 2y = 6$$
$$2y = 3$$
$$y = \frac{3}{2}$$

Thus, the completed ordered pair is $(1, \frac{3}{2})$.

Similarly, substitute 1 for y and solve for x to complete $(\ , 1)$, obtaining $(\frac{4}{3}, 1)$, and to complete $(-2, \ )$, substitute -2 for x and solve to obtain $(-2, 6)$. ◀

Practice Exercise 3

Given the equation $4x - 3y = 12$, complete the ordered pairs so that they are solutions to the equation.

$(0, \ ), (\ ,0), (2, \ ), (\ , -2), (-3, \ )$

Answer: $(0, -4), (3, 0), \left(2, -\frac{4}{3}\right), \left(\frac{3}{2}, -2\right), (-3, -8)$

EXAMPLE 4

Mr. Paducci has a small business that manufactures wood-burning stoves. He has determined that the cost, y, in dollars of producing a certain number, x, of stoves is given by the quation

$$y = 150x + 80.$$

Complete the ordered pairs $(1, \ ), (2, \ )$, and $(5, \ )$. Let $x = 1$ and solve for y.

$$y = 150(1) + 80$$
$$y = 230$$

Thus, $(1, \ )$ becomes $(1, 230)$. Similarly, $(2, \ )$ becomes $(2, 380)$ and $(5, \ )$ becomes $(5, 830)$. This means that it costs Mr. Paducci $230 to produce 1 stove, $380 to produce 2 stoves, and $830 to produce 5 stoves. ◀

Practice Exercise 4

Laura Hayes manufactures small appliances. She uses the equation $y = 45x + 60$ to determine the cost, y, of making x appliances. Find the cost to produce 2 appliances, the cost to produce 7, and the cost to produce 20.

Answer: $150, $375, $960

4.2 EXERCISES A

1. Plot the points associated with the given pairs of numbers: $A(2, 4)$, $B(4, -1)$, $C(-3, 4)$, $D(-3, 0)$, $E(2, 0)$, $F(-2, -2)$, $G(1, -4)$, and $H(4, -4)$.

2. Give the coordinates of the points A, B, C, D, E, F, G, and H.

3. Plot the points associated with the given pairs of numbers: $M(\frac{1}{2}, 2)$, $N(-\frac{3}{2}, 3)$, $P(-2, -\frac{3}{4})$, and $Q(3, -\frac{5}{2})$.

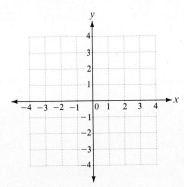
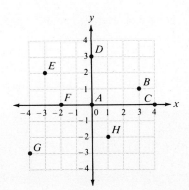
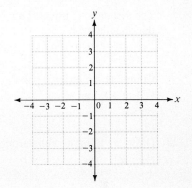

4. In which quadrant are the points M, N, P, and Q of Exercise 3 located?

Give the quadrant in which each of the following is located.

5. $(-8, -8)$
6. $(6, -3)$
7. $(4, 12)$
8. $(-2, 7)$

9. What are the four regions called into which a plane is separated by a Cartesian coordinate system?

10. What is the x-coordinate of the point named by the ordered pair $(-2, 5)$?

11. What is the y-coordinate of the point named by the ordered pair $(3, -8)$?

12. What are the coordinates of the origin?

13. What is another name for the horizontal axis in a Cartesian coordinate system?

14. If the first coordinate of a point is positive and the second coordinate is negative, in which quadrant is the point located?

15. If the x-coordinate of a point is negative and the y-coordinate of the point is positive, in which quadrant is the point located?

Using the given equation, complete each ordered pair so that it will be a solution to the equation.

16. $x - y = 2$
    (A) $(0,\ \ )$ (B) $(\ \ , 0)$
    (C) $(4,\ \ )$ (D) $(\ \ , -3)$

17. $x - 5 = 0$
    (A) $(0,\ \ )$ (B) $(\ \ , 0)$
    (C) $(5,\ \ )$ (D) $(\ \ , -10)$

172 GRAPHING

18. $x + y = 5$
   (A) (0,   )   (B) (, 0)
   (C) (3,   )   (D) (, −4)

19. $2x − y = 4$
   (A) (0,   )   (B) (, 0)
   (C) (−2,   )   (D) (, 6)

20. $x + 5y = 10$
   (A) (0,   )   (B) (, 0)
   (C) (−10,   )   (D) (, 3)

21. $5x − 2y = 10$
   (A) (0,   )   (B) (, 0)
   (C) (7,   )   (D) (, −4)

22. The distance, y, traveled by a car traveling at an average rate of 55 mph for a period of time, x, in hours is given by the equation

$$y = 55x.$$

Complete the ordered pairs (1,   ), (5,   ), and (10,   ). How far does the car travel (A) in 1 hour? (B) in 5 hours? (C) in 10 hours?

23. Draw the triangle with vertices (3, 4), (−3, 1), and (1, −2).

24. Draw the rectangle with corners (3, 1), (1, 3), (−3, −1), and (−1, −3).

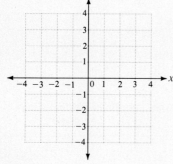
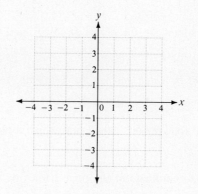

For Review

Graph the equation on a number line.

25. $a + 3 = 3(a − 1)$

26. $3x + 8 = 8 + 3x$

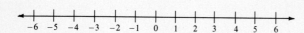

Graph the inequality on a number line.

27. $2x \geq 7$

28. $3(y − 1) < 4 − (y − 1)$

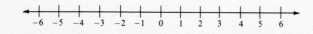
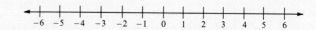

ANSWERS: 1. 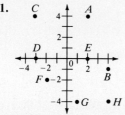   2. $A(0, 0)$, $B(3, 1)$, $C(4, 0)$, $D(0, 3)$, $E(−3, 2)$, $F(−2, 0)$, $G(−4, −3)$, $H(1, −2)$.   3.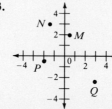

4.2 THE CARTESIAN COORDINATE SYSTEM   173

4. M:I, N:II, P:III, Q:IV   5. III   6. IV   7. I   8. II   9. quadrants   10. −2   11. −8   12. (0, 0)   13. x-axis
14. IV   15. II   16. (A) (0, −2)   (B) (2, 0)   (C) (4, 2)   (D) (−1, −3)   17. (A) x cannot be 0   (B) (5, 0)
(C) (5, any number)   (D) (5, −10)   18. (A) (0, 5)   (B) (5, 0)   (C) (3, 2)   (D) (9, −4)   19. (A) (0, −4)   (B) (2, 0)
(C) (−2, −8)   (D) (5, 6)   20. (A) (0, 2)   (B) (10, 0)   (C) (−10, 4)   (D) (−5, 3)   21. (A) (0, −5)   (B) (2, 0)
(C) $\left(7, \frac{25}{2}\right)$ (D) $\left(\frac{2}{5}, -4\right)$ 22. (1, 55), (5, 275), (10, 550)   (A) 55 miles   (B) 275 miles   (C) 550 miles

23.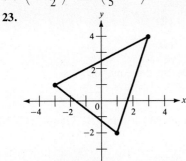

24.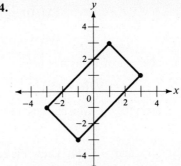

25.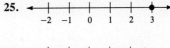

26.

27.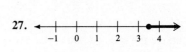

28.

4.2 EXERCISES B

1. Plot the points associated with the given pairs of numbers: $A(1, 3)$, $B(2, -1)$, $C(-3, 4)$, $D(-1, 1)$, $E(-3, -3)$, $F(2, -3)$.

2. Give the coordinates of the points A, B, C, D, E, and F in the figure at right.

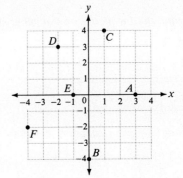

3. Plot the points associated with the given pairs of numbers: $M(-\frac{1}{2}, 1)$, $N(\frac{3}{2}, 2)$, $P(2, -\frac{3}{4})$, and $Q(-3, -\frac{5}{2})$.

4. In which quadrants are the points M, N, P, and Q of Exercise 3 located?

Give the quadrant in which each of the following is located.

5. (4, −1)   6. (−4, 3)   7. (6, 6)   8. (−12, −4)

9. What is another name for a Cartesian coordinate system?

10. What is the y-coordinate of the point named by the ordered pair (−2, 5)?

11. What is the x-coordinate of the point named by the ordered pair (3, −8)?

12. What is the name of the point with coordinates (0, 0)?

13. What is another name for the vertical axis in a Cartesian coordinate system?

14. If the first coordinate of a point is negative and the second coordinate of the point is positive, in which quadrant is the point located?

15. If the x-coordinate of a point is negative and the y-coordinate of the point is also negative, in which quadrant is the point located?

Using the given equation, complete each ordered pair so that it will be a solution to the equation.

16. $x + y = 2$

   (A) (0,   )   (B) (, 0)
   (C) (3,   )   (D) (, −2)

17. $y + 1 = 0$

   (A) (0,   )   (B) (, 0)
   (C) (3,   )   (D) (, −1)

18. $x - y = 3$

(A) $(0,)$ (B) $(, 0)$

(C) $(2,)$ (D) $(, -4)$

19. $2x + y = 4$

(A) $(0,)$ (B) $(, 0)$

(C) $(-2,)$ (D) $(, 6)$

20. $x - 5y = 10$

(A) $(0,)$ (B) $(, 0)$

(C) $(-10,)$ (D) $(, 3)$

21. $5x + 2y = 10$

(A) $(0,)$ (B) $(, 0)$

(C) $(7,)$ (D) $(, -4)$

22. The cost, y, in dollars of producing a number, x, of items has been estimated by the equation

$$y = 300x + 15.$$

Complete the ordered pairs $(1,)$, $(3,)$, and $(10,)$. Give the cost of producing **(A)** 1 item. **(B)** 3 items. **(C)** 10 items.

23. Draw the triangle with vertices $(-2, 7)$, $(1, 1)$, and $(-4, -5)$.

24. Draw the rectangle with corners $(-4, 2)$, $(-1, 5)$, $(5, -1)$, and $(2, -4)$.

For Review

Graph the equation on a number line.

25. $2a + 1 = 3(a - 2)$

26. $3x - 8 = 3x + 8$

Graph the inequality on a number line.

27. $3x < -5$

28. $2(y - 3) \geq 3 - (y - 6)$

4.2 EXERCISES C

Using the given equation, complete each ordered pair so that it will be a solution to the equation.

1. $0.012x - 1.565y = 1.878$

(A) $(0,)$ (B) $(, 0)$

(C) $(234.75,)$ (D) $(, 2.4)$

[Answer: **(A)** $(0, -1.2)$; **(C)** $(234.75, 0.6)$]

2. $-\dfrac{3}{16}x + \dfrac{7}{12}y = 3\dfrac{5}{8}$

(A) $(0,)$ (B) $(, 0)$

(C) $\left(-\dfrac{5}{9}, \right)$ (D) $\left(, 1\dfrac{3}{7}\right)$

4.3 GRAPHING LINEAR EQUATIONS

STUDENT GUIDEPOSTS

1. Method for graphing
2. General form of linear equations
3. Intercepts
4. Graphing $ax + by + c = 0$

The graph of an equation with two variables x and y is the graph of the ordered pairs of numbers that make the equation true. Since there are usually infinitely many such ordered pairs, we cannot find and plot each pair. Generally, we plot enough points to see a pattern and then connect these points with a line or curve to graph the equation.

An excellent way to display a collection of ordered-pair solutions to an equation such as

$$y = 2x + 5$$

is to make a table of values. Choose several values for x and substitute these values into the equation to compute the corresponding value for y. Place each y-value beside the x-value used to calculate it. (Calculations are usually done mentally or as scratch work.)

Substitution	Result in $y = 2x + 5$		x	y
$x = 0$	$2(0) + 5 = 5$		0	5
$x = 1$	$2(1) + 5 = 7$		1	7
$x = -1$	$2(-1) + 5 = 3$		-1	3
$x = 2$	$2(2) + 5 = 9$		2	9
$x = -2$	$2(-2) + 5 = 1$		-2	1
$x = 3$	$2(3) + 5 = 11$		3	11
$x = -3$	$2(-3) + 5 = -1$		-3	-1

This table contains seven (of the infinitely many) solutions to the equation $y = 2x + 5$. These seven solutions represent the ordered pairs

$(0, 5), \quad (1, 7), \quad (-1, 3), \quad (2, 9), \quad (-2, 1), \quad (3, 11), \quad (-3, -1).$

Now plot the points that correspond to these ordered-pair solutions in a Cartesian coordinate system, as in Figure 4.16.

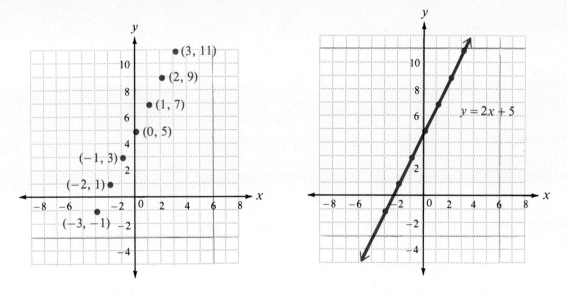

Figure 4.16   **Figure 4.17**

It appears that all seven points lie on a straight line. It is reasonable to assume that the graph of this equation is the straight line passing through these seven points, as in Figure 4.17.

To graph an equation in the two variables x and y

1. Make a table of values. These values represent the ordered-pair solutions of the equation.
2. Plot the points that correspond to the ordered-pair solutions in a Cartesian coordinate system.
3. Connect the points with a line or curve.

EXAMPLE 1

Graph $y + x = 3$.

Before making a table, it is often helpful to solve the equation for one of the variables (usually for y):

$$y = -x + 3.$$

To make a table of values, choose several values for x and substitute each into the equation to compute the corresponding value of y. See the table at the side. Plot the points that correspond to the ordered pairs from the table:

(0, 3), (1, 2), (−1, 4), (2, 1), (−2, 5), (3, 0), (−3, 6).

Connecting these points gives us a straight line, as in Figure 4.18. ◀◀

x	y
0	3
1	2
−1	4
2	1
−2	5
3	0
−3	6

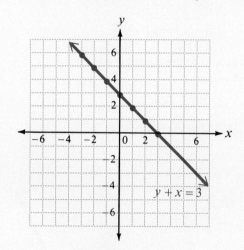

Figure 4.18

Practice Exercise 1

Graph $2x + y = -2$.

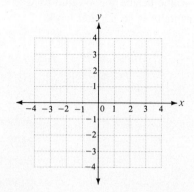

Answer: Straight line passing through (0, −2) and (−1, 0).

In the first part of this chapter, we graphed an equation in one variable on a number line. Usually, for graphing purposes, an equation in one variable such as

$$x = 5 \quad \text{or} \quad y + 2 = 0$$

is thought of as an equation in two variables with the coefficient of the missing variable equal to zero. That is,

$$x = 5 \text{ is the same as } x + 0 \cdot y = 5$$

and

$$y + 2 = 0 \text{ is the same as } 0 \cdot x + y + 2 = 0.$$

With this in mind, such equations can be graphed in a Cartesian coordinate system.

EXAMPLE 2

(A) Graph $x = 5$ in a Cartesian coordinate system.

Solutions to this equation always have an x-coordinate of 5 and can have any number as y-coordinate. For example,

(5, 0), (5, −1), (5, 1), (5, 2)

are all solutions, since $x = 5$ is the same as $x + 0 \cdot y = 5$, and we know that

$$5 + 0 \cdot (\text{any number}) = 5 + 0 = 5.$$

Practice Exercise 2

Plot the points from the table and draw a line through them. The graph of $x = 5$ is a straight line parallel to the y-axis, 5 units to the right of the y-axis. See Figure 4.19.

(A) Graph $x + 3 = 0$.

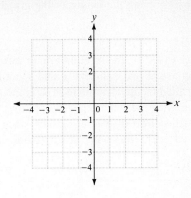

x	y
5	0
5	1
5	−1
5	2
5	−2
5	3
5	−3

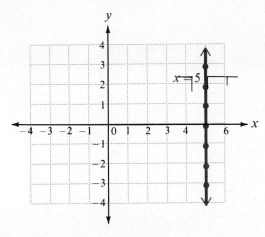

Figure 4.19

(B) Graph $y + 2 = 0$ in a Cartesian coordinate system.

We can write this equation as $y = -2$. Thus, solutions to this equation always have a y-coordinate of -2 and can have any number as x-coordinate. For example,

$$(0, -2), \quad (1, -2), \quad (-1, -2), \quad (2, -2)$$

are all solutions since $y + 2 = 0$ is the same as $0 \cdot x + y + 2 = 0$, and we know that

$$0 \cdot (\text{any number}) + (-2) + 2 = -2 + 2 = 0.$$

(B) Graph $y = \frac{5}{2}$.

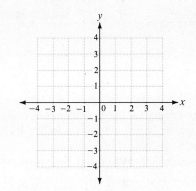

x	y
0	−2
1	−2
−1	−2
2	−2
−2	−2
3	−2
−3	−2

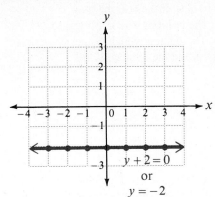

Figure 4.20

Plot the points given in the table and draw a line through them. The graph of $y + 2 = 0$ ($y = -2$) is a straight line parallel to the x-axis, 2 units below the x-axis. (See Figure 4.20.) ◀◀

Answers: **(A)** A straight line parallel to the y-axis, 3 units to the left of the y-axis. **(B)** A straight line parallel to the x-axis, $\frac{5}{2}$ units above the x-axis.

If we are asked to graph an equation such as $x = 5$ or $y + 2 = 0$, we must first determine whether it is to be considered an equation in one variable or as an equation in two variables having a zero coefficient on the missing variable. In the first case, the graph would be a point on a number line, while in the second, the graph would be a straight line in a Cartesian coordinate system. From this point on, we will only consider graphing equations in two variables, and all graphs will be in the coordinate plane.

178 GRAPHING

2 In graphing we can sometimes minimize the number of points that we plot by knowing the nature of the graph. An equation of the form

$$ax + by + c = 0$$

is called a **linear equation in two variables.** The numbers a, b, and c are constant real numbers, a and b not both zero, and this form of the equation is called the **general form.** The graph of a **linear** equation is always a straight **line** and can be determined by plotting just two points. Linear equations are also called **first-degree equations** since the variables are raised to the first power only. The following table provides practice at identifying linear equations.

Equation	Nature	General form	Constants
$2x + y = 7$	linear	$2x + y - 7 = 0$	$a = 2, b = 1, c = -7$
$x = -y - 8$	linear	$x + y + 8 = 0$	$a = 1, b = 1, c = 8$
$2y = 3x - 5$	linear	$3x - 2y - 5 = 0$	$a = 3, b = -2, c = -5$
$x + 5 = 0$	linear	$x + 0 \cdot y + 5 = 0$	$a = 1, b = 0, c = 5$
$2y = 3$	linear	$0 \cdot x + 2y - 3 = 0$	$a = 0, b = 2, c = -3$
$y = x^2 + 3$	not linear (x to second power)		
$x - y^3 + 7 = 0$	not linear (y to third power)		
$xy = 5$	not linear (cannot be put in the form $ax + by + c = 0$)		

Suppose we are given a linear equation such as

$$2x - 3y = 6.$$

Knowing that the graph of a linear equation is a straight line and that only two points are needed to determine a straight line, our work graphing linear equations can be shortened considerably. Rather than making a table of values which includes many solutions, we need only two solutions. In most instances, the two

3 pairs that are easiest to determine are the **intercepts.** The point at which a line crosses the x-axis, where y is zero, is the **x-intercept.** The point at which a line crosses the y-axis, where x is zero, is the **y-intercept.** To find the intercepts, fill in the following table.

x	y
0	
	0

The x-intercept, which is a point on the x-axis, has y-coordinate 0 while the y-intercept, which is a point on the y-axis, has x-coordinate 0. We substitute 0 for x in $2x - 3y = 6$ and solve for y to find the y-intercept.

$$2(0) - 3y = 6$$
$$0 - 3y = 6$$
$$-3y = 6$$
$$y = -2$$

Find the x-intercept by substituting 0 for y in $2x - 3y = 6$ and solving.

$$2x - 3(0) = 6$$
$$2x = 6$$
$$x = 3$$

The completed table

x	y
0	−2
3	0

displays the y-intercept (0, −2) and the x-intercept (3, 0). Plot these two intercepts and draw the straight line through them for the graph of $2x - 3y = 6$ in Figure 4.21.

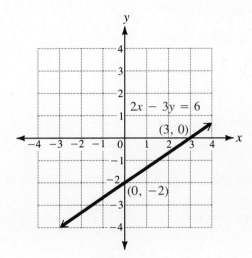

Figure 4.21

4 ▶▶ To graph a linear equation of the form $ax + by + c = 0$

1. If $a \neq 0$ and $b \neq 0$, find the x- and y-intercepts, plot them, and draw the line through them. If both intercepts are (0, 0), find and plot another point.
2. If $a = 0$, the equation becomes $by + c = 0$ or $y = -c/b$. The graph is a line through y-intercept $(0, -c/b)$ parallel to the x-axis.
3. If $b = 0$, the equation becomes $ax + c = 0$ or $x = -c/a$. The graph is a line through x-intercept $(-c/a, 0)$ parallel to the y-axis.

EXAMPLE 3

(A) Graph $3x + 4y = 12$.

First find the x- and y-intercepts by completing the following table.

x	y
0	
	0

When $x = 0$, $4y = 12$, so that $y = 3$. When $y = 0$, $3x = 12$, so that $x = 4$. The completed table,

x	y
0	3
4	0

Practice Exercise 3

(A) Graph $-6x + y = 6$.

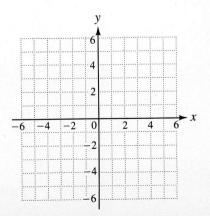

shows that the *x*-intercept is (4, 0) and the *y*-intercept is (0, 3). Plot (0, 3) and (4, 0) and connect the points to obtain the desired graph. See Figure 4.22.

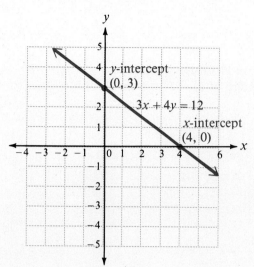

Figure 4.22

(B) Graph $x + 3y = 0$.

First find the *x*- and *y*-intercepts by completing the table.

x	y
0	
	0

This time both intercepts are (0, 0), so we must find another point on the line. When $x = 3$, $3 + 3y = 0$, so that $y = -1$. Thus, the table we use is given below.

x	y
0	0
3	-1

Plot (0, 0) and (3, -1) and connect the points to obtain the graph given in Figure 4.23. ◀◀

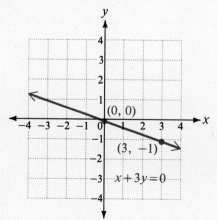

Figure 4.23

(B) Graph $4x - 3y = 0$.

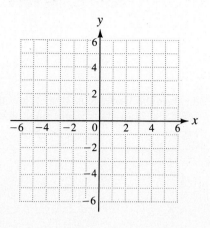

Answers: **(A)** Straight line through *x*-intercept (-1, 0) and *y*-intercept (0, 6). **(B)** Straight line through *x*- and *y*-intercept (0, 0) and the point (3, 4).

EXAMPLE 4

(A) Graph $x = 2$.

In this case, with no y term, the solutions will be $(2, y)$ for any number y. Thus, the graph is the line through the x-intercept $(2, 0)$ parallel to the y-axis, as in Figure 4.24.

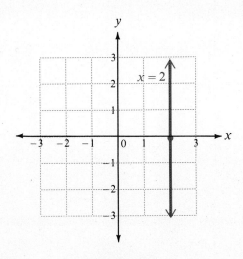

Figure 4.24

(B) Graph $2y = -2$.

The equation can be simplified to $y = -1$. In this case, with no x term, the solutions will be $(x, -1)$ for any number x. Thus, the graph is the line through the y-intercept $(0, -1)$ parallel to the x-axis, as shown in Figure 4.25. ◀◀

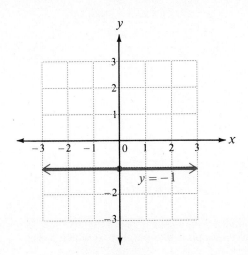

Figure 4.25

Practice Exercise 4

(A) Graph $2x = -10$.

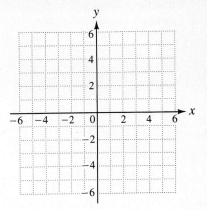

(B) Graph $y = -\frac{7}{2}$.

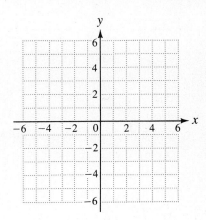

Answers: **(A)** Line through x-intercept $(-5, 0)$ parallel to the y-axis. **(B)** Line through y-intercept $\left(0, -\frac{7}{2}\right)$ parallel to the x-axis.

4.3 EXERCISES A

Each of the following is an equation in the two variables x and y. Make a table of values and graph each equation in a Cartesian coordinate system.

1. $y = x + 2$

2. $x - y = 1$

3. $y = 3x + 1$

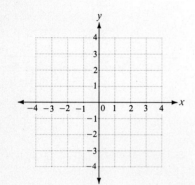
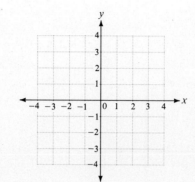
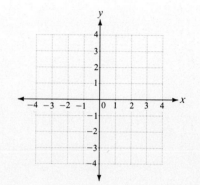

4. $y = \dfrac{1}{2}x - 1$

5. $y = 2$

6. $x = -1$

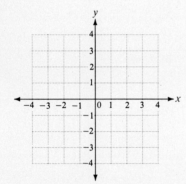
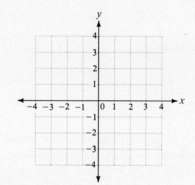
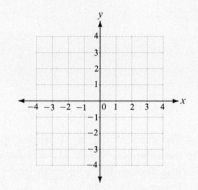

7. Which of the following are linear equations? Explain.

   (A) $2x + y = -4$
   (B) $x - y^3 = 5$
   (C) $2xy = 7$

   (D) $\dfrac{3}{x} + y = 7$
   (E) $x^2 + y^2 = 5$
   (F) $x = y - 8$

8. Why is a linear equation also called a first-degree equation?

9. What is an equation of the form $ax + by + c = 0$ ($a \neq 0$ or $b \neq 0$) called?

10. An equation of the type $x = c$ (c a constant) has as its graph a line parallel to which axis?

11. An equation of the type $y = c$ (c a constant) has as its graph a line parallel to which axis?

12. What is a point where a graph crosses the *x*-axis called?

13. What is a point where a graph crosses the *y*-axis called?

14. To graph a general linear equation, only two points are necessary. What are the best points to use?

Determine the intercepts and graph the following.

15. $y + 2x - 4 = 0$ **16.** $3y - 2x = 12$ **17.** $y + x = 0$

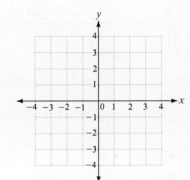 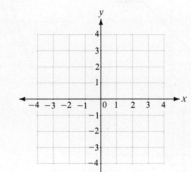 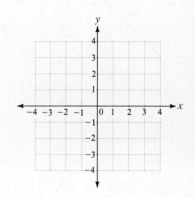

18. $y = 3$ **19.** $2x - 1 = 0$ **20.** $y + 2x = 0$

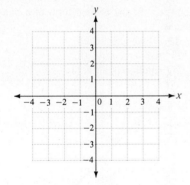 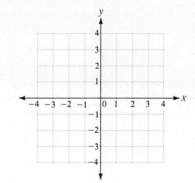 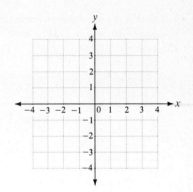

21. What are the intercepts of the line $x = 0$? What is the graph?

For Review

Using the given equation, complete each ordered pair so that it will be a solution to the equation.

22. $5x - 3y = 15$
    (A) (0,   )   **(B)** (, 0)
    (C) (, −1)   **(D)** (−1,   )

23. $y = 3$
    (A) (1,   )   **(B)** (−1,   )
    (C) (5,   )   **(D)** (−3,   )

24. The annual salary, y, of a salesman is given in terms of his total sales, x, by the equation

$$y = \$10{,}000 + (0.1)x.$$

Complete the ordered pairs (10000,), (20000,), and (50000,). How much does the salesman earn if he sells **(A)** 10,000 items; **(B)** 20,000 items? **(C)** 50,000 items?

25. Find the area of the rectangle with corners $(4, -3)$, $(4, 5)$, $(-3, -3)$, and $(-3, 5)$.

ANSWERS:

1. 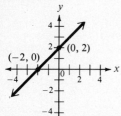   **2.** 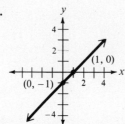   **3.** 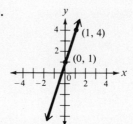   **4.**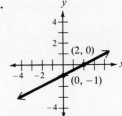

5. 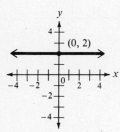   **6.**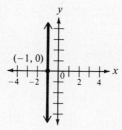

7. **(A)** and **(F)** are the only linear equations   **8.** the variables appear only to the first power   **9.** the general form of a linear equation   **10.** y-axis   **11.** x-axis   **12.** an x-intercept   **13.** a y-intercept   **14.** intercepts
15. x-intercept $(2, 0)$; y-intercept $(0, 4)$ **16.** y-intercept $(-6, 0)$; y-intercept $(0, 4)$ **17.** x-intercept $(0, 0)$; y-intercept $(0, 0)$; another point is $(1, -1)$

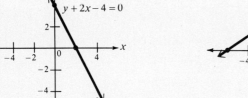 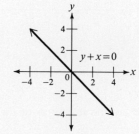

18. no x-intercept; y-intercept $(0, 3)$ **19.** x-intercept $(\frac{1}{2}, 0)$; no y-intercept   **20.** x-intercept $(0, 0)$; y-intercept $(0, 0)$; another point is $(1, -2)$

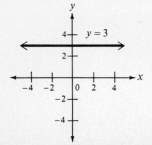 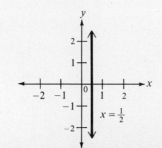 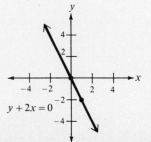

21. x-intercept (0, 0); y-intercept; every point on the y-axis; the graph is the y-axis   **22. (A)** (0, −5) **(B)** (3, 0) **(C)** $\left(\frac{12}{5}, -1\right)$ **(D)** $\left(-1, \frac{-20}{3}\right)$ **23. (A)** (1, 3) **(B)** (−1, 3) **(C)** (5, 3) **(D)** (−3, 3)   **24.** (10000, 11000), (20000, 12000), (50000, 15000)   **(A)** $11,000 **(B)** $12,000 **(C)** $15,000
25. 56 square units

4.3 EXERCISES B

Each of the following is an equation in the two variables x and y. Make a table of values and graph each equation in a Cartesian coordinate system.

1. $y = x + 1$

2. $y = 3x - 1$

3. $y = \frac{1}{2}x + 1$

4. $x + y = -1$

5. $x = 4$

6. $y = -2$

7. Which of the following are linear equations? Explain.

  (A) $3x - y = 8$

  (B) $x + y^2 = 7$

  (C) $3xy = -1$

  (D) $\frac{5}{y} + x = 3$

  (E) $x^2 + y^2 = 16$

  (F) $x = y - 3$

8. Why is a first-degree equation also called a linear equation?

9. Give the general form of a linear equation.

10. Give the general form of the equation of a line parallel to the x-axis.

11. Give the general form of the equation of a line parallel to the y-axis.

12. The x-intercept of the graph of a line is a point on which axis?

13. The y-intercept of the graph of a line is a point on which axis?

14. In general, what are the two best points to use when graphing a linear equation?

Determine the intercepts and graph the following.

15. $y - 2x + 4 = 0$

16. $3y + 2x = -12$

17. $2y + x = 0$

18. $y = -4$

19. $2x + 3 = 0$

20. $x - 3y = 0$

21. What are the intercepts of the line $y = 0$? What is the graph?

For Review

Using the given equation, complete each ordered pair so that it will be a solution to the equation.

22. $x - 3y = 6$

  (A) (0,   )   **(B)** (, 0)

  (C) (, −1)   **(D)** (−1,   )

23. $x = -1$

  (A) (, 1)   **(B)** (, −1)

  (C) (, 5)   **(D)** (, −3)

24. The number, y, of deer that can live in a forest preserve is related to the number of acres, x, in the preserve. If this relationship is given by the equation

$$y = 15x + 50,$$

complete the ordered pairs (20,), (100,), and (200,). How many deer can the preserve sustain if it contains **(A)** 20 acres? **(B)** 100 acres? **(C)** 200 acres?

25. Find the area of the triangle with vertices (1, 2), (−3, −2), and (5, −2).

4.3 EXERCISES C

Graph.

1. $0.05x - 0.02y = 0.10$

2. $\dfrac{2}{5}x + \dfrac{3}{10}y = \dfrac{6}{5}$

4.4 SLOPE OF A LINE AND EQUATIONS OF A LINE

▶▶ STUDENT GUIDEPOSTS

1 Definition of slope
2 Positive, negative, zero, and no slope
3 Slope of parallel lines
4 Slope-intercept form
5 Point-slope form

The graph of a given linear equation may be horizontal (parallel to the x-axis), vertical (parallel to the y-axis), may "slope" upward from lower left to upper right, or may "slope" downward from upper left to lower right. The *slope* of a line (or lack of slope) can be precisely defined.

Suppose that two points P and Q with coordinates (x_1, y_1) and (x_2, y_2), respectively, lie on the same straight line. (See Figure 4.26.) The difference $y_2 - y_1$ is the change in y-coordinates, while $x_2 - x_1$ is the change in x-coordinates of the two points.

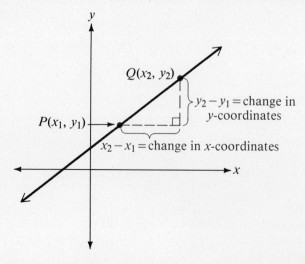

Figure 4.26

4.4 SLOPE OF A LINE AND EQUATIONS OF A LINE

The **slope** of the line (often denoted by the letter m) is given by

$$m = \frac{y_2 - y_1}{x_2 - x_1} = \frac{\text{change in } y\text{-coordinates}}{\text{change in } x\text{-coordinates}}.$$

EXAMPLE 1

Find the slope of the line passing through the given points.

(A) (4, 3) and (1, 2).

See the graph in Figure 4.27. Suppose we identify point $P(x_1, y_1)$ with (1, 2) and point $Q(x_2, y_2)$ with (4, 3). The slope will then be given by

$$m = \frac{y_2 - y_1}{x_2 - x_1} = \frac{3 - 2}{4 - 1} = \frac{1}{3}.$$

What happens if we identify $P(x_1, y_1)$ with (4, 3) and $Q(x_2, y_2)$ with (1, 2)? In this case we have

$$m = \frac{y_2 - y_1}{x_2 - x_1} = \frac{2 - 3}{1 - 4} = \frac{-1}{-3} = \frac{1}{3}.$$

Thus, we see that *the slope is the same regardless of how the two points are identified.*

Practice Exercise 1

Find the slope of the line passing through the given points.

(A) (5, 2) and (3, 7)

(B) (6, −3) and (−1, 8)

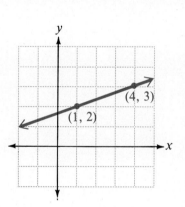

Figure 4.27

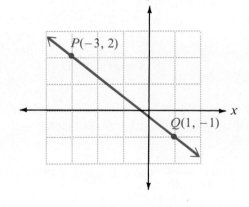

Figure 4.28

(B) (−3, 2) and (1, −1).

Let us identify $P(x_1, y_1)$ with (−3, 2) and $Q(x_2, y_2)$ with (1, −1) as in Figure 4.28. The slope is given by

$$m = \frac{y_2 - y_1}{x_2 - x_1} = \frac{(-1) - (2)}{(1) - (-3)} \quad \text{Watch signs}$$

$$= \frac{-3}{1 + 3} = -\frac{3}{4}.$$

Answers: **(A)** $-\dfrac{5}{2}$ **(B)** $-\dfrac{11}{7}$

▶▶ **CAUTION** ◀◀ It is important to make sure that the coordinates are subtracted in the same order. That is, *do not compute*

$$\frac{y_2 - y_1}{x_1 - x_2}. \quad \text{This is wrong} \blacktriangleleft$$

188 GRAPHING

EXAMPLE 2

Find the slope of the line passing through the given points

(A) (3, 2) and (−1, 2).

Identifying $P(x_1, y_1)$ with (3, 2) and $Q(x_2, y_2)$ with (−1, 2) in Figure 4.29 we obtain

$$m = \frac{y_2 - y_1}{x_2 - x_1} = \frac{2 - 2}{-1 - 3} = \frac{0}{-4} = 0.$$

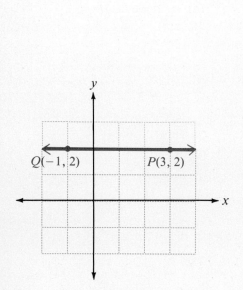

Figure 4.29

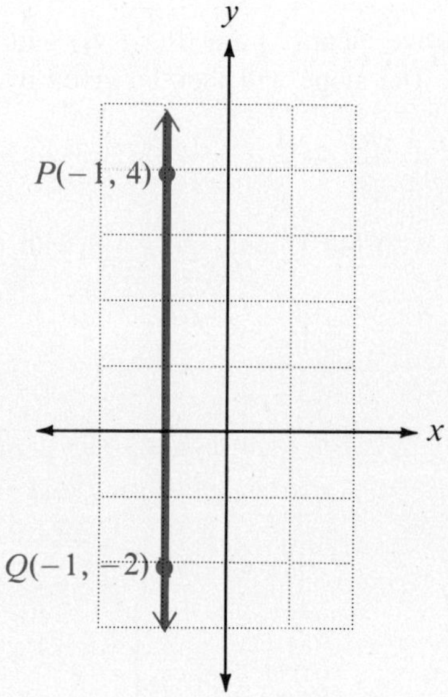

Figure 4.30

(B) (−1, 4) and (−1, −2).

Identifying $P(x_1, y_1)$ with (−1, 4) and $Q(x_2, y_2)$ with (−1, −2) in Figure 4.30, we obtain

$$m = \frac{y_2 - y_1}{x_2 - x_1} = \frac{-2 - 4}{-1 - (-1)} = \frac{-6}{-1 + 1} = \frac{-6}{0}, \text{ which is undefined.}$$

In this case we say that the line has **no slope**. ◀◀

Practice Exercise 2

Find the slope of the line passing through the given points.

(A) (−3, −4) and (5, −4)

(B) (3, −3) and (3, 0)

Answers: **(A)** 0   **(B)** no slope

In view of the above examples, the next rule seems reasonable.

2 ▶▶ The slope of a line

1. A line that "slopes" from lower left to upper right has a **positive slope**.
2. A line that "slopes" from upper left to lower right has a **negative slope**.
3. A horizontal line (parallel to the *x*-axis) has **zero slope**.
4. A vertical line (parallel to the *y*-axis) has **no slope**.

The graphs in Figure 4.31 show the four possibilities listed in the rule.

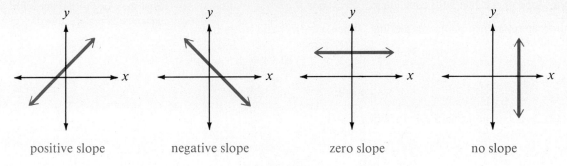

positive slope   negative slope   zero slope   no slope

Figure 4.31

3. The notion of the slope of a line can be used to determine when two lines are **parallel** (never intersect).

> Given two distinct lines with slopes m_1 and m_2, the lines are parallel if $m_1 = m_2$. (Equal slopes determine parallel lines.).

EXAMPLE 3

Verify that the line l_1 between (1, 8) and (−2, −1) and the line l_2 between (2, 4) and (−1, −5) in Figure 4.32 are parallel.

The slope of l_1 is $m_1 = \dfrac{8-(-1)}{1-(-2)} = \dfrac{8+1}{1+2} = \dfrac{9}{3} = 3.$

The slope of l_2 is $m_2 = \dfrac{4-(-5)}{2-(-1)} = \dfrac{4+5}{2+1} = \dfrac{9}{3} = 3.$

Since $m_1 = m_2$, the lines are parallel. ◀◀

Practice Exercise 3

Is the line l_1 between (4, −6) and (−3, −1) parallel to the line l_2 between (−2, 5) and (7, −1)?

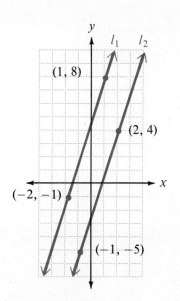

Figure 4.32

Answer: $m_1 = -\frac{5}{7}$, $m_2 = -\frac{2}{3}$, no

GRAPHING

EXAMPLE 4

Find the slope of the line with equation $x - 3y + 6 = 0$.

We must determine two points on the line. When $x = 0$,

$$0 - 3y + 6 = 0$$
$$-3y = -6$$
$$y = 2.$$

Thus, $(0, 2)$ is one point on the line. When $x = 3$,

$$3 - 3y + 6 = 0$$
$$-3y + 9 = 0$$
$$-3y = -9$$
$$y = 3.$$

Thus $(3, 3)$ is another point on the line. The slope of the line is

$$m = \frac{2 - 3}{0 - 3} = \frac{-1}{-3} = \frac{1}{3}.$$ ◀◀

Practice Exercise 4

Find the slope of the line with equation $3x + 4y - 8 = 0$.

Answer: $-\frac{3}{4}$

4 ▶▶ If the linear equation in Example 4 is solved for the variable y, we obtain a special form of that linear equation called the *slope-intercept form*:

$$x - 3y + 6 = 0$$
$$-3y = -x - 6$$
$$y = \frac{1}{3}x + 2. \quad \text{Divide through by } -3$$

Notice that the coefficient of x, $\frac{1}{3}$, is the slope of the line, while the constant term, 2, is the y-coordinate of the y-intercept of the line. This is true in general.

> ### Slope-intercept form of the equation of a line
> If the equation of a line is solved for y, we obtain the **slope-intercept form**
> $$y = mx + b,$$
> where m is the slope of the line and $(0, b)$ is the y-intercept.

EXAMPLE 5

What is the slope and y-intercept of the line with equation $4x + 2y + 1 = 0$?

First solve for y to obtain the slope-intercept form.

$$4x + 2y + 1 = 0$$
$$2y = -4x - 1$$
$$y = -2x - \frac{1}{2} \quad \text{Divide by 2}$$

$m = -2 =$ slope $\quad \left(0, -\frac{1}{2}\right) = y\text{-intercept} \quad \left[b = -\frac{1}{2}\right]$ ◀◀

Practice Exercise 5

What is the slope and y-intercept of the line with equation $3x - 6y + 10 = 0$?

Answer: slope is $\frac{1}{2}$, y-intercept is $\left(0, \frac{5}{3}\right)$

EXAMPLE 6

Find the general form of the equation of the line with slope -2 and y-intercept $(0, 5)$.

We first find the slope-intercept form of the equation by substituting -2 for m and 5 for b in

$$y = mx + b.$$
$$y = -2x + 5$$

By writing all terms on the left side of this equation we obtain the general form

$$2x + y - 5 = 0. \blacktriangleleft$$

Practice Exercise 6

Find the general form of the equation of the line with slope $\frac{2}{3}$ and y-intercept $(0, -3)$.

Answer: $2x - 3y - 9 = 0$

If we know that the slope of a line is m and that (x_1, y_1) is a point on the line, we can obtain the equation of the line. Suppose (x, y) is an arbitrary point on the desired line. Then using the formula for the slope of a line with (x, y) as (x_2, y_2), we have

$$m = \frac{y - y_1}{x - x_1}.$$

Multiplying both sides of this equation by $x - x_1$,

$$m(x - x_1) = y - y_1,$$

gives us another form of the equation of a line.

> **Point-slope form of the equation of a line**
>
> To find the equation of a line that has slope m and passes through the point (x_1, y_1), substitute these values into the **point-slope form**:
>
> $$y - y_1 = m(x - x_1).$$

EXAMPLE 7

Find the general form of the equation of the line with slope $\frac{1}{3}$ passing through the point $(-2, 1)$.

First find the point-slope form by substituting $\frac{1}{3}$ for m and $(-2, 1)$ for (x_1, y_1) in

$$y - y_1 = m(x - x_1).$$
$$y - 1 = \frac{1}{3}(x - (-2)) \quad \text{Watch the signs}$$
$$3(y - 1) = (x - (-2)) \quad \text{Multiply both sides by 3 to clear the fraction}$$
$$3y - 3 = x + 2$$

Writing all terms on one side results in the general form

$$x - 3y + 5 = 0. \blacktriangleleft$$

Practice Exercise 7

Find the general form of the equation of the line with slope -5 passing through the point $(2, -7)$.

Answer: $5x + y - 3 = 0$

To find the equation of a line passing through two points, we use both the slope formula and the point-slope form of the equation of a line. This is shown in the next example.

EXAMPLE 8

Find the general form of the equation of the line passing through two points $(3, -2)$ and $(-1, 5)$.

First find the slope of the line using $(3, -2) = (x_1, y_1)$ and $(-1, 5) = (x_2, y_2)$:

$$m = \frac{y_2 - y_1}{x_2 - x_1} = \frac{5 - (-2)}{-1 - 3} = \frac{7}{-4} = -\frac{7}{4}.$$

Next substitute $-\frac{7}{4}$ for m and $(3, -2)$ for (x_1, y_1) in the point-slope form:

$$y - y_1 = m(x - x_1).$$

$$y - (-2) = -\frac{7}{4}(x - 3)$$

$$4(y - (-2)) = -7(x - 3) \quad \text{Multiply both sides by 4}$$

$$4y + 8 = -7x + 21$$

$$7x + 4y - 13 = 0$$

This is the general form of the equation of the desired line. ◀

Practice Exercise 8

Find the general form of the equation of the line passing through the points $(1, 7)$ and $(-2, 3)$.

Answer: $4x - 3y + 17 = 0$

Remember that the names of the forms of the equations identify the values needed for each equation.

The *slope-intercept form* requires knowing the *slope* and *y-intercept*.

The *point-slope form* requires knowing a *point* and the *slope*.

A student should memorize these forms and be able to identify which should be used in a given problem.

4.4 EXERCISES A

Find the slope of the line passing through the given pair of points.

1. $(3, 5)$ and $(1, 3)$
2. $(-2, 3)$ and $(-4, 1)$
3. $(-7, 1)$ and $(3, 9)$
4. $(2, 7)$ and $(2, -3)$
5. $(3, 4)$ and $(-1, 4)$
6. $(0, 2)$ and $(5, 0)$

Answer Exercises 7–10 with one of the following phrases: **(A)** *positive slope* **(B)** *negative slope* **(C)** *zero slope* **(D)** *no slope.*

7. A line parallel to the *y*-axis has _____.

8. A line that "slopes" from lower left to upper right has _____.

9. A line perpendicular to the *y*-axis has _____.

10. The *x*-axis has _____.

11. Verify that the line l_1 between $(-2, -1)$ and $(1, 5)$ and the line l_2 between $(4, 3)$ and $(-1, -7)$ are parallel.

Determine the slope of the given line by first finding two points on the line then using the definition of slope.

12. $4x - y + 7 = 0$

13. $5x + 1 = 0$

14. $2 - y = 0$

15. $3x + 5y = 0$

16. $x + y = 5$

17. $2x - 3y + 1 = 0$

Write each equation in slope-intercept form and give the slope and y-intercept.

18. $5x + y - 12 = 0$

19. $2x - 5y + 10 = 0$

20. $4y + 7 = 0$

21. $5x + 1 = 0$

22. $x + y = 5$

23. $2x - 3y + 1 = 0$

Find the general form of the equation of the line with the given slope and y-intercept.

24. $m = 2$, $(0, 4)$

25. $m = 4$, $(0, -3)$

26. $m = -\dfrac{1}{2}$, $(0, 6)$

Find the general form of the equation of the line with the given slope passing through the given point.

27. $m = 3$, $(2, -4)$

28. $m = -2$, $(1, 5)$

29. $m = -\dfrac{1}{2}$, $(3, -2)$

Find the general form of the equation of the line passing through the given points.

30. $(2, 5)$ and $(4, 7)$

31. $(-1, 3)$ and $(2, 4)$

32. $(3, -2)$ and $(-1, -1)$

Some business problems can be described by a linear equation. For example, the cost y of producing a number of items x is given by y = mx + b, where b is the overhead cost (the cost when no items are produced) and m is the variable cost (the cost of producing a single item). Use this information to find the cost equation in 33–34.

33. Overhead cost: $25
Variable cost: $10

34. Overhead cost: $300
Variable cost: $10.50.

35. Do the three points $(2, 3)$, $(0, 2)$, and $(-2, 1)$ all lie on the same straight line? Explain using slopes.

For Review

Determine the intercepts and graph the following.

36. $x - 6y = 2$

37. $5x + y = 0$

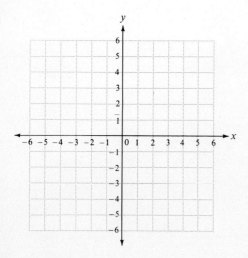
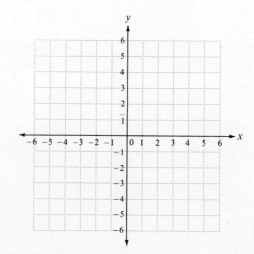

38. $x = -5$

39. $y = 6$

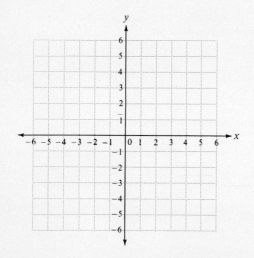
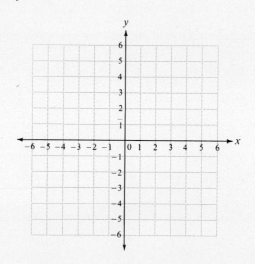

40. Is $\dfrac{3}{x} + \dfrac{2}{y} = 5$ a linear equation? Explain.

ANSWERS: **1.** 1 **2.** 1 **3.** $\dfrac{4}{5}$ **4.** no slope **5.** 0 **6.** $-\dfrac{2}{5}$ **7.** no slope **8.** positive slope **9.** zero slope **10.** zero slope **11.** They are parallel since both have slope 2. **12.** 4 **13.** no slope **14.** 0 (slope is zero) **15.** $-\dfrac{3}{5}$ **16.** -1 **17.** $\dfrac{2}{3}$ **18.** $y = -5x + 12$; slope is -5, y-intercept is $(0, 12)$ **19.** $y = \dfrac{2}{5}x + 2$; slope is $\dfrac{2}{5}$, y-intercept is $(0, 2)$ **20.** $y = -\dfrac{7}{4} = 0 \cdot x - \dfrac{7}{4}$; slope is 0, y-intercept is $\left(0, -\dfrac{7}{4}\right)$ **21.** The equation cannot be solved for y, hence it has no slope-intercept form. Also, it has no slope and no y-intercept. **22.** $y = -x + 5$; slope is -1, y-intercept is $(0, 5)$ **23.** $y = \dfrac{2}{3}x + \dfrac{1}{3}$; slope is $\dfrac{2}{3}$, y-intercept is $\left(0, \dfrac{1}{3}\right)$ **24.** $2x - y + 4 = 0$ **25.** $4x - y - 3 = 0$ **26.** $x + 2y - 12 = 0$ **27.** $3x - y - 10 = 0$ **28.** $2x + y - 7 = 0$ **29.** $x + 2y + 1 = 0$ **30.** $x - y + 3 = 0$ **31.** $x - 3y + 10 = 0$

32. $x + 4y + 5 = 0$ 33. $y = 10x + 25$ 34. $y = 10.50x + 300$ 35. Yes. The slope of the line through (2, 3) and (0, 2) is $\frac{1}{2}$, the same as the slope of the line through (0, 2) and (−2, 1).

36. x-intercept (2, 0); y-intercept $\left(0, -\frac{1}{3}\right)$

37. x-intercept (0, 0); y-intercept (0, 0); another point is (1, −5)

38. x-intercept (−5, 0); no y-intercept

39. no x-intercept; y-intercept (0, 6)

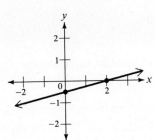
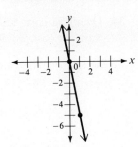
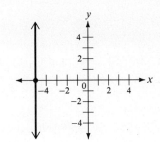
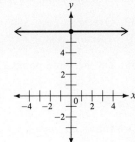

40. No. It cannot be written in the form $ax + by + c = 0$.

4.4 EXERCISES B

Find the slope of the line passing through the given pairs of points.

1. (4, 8) and (10, 2)
2. (−2, 1) and (−6, 5)
3. (6, 1) and (6, −5)
4. (4, −1) and (−2, −1)
5. (−3, −1) and (−2, −3)
6. (0, 1) and (−2, 0)

Answer Exercises 7–10 with one of the following phrases: **(A)** *positive slope* **(B)** *negative slope* **(C)** *zero slope* **(D)** *no slope.*

7. A line parallel to the x-axis has __?__.
8. A line that "slopes" from upper left to lower right has __?__.
9. A line perpendicular to the x-axis has __?__.
10. The y-axis has __?__.
11. Verify that the line l_1 between (4, −2) and (1, 1) and the line l_2 between (3, 3) and (7, −1) are parallel.

Determine the slope of the given line by first finding two points on the line then using the definition of slope.

12. $x + 4y − 9 = 0$
13. $3y − 7 = 0$
14. $4 + x = 0$
15. $2x − 7y = 0$
16. $y − x = 3$
17. $3x − 5y + 2 = 0$

Write each equation in slope-intercept form and give the slope and y-intercept.

18. $7x − y + 13 = 0$
19. $5x − 9y + 9 = 0$
20. $2x − 1 = 0$
21. $4y + 12 = 0$
22. $y − x = 3$
23. $3x − 5y + 2 = 0$

Find the general form of the equation of the line with the given slope and y-intercept.

24. $m = 3$, (0, 1)
25. $m = 5$, (0, −2)
26. $m = -\frac{1}{4}$, (0, 2)

Find the general form of the equation of the line with the given slope passing through the given point.

27. $m = 5$, (−3, 1)
28. $m = −4$, (2, 6)
29. $m = -\frac{1}{5}$, (−1, 4)

Find the general form of the equation of the line passing through the given points.

30. (2, 6) and (5, 9)   **31.** (−3, 2) and (2, 5)   **32.** (−4, 6) and (−2, −2)

Some business problems can be described by a linear equation. For example, the cost y of producing a number of items x is given by $y = mx + b$, where b is the overhead cost (the cost when no items are produced) and m is the variable cost (the cost of producing a single item). Use this information to find the cost equation in 33–34.

33. Overhead cost: $50
Variable cost: $30

34. Overhead cost: $500
Variable cost: $8.50

35. Do the three points (1, 2), (−1, −1), and (3, 6) all lie on the same straight line? Explain using slopes.

For Review

Determine the intercepts and graph the following.

36. $6x + y = 3$

37. $3x + y = 0$

38. $y = 5$

39. $x = -3$

40. Is $3x^2 + 2y^2 = 5$ a linear equation? Explain.

4.4 EXERCISES C

1. Find the general form of the equation of a line through the point (−1, 5) that is parallel to the line with equation $3x + 2y - 4 = 0$.
[Answer: $3x + 2y - 7 = 0$]

2. Find the slope of the line with equation $ax + by + c = 0$. [*Hint: a, b,* and *c* are constants.]

4.5 GRAPHING LINEAR INEQUALITIES

STUDENT GUIDEPOSTS

1 Linear inequalities in two variables
2 Solution to inequalities in two variables
3 Graph of solutions
4 Test point
5 Method of graphing

In Section 4.1 we graphed linear inequalities in one variable, such as

$$3x + 2 > 5 \quad \text{and} \quad x - 1 \le 3(x - 5) + 1,$$

on a number line. For example, if we solve

$3x + 2 > 5$
$3x > 3$ Subtract 2
$x > 1,$ Divide by 3

the solution $x > 1$ is graphed in Figure 4.33.

Figure 4.33

1 ▶◀ We now consider inequalities in two variables in which the variables are raised only to the first power. Such inequalities are called **linear inequalities in two variables.** For example,

$$2x + y < -1$$

2 ▶◀ is a linear inequality in the two variables x and y. A **solution** to such an inequality is an ordered pair of numbers which when substituted for x and y yields a true statement. Thus, $(-1, 0)$ is a solution to $2x + y < -1$ since by replacing x with -1 and y with 0 we obtain

$$2x + y < -1$$
$$2(-1) + 0 < -1$$
$$-2 < -1, \quad \text{which is true.}$$

On the other hand, $(4, -3)$ is not a solution since

$$2x + y < -1$$
$$2(4) + (-3) < -1$$
$$8 - 3 < -1$$
$$5 < -1 \quad \text{is false.}$$

EXAMPLE 1

Determine whether $(-1, -3)$, $(1, 1)$ and $(-2, 0)$, are solutions to $3x - 2y \geq 1$.

$$3(-1) - 2(-3) \geq 1$$
$$-3 + 6 \geq 1$$
$$3 \geq 1 \quad \text{True}$$

$(-1, -3)$ is a solution.

$$3(1) - 2(1) \geq 1$$
$$3 - 2 \geq 1$$
$$1 \geq 1 \quad \text{True}$$

$(1, 1)$ is a solution.

$$3(-2) - 2(0) \geq 1$$
$$-6 - 0 \geq 1$$
$$-6 \geq 1 \quad \text{False}$$

$(-2, 0)$ is not a solution. ◀

Practice Exercise 1

Determine whether $(3, 0)$, $(2, 3)$, and $(1, 1)$ are solutions to $-5x + 4y < -1$.

Answer: $(3, 0)$ is a solution, $(2, 3)$ is not a solution, $(1, 1)$ is not a solution.

3 ▶◀ The set of all solutions to a linear inequality in two variables can be displayed in a Cartesian coordinate system. Consider

$$2x + y > -3.$$

If we replace the inequality symbol with an equal sign, the resulting equation is

$$2x + y = -3.$$

To graph this equation we first plot the intercepts $(-\frac{3}{2}, 0)$ and $(0, -3)$ and draw the line as in Figure 4.34 on the following page. Notice that the graph of the line divides the plane into a region above the line, the line itself, and a region below the line.

198 GRAPHING

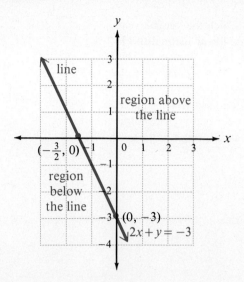

Figure 4.34

The graph of an inequality such as $2x + y > -3$ consists of all the points on one side of the boundary line $2x + y = -3$. The graph of $2x + y < -3$ is all the points on the other side of the line. To determine the correct side of the line to graph for an inequality such as $2x + y > -3$, we select any point not on the line as a **test point.** If $(0, 0)$ is used in this case the arithmetic is easy:

$$2x + y > -3$$
$$2(0) + 0 > -3$$
$$0 > -3. \quad \text{This is true}$$

Since this inequality is true, we graph the points on the side of the line containing the test point $(0, 0)$. If the inequality to be graphed had been $2x + y < -3$, a false inequality would have resulted using $(0, 0)$ as a test point:

$$2x + y < -3$$
$$2(0) + 0 < -3$$
$$0 < -3. \quad \text{This is false}$$

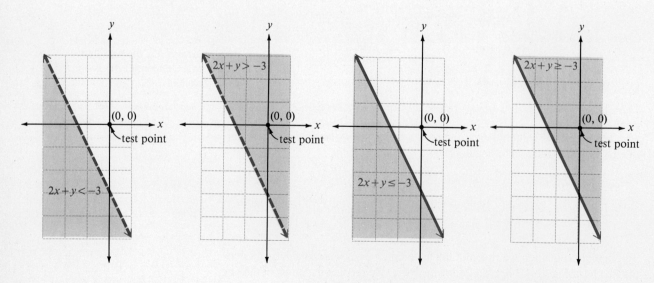

Figure 4.35

In the case when a false inequality is obtained, shade the region that does *not* contain the test point. The graphs of both $2x + y < -3$ and $2x + y > -3$ are shown in Figure 4.35 by shading the region that satisfies the inequality. The figure also has graphs of $2x + y \leq -3$ and $2x + y \geq -3$. We have used a dashed line for the boundary when the inequality is $<$ or $>$ to show that the boundary is *not* part of the graph. For the inequalities $\leq$ and $\geq$ a solid line is used to indicate that the boundary *is* part of the graph.

EXAMPLE 2

(A) Graph $x + 3y > 6$.

Graph the line $x + 3y = 6$ using the intercepts $(0, 2)$ and $(6, 0)$. Since the inequality is $>$, use a dashed line. Select the test point $(0, 0)$. (It is not on the line and the arithmetic is easy with $(0, 0)$.)

$$x + 3y > 6$$
$$0 + 3(0) > 6$$
$$0 > 6 \quad \text{This is false}$$

The inequality is false. Shade the region in Figure 4.36 that does not contain $(0, 0)$.

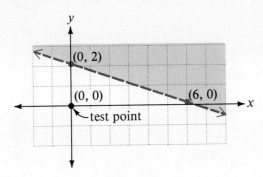

Figure 4.36

(B) Graph $2x - 3y \leq 0$.

Since both intercepts of $2x - 3y = 0$ are $(0, 0)$, we must find another point on the boundary line. If $x = 3$ then $y = 2$, and thus $(3, 2)$ is a second point. Draw a solid line through $(0, 0)$ and $(3, 2)$ as in Figure 4.37.

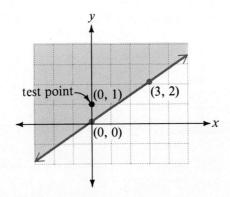

Figure 4.37

This time $(0, 0)$ cannot be used as a test point. Choosing $(0, 1)$ as a test point we obtain a true inequality:

Practice Exercise 2

(A) Graph $x + 3y \leq 6$.

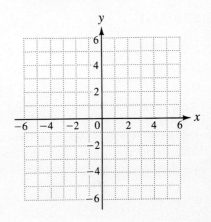

(B) Graph $2x - 3y > 0$.

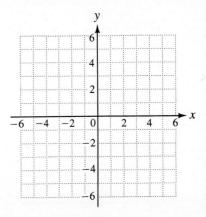

$$2x - 3y \leq 0$$
$$2(0) - 3(1) \leq 0$$
$$-3 \leq 0.$$

Thus, we shade the region containing the test point (0, 1).

Answers: **(A)** all points on and below the line in Figure 4.36 **(B)** all points below the line in Figure 4.37

Some linear inequalities in two variables have either the *x*-term or the *y*-term missing. The solution techniques are the same for these equations, but sometimes our work can be simplified by first solving the inequality for the remaining variable.

EXAMPLE 3

(A) Graph $2x - 4 \geq 0$.

$$2x - 4 \geq 0$$
$$2x \geq 4$$
$$x \geq 2$$

Replacing $\geq$ with $=$ we recognize the graph of $x = 2$ as a vertical line with *x*-intercept (2, 0). Since the inequality is $\geq$, we draw the boundary $x = 2$ as a solid line and use (0, 0) as a test point.

$$x \geq 2$$
$$0 \geq 2 \quad \text{This is false}$$

Since the inequality is false, shade the region not containing (0, 0) in Figure 4.38.

(B) Graph $3y + 3 < 0$.

$$3y + 3 < 0$$
$$3y < -3$$
$$y < -1$$

Replacing $<$ with $=$ we recognize the graph of $y = -1$ as a horizontal line with *y*-intercept (0, −1). Since the inequality is $<$, we draw the boundary $y = -1$ as a dashed line and use (0, 0) as a test point.

$$y < -1$$
$$0 < -1 \quad \text{This is false}$$

Since the inequality is false, shade the region not containing (0, 0) in Figure 4.39.

Practice Exercise 3

(A) Graph $2x - 4 < 0$.

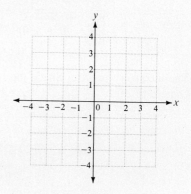

(B) Graph $3y + 3 \geq 0$.

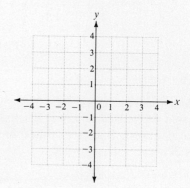

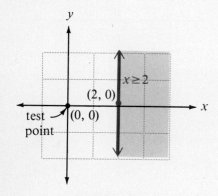

Figure 4.38

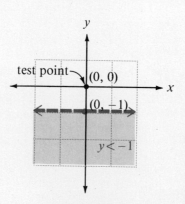

Figure 4.39

Answers: **(A)** all points to the left of the line in Figure 4.38 **(B)** all points on and above the line in Figure 4.39

We now summarize the techniques that we have learned.

> **To graph a linear inequality in two variables**
> 1. Graph the boundary line using a dashed line if the inequality is $<$ or $>$ and a solid line if it is $\leq$ or $\geq$.
> 2. Choose a test point that is not on the boundary line and substitute it into the inequality.
> 3. Shade the region that includes the test point if a true inequality is obtained, and shade the region that does not contain the test point if a false inequality results.

4.5 EXERCISES A

Graph the linear inequalities in two variables.

1. $x + y > 3$

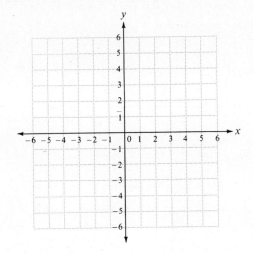

2. $x + y \leq 3$

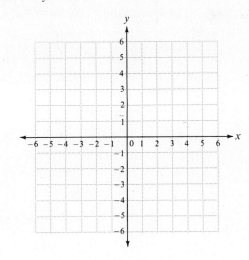

3. $x - y \geq -2$

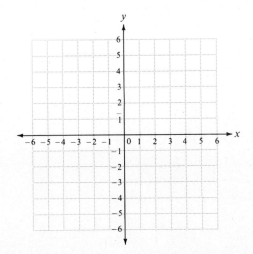

4. $x - y < -2$

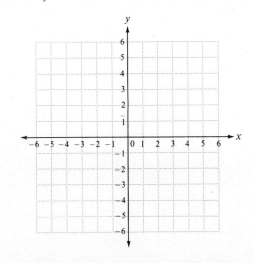

5. $x + 4y < 4$

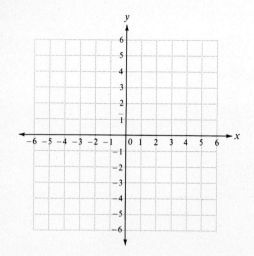

6. $x + 4y \geq 4$

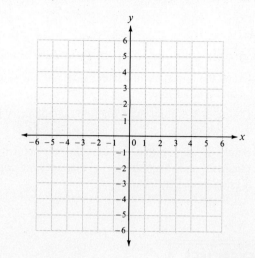

7. $4x + 12 > 0$

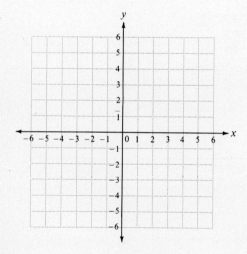

8. $4x + 12 \leq 0$

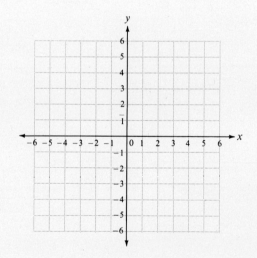

9. $2y - 8 \leq 0$

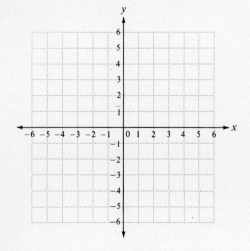

10. $2y - 8 > 0$

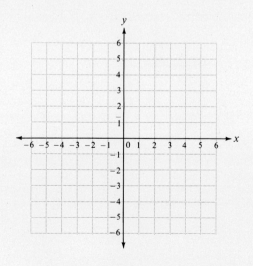

11. $x + y < 0$

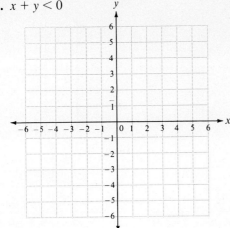

12. $x + y \geq 0$

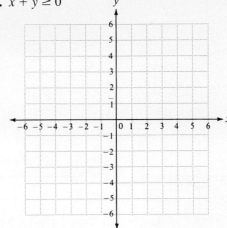

For Review

Find the slope of the line passing through the given pair of points.

13. $(-3, 7)$ and $(10, -1)$

14. $(6, 3)$ and $(6, -5)$

15. $(2, -5)$ and $(-8, -5)$

Write each equation in slope-intercept form and give the slope and y-intercept.

16. $12x - 9y + 36 = 0$

17. $5y - 15 = 0$

Find the general form of the equation of the line.

18. $m = \dfrac{1}{5}$, through $(-1, 4)$

19. through $(2, 6)$ and $(-4, -3)$

20. Verify that the line l_1 between $(-9, 2)$ and $(0, 5)$ and the line l_2 between $(1, -4)$ and $(19, 2)$ are parallel.

ANSWERS:

1.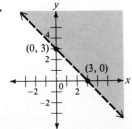
2.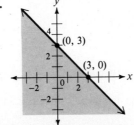
3.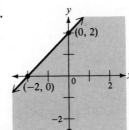
4.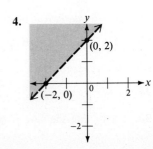

5. 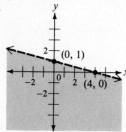 6. 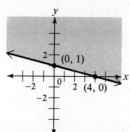 7. 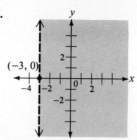 8.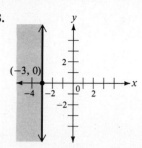

9. 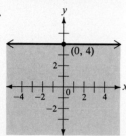 10. 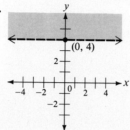 11. 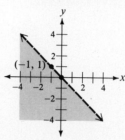 12.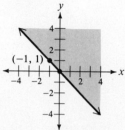

13. $-\dfrac{8}{13}$ 14. no slope 15. 0 (zero slope) 16. $y = \dfrac{4}{3}x + 4$; slope is $\dfrac{4}{3}$, y-intercept is $(0, 4)$ 17. $y = 3 = 0 \cdot x + 3$; slope is 0, y-intercept is $(0, 3)$ 18. $x - 5y + 21 = 0$ 19. $3x - 2y + 6 = 0$ 20. The lines are parallel since both have slope $\dfrac{1}{3}$.

4.5 EXERCISES B

Graph the linear inequalities in two variables.

1. $x + y > 2$
2. $x + y \leq 2$
3. $x - y \geq -1$
4. $x - y < -1$
5. $3x + 4y < 12$
6. $3x + 4y \geq 12$
7. $4x + 8 > 0$
8. $4x + 8 \leq 0$
9. $2y - 6 \leq 0$
10. $2y - 6 > 0$
11. $2x + y < 0$
12. $2x + y \geq 0$

For Review

Find the slope of the line passing through the given pair of points.

13. $(-5, 5)$ and $(3, -2)$
14. $(-3, 10)$ and $(2, 10)$
15. $(-9, 0)$ and $(-9, -7)$

Write each equation in slope-intercept form and give the slope and y-intercept.

16. $14x - 7y + 21 = 0$
17. $2y + 1 = 0$

Find the general form of the equation of the line.

18. $m = -3$, through $(-2, 5)$
19. through $(8, -1)$ and $(-4, 2)$

20. Verify that the line l_1 between $(3, 6)$ and $(5, -4)$ and the line l_2 between $(1, -2)$ and $(-1, 8)$ are parallel.

4.5 EXERCISES C

Determine whether the given point satisfies both of the inequalities.

1. $3x - y < 4$
   $y \geq -2$
2. $x + 4y \geq 5$
   $-2x + 3y < 0$

(A) (0, 2) (B) (2, 0)
(C) (−1, 5) (D) (−3, −4)
[Answer: (C) yes; (D) no]

(A) (0, −1) (B) (−1, 0)
(C) (5, 2) (D) (−3, 4)

CHAPTER 4 SUMMARY

Key Words and Phrases for Review

4.1 number line
origin
plotting points

4.2 rectangular (or Cartesian)
coordinate system
horizontal axis (x-axis)
vertical axis (y-axis)
ordered pair
x-coordinate
y-coordinate
quadrants
solution

4.3 linear (first-degree) equation in two variables
general form
x-intercept
y-intercept

4.4 slope
parallel
slope-intercept form
point-slope form

4.5 linear inequality in two variables
test point

Key Concepts

4.1 When graphing $x \geq c$ or $x \leq c$ on a number line, the point corresponding to c is part of the graph, which is indicated with a solid dot at c. When graphing $x > c$ or $x < c$, the point corresponding to c is *not* part of the graph, which is indicated with an ''open'' dot at c.

4.3 1. When graphing an equation in a Cartesian coordinate system, construct a table of values.

2. Two convenient points to use when graphing a linear equation are the intercepts (if they exist).

3. An equation of the form $x = c$ is parallel to the y-axis with x-intercept $(c, 0)$.

4. An equation of the form $y = c$ is parallel to the x-axis with y-intercept $(0, c)$.

4.4 1. The slope of the line passing through points (x_1, y_1) and (x_2, y_2) is given by

$$m = \frac{y_2 - y_1}{x_2 - x_1} = \frac{\text{change in } y\text{-coordinates}}{\text{change in } x\text{-coordinates}}.$$

2. Let m_1 and m_2 be the slopes of two distinct lines. If $m_1 = m_2$, the lines are parallel.

3. The slope-intercept form of the equation of a line is

$$y = mx + b,$$

where m is the slope and $(0, b)$ is the y-intercept.

4. The best way to determine the slope of a line quickly is to solve the equation for y (if possible) putting the equation into slope-intercept form. The coefficient of x is the slope.

5. The point-slope form of the equation of a line is

$$y - y_1 = m(x - x_1),$$

where m is the slope and (x_1, y_1) is any point on the line.

4.5 To graph a linear inequality in two variables, first graph the boundary line using a dashed line (for $<$ or $>$) or a solid line (for $\leq$ or $\geq$). Choose a test point not on the boundary line and substitute it into the inequality. Shade the region containing the test point if a true inequality results, and shade the region not containing the test point if a false inequality results.

206 GRAPHING

Review Exercises

4.1 *Graph the equation or inequality on a number line.*

1. $3x + 1 = 10$

2. $\dfrac{y}{\frac{1}{8}} = 16$

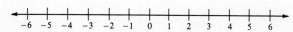
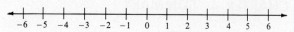

3. $4 - z = 8 - [1 - (z - 3)]$

4. $x > 3$

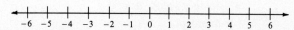
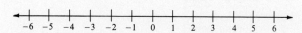

5. $-2a \leq -2$

6. $3(y - 1) \leq 2(y - 2)$

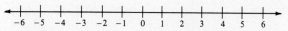
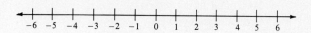

4.2 7. Plot the points associated with the pairs $A(-1, 3)$, $B(0, 0)$, $C(1, -2)$, $D(4, 2)$, and $E(-4, -3)$, and state which quadrant each is in.

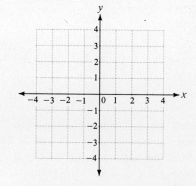

8. Complete the ordered pairs $(0, \ )$, $(\ , 0)$, and $(-2, \ )$ so that they will be solutions to the equation $2x + 3y = 6$.

9. The number, y, of bacteria in a culture is approximated using time x, in days, by the equation

$$y = 4000x + 5000.$$

Complete the ordered pair $(3, \ )$. After 3 days, what is the bacteria count?

4.3 10. Which of the following are linear equations?

(A) $2x + 3 = y$
(B) $xy = 5$
(C) $x + y^2 = 2$
(D) $x^3 + y = 3$
(E) $x = 4$
(F) $3y = -1$

Give the intercepts and graph in the Cartesian coordinate system.

11. $3x - 4y = 12$
x-intercept =
y-intercept =

12. $2x + 3 = 0$
x-intercept =
y-intercept =

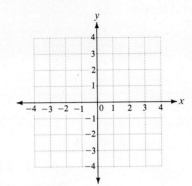
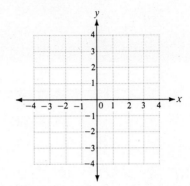

13. $x - 2y = 0$
x-intercept =
y-intercept =

14. $y - 3 = 0$
x-intercept =
y-intercept =

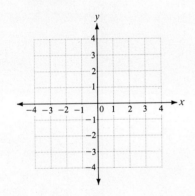
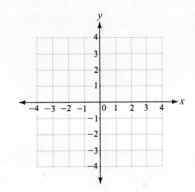

4.4 *Find the slope of the line through the given pair of points.*

15. $(-9, -2)$ and $(3, 6)$

16. $(0, 2)$ and $(-3, 0)$

17. $(-1, 8)$ and $(-1, 7)$

18. Write $8x + y - 3 = 0$ in slope-intercept form and give the slope and y-intercept.

19. What can be said about two distinct lines that both have slope $\frac{2}{3}$?

20. Find the general form of the equation of the line with slope $-\frac{2}{3}$ and y-intercept $(0, 4)$.

21. Find the general form of the equation of the line passing through $(-1, -2)$ and $(5, 3)$.

4.5 Graph the following in a Cartesian coordinate system.

22. $x - 2y > -2$

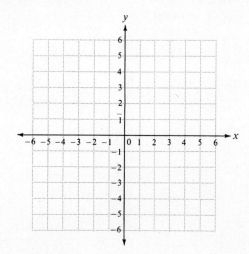

23. $2y + 4 \leq 0$

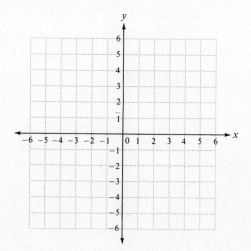

24. $3 - 3x < 0$

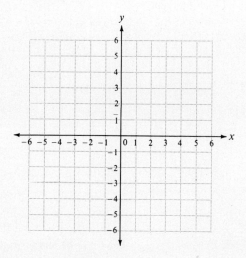

25. $3x + 2y \geq 6$

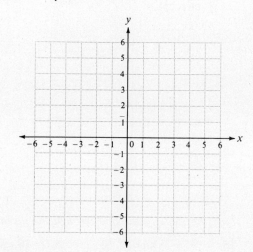

ANSWERS:

1.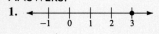
2.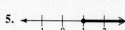
3.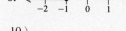
4.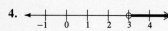
5.
6.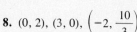

7. A is in II;
   B is origin;
   C is in IV;
   D is in I;
   E is in III

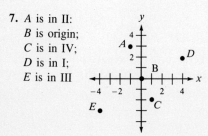

8. $(0, 2)$, $(3, 0)$, $\left(-2, \dfrac{10}{3}\right)$

9. $(3, 17000)$; $17,000$

10. **(A)**, **(E)**, and **(F)** are linear

11. x-intercept (4, 0); y-intercept (0, −3)

12. x-intercept $\left(-\frac{3}{2}, 0\right)$; no y-intercept

13. x-intercept (0, 0); y-intercept (0, 0); another point is (2, 1)

14. no x-intercept; y-intercept (0, 3)

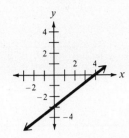 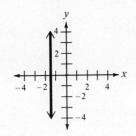 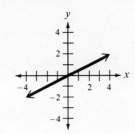 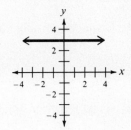

15. $\frac{2}{3}$ **16.** $\frac{2}{3}$ **17.** no slope   **18.** $y = -8x + 3$; slope is -8, y-intercept (0, 3)   **19.** They are parallel

20. $2x + 3y - 12 = 0$ **21.** $5x - 6y - 7 = 0$

22. 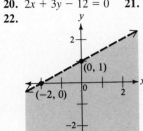   **23.** 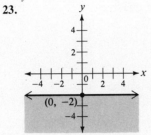   **24.** 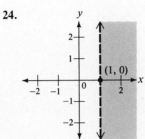   **25.**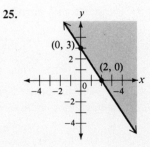

CHAPTER 4 TEST

1. Graph the equation $2x + 5 = 9$ on the number line.

    1.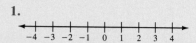

2. Graph the inequality $2(y + 1) \geq 1 - (y + 5)$ on the number line.

    2.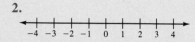

3. The point with coordinates $(2, -3)$ is located in which quadrant?

    3. _____

4. Complete the ordered pair $(-3,)$ so that it will be a solution to the equation $x + 5y = -1$.

    4. _____

5. The number, y, of items in a store during the week is approximated by using time x, in days, by the equation $y = -200x + 1000$. Complete the ordered pair $(3,)$ to determine the number of items in the store after three days?

    5. _____

6. True or False: $x - 2y^2 = 3$ is a linear equation.

    6. _____

7. Find the slope of the line through the points $(2, -3)$ and $(-4, -5)$.

    7. _____

8. Write $2x + 5y = 20$ in slope-intercept form and give the slope and the y-intercept.

    8. _____

9. Find the general form of the equation of the line with slope -6 and passing through $(4, -5)$.

    9. _____

Use the intercepts to graph each equation in the given Cartesian coordinate system.

10. $y = 2x + 4$

10.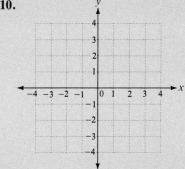

11. $3x + y = 0$

11.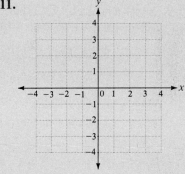

Graph each inequality in the given Cartesian coordinate system.

12. $2y + x - 3 > 0$

12.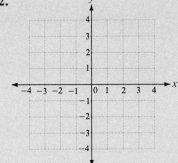

13. $y - 4 \leq 0$

13.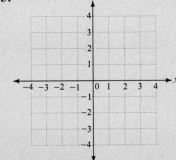

211

5

SYSTEMS OF LINEAR EQUATIONS

5.1 PARALLEL, COINCIDING, AND INTERSECTING LINES

STUDENT GUIDEPOSTS

1. System of equations
2. Solution to a system
3. Nature of the graphs of a system

In Chapter 4 we discovered that a solution to a linear equation in two variables such as

$$2x - y + 1 = 0$$

is an ordered pair of numbers that when substituted for x and y makes the equation true. Often in applied situations, two quantities are compared using two different equations. For example, a small businessman may discover that his total sales and his profit earned are related in two different ways that may be described by using two linear equations in two variables. A pair of linear equations such as

$$3x - 2y + 1 = 0$$
$$3x + y - 5 = 0$$

1, 2 is called **a system of two linear equations in two variables,** or simply a **system of equations.** A **solution to a system of equations** is an ordered pair of numbers (with x written first and y written second) that is a solution to *both* equations. We can verify that (1, 2) is a solution to the above system by direct substitution.

$$3x - 2y + 1 = 0 \qquad 3x + y - 5 = 0$$
$$3(1) - 2(2) + 1 \stackrel{?}{=} 0 \qquad 3(1) + (2) - 5 \stackrel{?}{=} 0$$
$$3 - 4 + 1 \stackrel{?}{=} 0 \qquad 3 + 2 - 5 \stackrel{?}{=} 0$$
$$0 = 0 \qquad 0 = 0$$

Finding solutions to systems of equations, that is, solving a system of equations, is easier to understand if we first examine the nature of the graphs of the system. Remember that a linear equation in x and y has a straight line for its graph. When the linear equations in a given system are graphed together in a Cartesian coordinate system, one of three possibilities occurs:

1. The lines coincide (the two equations represent the same line).
2. The lines are parallel (do not intersect).
3. The lines intersect in exactly one point.

The three following pairs of linear equations, graphed in Figure 5.1, illustrate these three possibilities.

(A) $3x - 2y + 1 = 0$ (C) $3x - 2y + 1 = 0$ (E) $3x - 2y + 1 = 0$
(B) $6x - 4y + 2 = 0$ (D) $3x - 2y - 4 = 0$ (F) $3x + y - 5 = 0$

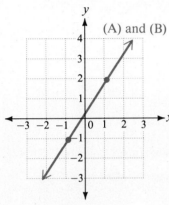
lines coincide

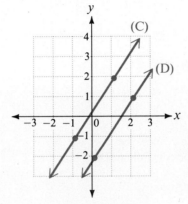
lines are parallel

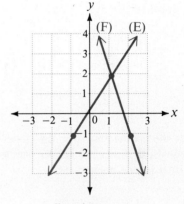
lines intersect in exactly one point

Figure 5.1

In the first case, lines (A) and (B) coincide, in the second case, lines (C) and (D) are parallel, and in the third case, lines (E) and (F) intersect in one point.

We need a method to determine quickly which case we have in a given situation. Let us solve each linear equation for y, writing it in slope-intercept form.

(A) $y = \dfrac{3}{2}x + \dfrac{1}{2}$ (C) $y = \dfrac{3}{2}x + \dfrac{1}{2}$ (E) $y = \dfrac{3}{2}x + \dfrac{1}{2}$

(B) $y = \dfrac{3}{2}x + \dfrac{1}{2}$ (D) $y = \dfrac{3}{2}x - 2$ (F) $y = -3x + 5$

In slope-intercept form, the equations can be compared more easily. Recall that the coefficient of x is the slope of the line, and that parallel lines must have the same slope.

> **To determine the nature of the graphs of a system of equations**
>
> 1. Write each equation in slope-intercept form and identify the slope and y-intercept of each.
> 2. If the slopes are equal and the y-intercepts are equal, the graphs (lines) coincide (the equations are equivalent).
> 3. If the slopes are equal and the y-intercepts are unequal, the graphs (lines) are parallel.
> 4. If the slopes are unequal, the graphs (lines) intersect in exactly one point.

214 SYSTEMS OF LINEAR EQUATIONS

The rule is not applicable when one (or both) of the equations in a system is of the form $ax = c$, since such equations cannot be solved for y. However, recalling that the graph of an equation of this form is always a line parallel to the y-axis, parallel or coinciding lines result only if both equations in the system are of this type.

EXAMPLE 1

Determine the nature of the graphs of the system.

$$2x + 3y + 1 = 0$$
$$x - 2y + 5 = 0$$

Solve for y.

$$3y = -2x - 1 \qquad\qquad -2y = -x - 5$$
$$\tfrac{1}{3} \cdot 3y = \tfrac{1}{3}(-2x - 1) \qquad \left(-\tfrac{1}{2}\right)(-2y) = \left(-\tfrac{1}{2}\right)(-x - 5)$$
$$y = -\tfrac{2}{3}x - \tfrac{1}{3} \qquad\qquad y = \tfrac{1}{2}x + \tfrac{5}{2}$$

Since the coefficients of x, the slopes of the lines, are $-\tfrac{2}{3}$ and $\tfrac{1}{2}$ (unequal), the lines are neither parallel nor coinciding. Thus, the lines intersect in one point. ◂◂

Practice Exercise 1

Determine the nature of the graphs of the system.

$$2x - 3y + 1 = 0$$
$$x + 5y - 10 = 0$$

Answer: intersecting

EXAMPLE 2

Determine the nature of the graphs of the system.

$$2x - 3y + 1 = 0$$
$$4x - 6y - 3 = 0$$

Solve for y.

$$-3y = -2x - 1 \qquad\qquad -6y = -4x + 3$$
$$\left(-\tfrac{1}{3}\right)(-3y) = \left(-\tfrac{1}{3}\right)(-2x - 1) \qquad \left(-\tfrac{1}{6}\right)(-6y) = \left(-\tfrac{1}{6}\right)(-4x + 3)$$
$$y = \tfrac{2}{3}x + \tfrac{1}{3} \qquad\qquad y = \tfrac{4}{6}x - \tfrac{3}{6}$$
$$\qquad\qquad\qquad\qquad y = \tfrac{2}{3}x - \tfrac{1}{2}$$

Since the coefficients of x, the slopes of the lines, are both $\tfrac{2}{3}$ (equal), the lines are either parallel or coinciding. Since the constants $\tfrac{1}{3}$ and $-\tfrac{1}{2}$ are unequal, that is the lines have two different y-intercepts, namely $(0, \tfrac{1}{3})$ and $(0, -\tfrac{1}{2})$, the lines do not coincide. Thus, the lines are parallel. ◂◂

Practice Exercise 2

Determine the nature of the graphs of the system.

$$5x - 10y + 20 = 0$$
$$-x + 2y - 4 = 0$$

Answer: coinciding

EXAMPLE 3

Determine the nature of the graphs of the system.

$$x - 2y + 2 = 0$$
$$-3x + 6y - 6 = 0$$

Solve for y.

$$-2y = -x - 2 \qquad\qquad 6y = 3x + 6$$

Practice Exercise 3

Determine the nature of the graphs of the system.

$$-3x + y - 5 = 0$$
$$6x - 2y + 3 = 0$$

$$\left(-\frac{1}{2}\right)(-2y) = \left(-\frac{1}{2}\right)(-x-2) \qquad \frac{1}{6}(6y) = \frac{1}{6}(3x+6)$$

$$y = \frac{1}{2}x + 1 \qquad\qquad y = \frac{3}{6}x + \frac{6}{6}$$

$$y = \frac{1}{2}x + 1$$

Since both coefficients of x are $\frac{1}{2}$ and both constants are 1, the lines coincide. ◀◀

Answer: parallel

EXAMPLE 4

Determine the nature of the graphs.

$$2x + 3y - 6 = 0$$
$$2x - 6 = 0$$

Since $2x - 6 = 0$ becomes $2x = 6$ or $x = 3$, its graph is a line parallel to the y-axis, 3 units to the right of the y-axis. Since there is a y term in $2x + 3y - 6 = 0$, the graph of this equation cannot be a line parallel to the y-axis. Thus, the two lines must intersect in one point. ◀◀

Practice Exercise 4

Determine the nature of the graphs of the system.

$$7x + 3 = 0$$
$$-2x + 5 = 0$$

Answer: parallel

EXAMPLE 5

Determine the nature of the graphs.

$$3y + 5 = 0$$
$$4x - 2y + 3 = 0$$

Solve for y.

$$3y = -5 \qquad\qquad -2y = -4x - 3$$

$$\left(\frac{1}{3}\right)(3y) = \left(\frac{1}{3}\right)(-5) \qquad \left(-\frac{1}{2}\right)(-2y) = \left(-\frac{1}{2}\right)(-4x - 3)$$

$$y = -\frac{5}{3} \qquad\qquad y = \frac{4}{2}x + \frac{3}{2}$$

$$y = 0 \cdot x - \frac{5}{3} \qquad\qquad y = 2x + \frac{3}{2}$$

Since the coefficients of x, 0 and 2, are not equal, the lines intersect in one point. ◀◀

Practice Exercise 5

Determine the nature of the graphs of the system.

$$2x - 3y + 6 = 0$$
$$5y - 1 = 0$$

Answer: intersecting

5.1 EXERCISES A

Complete Exercises 1–4 with one of the words (a) parallel, (b) coinciding, or (c) intersecting.

1. When two linear equations are both solved for y and the coefficients of x are unequal, the graphs are _____ lines.

2. When two linear equations are both solved for y and the coefficients of x are equal as are the constants, the graphs are _____ lines.

3. If the y term is missing in both of two linear equations, the graphs are _____ lines or lines _____ to the y-axis.

216 SYSTEMS OF LINEAR EQUATIONS

4. If the *y*-term is missing in only one of two linear equations, the graphs are _____ lines.

Determine the nature of the graphs of the system.

5. $-2x + 3y + 5 = 0$
   $-2x - 3y + 5 = 0$

6. $x - 5y + 2 = 0$
   $2x - 10y + 4 = 0$

7. $2x - 7y + 1 = 0$
   $-6x + 21y + 3 = 0$

8. $2y + x = 5$
   $2x = 10$

9. $x = 3y + 7$
   $y = 3x + 7$

10. $2y = 8$
    $2x = 8$

11. $2x + 1 = 3$
    $4x - 3 = 1$

12. $x + y = 5$
    $3y = 15$

13. $8y - 5 = -3x$
    $-6x + 10 = 16y$

Determine whether $(-3, 2)$ is a solution to the given system.

14. $2x + 3y = 0$
    $x + 8y = 19$

15. $3x + y + 7 = 0$
    $x + 4y - 5 = 0$

16. $3y = 6$
    $x - 5y = -13$

Determine whether $(4, 0)$ is a solution to the given system.

17. $2x - 3y = 8$
    $x - 4 = 0$

18. $x + y - 2 = 0$
    $2y + 2x = 4$

19. $2y + 1 = x$
    $2x - 8 = y$

ANSWERS: **1.** intersecting **2.** coinciding **3.** coinciding, parallel **4.** intersecting **5.** intersecting **6.** coinciding **7.** parallel **8.** intersecting **9.** intersecting **10.** intersecting **11.** coinciding **12.** intersecting **13.** coinciding **14.** no **15.** yes **16.** yes **17.** yes **18.** no **19.** no

5.1 EXERCISES B

Complete Exercises 1–4 with one of the following words: (a) parallel, (b) coinciding, or (c) intersecting.

1. When the slopes of the lines in a system of equations are equal but the *y*-intercepts are unequal, the graphs are _____ lines.

2. When the slopes of the lines in a system of equations are unequal, the graphs are _____ lines.

3. If the x term is missing in both equations in a system of equations, the graphs are **(A)** _____ lines or are lines **(B)** _____ to the x-axis.

4. If the x term is missing in only one of the equations in a system of equations, the graphs are _____ lines.

Determine the nature of the graphs of the system.

5. $-5x + y - 3 = 0$
$5x - y - 3 = 0$

6. $x + 2y + 3 = 0$
$-3x - 6y - 9 = 0$

7. $4x - y + 1 = 0$
$x - 4y - 1 = 0$

8. $2x + 3y = 1$
$y = 2$

9. $x = 3y - 1$
$y = -x - 1$

10. $3x = 1$
$3y = 1$

11. $3y + 1 = 4$
$-y + 2 = 1$

12. $y - 3x = 11$
$2x = 8$

13. $x + 2 = y$
$y + 1 = x$

Determine whether $(1, 0)$ is a solution to the given system.

14. $2x + 3y = 2$
$-x - 4y = -1$

15. $3x - y = 1$
$x + y = 1$

16. $3x = 3$
$-x + 9y = -1$

Determine whether $(-2, 1)$ is a solution to the given system.

17. $x - y = -3$
$y - 1 = 0$

18. $2x - 3y + 7 = 0$
$6y - 4x = 14$

19. $y + 2x = 3$
$2x + 5 = y$

5.1 EXERCISES C

1. Determine the value of a so that the line with equation $ax - 3y + 7 = 0$ will be parallel to the line with equation $5x + y - 2 = 0$. [*Hint:* Write each equation in slope-intercept form.]

2. Determine the value of c so that the line with equation $2x - 3y + c = 0$ will have the same y-intercept as the line with equation $4x + 2y + 3 = 0$.

5.2 SOLVING SYSTEMS OF EQUATIONS BY GRAPHING

STUDENT GUIDEPOSTS

1 Graphing method
2 Number of solutions

Consider the system of equations

$$x + y = 1$$
$$2x - y = 5.$$

Solve each equation for y.

$$y = -x + 1$$
$$y = 2x - 5$$

The slopes are different, so the lines intersect in exactly one point. Since every point on each line corresponds to an ordered pair of numbers that is a solution to that equation, the point of intersection must correspond to the ordered pair that solves both equations. Hence it is the solution to the system. If we graph both

218 SYSTEMS OF LINEAR EQUATIONS

equations in the same Cartesian coordinate system, as in Figure 5.2, it appears that the point of intersection has coordinates $(2, -1)$. We can check to see if $(2, -1)$ is indeed a solution by substitution.

$$x + y = 1 \qquad 2x - y = 5$$
$$2 + (-1) \stackrel{?}{=} 1 \qquad 2(2) - (-1) \stackrel{?}{=} 5$$
$$1 = 1 \qquad 4 + 1 \stackrel{?}{=} 5$$
$$5 = 5$$

1 ▶ This method for solving a system is the **graphing method.**

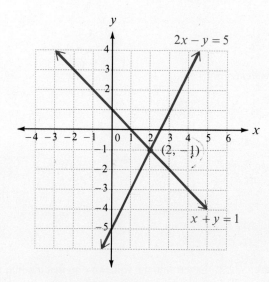

Figure 5.2

If the graphs of the lines in a system are parallel, there can be no point of intersection, so the system has no solution. If the graphs of the lines coincide, then any solution to one equation is also a solution to the other equation, and the system of equations has infinitely many solutions.

2 ▶ **A system of equations has**
1. Infinitely many solutions if the graphs of the equations coincide.
2. No solution if the graphs of the equations are parallel.
3. Exactly one solution if the graphs of the equations intersect in one point.

EXAMPLE 1

Determine the number of solutions to the system.

$$3x - 4y + 1 = 0$$
$$8y - 2 = 6x$$

Solve each equation for y.

$$-4y = -3x - 1 \qquad\qquad 8y = 6x + 2$$

Practice Exercise 1

Determine the number of solutions to the system.

$$x - 5y + 8 = 0$$
$$3y + 7 = -2x$$

$$\left(-\frac{1}{4}\right)(-4y) = \left(-\frac{1}{4}\right)(-3x - 1) \qquad \left(\frac{1}{8}\right)(8y) = \left(\frac{1}{8}\right)(6x + 2)$$

$$y = \frac{3}{4}x + \frac{1}{4} \qquad\qquad y = \frac{6}{8}x + \frac{2}{8}$$

$$y = \frac{3}{4}x + \frac{1}{4}$$

Since the lines coincide (why?), there are infinitely many solutions to the system. ◀◀

Answer: exactly one

EXAMPLE 2

Determine the number of solutions to the system.

$$x - 5y + 1 = 0$$
$$15y - 3x = 1$$

Solve for y.

$$-5y = -x - 1 \qquad 15y = 3x + 1$$
$$y = \frac{1}{5}x + \frac{1}{5} \qquad y = \frac{3}{15}x + \frac{1}{15}$$
$$\qquad\qquad\qquad y = \frac{1}{5}x + \frac{1}{15}$$

Since the lines are parallel (why?), there is no solution to the system. ◀◀

Practice Exercise 2

Determine the number of solutions to the system.

$$3x + 5y = -7$$
$$6x + 10y - 5 = 0$$

Answer: no solution

EXAMPLE 3

Determine the number of solutions to the system.

$$3x - 5 = 0$$
$$3x + y = -5$$

Since the y term is missing in the first equation, it cannot be solved for y. The graph of $3x - 5 = 0$, however, is a line parallel to the y-axis, so it intersects the line $3x + y = -5$ in only one point (why?) Thus, there is exactly one solution to the system. ◀◀

Practice Exercise 3

Determine the number of solutions to the system.

$$4x - 12 = 0$$
$$2x = 6$$

Answer: infinitely many

EXAMPLE 4

Solve the system using the graphing method.

$$x - 2y = 1$$
$$2x + y = 7$$

We graph each equation by finding its intercepts

$x - 2y = 1$

x	y
0	$-\frac{1}{2}$
1	0

$2x + y = 7$

x	y
0	7
$\frac{7}{2}$	0

Practice Exercise 4

Solve the system using the graphing method.

$$3x - 2y = 12$$
$$2x + 3y = -5$$

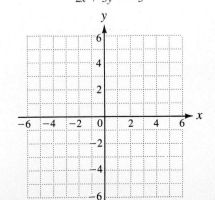

220 SYSTEMS OF LINEAR EQUATIONS

The point of intersection of the two lines, graphed in Figure 5.3, appears to have coordinates (3, 1). We check by substitution in both equations.

$$x - 2y = 1 \qquad 2x + y = 7$$
$$3 - 2(1) \stackrel{?}{=} 1 \qquad 2(3) + 1 \stackrel{?}{=} 7$$
$$3 - 2 \stackrel{?}{=} 1 \qquad 6 + 1 \stackrel{?}{=} 7$$
$$1 = 1 \qquad 7 = 7$$

Thus, the solution to the system is (3, 1). ◀◀

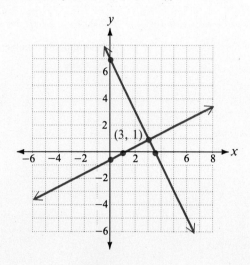

Figure 5.3

Answer: $(2, -3)$

5.2 EXERCISES A

1. How many solutions will a system of equations have if the lines in the system are parallel?

2. How many solutions will a system of equations have if the lines in the system are intersecting?

3. How many solutions will a system of equations have if the lines in the system are coinciding?

Determine the number of solutions to the given system. For those that have exactly one solution, find that solution by the graphing method. Check each solution.

4. $x + 3y = 4$
   $-x + y = 0$

5. $x + 3y = 4$
   $-2x - 6y = -8$

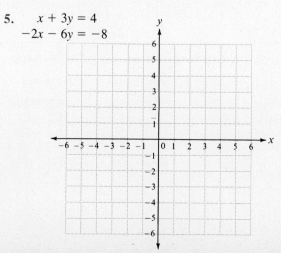

6. $x + 3y = 4$
   $-2x - 6y = 0$

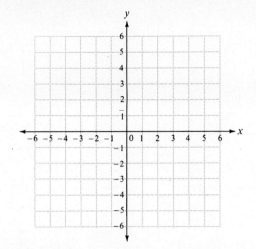

7. $2x + y = 5$
   $2x - y = 3$

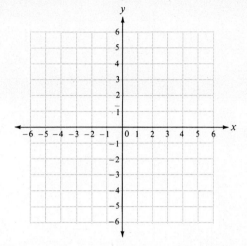

8. $x + y = 3$
   $x + 3y = -1$

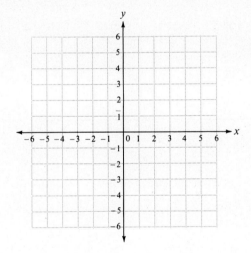

9. $2x + y = 3$
   $x + y = 3$

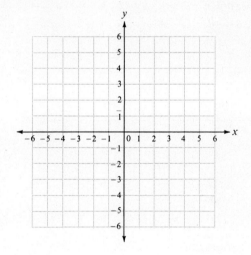

10. $3x - y = 1$
    $2y = 4$

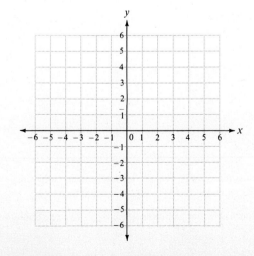

11. $2y = 3$
    $3x = 2$

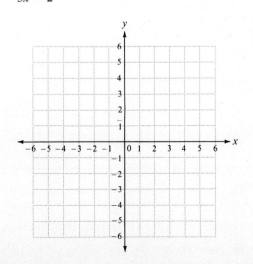

222 SYSTEMS OF LINEAR EQUATIONS

12. $x = y + 1$
$y = x - 1$

13. $x + 4y = 1$
$-x - 4y = 1$

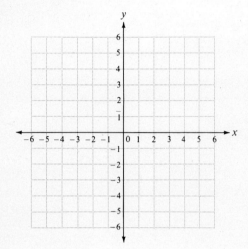

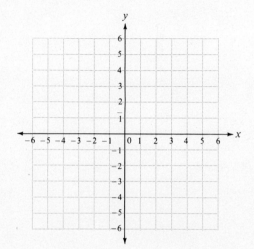

14. $4x - 3y = 2$
$-2x + y = -1$

15. $2x + y = 0$
$2x - y = 0$

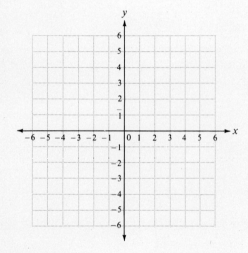

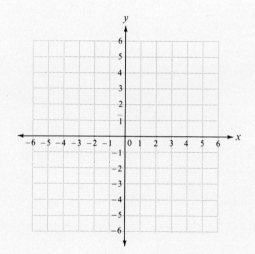

For Review

Determine whether $(0, 4)$ *is a solution to the given system.*

16. $2x + 3y - 12 = 0$
$x - 2y + 8 = 0$

17. $x - y = -4$
$2x + y = -4$

18. $y = 4$
$2x - 2y + 8 = 0$

ANSWERS: **1.** none   **2.** one (exactly one)   **3.** infinitely many   **4.** one solution; $(1, 1)$ **5.** infinitely many solutions   **6.** no solution   **7.** one solution; $(2, 1)$ **8.** one solution; $(5, -2)$ **9.** one solution; $(0, 3)$ **10.** one solution; $(1, 2)$ **11.** one solution; $\left(\frac{2}{3}, \frac{3}{2}\right)$ **12.** infinitely many solutions   **13.** no solution   **14.** one solution; $\left(\frac{1}{2}, 0\right)$ **15.** one solution; $(0, 0)$ **16.** yes   **17.** no   **18.** yes

5.2 EXERCISES B

1. If a system of equations has exactly one solution, what can you say about the graph of the system?
2. If a system of equations has no solution, what can you say about the graph of the system?
3. If a system of equations has infinitely many solutions, what can you say about the graph of the system?

Determine the number of the solutions to the given system. For those that have exactly one solution, find that solution by the graphing method. Check each solution.

4. $x + 2y = 1$
   $x - y = -2$

5. $x + 2y = 1$
   $-x - 2y = -1$

6. $x + 2y = 1$
   $-x - 2y = 1$

7. $x + 2y = 5$
   $x - 2y = -3$

8. $x - y = -4$
   $5x + y = 4$

9. $x + y = 2$
   $x + 2y = -1$

10. $2x + 3y = 6$
    $3x = 3$

11. $3x = 4$
    $4y = 3$

12. $x - 5y = 2$
    $-x + 5y = 2$

13. $y = x + 2$
    $x = y - 2$

14. $3x - 3y = -1$
    $x + 6y = 2$

15. $x + 5y = 0$
    $x - 5y = 0$

For Review

Determine whether $(1, -5)$ is a solution to the given system.

16. $2x - y - 7 = 0$
    $x + y = -4$

17. $x - 2y = 11$
    $2x - 7 = y$

18. $x + 5 = 0$
    $y - 1 = 0$

5.2 EXERCISES C

1. Determine the value of c so that the system will have infinitely many solutions.

   $x - 5y + c = 0$
   $-3x + 15y - 7 = 0$ [Answer: $c = \frac{7}{3}$]

2. Determine the value of a so that the system will have no solution.

   $ax - 5y + 8 = 0$
   $4x + y - 1 = 0$

5.3 SOLVING SYSTEMS OF EQUATIONS BY SUBSTITUTION

 STUDENT GUIDEPOSTS

1. Method of substitution
2. Contradiction
3. Identity

Solving a system by graphing takes too much time, and estimating the point of intersection depends on the accuracy of the graph. A better method for solving systems is the **method of substitution.** Consider the system

$$x - 2y = 1$$
$$2x + y = 7.$$

We will change the system into one equation in one unknown. Solve the first equation for x,

$$x - 2y = 1$$
$$x = 2y + 1,$$

and substitute this value of x into the second equation.

$$2(2y + 1) + y = 7$$

The result is an equation in the single variable y, which we can solve.

$$4y + 2 + y = 7$$
$$5y + 2 = 7$$
$$5y = 5$$
$$y = 1$$

Substitute this value of y into either of the original equations, say the first,

$$x - 2(1) = 1,$$

and solve for x.

$$x - 2 = 1$$
$$x = 3$$

We check the possible solution pair (3, 1), by substituting in both original equations.

$$x - 2y = 1 \qquad 2x + y = 7$$
$$3 - 2(1) \stackrel{?}{=} 1 \qquad 2(3) + 1 \stackrel{?}{=} 7$$
$$3 - 2 \stackrel{?}{=} 1 \qquad 6 + 1 \stackrel{?}{=} 7$$
$$1 = 1 \qquad 7 = 7$$

The solution to the system is thus (3, 1). Notice that we solve the same system with the same results in Example 4 of the preceding section.

> **To solve a system of equations using the substitution method**
>
> 1. Solve one of the equations for one of the variables.
> 2. Substitute that value of the variable in the *remaining* equation.
> 3. Solve this new equation and substitute the numerical solution into either of the two *original* equations to find the numerical value of the second variable.
> 4. Check your solution in both original equations.

EXAMPLE 1

Solve by the substitution method

$$5x + 3y = 17$$
$$x + 3y = 1$$

We could solve either equation for either variable. However, since the coefficient for x in the second equation is 1, we can avoid fractions if we solve for x using that equation.

$$x = 1 - 3y$$

Practice Exercise 1

Solve by the substitution method.

$$3x + y = 1$$
$$2x + 3y = 10$$

Substitute $1 - 3y$ for x in the first equation.
$$5(1 - 3y) + 3y = 17$$
$$5 - 15y + 3y = 17$$
$$5 - 12y = 17$$
$$-12y = 12$$
$$y = -1$$

Now substitute -1 for y in the second equation.
$$x + 3(-1) = 1$$
$$x - 3 = 1$$
$$x = 4$$

The solution is $(4, -1)$.

Check:
$$5(4) + 3(-1) \stackrel{?}{=} 17 \qquad 4 + 3(-1) \stackrel{?}{=} 1$$
$$20 - 3 \stackrel{?}{=} 17 \qquad 4 - 3 \stackrel{?}{=} 1$$
$$17 = 17 \qquad 1 = 1. \blacktriangleleft$$

Start by solving the first equation for y and substituting into the second equation.
$$y = -3x + 1$$
$$2x + 3(-3x + 1) = 10$$

Answer: $(-1, 4)$

EXAMPLE 2

Solve by the substitution method.
$$4x + 2y = 1$$
$$2x + y = 8$$

If we solve the second equation for y, we obtain
$$y = 8 - 2x.$$

Substitute this value into the first equation.
$$4x + 2(8 - 2x) = 1$$
$$4x + 16 - 4x = 1$$
$$16 = 1$$

But $16 \neq 1$. What went wrong? Return to the original system and solve both equations for y.

$$2y = 1 - 4x \qquad \text{and} \qquad y = 8 - 2x$$
$$y = \frac{1}{2} - \frac{4}{2}x \qquad\qquad y = -2x + 8$$
$$y = -2x + \frac{1}{2}$$

Both coefficients of x are -2 but the constant terms are different. Thus the two lines are parallel, and there is no solution. When solving a system it is wise first to determine the nature of the solutions. However, if there is no solution (the lines are parallel), substitution results in an equation that is a contradiction (such as $16 = 1$). ◀

Practice Exercise 2

Solve by the substitution method.
$$3x - 6y = 9$$
$$-x + 2y = -3$$

Answer: infinitely many solutions

EXAMPLE 3

Solve by the substitution method.
$$x + 2y = 1$$
$$2y = 1 - x$$

Practice Exercise 3

Solve by the substitution method.
$$10x + 5y = 2$$
$$-2x - y = 4$$

We solve the first equation for x.

$$x = 1 - 2y$$

Substitute this value into the second equation.

$$2y = 1 - (1 - 2y)$$
$$2y = 1 - 1 + 2y \quad \text{Watch the signs}$$
$$2y = 2y$$

Clearly, $2y = 2y$ no matter what number replaces y so that we end up with an equation that is an identity. Return to the original equations and solve them for y.

$$2y = 1 - x \qquad 2y = 1 - x$$
$$2y = -x + 1 \qquad 2y = -x + 1$$
$$y = \frac{1}{2}x - \frac{1}{2} \qquad y = \frac{1}{2}x - \frac{1}{2}$$

The coefficients of x are equal, as are the constants. The two lines coincide so there are infinitely many solutions to this system (any pair of numbers that is a solution to one equation is a solution to both, hence a solution to the system). ◀◀

Answer: no solution

Examples 2 and 3 lead to the following rules.

When solving a system of equations

1. If a **contradiction** (an equation that is never true) results, the lines are parallel and there is no solution to the system.
2. If an **identity** (an equation that is always true for every value of the variable) results, the lines coincide and there are infinitely many solutions to the system.

5.3 EXERCISES A

1. Suppose in the process of solving a system of equations using the substitution method we obtain the equation $3x = 3x$. What would this tell us about the system?

2. Suppose in the process of solving a system of equations using the substitution method we obtain the equation $5 = 0$. What would this tell us about the system?

Solve the following systems using the substitution method.

3. $y = 3x + 2$
   $5x - 2y = -7$

4. $2x - 3y = 14$
   $x = y + 10$

5. $3x + y = 35$
   $x - 2y = 7$

6. $13x - 4y = -66$
   $5x + 2y = 10$

7. $3x - 5y = 19$
   $2x - 4y = 16$

8. $3x - y = 7$
   $4x - 5y = 2$

9. $x - 3y = 14$
   $x - 2 = 0$
   [*Hint:* Start by solving for x in the second equation.]

10. $3x - 3y = 1$
    $x - y = -1$

11. $2x + 2y = -6$
    $-x - y = 3$

12. $y = 5$
    $4x - y = 19$

13. $2x + 3y = 5$
    $4x + 7y = 11$

14. $3x + 5y = 30$
    $5x + 3y = 34$

15. Which variable in which equation is easiest to solve for in Exercise 14? (The next section explains a method for solving equations that may be better for systems such as this.)

For Review

16. When graphing a system of equations, if the lines are parallel, how many solutions does the system have?

Solve by the graphing method.

17. $x + y = 5$
    $3x - y = -1$

18. $2x + y = 4$
    $x - 2 = 0$

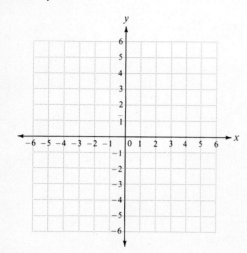
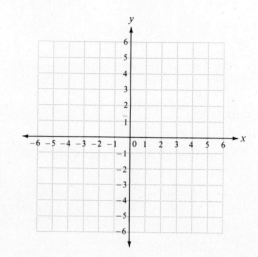

ANSWERS: **1.** There are infinitely many solutions; any solution to one equation is a solution to the other and therefore a solution to the system. **2.** There are no solutions. **3.** (3, 11); first equation is already solved for y **4.** (16, 6); second equation is already solved for x **5.** (11, 2); best to solve for x in second equation **6.** $(-2, 10)$ **7.** $(-2, -5)$ **8.** (3, 2) **9.** $(2, -4)$ **10.** no solution **11.** infinitely many solutions **12.** (6, 5) **13.** (1, 1); no one variable is better to solve for than any other **14.** (5, 3) **15.** no one variable is better to solve for than any other **16.** no solutions **17.** (1, 4) **18.** (2, 0)

5.3 EXERCISES B

1. Suppose in the process of solving a system of equations using the substitution method we obtain the equation $-3 = 0$. What would this tell us about the system?

2. Suppose in the process of solving a system of equations using the substitution method we obtain the equation $x + 1 = x + 1$. What would this tell us about the system?

Solve the following systems using the substitution method.

3. $y = -2x + 3$
   $3x - 4y = -1$

4. $5x - y = -2$
   $x = 3y - 6$

5. $x + y = 0$
   $5x + 2y = -3$

6. $2x - 8y = -10$
   $x - 3y = -2$

7. $2x - y = 3$
   $-4x + 2y = -6$

8. $4x - y = 13$
   $x + 2y = 1$

9. $3x - y = 2$
   $-6x + 2y = 2$

10. $2x - y = -2$
    $y - 4 = 0$

11. $4x + y = -2$
    $-2x - 3y = 1$

12. $3x - y = 14$
    $x - 4 = 0$

13. $2x + 3y = -4$
    $3x - 2y = 7$

14. $3x + 7y = 32$
    $7x + 3y = 8$

15. What is the difficulty in using the substitution method on a system like the one in Exercise 14?

For Review

16. When graphing a system of equations, if the lines are coinciding, how many solutions does the system have?

Solve by the graphing method.

17. $3x - y = -4$
    $-x + y = 0$

18. $x - 3y = 3$
    $y + 1 = 0$

5.3 EXERCISES C

Solve using the substitution method.

1. $0.03x - 0.01y = 0.12$
   $0.04x - 0.05y = 0.20$

2. $\dfrac{x}{10} + \dfrac{2y}{5} = 1$
   $\dfrac{2x}{5} + \dfrac{3y}{2} = 2$ [Answer: $(-70, 20)$]

5.4 SOLVING SYSTEMS OF EQUATIONS BY ADDITION-SUBTRACTION

An alternative method of solving systems of equations is the **addition-subtraction method.** It is based on the addition-subtraction rule for solving equations from Chapter 3: if the same expression is added to or subtracted from both sides of an equation, another equation with the same solution is obtained. Consider the system

$$x + y = 6$$
$$3x - y = 2.$$

In the second equation, $3x - y$ and 2 are both names for the same number, 2. Therefore we can add $3x - y$ to the left side of the first equation, and 2 to the right of it, to obtain another equation with the same solution.

$$(x + y) + (3x - y) = 6 + 2$$
$$4x = 8$$

Generally we add the equations vertically.

$$x + y = 6$$
$$\underline{3x - y = 2}$$
$$4x = 8$$
$$x = 2 \quad \text{Divide by 4}$$

By adding the two equations, we obtain one equation in one variable, x. Once x is known to be 2, we substitute this value into either one of the original equations (just as in the substitution method) to find the value of y.

$$2 + y = 6 \quad \text{Substitute in the first equation since it is simpler}$$
$$y = 4$$

Thus, the solution is $(2, 4)$. Check to see that this is correct by substituting in both equations.

EXAMPLE 1

Solve by the addition-subtraction method.

$$x + y = 5$$
$$-2x + y = -4$$

If we add both equations, the resulting equation still has both variables. (Try it.) In this case, we subtract instead of adding, to obtain one equation in one variable.

Practice Exercise 1

Solve by the addition-subtraction method.

$$4x + 3y = -1$$
$$2x - 3y = 13$$

$$x + y = 5$$
$$+2x - y = 4 \quad \text{Change all signs when subtracting}$$
$$3x = 9$$
$$x = 3$$

Substitute 3 for x in the first equation.

$$3 + y = 5$$
$$y = 2$$

The solution is the ordered pair (3, 2). Check by substitution. ◂◂

Answer: $(2, -3)$

Sometimes addition or subtraction does not yield one equation in only one variable. When this occurs, we may have to multiply one (or both) of the equations by a number so that the coefficients of one variable are negatives of each other. We then add to eliminate that variable.

EXAMPLE 2

Solve by the addition-subtraction method.

$$2x - 5y = 13$$
$$x + 2y = 11$$

We would not obtain one equation with one variable by simply adding or subtracting in this case. However, if we multiply both sides of the second equation by -2 the resulting equation is

$$-2x - 4y = -22.$$

Now add this equation to the first equation.

$$2x - 5y = 13$$
$$\underline{-2x - 4y = -22}$$
$$-9y = -9$$
$$y = 1$$

Substitute 1 for y in the second equation.

$$x + 2(1) = 11$$
$$x + 2 = 11$$
$$x = 9$$

The solution is (9, 1). Check by substitution. ◂◂

Practice Exercise 2

Solve by the addition-subtraction method.

$$6x + 5y = 9$$
$$4x + y = -1$$

Answer: $(-1, 3)$

Sometimes the multiplication rule has to be applied to both equations before adding or subtracting, as in the next example.

EXAMPLE 3

Solve by the addition-subtraction method.

$$3x + 5y = 30$$
$$5x + 3y = 34$$

Practice Exercise 3

Solve by the addition-subtraction method.

$$2x + 7y = -4$$
$$5x + 3y = 19$$

Multiply the first equation by -5 and the second equation by 3 (making the coefficients of x negatives of each other).

$$-15x - 25y = -150$$
$$15x + 9y = 102$$

Add to get one equation in one variable.

$$-16y = -48$$
$$y = 3$$

Substitute 3 for y in the first equation.

$$3x + 5(3) = 30$$
$$3x + 15 = 30$$
$$3x = 15$$
$$x = 5$$

The solution is (5, 3). Check by substitution. ◀◀

Answer: (5, −2)

Compare this example with your work on Exercise 14 **(A)** in the last section. In this case the addition-subtraction method is faster and easier to use than the substitution method.

To solve a system by the addition-subtraction method

1. Write both equations in the form $ax + by = c$.
2. If necessary, apply the multiplication rule to each equation so that addition or subtraction will eliminate one of the variables.
3. Solve the resulting single-variable equation. Substitute this value into one of the original equations. The resulting pair of numbers is the ordered-pair solution to the system.
4. Check your answer by substitution in both original equations.

EXAMPLE 4

Solve by the addition-subtraction method.

(A)
$$2x = 3y + 5$$
$$-2x + 3y = -1$$

Rewrite the first equation as $2x - 3y = 5$. Notice that by adding both equations, we can eliminate x.

$$2x - 3y = 5$$
$$\underline{-2x + 3y = -1}$$
$$0 = 4$$

We actually eliminated both variables and obtained a contradiction. Just as in the method of substitution, this indicates that there is no solution to the system. Verify that the lines are parallel by solving each equation for y.

Practice Exercise 4

Solve by the addition-subtraction method.

(A) $-3x + 9y = -18$
$\phantom{\text{(A)}} \quad x - 3y = 6$

(B)
$$2x + y = 3$$
$$-4x - 2y = -6$$

Multiply the first equation by 2 and add.
$$4x + 2y = 6$$
$$\underline{-4x - 2y = -6}$$
$$0 = 0$$

Again, both variables are eliminated by this process, but this time we obtain an identity. This indicates that there are infinitely many solutions to the system. Verify that the lines coincide by solving each equation for y. ◀

(B)
$$8x = -4y + 12$$
$$4x + 2y = -3$$

Answers: **(A)** infinitely many solutions **(B)** no solution

5.4 EXERCISES A

1. If we obtain an equation such as $0 = -3$ when using the addition-subtraction method, how many solutions does the system have?

2. If we obtain an equation such as $0 = 0$ when using the addition-subtraction method, how many solutions does the system have?

Solve the following systems using the addition-subtraction method. Check by substitution.

3. $x + y = 6$
   $x - y = 4$

4. $3x - 2y = 12$
   $2x + 2y = 13$

5. $6x + 5y = 11$
   $3x - 7y = -4$

6. $x - y = 16$
   $x - 3y = 2$

7. $9x - y = 19$
   $3x + 7y = -1$

8. $4x + y = -3$
   $-8x - 2y = 6$

9. $-2y = 2 - x$
   $3x - 6y = -4$

10. $2x + 3y = -4$
    $1 + 2y = -3x$

11. $3x + 7y = -21$
    $7x + 3y = -9$

12. $5x - 2y = 1$
    $3x + 7y = -24$

Decide whether the substitution or the addition-subtraction method is most appropriate, then use it to solve the system.

13. $x + 5y = 0$
    $x - 5y = 20$

14. $x - 4y = -6$
    $3x - 5y = -4$

15. $4x - 5y = 9$
    $5x + 4y = 1$

16. $3x + 3y = 3$
    $4x + 4y = -3$

17. $2x + 3y = -4$
    $5x + 7y = -10$

18. $3x = 9$
    $x - y = 7$

19. $y = 3x + 1$
    $4x + 5y = 24$

20. $y - 5 = 0$
    $x + 2 = 0$

ANSWERS: 1. no 2. infinitely many 3. (5, 1) 4. $\left(5, \dfrac{3}{2}\right)$ 5. (1, 1) 6. (23, 7) 7. (2, −1) 8. infinitely many solutions 9. no solution 10. (1, −2) 11. (0, −3) 12. (−1, −3) 13. (10, −2) 14. (2, 2) 15. (1, −1) 16. no solution 17. (−2, 0) 18. (3, −4) 19. (1, 4) 20. (−2, 5)

5.4 EXERCISES B

1. When solving a system of equations using the addition-subtraction method, if a contradiction results, how many solutions does the system have?

2. When solving a system of equations using the addition-subtraction method, if an identity results, how many solutions does the system have?

Solve the following systems using the addition-subtraction method. Check by substitution.

3. $x + y = 3$
   $x - y = -1$

4. $2x - 3y = -7$
   $3x + 3y = 27$

5. $x - y = 0$
   $x - 5y = -12$

6. $3x - 2y = 10$
   $5x + 3y = 4$

7. $x + 3y = 1$
   $3x + y = 11$

8. $2x + 3y = -4$
   $3x - 2y = 7$

9. $5x - 2y = 1$
   $7 + 10x = 4y$

10. $2x - 2y = 5$
    $4y = 3x - 7$

11. $4x - 2y = 2$
    $-2x + y = -1$

12. $3x + 2y = -10$
    $5x - 3y = 15$

Decide whether the substitution or the addition-subtraction method is most appropriate, then use it to solve the system.

13. $x + 3y = 4$
    $x + 8y = -6$

14. $x - 7y = 44$
    $5x + 3y = -8$

15. $5x + 7y = -2$
    $7x + 5y = 2$

16. $3x - 7y = -4$
    $7x + 3y = 10$

17. $2x = 8$
    $x - y = 1$

18. $2x + 5y = -10$
    $3x + 11y = -22$

19. $x = 4y - 3$
    $2x + 3y = 5$

20. $y + 7 = 0$
    $x - 3 = 0$

5.4 EXERCISES C

Solve using the addition-subtraction method.

1. $0.015x - 0.007y = 0.032$
   $-0.009x + 0.014y = -0.001$ [Answer: $(3, \tfrac{13}{7})$]

2. $-\dfrac{3x}{4} + \dfrac{7y}{10} = \dfrac{1}{5}$
   $\dfrac{5x}{2} - \dfrac{4y}{3} = -\dfrac{1}{12}$

5.5 APPLICATIONS USING SYSTEMS

▶▶ STUDENT GUIDEPOSTS
1. Mixture problems
2. Motion problems

Some applied problems can be simplified by translating them into a system of equations. The following examples illustrate this technique.

EXAMPLE 1

The sum of two numbers is 21 and their difference is 9. Find the numbers.

Let x = one number,
y = the other number.

Then

$$x + y = 21 \quad \text{Represents "their sum is 21"}$$
$$x - y = 9 \quad \text{Represents "their difference is 9"}$$

We should solve the system by the simplest method. In this case, one method is as easy as the other so we add the two equations.

$$2x = 30$$
$$x = 15$$

Now substitute 15 for x in $x + y = 21$.

$$15 + y = 21$$
$$y = 6$$

The numbers are 15 and 6. Check. ◀◀

Practice Exercise 1

If twice one number is added to a second number, the result is 25. If the first number is the larger of the two and their difference is 5, find the numbers.

Let x = the first number,
y = the second number.

Answer: 10, 5

EXAMPLE 2

The sum of Bob's and Joe's ages is 14. In 2 years Bob will be twice as old as Joe. What are their present ages?

Let x = Bob's age,
y = Joe's age.

Then

$$x + 2 = \text{Bob's age in 2 years}$$
$$y + 2 = \text{Joe's age in 2 years.}$$

In two years Bob will be twice as old as Joe
$(x+2)$ = 2 $(y+2)$

$$x + 2 = 2(y + 2)$$
$$x + 2 = 2y + 4$$
$$-2 = 2y - x \quad \text{or} \quad -x + 2y = -2$$

Practice Exercise 2

Annette is $\frac{3}{4}$ as old as Mona. How old is each if the sum of their ages is 84.

Let x = Annette's age,
y = Mona's age,
$\frac{3}{4}y$ = three fourths Mona's age.

Since the sum of their ages is 14, $x + y = 14$. This gives us the following system.

$$-x + 2y = -2$$
$$x + y = 14$$

Add these equations.

$$3y = 12$$
$$y = 4$$

Now substitute 4 for y in $x + y = 14$.

$$x + 4 = 14$$
$$x = 10$$

Bob is 10 and Joe is 4. Check. ◂◂

Answer: Annette is 36, and Mona is 48.

EXAMPLE 3

Two angles are **supplementary** (their sum is 180°). If one is 60° more than twice the other, find the angles.

Let $x =$ one angle,
$y =$ the other angle.

Then

$$x + y = 180.$$

If x is 60° more than $2y$, then we would add 60° to $2y$ to get x. Thus, we have the following system.

$$x + y = 180$$
$$2y + 60 = x$$

Substitute $2y + 60$ for x in $x + y = 180$.

$$(2y + 60) + y = 180$$
$$3y + 60 = 180$$
$$3y = 120$$
$$y = 40$$

Now substitute 40 for y in $x + y = 180$.

$$x + 40 = 180$$
$$x = 140$$

The angles are 40° and 140°. Check. ◂◂

Practice Exercise 3

Two angles are **complementary** (their sum is 90°). If one is 12° less than twice the other, find the angles.

Answer: 34°, 56°

An important type of word problem that yields a system of equations is known as a **mixture problem.** Generally, two quantities are mixed together, and the two equations we seek might be called a *quantity equation* and a *value equation.* The **total value** of something that is composed of a number of units each of equal value can be computed by

(total value) = (value per unit) · (number of units).

For example, the total value of 5 lb of candy worth $1.10 per pound would be

total value = (value per pound) · (number of pounds)
= ($1.10) · (5)
= $5.50.

5.5 APPLICATIONS USING SYSTEMS

Similarly, the value of a collection of 30 nickels can be determined.

$$\text{total value} = (\text{value per nickel}) \cdot (\text{number of nickels})$$
$$= (5¢)(30)$$
$$= 150¢, \text{ or } \$1.50$$

Calculation of values is important for solving mixture problems.

EXAMPLE 4

A man wishes to mix candy selling for 75¢ per pound with nuts selling for 50¢ per pound to obtain a party mix to be sold for 60¢ per pound. How many pounds of each must be used to obtain 50 pounds of the mixture?

Let x = the number of pounds of candy,
y = the number of pounds of nuts.

The first equation we obtain is the quantity equation

(no. lb of candy) + (no. lb of nuts) = (no. lb of mixture),

which translates to

$$x + y = 50.$$

Next we determine the value equation.

(total value of candy) + (total value of nuts) = (total value of mixture)

That is, the value of a quantity is equal to the sum of the values of its parts.

$$75 \cdot x = \text{value of candy}$$
$$50 \cdot y = \text{value of nuts}$$
$$60 \cdot 50 = \text{value of mixture}$$

Thus, we have the value equation

$$75x + 50y = 60 \cdot 50.$$

We must solve the following system of equations.

$$x + y = 50$$
$$75x + 50y = 3000$$

Solve the first equation for x,

$$x = 50 - y,$$

and substitute in the second equation.

$$75(50 - y) + 50y = 3000$$
$$3750 - 75y + 50y = 3000$$
$$-25y = -750$$
$$y = 30$$

Now substitute 30 for y in $x = 50 - y$.

$$x = 50 - 30 = 20$$

The man must use 20 lb of candy and 30 lb of nuts. ◂◂

Practice Exercise 4

Dr. Lynn Mowen wants to mix a 10% alcohol solution and a 70% alcohol solution to obtain 30 liters of 50% alcohol solution. How much of each must she use to obtain the desired mixture?

Let x = Number of liters of 10% solution to be used
y = Number of liters of 70% solution to be used

$x + y =$ _____
$0.10x + 0.70y = 0.50(30)$

Answer: 10 liters of 10% and 20 liters of 70%

EXAMPLE 5

A boy has 25 nickels and dimes. If the coins' total value is $2.10, find the number of nickels and the number of dimes in the collection.

Let x = the number of nickels,
y = the number of dimes.

The quantity equation

(no. of nickels) + (no. of dimes) = (no. coins in the collection)

translates to

$$x + y = 25.$$

The value equation

(value of nickels) + (value of dimes) = (value of collection)

becomes

$$5 \cdot x + 10 \cdot y = 210. \quad \text{Convert to cents}$$

Solve the first equation for y,

$$y = 25 - x,$$

and substitute into the second.

$$5x + 10(25 - x) = 210$$
$$5x + 250 - 10x = 210$$
$$-5x = -40$$
$$x = 8$$

Now substitute 8 for x in $y = 25 - x$.

$$y = 25 - 8 = 17$$

There are 8 nickels and 17 dimes in the collection. ◀◀

Practice Exercise 5

Jose Garcia invests a total of $8000 in two accounts, one paying 10% simple interest and the other 12% simple interest. How much did he invest in each account if his total interest for the year was $930?

Answer: $1500 at 10% and $6500 at 12%

---

2 Another type of problem that can be solved using systems is the **motion problem.**

EXAMPLE 6

A boat travels 60 miles upstream in 4 hours and returns to the starting place in 3 hours. What is the speed of the boat in still water, and what is the speed of the stream?

Let x = speed of the boat in still water,
y = speed of the stream,
$x - y$ = speed of the boat going upstream,
$x + y$ = speed of the boat going downstream.

Since $d = rt$ (distance equals rate times time), we have the following system.

$$60 = 4(x - y)$$
$$60 = 3(x + y)$$

Here we can simplify the system by dividing the first equation by 4 and the second equation by 3.

Practice Exercise 6

A boat can travel 30 miles upstream in 3 hours and return in 2 hours. What is the speed of the boat in still water and the speed of the stream?

$$x - y = 15$$
$$x + y = 20$$
$$2x = 35 \quad \text{Add the equations}$$
$$x = \frac{35}{2}$$
$$x = 17.5$$

Substituting 17.5 into $x + y = 20$ gives

$$17.5 + y = 20$$
$$y = 2.5$$

Thus, the speed of the boat is 17.5 mph, and the speed of the stream is 2.5 mph. ◂◂

Answer: speed of boat is 12.5 mph; speed of stream is 2.5 mph

5.5 EXERCISES A

Solve. (Some problems have been started.)

1. The sum of two numbers is 39 and their difference is 13. Find the numbers.

   Let x = the first number,
   y = the other number.

   $x + y =$
   $x - y =$

2. Mike is 2 years younger than Burford. Four years from now the sum of their ages will be 36. How old is each now?

   Let x = Burford's age,
   y = Mike's age,
   $x + 4$ = Burford's age in 4 years,
   $y + 4$ = Mike's age in 4 years.

3. Two angles are supplementary (their sum is 180°) and one is 5° more than six times the other. Find the angles.

4. If three times the smaller of two numbers is increased by five times the larger, the result is 195. Find the two numbers if their sum is 47.

   Let x = the smaller number,
   y = the larger number,

   $3x + 5y =$

5. Mr. Smith has two sons. If the sum of their ages is 19 and the difference between their ages is 3, how old are his sons?

6. Bill is three times as old as Jim. Find the age of each if the sum of their ages is 24.

7. One third of one number is the same as twice another. The sum of the first and five times the second is 33. Find the numbers.

8. Two angles are *complementary* (their sum is 90°) and their difference is 66°. Find the angles.

9. A 30-ft rope is cut into two pieces. One piece is 8 ft longer than the other. How long is each?

10. The starting players on a basketball team scored 40 points more than the reserves. If the team scored a total of 90 points, how many points did the starters score?

11. A candy mix sells for $1.10 per pound. If it is composed of two kinds of candy, one worth 90¢ per lb and the other worth $1.50 per pound, how many pounds of each would be in 30 pounds of the mixture?

12. A collection of nickels and dimes is worth $2.85. If there are 34 coins in the collection, how many of each are there?

   Let x = the number of nickels,
        y = the number of dimes,
       $5x$ = value of nickels (in cents),
      $10y$ = value of dimes (in cents).

   $5x + 10y =$
   $x + y =$

13. A collection of 13 coins consists of dimes and quarters. If the value of the collection is $2.80, how many of each are there?

14. A man wishes to mix two grades of nuts selling for 48¢ and 60¢ per pound. How many pounds of each must he use to get a 20-pound mixture that sells for 54¢ per pound? (Does your answer seem reasonable?)

15. There were 450 people at a play. If the admission price was $2.00 for adults and 75¢ for children and the receipts were $600.00, how many adults and how many children were in attendance?

16. A painter needs to have 24 gallons of paint that contains 30% solvent. How many gallons of paint containing 20% solvent and how many gallons containing 60% solvent should be mixed to obtain the desired paint?

17. When riding his bike against the wind, Terry Larsen can go 40 miles in 4 hours. He takes 2 hours to return riding with the wind. What is Terry's average speed in still air and the average wind speed?

18. Victoria Harris can ride her bike 120 miles with the wind in 5 hr and return against the wind in 10 hours. What is her speed in still air and the speed of the wind?

19. Vince McKee has $10,000 to invest, part at 10% simple interest, and the rest at 12% simple interest. If he wants to have $1160 in interest at the end of the year, how much should he invest at each rate?

20. Wendy Young has $20,000 invested. Part of the money earns 9% simple interest and the rest earns 11% simple interest. If her interest income for the year was $2060, how much did she have invested at each rate?

21. International Car Rentals charges a daily fee plus a mileage fee. Carl Tyson was charged $82.00 for 3 days and 400 miles, while Brent Alderman was charged $120.00 for 5 days and 500 miles. What is the daily rate and the mileage rate?

22. Harder-Than-Ever Car Rentals charged Terry McGinnis $76.00 for 2 days and 300 miles. For the same type of car, Linda Youngman was charged $132 for 6 days and 100 miles. What are the daily fee and the mileage fee?

For Review

Solve each system using either the addition-subtraction or the substitution method, whichever seems most appropriate.

23. $2x - 3y = 20$
$7x + 5y = -54$

24. $x - 11y = -5$
$7x + 13y = -35$

ANSWERS: **1.** 13, 26 **2.** Burford is 15, Mike is 13 **3.** 25°, 155° **4.** 20, 27 **5.** 11 yr, 8 yr **6.** 6 yr, 18 yr **7.** 18, 3 **8.** 12°, 78° **9.** 11 ft, 19 ft **10.** 65 **11.** 10 lb ($1.50), 20 lb (90¢) **12.** 11 nickels, 23 dimes **13.** 3 dimes, 10 quarters **14.** 10 lb (48¢), 10 lb (60¢) **15.** 210 adults, 240 children **16.** 18 gal of 20%; 6 gal of 60% **17.** Terry: 15 mph; wind: 5 mph **18.** bike: 18 mph; wind: 6 mph **19.** $2000 at 10%, $8000 at 12% **20.** $7000 at 9%; $13,000 at 11% **21.** $14 per day, 10¢ per mile **22.** $20 per day; 12¢ per mile **23.** (−2, −8) **24.** (−5, 0)

5.5 EXERCISES B

Solve.

1. If three times the smaller of two numbers is increased by two times the larger, the result is 16. Find the two numbers if their sum is 7.

2. Julie is 4 years older than Randy. Five years from now the sum of their ages will be 24. How old is each now?

3. Two angles are supplementary (their sum is 180°) and one is 15° more than twice the other. Find the angles.

4. If four times the larger of two numbers is decreased by six times the smaller, the result is 4. Find the two numbers if they sum up to 16.

5. Tim Chandler has two sons. If the sum of their ages is 30 and the difference between their ages is 4, how old are his sons?

6. Sherry Maducci is five times as old as Mac Davis. Find the age of each if the sum of their ages is 42.

7. One half of one number is the same as three times another. The sum of twice the first and the second is 80. Find the numbers.

8. Two angles are complementary (their sum is 90°) and one is 15° more than four times the other. Find each angle.

9. A 45-inch wire is cut into two pieces. One piece is 11 inches more than the other. How long is each?

10. The starting five on the Memphis State basketball team scored 35 more points than the reserves. If the team scored a total of 85 points, how many points did the starters score?

11. A collection of nickels and dimes is worth $2.00. If the number of dimes minus the number of nickels is 5, how many of each are there?

12. A man wishes to make a party mix consisting of nuts worth 50¢ a pound and candy worth 80¢ a pound. How many pounds of each must he use to make 100 pounds of the mixture that will sell for 68¢ a pound?

13. Mark Bonnett has 40 coins in his piggy bank, consisting of dimes and quarters. If the value of the collection is $5.80, how many of each are in the bank?

14. A grocer mixes two types of candy selling for 50¢ and 70¢ per pound. How many pounds of each must he use to obtain a 30-pound mixture that sells for 60¢ per pound?

15. There were 12,000 people at a rock concert at Colorado State University. If the admission price was $7.00 for a student with an I.D. and $10.00 for a non-student, how many of each were in attendance if the total receipts amounted to $94,500?

16. Dr. Karen Winter needs 20 liters of 70% alcohol solution in her clinic. How many liters of 40% solution and how many liters of 90% solution must be mixed to obtain the desired solution?

17. While driving his boat upstream, Charlie Moore can go 75 miles in 5 hours. It takes him 3 hours to make the return trip downstream. What is the average speed of the boat in still water and the average stream speed?

18. Maria Lopez traveled 60 miles downstream in 2 hours but required 5 hours to go back upstream to her starting point. What is the average speed of the boat in still water and the average speed of the stream?

19. Grady and Barbara wish to invest $5000, part at 8% simple interest, and the rest at 14% simple interest. If they must earn $610 by the end of one year, how much should they invest at each rate?

20. A student has $1200 invested. Part of this money earns 12% simple interest, and the rest earns 10% simple interest. How much is invested at each rate if the interest income for the year will be $140?

21. Horn Rentals charges a daily fee and a mileage fee to rent a truck. Syd was charged $375 for 5 days and 1500 miles, while Barry was charged $165 for 3 days and 600 miles. What is the daily rate and the mileage rate?

22. Better-Than-Ever Car Rentals charges a daily fee and a mileage fee. Renee paid $110 for ten days and 200 miles, while Pamela rented the same type of car and paid $168 for six days and 800 miles. What is the daily fee and the mileage fee?

For Review

Solve the system using either the addition-subtraction or the substitution method, whichever seems most appropriate.

23. $3x + 2y = 2$
$5x - 2y = 30$

24. $4x - y = 6$
$3x + 5y = -30$

5.5 EXERCISES C

Solve.

1. On a Pacific island Angie is charged $157.92 after driving a rented car for 420 kilometers in 5 days. Andy paid $217.56 for 7 days and 560 kilometers. If each bill includes a 5% pleasure tax, what was the daily fee and what was the fee charged per kilometer? [*Hint:* Determine the fees before tax first.]

2. John and Pat Woods invested $22,000, part at 14% simple interest and the rest at 9% simple interest. How much was invested at 14% if the total interest received in three years was $8040. [Answer: $14,000]

CHAPTER 5 SUMMARY

Key Words and Phrases for Review

5.1 system of equations
solution to a system
coinciding lines
parallel lines
intersecting lines

5.2 graphing method
5.3 substitution method
5.4 addition-subtraction method
5.5 mixture problem
total value

Key Concepts

5.1 A system of equations has:
5.2 **(1)** infinitely many solutions if the lines coincide.

(2) no solution if the lines are parallel.

(3) exactly one solution if the lines intersect.

5.2 The method of graphing is not usually used to solve a system since it is too time-consuming and depends on our ability to graph accurately.

5.3 **1.** To solve a system of equations, use either
5.4 the substitution method or the addition-subtraction method, whichever seems most appropriate.

2. If an identity (such as 0 = 0) results when solving a system, the system has infinitely many solutions.

3. If a contradiction (such as 3 = 0) results when solving a system, the system has no solution.

4. When solving a system, do not forget to find the value of both variables, not just one of them. For example, when solving

$x + y = 3$
$x - y = 1$
$\overline{2x = 4}$ Adding the two
$x = 2$

do not stop here. Be sure to continue and solve for *y* also.

5.5 1. The total value of a quantity that is composed of a number of units each of equal value can be found by the following equation.

(total value) = (value per unit) · (number of units)

For example, 25 dimes has total value

$$250¢ = (10¢)(25).$$

2. The value of a quantity is equal to the sum of the values of its parts.

Review Exercises

5.1 *Determine the nature of the graphs of the system.*

1. $2x + y = -3$
   $-4x - 2y = 6$

2. $x - 3y = 7$
   $4x - 12y = -3$

3. $x + y = 3$
   $2x + y = -4$

5.2
5.3 *Solve by whichever method seems most appropriate.*
5.4

4. $3x - y = 9$
   $2x + 5y = -11$

5. $2x - 3y = -14$
   $5x + 2y = 3$

6. $8x - 2y = 4$
   $-4x + y = -2$

7. $x + 3 = 0$
   $3x + 2y = -7$

8. $3x + y = -2$
   $-6x - 2y = -2$

9. $x - 8 = 0$
   $2y + 1 = 0$

5.5 *Solve.*

10. The sum of two numbers is 37 and twice one minus the other is 14. Find the numbers.

11. The sum of Barb's and Cindy's ages is 17. In 4 years Barb will be twice as old as Cindy is now. What is the present age of each?

12. Two angles are complementary (their sum is 90°) and one is 6° more than five times the other. Find each angle.

13. Mrs. Carlson has two sons. If the sum of their ages is 17 and the difference between their ages is 7, how old are her sons?

14. A collection of nickels and dimes is worth $2.00. If the number of dimes minus the number of nickels is 5, how many of each are there?

15. A dealer wishes to mix tea worth 80¢ per pound with tea worth $1.00 per pound to obtain a 50-pound mixture worth 94¢ per pound. How many pounds of each must he use?

16. A boat travels 96 miles downstream in 4 hours and returns to the starting point in 6 hours. What is the speed of the boat in still water?

17. Tina invested $7000, part at 8% simple interest and the rest at 12% simple interest. How much did she have invested at 12% if her total interest income for the year was $640?

ANSWERS: **1.** coinciding   **2.** parallel   **3.** intersecting   **4.** (2, −3)   **5.** (−1, 4)   **6.** infinitely many solutions   **7.** (−3, 1)   **8.** no solution   **9.** $\left(8, -\frac{1}{2}\right)$ **10.** 17, 20   **11.** Barb is 10, Cindy is 7   **12.** 14°, 76°   **13.** 12 years, 5 years   **14.** 10 nickels, 15 dimes   **15.** 15 pounds (80¢), 35 pounds ($1.00)   **16.** 20 mph   **17.** $2000

CHAPTER 5   TEST

1. Without graphing, determine the nature of the graphs (intersecting, parallel, or coinciding) of the pair of linear equations.

    $x + 3y = 1$
    $2x - y = 5$

    1. _____

2. Determine whether $(4, -2)$ is a solution to the following system.

    $3x + 4y = 4$
    $-2x - y = -10$

    2. _____

Solve the system of equations by whichever method seems most appropriate.

3. $2x - 5y = 11$
   $x + 3y = 0$

    3. _____

4. $3x - 2y = 1$
   $-6x + 4y = -1$

    4. _____

5. In a system of equations, if the lines are coinciding, how many solutions will the system have?

    5. _____

6. The sum of two numbers is 18 and their difference is 12. Find the two numbers.

6. _____

7. Two angles are supplementary and one is 12° more than six times the other. Find each angle.

7. _____

8. A collection of nickels and dimes has a value of $3.70. If there are 52 coins in the collection, how many of each are there?

8. _____

9. An airplane can fly 1000 miles with the wind in 2 hours and return against the wind in 2.5 hours. What is the speed of the plane in still air?

9. _____

6

POLYNOMIALS

6.1 BASIC CONCEPTS OF POLYNOMIALS

STUDENT GUIDEPOSTS

1 Polynomials
2 Coefficients
3 Monomials, binomials, and trinomials
4 Polynomials in one and in several variables
5 Degree of a polynomial
6 Ascending and descending order
7 Like terms
8 Collecting like terms

In Chapter 1 we defined a **variable** as a letter used to represent an unspecified number and said that an **algebraic expression** involved sums, differences, products, or quotients of numbers and variables. The **terms** in an algebraic expression are the parts that are separated by plus and minus signs.

1 A **polynomial** is an algebraic expression having its variables raised to whole number powers only. The following are examples of polynomials:

$$2x^2 + x - 1, \quad 2a - 3, \quad 7b, \quad \frac{7}{4}x - 3x^4, \quad \text{and} \quad -2x^2y + xy^2 + y.$$

Since the polynomial $2x^2 + x - 1$ can be written as $2x^2 + 1 \cdot x + (-1)$, its terms are

$$2x^2, \quad 1 \cdot x \text{ (or } x\text{)}, \quad -1$$

and the coefficients of these terms are

$$2, \quad 1, \quad -1,$$

2 respectively. Remember that the **(numerical) coefficient** of a term is its numerical factor. Also note that the minus sign goes with the term. Thus, in $\frac{7}{4}x - 3x^4$, $-3x^4$ is a term and -3 is its coefficient.

3 A polynomial with one term is a **monomial.** A **binomial** has two terms, and a **trinomial** involves three terms. When a polynomial has more than three terms, no special name is used and we simply call it a polynomial. Study the polynomials given in the following table.

Polynomial	Type	Terms	Coefficients
$5x - 2$	binomial	$5x, -2$	$5, -2$
6	monomial	6	6
$\frac{1}{3}y^2 + 2y - 3$	trinomial	$\frac{1}{3}y^2, 2y, -3$	$\frac{1}{3}, 2, -3$
$7a^3 + 2a^2 - 6a + 5$	polynomial	$7a^3, 2a^2, -6a, 5$	$7, 2, -6, 5$
$0.5x^3 + 0.1$	binomial	$0.5x^3, 0.1$	$0.5, 0.1$
$-4b^2$	monomial	$-4b^2$	-4

4 A polynomial such as $3x^3 + 4x^2 - x + 1$, which has the same variable in each term, is called a **polynomial in one variable.** If two or more variables are involved, such as in $-2a^2b + 3ab - b$, we have a **polynomial in several variables.** Our primary interest will be in polynomials in one variable, but we will discuss some of the properties of polynomials with several variables.

5 For polynomials involving only one variable, the **degree of a term** is the exponent on the variable. The **degree of a polynomial** is the degree of the term of highest degree.

EXAMPLE 1

For the polynomial $-3y^2 + 9y^4 - 6y^7 + y + 8$, list the terms and their degree. Also give the degree of the polynomial.

Term	Degree of the term	Reason
$-3y^2$	2	Exponent on y is 2
$9y^4$	4	Exponent on y is 4
$-6y^7$	7	Exponent on y is 7
y	1	Exponent on y is 1, since $y = y^1$
8	0	Exponent on y is 0, since $8 = 8 \cdot 1 = 8y^0$ (Remember that $y^0 = 1$)

Since $-6y^7$ has the highest degree, 7, the degree of the polynomial is 7. ◀◀

Practice Exercise 1

For the polynomial $-2x^3 + 9x^5 + 11$, list the terms and their degree. Also give the degree of the polynomial.

Answer: The term $-2x^3$ has degree 3, the term $9x^5$ has degree 5, and the term 11 has degree 0. The degree of the polynomial is 5.

6 If the polynomial in Example 1 is written in the order

$$-6y^7 + 9y^4 - 3y^2 + y + 8,$$

we say that it is in **descending order.** Note that the highest degree term is first, followed by the next highest, and so forth. The constant term is last since 8 can be written as $8y^0$. Written in **ascending order** the same polynomial would appear as

$$8 + y - 3y^2 + 9y^4 - 6y^7.$$

Most of the time we will use descending order.

EXAMPLE 2

Write $-3x + 5 - x^6 + x^3$ in descending order and in ascending order.

$$-x^6 + x^3 - 3x + 5 \quad \text{Descending order}$$

$$5 - 3x + x^3 - x^6 \quad \text{Ascending order} \blacktriangleleft\blacktriangleleft$$

Practice Exercise 2

Write $y^3 - 3y^5 + 12 - y$ in descending order and in ascending order.

Answer: Descending order:
$-3y^5 + y^3 - y + 12$
Ascending order: $12 - y + y^3 - 3y^5$

7   Terms that involve the variable raised to the same power are called **like terms**.

EXAMPLE 3

For the polynomial $7a^2 - 3a^3 + 2a - 5a^2 + 1 - 2a - 5$, give the like terms.

$$7a^2 \quad \text{and} \quad -5a^2$$
$$2a \quad \text{and} \quad -2a$$
$$1 \quad \text{and} \quad -5 \blacktriangleleft\blacktriangleleft$$

Practice Exercise 3

For the polynomial $6w^3 - 9 + 2w - w^3 + 8w + 4$, give the like terms.

Answer: $6w^3$ and $-w^3$, $2w$ and $8w$, and -9 and 4 are like terms.

8   When we have a polynomial like the one in Example 3, we may simplify it by **collecting like terms**.

EXAMPLE 4

Collect like terms and write each polynomial in descending order.

(A) $7a^2 - 3a^3 + 2a - 5a^2 + 1 - 2a - 5$
$= 7a^2 - 5a^2 - 3a^3 + 2a - 2a + 1 - 5$ Commute so that like terms are adjacent
$= (7 - 5)a^2 - 3a^3 + (2 - 2)a + (1 - 5)$ Use distributive law
$= 2a^2 - 3a^3 + 0 \cdot a - 4$
$= 2a^2 - 3a^3 - 4$
$= -3a^3 + 2a^2 - 4$ Descending order

(B) $-8x^3 + x^3 - 3x + 2 + 5x - 7$
$= -8x^3 + x^3 - 3x + 5x + 2 - 7$ Commute
$= (-8 + 1)x^3 + (-3 + 5)x + (2 - 7)$ Distributive law
$= -7x^3 + 2x - 5$

(C) $-3x^3 + x^7 - 7x - 4x^7 + x^3 + 1$
$= x^7 - 4x^7 - 3x^3 + x^3 - 7x + 1$
$= (1 - 4)x^7 + (-3 + 1)x^3 - 7x + 1$
$= -3x^7 - 2x^3 - 7x + 1 \blacktriangleleft\blacktriangleleft$

Practice Exercise 4

Collect like terms and write each polynomial in descending order.

(A) $3z^4 + z - 2z^4 - 5z + 1$

(B) $-9w^6 + 2w + 3w^6 - 5 - w^6 + 11$

(C) $y^3 - 3y + 6y^3 + 1 + 3y - 1 - 5y^3$

Answers: (A) $z^4 - 4z + 1$
(B) $-7w^6 + 2w + 6$ (C) $2y^3$

Like terms in polynomials with several variables must contain the same variables raised to the same power. The following table lists several terms along with like and unlike terms.

Term	Like term	Unlike term
$-2xy$	$5xy$	$6y$
$8a^2b^3$	$-2a^2b^3$	$3a^3b^2$
$9xy^4z$	$4xy^4z$	$-7x^2y^4z$
2	-10	$5a^2b^2$

EXAMPLE 5

Collect like terms.

$3x^2y - 5xy - 2x - 4x^2y + 4x + 2xy^2$
$= 3x^2y - 4x^2y - 5xy - 2x + 4x + 2xy^2$
$= (3 - 4)x^2y - 5xy + (-2 + 4)x + 2xy^2$
$= -x^2y - 5xy + 2x + 2xy^2$ ◀◀

Practice Exercise 5

Collect like terms.

$6u^2v - 3uv + 5u + 9uv - 5u^2v + uv^2$

Answer: $u^2v + uv^2 + 6uv + 5u$

Since variables stand for numbers, we may evaluate a polynomial if the values of the variables are given.

EXAMPLE 6

Evaluate the polynomials for the given values of the variables.

(A) $5x^2 - 3x + 1$ for $x = -2$

$5x^2 - 3x + 1 = 5(-2)^2 - 3(-2) + 1$ Substitute -2 for x
$= 5(4) + 6 + 1$
$= 20 + 6 + 1 = 27$

(B) $3ab + 2a - 5b$ for $a = -1$ and $b = 0$

$3ab + 2a - 5b = 3(-1)(0) + 2(-1) - 5(0)$
$= 0 - 2 - 0 = -2$ ◀◀

Practice Exercise 6

Evaluate the polynomials for the given values of the variables.

(A) $6y^3 - y + 5$ for $y = -1$

(B) $-2uv + v - 7u + 1$ for $u = -2$ and $v = 0$

Answers: (A) 0 (B) 15

EXAMPLE 7

The profit in dollars made by a machine manufacturer who sells x machines per week is $100x^3 - 2500$.

(A) Find the profit when 5 machines are sold.

$100x^3 - 2500 = 100(5)^3 - 2500$ Substitute 5 for x
$= 100(125) - 2500$
$= 12{,}500 - 2500 = 10{,}000$

Thus, the profit for the week was $10,000.

(B) Find the profit when 2 machines are sold.

$100x^3 - 2500 = 100(2)^3 - 2500$
$= 100(8) - 2500$
$= 800 - 2500 = -1700$

Since the profit is negative, the company lost $1700 that week. ◀◀

Practice Exercise 7

The cost in dollars of manufacturing a particular type of circuit board is given by $0.45n^2 + 180$, where n is the number of boards produced.

(A) Find the cost when 10 boards are made.

(B) Find the overhead cost that results when no boards are made ($n = 0$).

Answers: (A) $225 (B) $180

6.1 EXERCISES A

Fill in the following table.

	Polynomial	Type	Terms	Coefficients
1.	$-3x^2 + 2$	(A) _____	(B) _____	(C) _____
2.	$2a$	(A) _____	(B) _____	(C) _____
3.	$-4y^2 + 2y^3 - 7y + 1$	(A) _____	(B) _____	(C) _____
4.	$a^2 - 2a + \frac{1}{3}$	(A) _____	(B) _____	(C) _____
5.	-6	(A) _____	(B) _____	(C) _____
6.	$3x - 4y$	(A) _____	(B) _____	(C) _____
7.	$7a^2b^3$	(A) _____	(B) _____	(C) _____
8.	$x^2 - 2xy + y^2$	(A) _____	(B) _____	(C) _____

Determine the degree of each of the following terms.

9. $3x^2$ 10. $9b$ 11. -5 12. $-7b^{14}$

Determine the degree of each of the following polynomials.

13. $x^2 - 6x^3 + x^4 - 9x$ 14. $-3y^2 + 9y^5 + 2y - y^7$ 15. $a - 2$

16. 6 17. $2x^3 - x$ 18. $-y^2 - 8y + y^4 - 3y^3 + 5$

Collect like terms and write in descending order.

19. $8x - 3x$ 20. $-5y + 2y$

21. $10a^2 - 7a + 5a^2 + 3a$ 22. $-4b^2 - 9 + 12 - 6b^2$

23. $2x^4 - 2x^3 + x^3 - 3x^4 + 7 - 5$ 24. $3x^3 - 7 + 4x^2 - 3x^3 + 6 + 2x^3$

25. $7y^3 + 3y^2 - 7y^3 - 3y^2$ 26. $4y^4 - 2y^3 + 2y^3 - 4y^4 + 8$

27. $6a^2 + 7a + 8a^2 - 3 - 7a - 9a^2 + 1$ 28. $-8x^{10} + x^5 - 2x^{10} - 7x^5 + 1 - x^{10} + 3$

29. $\frac{3}{4}y^2 - \frac{1}{8}y^2$ 30. $-0.5b^3 + 0.77b^3$

Collect like terms.

31. $2xy - 5xy$ 32. $5x^2 + 2y^2$ 33. $2xy - 3y^2 + 5xy + y^2$

34. $a^2b^2 - ab + a^2b^2 + ab$ **35.** $-4x^2y + 2xy^2 + x^2 - 3x^2y$ **36.** $5ab^2 - 3ab + 3ab - 5ab^2$

Evaluate the polynomials for the given value of the variables.

37. $3x + 2$ for $x = 5$

38. $7y^2 - 2y - 5$ for $y = -2$

39. $8a^3 - 5$ for $a = -3$

40. $2a^2 + 5b^2$ for $a = 1$ and $b = -1$

41. The cost in dollars of manufacturing x bolts is given by the expression $0.05x + 15.5$. Find the cost when 440 bolts are made.

42. The number of grams of sugar required when x liters of milk are used in an ice cream mix is given by the expression $x^2 + 200x - 50$. Find the number of grams of sugar required for 5 liters of milk.

ANSWERS: **1.** (A) binomial (B) $-3x^2, 2$ (C) $-3, 2$ **2.** (A) monomial (B) $2a$ (C) 2 **3.** (A) polynomial (B) $-4y^2, 2y^3, -7y, 1$ (C) $-4, 2, -7, 1$ **4.** (A) trinomial (B) $a^2, -2a, \frac{1}{3}$ (C) $1, -2, \frac{1}{3}$ **5.** (A) monomial (B) -6 (C) -6 **6.** (A) binomial (B) $3x, -4y$ (C) $3, -4$ **7.** (A) monomial (B) $7a^2b^3$ (C) 7 **8.** (A) trinomial (B) $x^2, -2xy, y^2$ (C) $1, -2, 1$ **9.** 2 **10.** 1 **11.** 0 **12.** 14 **13.** 4 **14.** 7 **15.** 1 **16.** 0 **17.** 3 **18.** 4 **19.** $5x$ **20.** $-3y$ **21.** $15a^2 - 4a$ **22.** $-10b^2 + 3$ **23.** $-x^4 - x^3 + 2$ **24.** $2x^3 + 4x^2 - 1$ **25.** 0 **26.** 8 **27.** $5a^2 - 2$ **28.** $-11x^{10} - 6x^5 + 4$ **29.** $\frac{5}{8}y^2$ **30.** $0.27b^3$ **31.** $-3xy$ **32.** $5x^2 + 2y^2$ (no like terms) **33.** $7xy - 2y^2$ **34.** $2a^2b^2$ **35.** $-7x^2y + 2xy^2 + x^2$ **36.** 0 **37.** 17 **38.** 27 **39.** -221 **40.** 7 **41.** \$37.50 **42.** 975 g

---

6.1 EXERCISES B

Give the type and list the terms and coefficients of the following polynomials.

1. $7x - 5$ **2.** $3y^2 - 2y + 3$ **3.** $-4a^3 + 6a + a^4 - 7$ **4.** 22

5. $-12b^{15}$ **6.** $-8a + b$ **7.** $4x^2 + 4xy - y^2$ **8.** $9x^4y^6$

Determine the degree of each of the following terms.

9. $5y^7$ **10.** $-2x^{32}$ **11.** $4a^0$ **12.** $44b^{12}$

Determine the degree of each of the following polynomials.

13. $3 + 4x^{10}$ **14.** $6y^4 + 12y^2 - 8y + 7y^5$ **15.** $-2 - 3a$

16. -9 **17.** $-x^4 - x^2 + 9x$ **18.** $5y^{10} + 14y^{20} - 2y^{15} + 3$

Collect like terms and write in descending order.

19. $-2x + 10x$ **20.** $y - 9y$

21. $a^3 - 3a^2 + 4a^3 - a^2$ **22.** $5 - b^4 + 3b^4 - 8$

23. $3x - 8x^2 + 5x^3 - 2x^3 + 4x^2 - 2x$ **24.** $3y + 2 - 7y^2 - 2y - 5$

254 POLYNOMIALS

25. $11a^3 - 17a^4 + 8a^2 - 5 + a^3 - 6 + a^4$

26. $-3b + 21b^7 - 5b^3 + 12b - b^7$

27. $10 - 3y + 7y^2 - 8y^3$

28. $-22x^3 + 17x^5 - 4x^3 + 2$

29. $0.2b - 0.8b^2 + 0.7b^2$

30. $\frac{3}{2}a^3 - \frac{1}{3}a^4 + \frac{1}{4}a^3 + \frac{2}{9}a^4$

Collect like terms.

31. $-9xy + 7xy$

32. $5x^2y^2 - 5$

33. $x^2 - 3xy + 8xy - 4x^2$

34. $2a^2b - a^2b^2 - 2a^2b + a^2b^2$

35. $2x^2y^2 - 3xy^2 + 5x^2y^2 + xy^2$

36. $5ab - 2a^2 - 2a^2 + 5ab + 6$

Evaluate the polynomials for the given values of the variables.

37. $-8x + 5$ for $x = 4$

38. $-2y^2 + 3y - 2$ for $y = -3$

39. $2a^3 + a^2$ for $a = 5$

40. $5a^2b^2 - 2$ for $a = 9$ and $b = 0$

41. The profit in dollars when x pairs of shoes are sold is given by the expression $2x^2 - 120$. Find the profit when 10 pairs are sold.

42. The cost of making dresses is described as ten times the number of dresses plus eight. Use x as the number of dresses and write a polynomial to describe this cost. Find the cost of making twenty dresses.

6.1 EXERCISES C

The degree of a term of polynomial in several variables is found by adding the exponents on the variables. Find the degree of each term.

1. $3x^2y^3$
2. $-8xy^6$
3. $6xyz$ [Answer: 3]
4. $-9a^4b^2c^5$

Write each polynomial in descending powers of x.

5. $7xy^3 - 4x^2y + 8x^4y^4$
   [Answer: $8y^4y^4 - 4x^2y + 7xy^3$]

6. $-3x^2yz^4 + 4yz - 8xyz$

6.2 ADDITION AND SUBTRACTION OF POLYNOMIALS

▶▶ STUDENT GUIDEPOSTS

1 Adding polynomials
2 Subtracting polynomials

When adding polynomials we proceed as follows.

 To add polynomials

1. Indicate the addition with a plus sign.
2. Remove parentheses and collect like terms.

EXAMPLE 1

Add $2x^2 - 3x + 5$ and $-x^2 + 6x + 2$.

$(2x^2 - 3x + 5) + (-x^2 + 6x + 2)$

Practice Exercise 1

Add $3y^3 - y + 5$ and $-2y^3 + 6y - 3$.

$$= 2x^2 - 3x + 5 - x^2 + 6x + 2$$ Remove parentheses
$$= (2-1)x^2 + (-3+6)x + (5+2)$$ Collect like terms
$$= x^2 + 3x + 7 \;\blacktriangleleft$$

Answer: $y^3 + 5y + 2$

When adding polynomials in one variable, we usually arrange the terms in descending order before collecting like terms.

EXAMPLE 2

Add $-3x + x^4 - 5x^3 + 7$ and $3 - 2x - 5x^4 + 8x^2$.

$(x^4 - 5x^3 - 3x + 7) + (-5x^4 + 8x^2 - 2x + 3)$ Arrange in descending order and indicate addition

$$= x^4 - 5x^3 - 3x + 7 - 5x^4 + 8x^2 - 2x + 3$$ Remove parentheses
$$= (1-5)x^4 - 5x^3 + 8x^2 + (-3-2)x + (7+3)$$ Collect like terms
$$= -4x^4 - 5x^3 + 8x^2 - 5x + 10 \;\blacktriangleleft$$

Practice Exercise 2

Add $2z - 5z^6 + z^3 - 5$ and $4 + z^6 - 8z + 3z^2$.

Answer: $-4z^6 + z^3 + 3z^2 - 6z - 1$

In some cases it is easier to add polynomials by arranging like terms in vertical columns and then adding. This technique, called the column method, will be used later when we multiply polynomials.

EXAMPLE 3

Add $2 - 3x^2$, $-x^4 + 7x - 3x^3 - 5$, and $-8x^3 + 4x^2 - 7x$ using the column method.

$$\begin{array}{r} -3x^2 + 2 \\ -x^4 - 3x^3 + 7x - 5 \\ -8x^3 + 4x^2 - 7x \\ \hline -x^4 - 11x^3 + x^2 + 0x - 3 = -x^4 - 11x^3 + x^2 - 3 \;\blacktriangleleft \end{array}$$

Note the spaces where terms are missing

Practice Exercise 3

Add $y^5 + 1 - 3y$, $y - 2y^4$, and $2y - y^3 + 3y^5 - 5$ using the column method.

Answer: $4y^5 - 2y^4 - y^3 - 4$

When adding polynomials it does not matter in which order we write them. However, when subtracting we must be careful to indicate the subtraction correctly. Thus, when the instructions are to subtract

$$3y - 4 \quad \text{from} \quad 2y + 5$$

we write

$$(2y + 5) - (3y - 4)$$

In this case $3y - 4$ is being subtracted. However, if we see

$$(y^2 + 3) - (-2y + 1)$$

the problem is already set up for us.

To subtract polynomials

1. Indicate the subtraction by putting a minus sign before the polynomial to be subtracted.
2. Remove parentheses, changing all signs in the polynomial being subtracted.
3. Collect like terms as in addition.

256 POLYNOMIALS

It helps to arrange terms in descending order when subtracting polynomials of one variable.

EXAMPLE 4

Subtract $7x - 3x^2$ from $4 - 2x - 4x^2$.

$(-4x^2 - 2x + 4) - (-3x^2 + 7x)$ Arrange in descending order and indicate the subtraction

$= -4x^2 - 2x + 4 + 3x^2 - 7x$ Remove parentheses, changing signs

$= -x^2 - 9x + 4$ Combine like terms

To do the same problem by the column method, we change the signs in $7x - 3x^2$ and it becomes $-7x + 3x^2$. We then arrange like terms in vertical columns and add.

$$\begin{array}{r} -4x^2 - 2x + 4 \\ \underline{+3x^2 - 7x} \\ -x^2 - 9x + 4 \end{array}$$ Change signs   Add ◀◀

Practice Exercise 4

Subtract $5z^2 + 2$ from $1 - z^2 + 8z$.

Answer: $-6z^2 + 8z - 1$

EXAMPLE 5

Subtract $-3x + 7x^4 + 5x^5 - 2$ from $2x^4 - 8x^5 - 4x + 12x^2 - x^3$.

$(-8x^5 + 2x^4 - x^3 + 12x^2 - 4x) - (5x^5 + 7x^4 - 3x - 2)$

$= -8x^5 + 2x^4 - x^3 + 12x^2 - 4x - 5x^5 - 7x^4 + 3x + 2$ Change signs

$= (-8 - 5)x^5 + (2 - 7)x^4 - x^3 + 12x^2 + (-4 + 3)x + 2$ Commute and distribute

$= -13x^5 - 5x^4 - x^3 + 12x^2 - x + 2$

By the column method we would proceed as follows:

$$\begin{array}{r} -8x^5 + 2x^4 - x^3 + 12x^2 - 4x \\ \underline{-5x^5 - 7x^4 + 3x + 2} \\ -13x^5 - 5x^4 - x^3 + 12x^2 - x + 2. \end{array}$$ Change signs   Add ◀◀

Practice Exercise 5

Subtract $1 - 6w^6 + 2w - 5w^5$ from $4w + 2w^6 - 7 - 8w^5 + w^3$.

Answer: $8w^6 - 3w^5 + w^3 + 2w - 8$

EXAMPLE 6

Perform the indicated operations.

$(5x^3 - 6 + x^2) + (7x^2 - 3) - (-4x^3 + 7x - 5)$

$= (5x^3 + x^2 - 6) + (7x^2 - 3) - (-4x^3 + 7x - 5)$

$= 5x^3 + x^2 - 6 + 7x^2 - 3 + 4x^3 - 7x + 5$ Change signs on polynomial being subtracted

$= (5 + 4)x^3 + (1 + 7)x^2 - 7x + (5 - 6 - 3)$

$= 9x^3 + 8x^2 - 7x - 4$

By the column method we proceed as follows:

$$\begin{array}{r} 5x^3 + x^2 - 6 \\ 7x^2 - 3 \\ \underline{+4x^3 - 7x + 5} \\ 9x^3 + 8x^2 - 7x - 4. \end{array}$$ Change all signs   Add ◀◀

Practice Exercise 6

Perform the indicated operations.

$(6y^4 + 5 - y^3) - (4 - y^3)$
$+ (y^2 - 5y^4 + 2)$

Answer: $y^4 + y^2 + 3$

For polynomials in several variables, the procedures are exactly the same. Care must be taken to insure that only like terms are combined.

EXAMPLE 7

Add $8x^2 + 2y^2 - 3xy$ and $6x^2 - 7y^2 - 9xy + 3$.

$$(8x^2 + 2y^2 - 3xy) + (6x^2 - 7y^2 - 9xy + 3)$$
$$= 8x^2 + 2y^2 - 3xy + 6x^2 - 7y^2 - 9xy + 3$$
$$= (8 + 6)x^2 + (2 - 7)y^2 + (-3 - 9)xy + 3$$
$$= 14x^2 - 5y^2 - 12xy + 3$$ ◀◀

Practice Exercise 7

Add $6a^3 + b^2 - 4ab$ and $2b^2 - 3a^3 + 2ab - 5$.

Answer: $3a^3 + 3b^2 - 2ab - 5$

EXAMPLE 8

Subtract $3ab - 5a + 9$ from $-6ab + 2a + b$.

$$(-6ab + 2a + b) - (3ab - 5a + 9)$$
$$= -6ab + 2a + b - 3ab + 5a - 9 \quad \text{Change signs}$$
$$= (-6 - 3)ab + (2 + 5)a + b - 9 \quad \text{Collect like terms}$$
$$= -9ab + 7a + b - 9$$ ◀◀

Practice Exercise 8

Subtract $6uv - 3v^2 + 12$ from $v^2 - 2uv + 5$.

Answer: $4v^2 - 8uv - 7$

EXAMPLE 9

Perform the indicated operations.

$$(7x^2y^2 - 2xy) - (5x^2y + 9xy) - (-x^2y^2 + 3xy^2 - 4xy)$$
$$= 7x^2y^2 - 2xy - 5x^2y - 9xy + x^2y^2 - 3xy^2 + 4xy \quad \text{Change signs}$$
$$= (7 + 1)x^2y^2 + (-2 - 9 + 4)xy - 5x^2y - 3xy^2 \quad \text{Collect like terms}$$
$$= 8x^2y^2 - 7xy - 5x^2y - 3xy^2$$

Note that $-5x^2y$ and $-3xy^2$ are not like terms. ◀◀

Practice Exercise 9

Perform the indicated operations.

$$(9a^3b^3 + ab) - (3ab - a^3b^3) + (8a - 5a^3b^3 + 2ab)$$

Answer: $5a^3b^3 + 8a$

6.2 EXERCISES A

Add.

1. $3x - 5$ and $-8x + 4$

2. $7x^2 + 6$ and $x^2 - 2$

3. $y^2 + 3y - 5$ and $-8y^2 - 5y + 9$

4. $2y - 3$ and $-4y^2 + 2$

5. $3x^5 - 2x^3 + 5x^2$ and $-5x^5 + 8x^4 - 7x^2$

6. $-x^6 + 2x^4$ and $2x^7 - x^6 + 5x^2 + 2$

7. $8y - 6y^2 + 2$ and $-5 + 2y + 8y^2$

8. $6y^5 - 2y + y^4$ and $3 - 5y^2 + 21y + 2y^5$

9. $2a^2 + 3a$, $-5a^2 + 6$, and $-9a + 2$

10. $-4a^3 + 7a^4 + 3a + 2$, $5 - 3a + 7a^3$, and $17a^4 - 5 + 12a^3$

11. $2 - x^2 + 7x$, $9x - 7 + x^3$, $-6x$, and $12 - 3x^3$

12. $25x^4 - 7x^5$, $12x - 17x^2 + 40x^3 - 18x^4$, and $x - 21 + 18x^5$

13. $\quad 3y^2 - 2y + 1$
    $\quad \underline{-2y^2 + 2y + 8}$

14. $\quad 5x^5 \quad\quad + x^3 \quad\quad - 2x$
    $\quad \underline{2x^5 - 8x^4 - 7x^3 + 6x^2}$

15. $\quad\quad\quad\quad 5x^3 + 6x^2 \quad\quad - 7$
    $\quad\quad -4x^4 + 3x^3 - 3x^2 - 8x$
    $\quad\quad \underline{3x^4 - 2x^3 + 4x^2 \quad\quad + 1}$

16. $\quad 0.03y^3 - 0.75y^2 - 3y + 2$
    $\quad -0.15y^3 \quad\quad\quad + 5y - 0.3$
    $\quad \underline{0.21y^3 - 0.13y^2 \quad\quad + 0.6}$

Subtract.

17. $2x + 5$ from $3x - 6$

18. $6x^2 - 2$ from $4x^2 + 5$

19. $-3y^2 + 2y - 7$ from $-8y^2 + y - 9$

20. $-3y + 2$ from $y^2 - 4y - 5$

21. $a^4 + 5a^3 - 3a^2$ from $7a^5 - 2a^4 - a^2 + 3a$

22. $3a - 2a^2 + 7a^3$ from $a^4 - 2a^2 + a^3 - 2$

23. $3x^4 - 7x^5 + 15x - 32$ from
    $56x - 93x^3 + 21x^4 + 32x^5$

24. $7x^{10} - 4x^5 + 1$ from $3x^5 + 1 - x^6 - 3x^2$

Perform the indicated operations.

25. $(-8y^2 + 4) - (7y^2 - 3)$

26. $(-8a^2 - 2a + 5) - (-8a^2 + a + 1)$

27. $(6y^7 - y + y^5 + 2) - (2y + 3y^2 - 4y^5 - 2)$

28. $(8y^{10} - y^8) - (3y^{12} + 2y^{10} - y^8)$

29. $(3y^2 + 2) + (-5y - 5) - (y^2 - 2y + 10)$

30. $(-2a^2 + 3a) - (a^2 + a + 1) - (a^2 - 2a - 5)$

31. $(9x^4 + 3x^3 + 8x) + (3x^4 + x^3 - 7x^2) - (12x^4 - 3x^2 + x)$

32. $(9y^4 + 3y^3 + 8y) - (3y^4 + y^3 - 7y^2) - (12y^4 - 3y^2 + y)$

33. $(4a^2b^2 - 2ab) + (5a^2b^2 + 9ab)$

34. $(3a^2 + 2b^2 + 3) + (-6a^2 + 2b^2 - 2)$

35. $(4x^2y^2 - 2xy) - (5x^2y^2 + 9xy)$

36. $(3x^2 + 2y^2 + 3) - (-6x^2 + 2y^2 - 2)$

37. $(-2a^2b + ab - 4ab^2) + (6a^2b + 4ab^2)$

38. $(-2a^2b + ab - 4ab^2) - (6a^2b + 4ab^2)$

39. $(6x^2y - xy) + (3x^2y - 7xy^2) - (4xy - 5xy^2)$

40. $(6x^2y - xy) - (3x^2y - 7xy^2) - (4xy - 5xy^2)$

For Review

Determine the degree of each polynomial.

41. $-6y + 8y^5 + y^4 - 2$ **42.** 14 **43.** $x^{10} - 6x^{20} + x^{30}$

Evaluate the polynomial for the given value of the variable.

44. $-7a + 3$ for $a = -2$ **45.** $3y^3 + 2y^2$ for $y = -1$

46. The profit in dollars when x suits are sold is given by the expression $7x - 50$. Find the profit when 60 suits are sold.

ANSWERS: **1.** $-5x - 1$ **2.** $8x^2 + 4$ **3.** $-7y^2 - 2y + 4$ **4.** $-4y^2 + 2y - 1$ **5.** $-2x^5 + 8x^4 - 2x^3 - 2x^2$ **6.** $2x^7 - 2x^6 + 2x^4 + 5x^2 + 2$ **7.** $2y^2 + 10y - 3$ **8.** $8y^5 + y^4 - 5y^2 + 19y + 3$ **9.** $-3a^2 - 6a + 8$ **10.** $24a^4 + 15a^3 + 2$ **11.** $-2x^3 - x^2 + 10x + 7$ **12.** $11x^5 + 7x^4 + 40x^3 - 17x^2 + 13x - 21$ **13.** $y^2 + 9$ **14.** $7x^5 - 8x^4 - 6x^3 + 6x^2 - 2x$ **15.** $-x^4 + 6x^3 + 7x^2 - 8x - 6$ **16.** $0.09y^3 - 0.88y^2 + 2y + 2.3$ **17.** $x - 11$ **18.** $-2x^2 + 7$ **19.** $-5y^2 - y - 2$ **20.** $y^2 - y - 7$ **21.** $7a^5 - 3a^4 - 5a^3 + 2a^2 + 3a$ **22.** $a^4 - 6a^3 - 3a - 2$ **23.** $39x^5 + 18x^4 - 93x^3 + 41x + 32$ **24.** $-7x^{10} - x^6 + 7x^5 - 3x^2$ **25.** $-15y^2 + 7$ **26.** $-3a + 4$ **27.** $6y^7 + 5y^5 - 3y^2 - 3y + 4$ **28.** $-3y^{12} + 6y^{10}$ **29.** $2y^2 - 3y - 13$ **30.** $-4a^2 + 4a + 4$ **31.** $4x^3 - 4x^2 + 7x$ **32.** $-6y^4 + 2y^3 + 10y^2 + 7y$ **33.** $9a^2b^2 + 7ab$ **34.** $-3a^2 + 4b^2 + 1$ **35.** $-x^2y^2 - 11xy$ **36.** $9x^2 + 5$ **37.** $4a^2b + ab$ **38.** $-8a^2b + ab - 8ab^2$ **39.** $9x^2y - 5xy - 2xy^2$ **40.** $3x^2y - 5xy + 12xy^2$ **41.** 5 **42.** 0 **43.** 30 **44.** 17 **45.** -1 **46.** \$370

6.2 EXERCISES B

Add.

1. $9x + 3$ and $-4x - 2$

2. $-8x^2 + 5$ and $3x^2 + 2$

3. $8a^2 + a - 7$ and $-a + 3$

4. $2a^4 - 5a^2 + 6$ and $3a^2$

5. $6x^4 + 2x^3 - 7x^2$ and $-9x^5 + 3x^4 - x^3 + 7$

6. $6x^9 - 3x^6 + 2x^3$ and $2x^8 - 4x^6 - 9x^3$

7. $-8y^3 + 5 - 6y^2$ and $3 - 2y^2 + y^3$

8. $3y - 5y^3 - 2y^2 + 2$ and $9 + 2y^2 - y^3$

9. $-12a^2 - 6$, $2a^2 + 5a$, and $a - 3$

10. $-9a^3 + a^4 - 6$, $2a - 5a^4 + a^3$, and $8a^4 - 2a^3 + 5$

11. $x^2 - x^3 + x$, $5x - 6x^3 + 3$, 9, and $3 + 2x^3$

12. $-3x^5 + 9x^2$, $6 - 12x^5 + 6x^2$, and $3 - 6x + 5x^5 + x^2$

13. $-9y^2 + 2y - 6$
    $6y^2 + 3y + 9$

14. $-8x^4 - 2x^2 + 6x - 8$
    $7x^4 + 12x^3 + 2x^2 + 4$

6.2 ADDITION AND SUBTRACTION OF POLYNOMIALS

15.
$$\begin{array}{r} -2x^3 + 8x^2 - 9x + 9 \\ 7x^4 + 2x^2 + 9x - 8 \\ -6x^4 + 6x^3 - 4x^2 + 7 \\ \hline \end{array}$$

16.
$$\begin{array}{r} 0.23y^3 + 0.98y^2 - 0.6y + 0.8 \\ 0.54y^3 - 0.82y^2 + 0.1y \\ -0.77y^3 + 0.9y - 0.3 \\ \hline \end{array}$$

Subtract.

17. $3x - 5$ from $2x + 8$

18. $2x^2 - 9$ from $-5x^2 + 7$

19. $6y^2 - 2y + 8$ from $-4y^2 - 2y + 1$

20. $-3y^2 + 9$ from $2y^2 - 9y + 2$

21. $2a^4 - 3a^3 + 5a^2$ from $-9a^5 + 2a^4 - 3a^2 + 2a$

22. $5a - 3a^2 + 5a^3$ from $a^4 + 5a^2 + 5a^3 + 3$

23. $5x^4 + 7x^5 - 8x + 20$ from $19x + 20x^3 + 72x^4 - 10x^5$

24. $10x^8 + 10x^6 - 13$ from $2x^4 - 13 + 8x^8 - x^2$

Perform the indicated operations.

25. $(7y^2 - 8) - (-2y^2 + 7)$

26. $(9a^2 + 7a - 4) - (-2a^2 + 7a - 7)$

27. $(-6y^7 + 4y - y^5 + 8) - (5y - 4y^2 + 7y^5 - 2)$

28. $(3y^{14} - 2y^{10}) - (-y^{14} + 2y^{10} + y^6)$

29. $(5y + 6) + (-2y^2 + 5) - (2y^2 - 3y - 8)$

30. $(-8a^2 - 2a) - (2a^2 + 3a - 1) - (a^2 - 9a + 6)$

31. $(3x^4 - 8x^3 - 4x) + (2x^4 - 8x^3 + 5x^2) - (5x^4 + 5x^2 - 6x)$

32. $(3y^4 - 8y^3 - 4y) - (2y^4 - 8y^3 + 5y^2) - (5y^4 + 5y^2 - 6y)$

33. $(2a^2b^2 + 5ab) + (9a^2b^2 - 2ab)$

34. $(4a^2 - 3b^2 + 9) + (-4a^2 - 5b^2 + 3)$

35. $(8x^2y^2 - 9xy) - (-4x^2y^2 + 7xy)$

36. $(5x^2 + 3y^2 - 7) - (-2x^2 + 3y^2 + 7)$

37. $(5a^2b - 2ab - 2ab^2) + (2a^2b - 6ab^2)$

38. $(5a^2b - 2ab - 2ab^2) - (2a^2b - 6ab^2)$

39. $(11x^2y - 2xy) + (4x^2y - 9xy^2) - (8xy - 10xy^2)$

40. $(11x^2y - 2xy) - (4x^2y - 9xy^2) - (8xy - 10xy^2)$

For Review

Determine the degree of each polynomial.

41. $8y^2 - 6y^{10} - y^{14}$

42. $22a$

43. $x^{20} + 2x^{40}$

Evaluate the polynomial for the given value of the variable.

44. $6a^2 - 2a$ for $a = 3$

45. $y^4 - 6y^2 + 8$ for $y = -2$

46. The cost in dollars when x suits are made is given by the expression $4x^2 + 5x - 100$. Find the cost when 15 suits are made.

6.2 EXERCISES C

Perform the indicated operations.

1. $\left(\dfrac{1}{3}x^4 - \dfrac{1}{2}x^3 + \dfrac{7}{8}\right) + \left(-\dfrac{1}{9}x^4 + \dfrac{3}{4}x^3 - \dfrac{2}{3}x^2 + x\right) + \left(\dfrac{2}{9}x^4 - \dfrac{3}{2}x^3 + \dfrac{4}{3}x^2 + \dfrac{1}{3}x + \dfrac{3}{8}\right)$
   $\left[\text{Answer: } \dfrac{4}{9}x^4 - \dfrac{5}{4}x^3 + \dfrac{2}{3}x^2 + \dfrac{4}{3}x + \dfrac{5}{4}\right]$

2. $(100a^3 - 97a^2 + 21a - 105) + (16a^3 + 45a^2 - 115a) - (88a^3 - 19a^2 - 47)$

3. $(-2.13a^2 + 3.25a + 7.98) - (0.32a^3 - 2.10a + 1.92)$ [Answer: $-0.32a^3 - 2.13a^2 + 5.35a + 6.06$]

4. $\left(-\dfrac{1}{9}x^5 + \dfrac{1}{3}x^4 - \dfrac{2}{3}x + \dfrac{1}{3}\right) - \left(-\dfrac{5}{9}x^5 - \dfrac{2}{9}x^4 + \dfrac{8}{9}x + \dfrac{7}{9}\right)$

6.3 MULTIPLICATION OF POLYNOMIALS

STUDENT GUIDEPOSTS

1. Multiplying monomials
2. Multiplying a binomial by a monomial
3. Multiplying two polynomials
4. The FOIL method

Recall that when we multiply variables that are raised to powers we add the exponents. Thus, the rule $x^m x^n = x^{m+n}$ is used in multiplying polynomials. We begin by multiplying monomials.

To multiply monomials
1. Multiply the coefficients.
2. Use the rule $x^m x^n = x^{m+n}$ to multiply the variables.

EXAMPLE 1

Multiply.

(A) $(-3x^2)(7x^5) = (-3)(7)x^2 x^5$ The order of the factors can be changed
$= -21x^{2+5}$ Use $x^m x^n = x^{m+n}$
$= -21x^7$

(B) $(5x^4)(-x) = (5)(-1)x^4 x$ $-x = (-1) \cdot x$
$= -5x^{4+1}$ $x = x^1$
$= -5x^5$ ◀◀

Practice Exercise 1

Multiply.

(A) $(4y^3)(-5y^7)$

(B) $(-8y^2)(-y)$

Answers: **(A)** $-20y^{10}$ **(B)** $8y^3$

To multiply a binomial by a monomial
1. Use the distributive property of multiplication over addition to multiply both terms of the binomial by the monomial.
2. Use the rules for multiplying monomials.

EXAMPLE 2

Multiply.

(A) $a(3a^2 + 2) = (a)(3a^2) + (a)(2)$ Distributive law
$= 3aa^2 + 2a$
$= 3a^3 + 2a$

(B) $-3a^2(8a^3 - 5a) = (-3a^2)(8a^3) + (-3a^2)(-5a)$ Distributive law
$= (-3)(8)a^2 a^3 + (-3)(-5)a^2 a$
$= -24a^{2+3} + 15a^{2+1}$
$= -24a^5 + 15a^3$ ◀◀

Practice Exercise 2

Multiply.

(A) $x(5 + 3x^3)$

(B) $-4x^3(2x^2 - 7x)$

Answers: **(A)** $5x + 3x^4$
(B) $-8x^5 + 28x^4$

To multiply two polynomials, neither of which is a monomial

1. Multiply each term in one by each term in the other.
2. Collect like terms.

One way to do this is to distribute one polynomial over the terms of the other.

EXAMPLE 3

Multiply.

(A) $(3x + 2)(x^2 - 5x) = (3x + 2)x^2 + (3x + 2)(-5x)$ Distributive law
$= (3x)(x^2) + (2)(x^2) + (3x)(-5x) + (2)(-5x)$ Distributive law
$= 3x^3 + 2x^2 - 15x^2 - 10x$
$= 3x^3 - 13x^2 - 10x$

(B) $(4x + 1)(3x^2 - x + 5) = (4x + 1)3x^2 + (4x + 1)(-x) + (4x + 1)5$
$= (4x)(3x^2) + (1)(3x^2) + (4x)(-x) +$
$(1)(-x) + (4x)(5) + (1)(5)$
$= 12x^3 + 3x^2 - 4x^2 - x + 20x + 5$
$= 12x^3 - x^2 + 19x + 5$

Practice Exercise 3

Multiply.

(A) $(2z + 1)(z^3 - 3z)$

(B) $(5z - 3)(2z^2 + z - 7)$

Answers: (A) $2z^4 + z^3 - 6z^2 - 3z$
(B) $10z^3 - z^2 - 38z + 21$

We can do the same problem in Example 3(A) without going through the first step above.

$(3x + 2)(x^2 - 5x) = (3x)(x^2) + (3x)(-5x) + (2)(x^2) + (2)(-5x)$
$= 3x^3 - 15x^2 + 2x^2 - 10x$
$= 3x^3 - 13x^2 - 10x$

Notice that we added together the products of the Ⓕirst terms, the Ⓞutside terms, the Ⓘnside terms, and the Ⓛast terms. These letters, F, O, I, L spell "FOIL." Remember this word, and you will not omit terms when multiplying. We call this the **FOIL method**.

▶▶ **CAUTION** ◀◀   The FOIL method applies only to the multiplication of two binomials. ◀◀

Foil method

$(a + b)(c + d) = ac + ad + bc + bd$

EXAMPLE 4

Use the FOIL method to multiply the binomials.

(A) $(x - 3)(x + 4) = x^2 + 4x + 3x - 12$

(B) $(x + 7)(x + 5) = x^2 + 5x + 7x + 35$
$= x^2 + 12x + 35$ ◂◂

Practice Exercise 4

Use the FOIL method to multiply the binomials.

(A) $(a - 2)(a + 8)$

(B) $(a + 2)(a + 9)$

Answers: (A) $a^2 + 6a - 16$
(B) $a^2 + 11a + 18$

When we study factoring in Chapter 7, we will start with trinomials such as $x^2 + x - 12$ in Example 4(A) and be expected to find the factors $(x - 3)$ and $(x + 4)$. Practicing the FOIL method now will help us understand factoring later.

EXAMPLE 5

Use the FOIL method to multiply the binomials.

(A) $(2x - 3)(3x + 7) = 6x^2 + 14x - 9x - 21$
$= 6x^2 + 5x - 21$

(B) $(2a - 5)(2a + 5) = 4a^2 + 10a - 10a - 25$
$= 4a^2 + 0 \cdot a - 25$
$= 4a^2 - 25$

(C) $(5x + 3)(x + 8) = 5x^2 + 40x + 3x + 24$
$= 5x^2 + 43x + 24$

(D) $(7y - 2)(3y - 8) = 21y^2 - 56y - 6y + 16$
$= 21y^2 - 62y + 16$

(E) $(10a + 5)(3a - 5) = 30a^2 - 50a + 15a - 25$
$= 30a^2 - 35a - 25$ ◂◂

Practice Exercise 5

Use the FOIL method to multiply the binomials.

(A) $(3y - 1)(2y + 4)$

(B) $(6x - 7)(6x + 7)$

(C) $(8z + 3)(z - 1)$

(D) $(4a - 5)(2a - 3)$

(E) $(12x + 1)(5x - 2)$

Answers: (A) $6y^2 + 10y - 4$
(B) $36x^2 - 49$ (C) $8z^2 - 5z - 3$
(D) $8a^2 - 22a + 15$
(E) $60x^2 - 19x - 2$

Another way to multiply is to write one polynomial above the other and multiply term by term, arranging like terms in vertical columns. For example, to multiply the binomials in Example 3(A), we would proceed as follows.

$$
\begin{array}{r}
x^2 - 5x \\
3x + 2 \\
\hline
3x^3 - 15x^2 \\
2x^2 - 10x \\
\hline
3x^3 - 13x^2 - 10x
\end{array}
$$

$3x$ multiplied by each term of the top polynomial
2 multiplied by each term of the top polynomial
Sum

This method is especially good when trinomials or larger polynomials are involved in the multiplication.

EXAMPLE 6

Multiply.

(A) $\quad 5x^3 - 2x + 3$
$\quad\quad \underline{-7x^2 + 4x}$
$\quad -35x^5 \quad\quad + 14x^3 - 21x^2 \quad\quad$ $-7x^2$ multiplied by top polynomial
$\quad\quad\quad\quad 20x^4 \quad\quad - 8x^2 + 12x \quad\quad$ $4x$ multiplied by top polynomial
$\quad \overline{-35x^5 + 20x^4 + 14x^3 - 29x^2 + 12x} \quad\quad$ Sum

We leave spaces for missing terms so that we can add like terms in columns.

(B) $\quad a^2 + 2a + 4$
$\quad\quad \underline{a - 2}$
$\quad a^3 + 2a^2 + 4a \quad\quad$ a multiplied by top polynomial
$\quad\quad \underline{- 2a^2 - 4a - 8} \quad\quad$ -2 multiplied by top polynomial
$\quad a^3 \quad\quad\quad\quad - 8 = a^3 - 8 \quad$ Sum

(C) $\quad 5x^2 - 3x + 7$
$\quad\quad \underline{-4x^2 + x + 8}$
$\quad -20x^4 + 12x^3 - 28x^2 \quad\quad$ $-4x^2$ multiplied by top polynomial
$\quad\quad\quad\quad 5x^3 - 3x^2 + 7x \quad\quad$ x multiplied by top polynomial
$\quad\quad\quad\quad\quad\quad\quad 40x^2 - 24x + 56 \quad\quad$ 8 multiplied by top polynomial
$\quad \overline{-20x^4 + 17x^3 + 9x^2 - 17x + 56} \quad$ Sum ◀◀

All the procedures used above can be applied to the multiplication of polynomials in several variables.

EXAMPLE 7

Use the FOIL method to multiply the binomials in two variables.

(A) $(2x + y)(x - 3y) = 2x^2 - 6xy + xy - 3y^2$
$\quad\quad\quad\quad\quad\quad\quad\quad = 2x^2 - 5xy - 3y^2$

(B) $(2a - 5b)(2a + 5b) = 4a^2 + 10ab - 10ab - 25b^2$
$\quad\quad\quad\quad\quad\quad\quad\quad\quad = 4a^2 + 0 \cdot ab - 25b^2$
$\quad\quad\quad\quad\quad\quad\quad\quad\quad = 4a^2 - 25b^2$

(C) $(5x + 3y)(x + 8y) = 5x^2 + 40xy + 3xy + 24y^2$
$\quad\quad\quad\quad\quad\quad\quad\quad = 5x^2 + 43xy + 24y^2$ ◀◀

EXAMPLE 8

Use the column method to multiply.

$\quad 3x^2 - 4xy + 5y^2$
$\quad \underline{6x - 7y}$
$\quad 18x^3 - 24x^2y + 30xy^2 \quad\quad$ $6x$ multiplied by top polynomial
$\quad\quad \underline{- 21x^2y + 28xy^2 - 35y^3} \quad\quad$ $-7y$ multiplied by top polynomial
$\quad 18x^3 - 45x^2y + 58xy^2 - 35y^3 \quad\quad$ Sum ◀◀

Practice Exercise 6

Multiply.

(A) $\quad 3y^2 - y + 2$
$\quad\quad \underline{y^2 + 2y}$

(B) $\quad 4x^2 + 6x + 9$
$\quad\quad \underline{2x - 3}$

(C) $\quad 2a^2 + 5a - 6$
$\quad\quad \underline{-3a^2 + a + 5}$

Answers: (A) $3y^4 + 5y^3 + 4y$
(B) $8x^3 - 27$ (C) $-6a^4 - 13a^3 + 33a^2 + 19a - 30$

Practice Exercise 7

Use the FOIL method to multiply the binomials in two variables.

(A) $(5a + 2b)(a - b)$

(B) $(3x + 7y)(3x - 7y)$

(C) $(6z - 4w)(5z - 3w)$

Answers: (A) $5a^2 - 3ab - 2b^2$
(B) $9x^2 - 49y^2$ (C) $30z^2 - 38zw + 12w^2$

Practice Exercise 8

Use the column method to multiply.

$\quad 4a^2 + 3ab + b^2$
$\quad \underline{5a - 2b}$

Answer: $20a^3 + 7a^2b - ab^2 - 2b^3$

6.3 EXERCISES A

Multiply the following monomials.

1. $(2)(6x)$
2. $(-8y)(4)$
3. $(-7a^2)(-4a)$
4. $(4y^2)(2y^2)$
5. $(-3x^4)(2x)$
6. $(8y^3)(-9y^2)$
7. $(-13x^7)(10x^3)$
8. $(-8y^6)(-7y^5)$

Multiply using the distributive law.

9. $2(x + 3)$
10. $3x(x - 4)$
11. $-4y(y + 8)$
12. $-6a^2(4a + 5)$
13. $3(2x^2 - 3x + 1)$
14. $4y(y^2 - 8y - 5)$
15. $2y^3(-3y^2 - 2y + 5)$
16. $-10x^2(x^5 - 6x^3 + 7x^2)$
17. $-9a^3(4a^4 - 3a^2 + 2)$
18. $(x + 8)(x + 4)$
19. $(2a - 3)(3a - 2)$
20. $(x - 5)(x^2 - 3x + 2)$
21. $(2y + 7)(3y^2 + y - 1)$
22. $(a^2 + 1)(a^4 - a^2 + 1)$
23. $(x^2 + 4x - 2)(2x^2 - x + 3)$

Use the FOIL method to multiply the following binomials.

24. $(x + 10)(x + 2)$
25. $(a - 6)(a + 6)$
26. $(y + 5)(y - 3)$
27. $(x - 7)(x - 3)$
28. $(a + 8)(a + 8)$
29. $(x - 12)(x + 4)$
30. $(a - 8)(a - 8)$
31. $(3x + 2)(x + 5)$
32. $(5a - 3)(a - 7)$
33. $(2y - 7)(y + 10)$
34. $(7x + 3)(x - 5)$
35. $(2a + 3)(3a + 5)$
36. $(4y - 3)(2y - 5)$
37. $(5x - 2)(3x + 4)$
38. $(9a + 2)(2a - 7)$
39. $(10y - 1)(4y + 5)$
40. $(7x + 5)(3x - 8)$
41. $(2z^2 + 1)(z^2 - 2)$

Multiply.

42. $\begin{array}{r} 3a^2 + 5 \\ \underline{a - 4} \end{array}$

43. $\begin{array}{r} 2a^3 - 3a \\ \underline{7a^2 - 5} \end{array}$

44. $\begin{array}{r} 3y^3 - 2 \\ \underline{5y^2 + 4y} \end{array}$

45. $\begin{array}{r} 3a^2 - 5 \\ \underline{3a^2 - 5} \end{array}$

46. $\begin{array}{r} 7y^3 - 2 \\ \underline{7y^3 + 2} \end{array}$

47. $\begin{array}{r} 5x^4 + 3x^2 \\ \underline{8x^2 - 7} \end{array}$

48. $3x^2 - 5x + 2$
$\underline{x + 4}$

49. $-7y^2 - 3y + 10$
$\underline{2y - 1}$

50. $12a^3 - 3a + 8$
$\underline{4a^2 + 7a}$

51. $12a^3 - 6a + 4$
$\underline{7a^2 - 3}$

52. $0.3x^2 + 0.2$
$\underline{0.5x - 0.7}$

53. $\dfrac{2}{3}y^2 - \dfrac{3}{4}$
$\underline{\dfrac{1}{2}y + 5}$

54. $a^2 - 2a + 2$
$\underline{a^2 + 2a - 2}$

55. $3y^2 + 5y - 6$
$\underline{y^2 - 3y + 2}$

56. $-4x^3 + 2x - 5$
$\underline{6x^2 - x + 1}$

Multiply the polynomials in two variables.

57. $2xy(x^2 + 2xy)$

58. $3a^2b(a^2b + 2ab - ab^2)$

59. $(2a - 3b)(5a + b)$

60. $(x + 3y)(x + 2y)$

61. $(4x - y)(4x + y)$

62. $(7a - 2b)(2a - 5b)$

63. $3x^2 - 2xy + 4y^2$
$\underline{2x + y}$

64. $a^2b^2 + ab - 3$
$\underline{a^2 - b^2}$

Perform the following operations.

65. $(x - 1)(x + 1)(x + 2)$

66. $(y + 3)(y - 2)(y + 2)$

67. The length of a rectangle measures $2x + 1$ feet and the width is $3x - 2$ feet. Find the area of the rectangle.

68. The dimensions of a box, given in centimeters, are y, $y + 1$, and $2y + 3$. Find the volume of the box.

For Review

Perform the indicated operations.

69. $(3x^2 + 2x - 1) + (2x + 1) - (x^2 - 2)$

70. $(-4y^2 + 2y) + (y^3 - 2y) - (y^2 + 3y - 5)$

71. $(3a^2 - 5a + 5) - (-a^5 + 4a^2 - 2) - (6a - 7)$

72. $(x^2 - y^2) - (2x^2 - 3y^2) - (5x^2 - y^2)$

268 POLYNOMIALS

ANSWERS: 1. $12x$ 2. $-32y$ 3. $28a^3$ 4. $8y^4$ 5. $-6x^5$ 6. $-72y^5$ 7. $-130x^{10}$ 8. $56y^{11}$ 9. $2x + 6$
10. $3x^2 - 12x$ 11. $-4y^2 - 32y$ 12. $-24a^3 - 30a^2$ 13. $6x^2 - 9x + 3$ 14. $4y^3 - 32y^2 - 20y$
15. $-6y^5 - 4y^4 + 10y^3$ 16. $-10x^7 + 60x^5 - 70x^4$ 17. $-36a^7 + 27a^5 - 18a^3$ 18. $x^2 + 12x + 32$ 19. $6a^2 - 13a + 6$
20. $x^3 - 8x^2 + 17x - 10$ 21. $6y^3 + 23y^2 + 5y - 7$ 22. $a^6 + 1$ 23. $2x^4 + 7x^3 - 5x^2 + 14x - 6$ 24. $x^2 + 12x + 20$
25. $a^2 - 36$ 26. $y^2 + 2y - 15$ 27. $x^2 - 10x + 21$ 28. $a^2 + 16a + 64$ 29. $x^2 - 8x - 48$ 30. $a^2 - 16a + 64$
31. $3x^2 + 17x + 10$ 32. $5a^2 - 38a + 21$ 33. $2y^2 + 13y - 70$ 34. $7x^2 - 32x - 15$ 35. $6a^2 + 19a + 15$
36. $8y^2 - 26y + 15$ 37. $15x^2 + 14x - 8$ 38. $18a^2 - 59a - 14$ 39. $40y^2 + 46y - 5$ 40. $21x^2 - 41x - 40$
41. $2z^4 - 3z^2 - 2$ 42. $3a^3 - 12a^2 + 5a - 20$ 43. $14a^5 - 31a^3 + 15a$ 44. $15y^5 + 12y^4 - 10y^2 - 8y$
45. $9a^4 - 30a^2 + 25$ 46. $49y^6 - 4$ 47. $40x^6 - 11x^4 - 21x^2$ 48. $3x^3 + 7x^2 - 18x + 8$ 49. $-14y^3 + y^2 + 23y - 10$
50. $48a^5 + 84a^4 - 12a^3 + 11a^2 + 56a$ 51. $84a^5 - 78a^3 + 28a^2 + 18a - 12$ 52. $0.15x^3 - 0.21x^2 + 0.10x - 0.14$
53. $\frac{1}{3}y^3 + \frac{10}{3}y^2 - \frac{3}{8}y - \frac{15}{4}$ 54. $a^4 - 4a^2 + 8a - 4$ 55. $3y^4 - 4y^3 - 15y^2 + 28y - 12$
56. $-24x^5 + 4x^4 + 8x^3 - 32x^2 + 7x - 5$ 57. $2x^3y + 4x^2y^2$ 58. $3a^4b^2 + 6a^3b^2 - 3a^3b^3$ 59. $10a^2 - 13ab - 3b^2$
60. $x^2 + 5xy + 6y^2$ 61. $16x^2 - y^2$ 62. $14a^2 - 39ab + 10b^2$ 63. $6x^3 - x^2y + 6xy^2 + 4y^3$
64. $a^4b^2 + a^3b - 3a^2 - a^2b^4 - ab^3 + 3b^2$ 65. $x^3 + 2x^2 - x - 2$ 66. $y^3 + 3y^2 - 4y - 12$ 67. $(6x^2 - x - 2)$ ft^2
68. $(2y^3 + 5y^2 + 3y)$ cm^3 69. $2x^2 + 4x + 2$ 70. $y^3 - 5y^2 - 3y + 5$ 71. $a^5 - a^2 - 11a + 14$ 72. $-6x^2 + 3y^2$

6.3 EXERCISES B

Multiply the following monomials.

1. $(3)(5x)$
2. $(-4y)(8)$
3. $(9a)(2a^3)$
4. $(-3a^3)(5a^2)$
5. $(-10y)(-8y^5)$
6. $(a^7)(a^7)$
7. $(18x^8)(-2x^3)$
8. $(-10y^7)(-20y^8)$

Multiply using the distributive law.

9. $5x(x - 6)$
10. $2x(-3x + 5)$
11. $-8x(3x - 4)$
12. $-9a(2a^2 - 3)$
13. $5(3x^2 - 2x - 3)$
14. $3y(y^2 - 5y + 4)$
15. $6y^2(-8y^2 + 5y - 3)$
16. $6x^3(2x^7 - 3x^6 + 5x^4)$
17. $-10a^5(-2a^5 + 2a^4 - 8a^2)$
18. $(x + 5)(x + 10)$
19. $(5a - 2)(2a - 7)$
20. $(x - 3)(x^2 + 5x - 4)$
21. $(3y + 8)(2y^2 - 2y + 3)$
22. $(a^2 - 1)(a^4 + a^2 + 1)$
23. $(2x^2 + 3x - 2)(x^2 - 5x + 1)$

Use the FOIL method to multiply the following binomials.

24. $(x + 3)(x - 9)$
25. $(a + 8)(a - 8)$
26. $(x - 4)(x - 5)$
27. $(y - 12)(y + 3)$
28. $(x + 2)(x - 8)$
29. $(a - 9)(a - 9)$
30. $(y - 9)(y + 4)$
31. $(5x + 3)(x + 2)$
32. $(4a - 5)(a - 6)$
33. $(7y - 2)(y + 5)$
34. $(8x + 1)(x - 10)$
35. $(4a + 5)(5a + 4)$
36. $(3y - 5)(4y - 7)$
37. $(2x - 9)(5x + 3)$
38. $(10a + 3)(5a - 9)$
39. $(7y - 3)(8y + 3)$
40. $(9x + 7)(2x - 1)$
41. $(3z^2 - 1)(z^2 + 1)$

Multiply.

42. $3x + 4$
    $\underline{5x - 8}$

43. $3a^3 - 5a$
    $\underline{4a^2 - 7}$

44. $2y^2 - 9$
    $\underline{3y^2 + 5y}$

45. $4x^2 + 9$
    $\underline{4x^2 + 9}$

46. $8y^3 + 5$
    $\underline{8y^3 - 5}$

47. $10x^4 + 7x$
    $\underline{5x^2 - 2x}$

48. $2x^2 - 8x + 3$
    $\underline{x + 3}$

49. $-6y^2 + 2y - 5$
    $\underline{3y - 2}$

50. $8a^4 - 5a + 1$
    $\underline{3a^2 + 5}$

51. $15a^3 + 5a^2 - 3$
    $\underline{6a^2 - 7}$

52. $1.2x^2 + 4.5$
    $\underline{0.1x - 1.2}$

53. $\frac{1}{2}y^2 - \frac{5}{4}$
    $\underline{\frac{4}{3}y + 7}$

54. $a^2 + 3a - 1$
    $\underline{a^2 - 3a + 1}$

55. $4y^2 + 3y - 5$
    $\underline{y^2 - 6y + 4}$

56. $-2x^4 + 5x^2 - 3$
    $\underline{8x^2 - 3x + 2}$

Multiply the polynomials in two variables.

57. $5xy(3xy + y^2)$

58. $6a^3b^2(2a^4b - 3a^2b^2 + ab)$

59. $(x + 4y)(x + 6y)$

60. $(3a - 5b)(7a + 8b)$

61. $(6x - y)(6x + y)$

62. $(3a - 8b)(10a - 3b)$

63. $5x^2 - 5xy + 6y^2$
    $\underline{4x + 3y}$

64. $2a^2b^2 - 3ab + 9$
    $\underline{ab - 2}$

Perform the following operations.

65. $(y - 3)(y + 3)(y - 2)$

66. $(x + 4)(x + 2)(x - 4)$

67. The base of a triangle measures $3y + 2$ inches and the height is $2y - 1$ inches. Find the area of the triangle.

68. The dimensions of a box, given in feet, are x, $x + 2$, and $3x + 5$. Find the volume of the box.

For Review

Perform the indicated operations.

69. $(-4x^2 + 5x - 2) + (5x - 2) - (3x^2 - 2x)$

70. $(9y^2 - 6) + (y^4 - 5y^2) - (2y^2 - 5y + 7)$

71. $(7a^2 - 4a + 10) - (a^3 - 3a^2 + 5) - (-a + 4)$

72. $(x^2y^2 - xy + 2) - (5x^2y^2 - 3) - (4xy - 11)$

6.3 EXERCISES C

Perform the indicated operations.

1. $(x + y)(x^2 - xy + y^2)$ [Answer: $x^3 + y^3$]

2. $(x - y)(x^2 + xy + y^2)$

3. $(a - 5b)(2a + b)(2a - 3b)$

4. $(a - b)[3a^2 - (a + b)(a - 2b)]$
   [Answer: $2a^3 - a^2b + ab^2 - 2b^3$]

6.4 SPECIAL PRODUCTS

STUDENT GUIDEPOSTS

1 Difference of squares
2 Perfect square trinomials

Suppose we use the FOIL method to multiply the two binomials $x + 6$ and $x - 6$.

$$(x + 6)(x - 6) = x^2 - 6x + 6x - 6^2$$
$$= x^2 - 36$$

Notice that the middle terms add to zero. This will always happen when we multiply two binomials one of which is of the form $a + b$ and the other $a - b$. We can use the FOIL method to obtain products of this nature, but knowing the following formula is helpful when we factor in Chapter 7.

270   POLYNOMIALS

1 ▶ Difference of two squares

$$(a + b)(a - b) = a^2 - b^2$$

Remember this rule as: "The product of a sum and a difference of two terms is the square of the first minus the square of the second." Since the order of multiplication can be changed, we also know that $(a - b)(a + b) = a^2 - b^2$.

EXAMPLE 1

Find each product first using FOIL, then using the difference of squares formula.

(A) $(x - 7)(x + 7)$.

FOIL method

$(x - 7)(x + 7) = x^2 - 7x - 49$
$ = x^2 - 49$

Using $(a - b)(a + b) = a^2 - b^2$

$(x - 7)(x + 7) = x^2 - 7^2$
$ = x^2 - 49$

(B) $(3z + 4)(3z - 4)$.

FOIL method

$(3z + 4)(3z - 4) = 9z^2 - 12z + 12z - 16$
$ = 9z^2 - 16$

Using $(a + b)(a - b) = a^2 - b^2$

$(3z + 4)(3z - 4) = (3z)^2 - 4^2$
$ = 9z^2 - 16$

(C) $(7x^2 - 2x)(7x^2 + 2x)$.

FOIL method

$(7x^2 - 2x)(7x^2 + 2x)$
$= 49x^4 + 14x^3 - 14x^3 - 4x^2$
$= 49x^4 - 4x^2$ ◀◀

Using $(a - b)(a + b) = a^2 - b^2$

$(7x^2 - 2x)(7x^2 + 2x) = (7x^2)^2 - (2x)^2$
$ = 49x^4 - 4x^2$

Practice Exercise 1

Find each product first using FOIL, then using the difference of squares formula.

(A) $(a - 5)(a + 5)$

(B) $(4x + 7)(4x - 7)$

(C) $(2y^2 - y)(2y^2 + y)$

Answers: **(A)** $a^2 - 25$
(B) $16x^2 - 49$ **(C)** $4y^4 - y^2$

Suppose we use the FOIL method to multiply $x + 6$ and $x + 6$ or $(x + 6)^2$.

$(x + 6)^2 = (x + 6)(x + 6) = x^2 + 6x + 6x + 6^2$
$ = x^2 + 2(6x) + 6^2$
$ = x^2 + 12x + 36$

This is an example of a product that can be determined using the perfect square formula.

2 ▶ Perfect square trinomial

$$(a + b)^2 = (a + b)(a + b) = a^2 + 2ab + b^2$$

Remember this rule as: "The square of the sum of two terms is the square of the first plus twice their product plus the square of the second."

EXAMPLE 2

Find each product first using FOIL, then using the perfect square formula.

(A) $(y + 3)(y + 3)$.

FOIL method

$(y + 3)(y + 3) = y^2 + 3y + 3y + 9$
$ = y^2 + 6y + 9$

Using $(a + b)^2 = a^2 + 2ab + b^2$

$(y + 3)^2 = y^2 + 2(y)(3) + 3^2$
$ = y^2 + 6y + 9$

Practice Exercise 2

Find each product first using FOIL, then using the perfect square formula.

(A) $(x + 8)(x + 8)$

(B) $(5x + 4)(5x + 4)$.

FOIL method Using $(a + b)^2 = a^2 + 2ab + b^2$

$(5x + 4)(5x + 4)$ $(5x + 4)^2 = (5x)^2 + 2(5x)(4) + 4^2$
$= 25x^2 + 20x + 20x + 16$
$= 25x^2 + 40x + 16$ $= 25x^2 + 40x + 16$ ◀◀

(B) $(7y + 2)(7y + 2)$

Answers: **(A)** $x^2 + 16x + 64$
(B) $49y^2 + 28y + 4$

Next we consider products like $(x - 6)^2$.

$(x - 6)^2 = (x - 6)(x - 6) = x^2 - 6x - 6x + 6^2$
$= x^2 - 2(6x) + 6^2$
$= x^2 - 12x + 36$

Here we may use the formula for the perfect square $(a - b)^2$.

> **Perfect square trinomial**
> $(a - b)^2 = (a - b)(a - b) = a^2 - 2ab + b^2$

Remember this rule as: "The square of the difference of two terms is the square of the first minus twice their product plus the square of the second."

▶▶ **CAUTION** ◀◀ A common mistake when squaring a binomial is to write

$(a - b)^2 = a^2 - b^2$ THIS IS WRONG

or

$(a + b)^2 = a^2 + b^2$.   THIS IS WRONG

To see that $(a + b)^2 \neq a^2 + b^2$, substitute 1 for a and 2 for b. Then,

$(a + b)^2 = (1 + 2)^2 = 3^2 = 9$.

However,

$a^2 + b^2 = (1)^2 + (2)^2 = 1 + 4 = 5$.

Don't forget the middle term $2ab$ when finding $(a + b)^2$, or $-2ab$ when finding $(a - b)^2$. ◀◀

EXAMPLE 3

Find each product first using FOIL, then using the perfect square formula.

(A) $(y - 2)(y - 2)$

FOIL method Using $(a - b)^2 = a^2 - 2ab + b^2$

$(y - 2)(y - 2)$ $(y - 2)^2 = y^2 - 2(y)(2) + 2^2$
$= y^2 - 2y - 2y + 4$ $= y^2 - 4y + 4$
$= y^2 - 4y + 4$

(B) $(5x - 4)(5x - 4)$.

FOIL method Using $(a - b)^2 = a^2 - 2ab + b^2$

$(5x - 4)(5x - 4)$ $(5x - 4)^2 = (5x)^2 - 2(5x)(4) + 4^2$
$= 25x^2 - 20x - 20x + 16$ $= 25x^2 - 40x + 16$
$= 25x^2 - 40x + 16$ ◀◀

Practice Exercise 3

Find each product first using FOIL, then using the perfect square formula.

(A) $(x - 9)(x - 9)$

(B) $(7a - 1)(7a - 1)$

Answers: **(A)** $x^2 - 18x + 81$
(B) $49a^2 - 14a + 1$

Remember that when we square a binomial we will always have the middle term, $2ab$ or $-2ab$. It is only when we have a binomial product of the form $(a + b)(a - b)$ that the middle term is not present.

EXAMPLE 4

Use the special formulas in this section to find the following products involving two variables.

(A) $(2x - 3y)(2x + 3y) = (2x)^2 - (3y)^2$ $(a - b)(a + b) = a^2 - b^2$
$= 4x^2 - 9y^2$

(B) $(2x + 3y)^2 = (2x)^2 + 2(2x)(3y) + (3y)^2$ $(a + b)^2 = a^2 + 2ab + b^2$
$= 4x^2 + 12xy + 9y^2$

(C) $(2x - 3y)^2 = (2x)^2 - 2(2x)(3y) + (3y)^2$ $(a - b)^2 = a^2 - 2ab + b^2$
$= 4x^2 - 12xy + 9y^2$ ◀◀

Practice Exercise 4

Use the special formulas to find each product.

(A) $(5a - b)(5a + b)$

(B) $(9a + 2b)^2$

(C) $(9a - 2b)^2$

Answers: (A) $25a^2 - b^2$ (B) $81a^2 + 36ab + 4b^2$ (C) $81a^2 - 36ab + 4b^2$

6.4 EXERCISES A

Multiply using special formulas.

1. $(x - 3)(x + 3)$

2. $(y + 7)(y - 7)$

3. $(x + 5)^2$

4. $(x - 5)^2$

5. $(y - 8)(y + 8)$

6. $(a + 10)(a - 10)$

7. $(x - 12)^2$

8. $(x + 12)^2$

9. $(2y + 5)(2y - 5)$

10. $(7a - 1)(7a + 1)$

11. $(2x + 7)^2$

12. $(2x - 7)^2$

13. $(4y - 9)(4y + 9)$

14. $(5a + 7)(5a - 7)$

15. $(3a + 1)^2$

16. $(3a - 1)^2$

17. $(5x - 3)(5x + 3)$

18. $(5y + 1)(5y - 1)$

19. $(6a + 7)^2$

20. $(6a - 7)^2$

21. $(4x + 8)^2$

22. $(3y - 9)^2$

23. $(5a - 10)(5a + 10)$

24. $(3x + 12)(3x - 12)$

25. $(5y^2 + 1)^2$

26. $(5y^2 - 1)^2$

27. $(5a^2 - 7)(5a^2 + 7)$

28. $(2x^2 - 3x)(2x^2 + 3x)$

29. $(2y^2 + 3y)^2$

30. $(2y^2 - 3y)^2$

31. $(0.7y - 3)^2$

32. $\left(\dfrac{1}{2}a - \dfrac{1}{3}\right)\left(\dfrac{1}{2}a + \dfrac{1}{3}\right)$

33. $(5x - 2y)(5x + 2y)$

34. $(5x + 2y)^2$

35. $(5x - 2y)^2$

36. $(x^2 + y^2)(x^2 - y^2)$

37. $(2x^2 - y)^2$

38. $(8x + 3y)^2$

39. $(a^2 + 2b)^2$

Perform the following operations.

40. $(x + 1)^2 - (x - 1)^2$

41. $(a - 2b)^2 + (a + 2b)^2$

42. $(z - 4)(z + 4) - (z + 4)^2$

43. $(w + 2)^2 - (w + 2)(w - 2)$

44. A field that is rectangular in shape is $(2a + 7)$ miles long and $(a - 1)$ miles wide. Find the area of the field.

For Review

Multiply.

45. $-3a^2(ab^2 - ab + 7)$

46. $(5a + 7)(3a - 8)$

47. $4x^2 - 9$
    $\underline{2x + 3}$

48. $2x^2 - xy + y^2$
    $\underline{x - y}$

ANSWERS: 1. $x^2 - 9$ 2. $y^2 - 49$ 3. $x^2 + 10x + 25$ 4. $x^2 - 10x + 25$ 5. $y^2 - 64$ 6. $a^2 - 100$ 7. $x^2 - 24x + 144$ 8. $x^2 + 24x + 144$ 9. $4y^2 - 25$ 10. $49a^2 - 1$ 11. $4x^2 + 28x + 49$ 12. $4x^2 - 28x + 49$ 13. $16y^2 - 81$ 14. $25a^2 - 49$ 15. $9a^2 + 6a + 1$ 16. $9a^2 - 6a + 1$ 17. $25x^2 - 9$ 18. $25y^2 - 1$ 19. $36a^2 + 84a + 49$

20. $36a^2 - 84a + 49$ **21.** $16x^2 + 64x + 64$ **22.** $9y^2 - 54y + 81$ **23.** $25a^2 - 100$ **24.** $9x^2 - 144$ **25.** $25y^4 + 10y^2 + 1$ **26.** $25y^4 - 10y^2 + 1$ **27.** $25a^4 - 49$ **28.** $4x^4 - 9x^2$ **29.** $4y^4 + 12y^3 + 9y^2$ **30.** $4y^4 - 12y^3 + 9y^2$ **31.** $0.49y^2 - 4.2y + 9$ **32.** $\frac{1}{4}a^2 - \frac{1}{9}$ **33.** $25x^2 - 4y^2$ **34.** $25x^2 + 20xy + 4y^2$ **35.** $25x^2 - 20xy + 4y^2$ **36.** $x^4 - y^4$ **37.** $4x^4 - 4x^2y + y^2$ **38.** $64x^2 + 48xy + 9y^2$ **39.** $a^4 + 4a^2b + 4b^2$ **40.** $4x$ **41.** $2a^2 + 8b^2$ **42.** $-8z - 32$ **43.** $4w + 8$ **44.** $(2a^2 + 5a - 7)$ mi^2 **45.** $-3a^3b^2 + 3a^3b - 21a^2$ **46.** $15a^2 - 19a - 56$ **47.** $8x^3 + 12x^2 - 18x - 27$ **48.** $2x^3 - 3x^2y + 2xy^2 - y^3$

6.4 EXERCISES B

Multiply using special formulas.

1. $(x - 4)(x + 4)$ **2.** $(y + 5)(y - 5)$ **3.** $(x + 10)^2$ **4.** $(x - 10)^2$

5. $(y - 9)(y + 9)$ **6.** $(a + 12)(a - 12)$ **7.** $(x - 4)^2$ **8.** $(x + 4)^2$

9. $(3y - 7)(3y + 7)$ **10.** $(5a - 2)(5a + 2)$ **11.** $(3x + 4)^2$ **12.** $(3x - 4)^2$

13. $(8y - 1)(8y + 1)$ **14.** $(6a - 5)(6a + 5)$ **15.** $(9a + 1)^2$ **16.** $(9a - 1)^2$

17. $(10x - 3)(10x + 3)$ **18.** $(3y + 5)(3y - 5)$ **19.** $(4a + 5)^2$ **20.** $(4a - 5)^2$

21. $(2x + 10)^2$ **22.** $(6y - 3)^2$ **23.** $(2a - 10)(2a + 10)$ **24.** $(6x + 3)(6x - 3)$

25. $(3y^2 + 2)^2$ **26.** $(3y^2 - 2)^2$ **27.** $(2a^2 - 5)(2a^2 + 5)$ **28.** $(3x^2 + 4x)(3x^2 - 4x)$

29. $(3y^2 + 4y)^2$ **30.** $(3y^2 - 4y)^2$ **31.** $(0.4y - 5)^2$ **32.** $\left(\frac{2}{3}a - \frac{1}{4}\right)\left(\frac{2}{3}a + \frac{1}{4}\right)$

33. $(7x - 3y)(7x + 3y)$ **34.** $(7x + 3y)^2$ **35.** $(7x - 3y)^2$ **36.** $(x^3 + y^3)(x^3 - y^3)$

37. $(3x - y^2)^2$ **38.** $(8x - 3y)^2$ **39.** $(a^2 - 2b)^2$

Perform the following operations.

40. $(x - 1)^2 - (x + 1)^2$

41. $(a + 3b)^2 + (a - 3b)^2$

42. $(y - 2)(y + 2) - (y - 2)^2$

43. $(z + 3)^2 - (z + 3)(z - 3)$

44. A picture frame is in the shape of a square with sides $(3z + 2)$ feet in length. Find the area of the frame.

For Review

Multiply.

45. $4ab(-2a^2b - 6a + 9b)$

46. $(2a - 5)(4a - 3)$

47. $3x^2 + 11$
     $\underline{3x - 2}$

48. $5x^2y^2 - 2xy + 3$
     $\underline{2xy - 1}$

6.4 EXERCISES C

Perform the indicated operations. Assume all exponents are positive integers.

1. $[2x^2 - (x - 2y)][2x^2 + (x - 2y)]$
  [*Hint:* Multiply before removing parentheses.]

2. $(2x + y - z)^2$

3. $(a^n + b^n)(a^n - b^n)$

4. $(a^n - b^n)^2$ [Answer: $a^{2n} - 2a^nb^n + b^{2n}$]

6.5 DIVISION OF POLYNOMIALS

STUDENT GUIDEPOSTS
1. Dividing a polynomial by a monomial
2. Dividing a polynomial by a binomial

When dividing a polynomial by a monomial, we must divide term by term using the rules of exponents.

1 | **To divide a polynomial by a monomial**
1. Divide each term of the polynomial by the monomial.
2. Use the rule
$$\frac{x^m}{x^n} = x^{m-n}$$
to divide the variables.

EXAMPLE 1
Divide.

(A) $\dfrac{14x^3 - 7x^2 + 28x - 7}{7} = \dfrac{14x^3}{7} - \dfrac{7x^2}{7} + \dfrac{28x}{7} - \dfrac{7}{7}$
$= 2x^3 - x^2 + 4x - 1$

(B) $\dfrac{8x^4 - 6x^2 + 12x}{-2x} = \dfrac{8x^4}{-2x} - \dfrac{6x^2}{-2x} + \dfrac{12x}{-2x}$
$= -4x^3 + 3x - 6$

$\dfrac{x^4}{x} = x^{4-1} = x^3,$

$\dfrac{x^2}{x} = x^{2-1} = x,$

$\dfrac{x}{x} = x^{1-1} = x^0 = 1$

(C) $\dfrac{7x^5 - 35x^4 + 40x}{5x^2} = \dfrac{7x^5}{5x^2} - \dfrac{35x^4}{5x^2} + \dfrac{40x}{5x^2}$
$= \dfrac{7}{5}x^3 - 7x^2 + \dfrac{8}{x}$

$\dfrac{x}{x^2} = \dfrac{1}{x}$ ◀

Practice Exercise 1
Divide.

(A) $\dfrac{6a^4 - 9a^3 + 12a^2 + 3a - 15}{3}$

(B) $\dfrac{25y^5 - 15y^3 + 30y}{-5y}$

(C) $\dfrac{18w^6 - 27w^4 - 9w^2}{9w^2}$

Answers: (A) $2a^4 - 3a^3 + 4a^2 + a - 5$ (B) $-5y^4 + 3y^2 - 6$ (C) $2w^4 - 3w^2 - 1$

If the division is expressed using the sign ÷, we change to the notation used in Example 1 before proceeding. Thus, to solve the division problem
$$(3a^3b^3 - 9a^2b + 27ab) \div 9ab,$$
we write
$$\dfrac{3a^3b^3 - 9a^2b + 27ab}{9ab} = \dfrac{3a^3b^3}{9ab} - \dfrac{9a^2b}{9ab} + \dfrac{27ab}{9ab}$$
$$= \dfrac{1}{3}a^2b^2 - a + 3.$$

276 POLYNOMIALS

In order for us to understand the procedures for dividing a polynomial by a binomial, we first review division of whole numbers. Refer to the following numerical division problem as you study the rule for division of a polynomial by a binomial.

$$43 \div 37$$
$$61 \div 37$$
$$247 \div 37$$
$$252 \div 37$$

$$\begin{array}{r} 1166 \\ 37\overline{)43172} \\ \underline{37} \\ 61 \\ \underline{37} \\ 247 \\ \underline{222} \\ 252 \\ \underline{222} \\ 30 \end{array}$$ Remainder 30

⟵ First digit of quotient times divisor, 37
   Subtract and bring down next digit of dividend, 43172
⟵ Second digit of quotient times 37
   Subtract and bring down next digit of dividend
⟵ Third digit of quotient times 37
   Subtract and bring down next digit
⟵ Fourth digit of quotient times 37
   Subtract; no more digits in dividend

The answer is 1166 with a remainder of 30, or $1166 + \frac{30}{37}$.

> **To divide a polynomial by a binomial**
>
> 1. Arrange the terms of both polynomial and binomial in descending order.
>
> 2. Divide the first term of the polynomial (the dividend) by the first term of the binomial (the divisor) to obtain the first term of the quotient.
>
> 3. Multiply the first term of the quotient by the binomial and subtract the results from the dividend. Bring down the next terms to obtain a new polynomial which becomes the new dividend.
>
> 4. Divide the new dividend polynomial by the binomial. Continue the process until the variable in the first term of the remainder dividend is raised to a lower power than the variable in the first term of the divisor.

EXAMPLE 2

Divide $6 - 5x + x^2$ by $x - 2$.

1. Arrange terms in descending order.

$$x - 2\overline{)x^2 - 5x + 6}$$

2. Divide the first term of the polynomial by the first term of the binomial.

   equals ⟶ x
   $x - 2\overline{)x^2 - 5x + 6}$ $x^2 \div x = x$
   divided by

3. Multiply the first term of the quotient by the binomial and subtract the results from the dividend. Bring down the next terms to obtain a new dividend.

Practice Exercise 2

Divide $8x - 9 + x^2$ by $x - 1$.

equals ⟶ x
$x - 1\overline{)x^2 + 8x - 9}$ ⟵ Descending order
$\underline{x^2 - x}$ ⟵ $x(x - 1)$
$9x$ ⟵ Remember to subtract like terms: $x^2 - x^2 = 0$ and $8x - (-x) = 8x + x = 9x$

Bring down the -9 and continue by dividing x into $9x$.

$$\begin{array}{r}\overset{\text{times}\ \ x}{x-2\overline{)x^2-5x+6}}\\ \underset{\text{equals}}{x^2-2x}\\ \hline -3x+6\end{array}$$

$x(x-2) = x^2 - 2x$
Subtract $x^2 - 2x$ from $x^2 - 5x$ and bring down $+6$

4. Divide the new dividend polynomial by the binomial, using Steps 2 and 3.

$$\begin{array}{r}x-3\\ x-2\overline{)x^2-5x+6}\\ x^2-2x\\ \hline -3x+6\\ -3x+6\\ \hline 0\end{array}$$

Divide $-3x + 6$ by $x - 2$

Multiply $x - 2$ by -3
No variable in the new dividend; the process terminates ◀◀

Answer: $x + 9$

EXAMPLE 3

Divide $a^3 + 27$ by $a + 3$.

$$\begin{array}{r}a^2-3a+9\\ a+3\overline{)a^3+27}\\ a^3+3a^2\\ \hline -3a^2+27\\ -3a^2-9a\\ \hline 9a+27\\ 9a+27\\ \hline 0\end{array}$$

The quotient of a^3 and a
The quotient of $-3a^2$ and a
The quotient of $9a$ and a
Leave space for missing terms
The quotient a^2 times the divisor $a + 3$
Subtract $a^3 + 3a^2$ from a^3 and bring down 27
The quotient $-3a$ times $a + 3$
Subtract $-3a^2 - 9a$ from $3a^2 + 27$
The quotient 9 times $a + 3$
Subtract; no variable in new dividend

The answer is $a^2 - 3a + 9$ with a remainder of 0. ◀◀

Practice Exercise 3

Divide $x^3 - 8$ by $x - 2$.

Answer: $x^2 + 2x + 4$

EXAMPLE 4

Divide $3x^4 - 19x^3 + 27x^2 - 41x + 32$ by $x - 5$.

$$\begin{array}{r}3x^3-4x^2+7x-6\\ x-5\overline{)3x^4-19x^3+27x^2-41x+32}\\ 3x^4-15x^3\\ \hline -4x^3+27x^2-41x+32\\ -4x^3+20x^2\\ \hline 7x^2-41x+32\\ 7x^2-35x\\ \hline -6x+32\\ -6x+30\\ \hline 2\end{array}$$

Divide $3x^4$ by x

No variable in new dividend; process ends

The answer is $3x^3 - 4x^2 + 7x - 6$ with remainder 2, or $3x^3 - 4x^2 + 7x - 6 + \dfrac{2}{x-5}$. ◀◀

Practice Exercise 4

Divide $2y^4 + 9y^3 - 3y^2 - 25y + 10$ by $y + 4$.

Answer: $2y^3 + y^2 - 7y + 3 - \dfrac{2}{y+4}$

EXAMPLE 5

Divide $20x^3 + 18x^2 + 21x + 40$ by $5x + 7$.

$$\begin{array}{r} 4x^2 - 2x + 7 \\ 5x+7\overline{)20x^3 + 18x^2 + 21x + 40} \\ \underline{20x^3 + 28x^2} \\ -10x^2 + 21x + 40 \\ \underline{-10x^2 - 14x} \\ 35x + 40 \\ \underline{35x + 49} \\ -9 \end{array}$$

No variable in new dividend

The answer is $4x^2 - 2x + 7$ with remainder -9, or $4x^2 - 2x + 7 - \dfrac{9}{5x + 7}$. ◀◀

Practice Exercise 5

Divide $16a^3 - 8a^2 + 5a - 3$ by $4a - 3$.

Answer: $4a^2 + a + 2 + \dfrac{3}{4a - 3}$

EXAMPLE 6

Divide $4x^4 - 8x^3 - 12x^2 + 44x - 15$ by $4x^2 - 8$.

$$\begin{array}{r} x^2 - 2x - 1 \\ 4x^2-8\overline{)4x^4 - 8x^3 - 12x^2 + 44x - 15} \\ \underline{4x^4 - 8x^2} \\ -8x^3 - 4x^2 + 44x - 15 \\ \underline{-8x^3 + 16x} \\ -4x^2 + 28x - 15 \\ \underline{-4x^2 + 8} \\ 28x - 23 \end{array}$$

Keep like terms in same column

x is raised to first power; process ends

The answer is $x^2 - 2x - 1 + \dfrac{28x - 23}{4x^2 - 8}$. ◀◀

Practice Exercise 6

Divide $9x^4 + 6x^3 + 3x^2 + 7x - 1$ by $3x^2 - 1$.

Answer: $3x^2 + 2x + 2 + \dfrac{9x + 1}{3x^2 - 1}$

6.5 EXERCISES A

Divide.

1. $\dfrac{3x^3 - 9x^2 + 27x + 3}{3}$

2. $\dfrac{14x^4 - 7x^2 + 28}{7}$

3. $(25a^5 - 20a^4 + 15a^3) \div (5a^3)$

4. $(y^8 - y^6 + y^5) \div (y^3)$

5. $\dfrac{7x^5 - 35x^4 + 14x^2}{14x^2}$

6. $\dfrac{8x^3 - 64x^2}{8x^2}$

7. $(3a^{12} - 9a^6 + 27a^5 + 81a^4) \div (9a^2)$

8. $(y^6 - 3y^2) \div (2y^2)$

9. $\dfrac{-8x^3 + 6x^2 - 4x}{-2x}$

10. $\dfrac{-8x^3 + 6x^2 - 4x}{0.2x}$

11. $\dfrac{y^3 + 2y^2 + y}{y^2}$

12. $\dfrac{-8y^4 + 16y^3 - 4y}{2y^2}$

13. $\dfrac{-5a^2b + 3ab - 2a}{-a}$

14. $\dfrac{-6x^2y^3 + 9x^2y}{3xy}$

15. $\dfrac{27x^2y^4 - 18xy^6 + 36xy^3}{9xy^3}$

16. $\dfrac{12x^4y^4 - 42x^2y^2}{6x^3y^3}$

17. $(x^2 - 5x + 6) \div (x - 3)$

18. $(a^2 + 5a + 6) \div (a + 3)$

19. $(y^2 - 2y - 3) \div (y + 1)$

20. $(x^2 - 9x - 5) \div (x - 7)$

21. $(3y + 2 + y^2) \div (y + 2)$
[*Hint:* Remember descending order.]

22. $(6 + 8y - y^2) \div (4 - y)$

23. $(3a^2 - 23a + 40) \div (a - 5)$

24. $(14y - 6y^2 + 80) \div (3y + 8)$

25. $4x + 6 \overline{) 28x^3 + 26x^2 - 44x - 36}$

26. $2x - 3 \overline{) 10x^4 - 15x^3 - 8x + 12}$

27. $3a - 12 \overline{) 6a^3 - 18a^2 + 33a - 80}$

28. $y + 2 \overline{) y^5 + 32}$

29. $3x^2 + 2 \overline{) 9x^3 + 30x^2 + x - 10}$

30. $x^2 - 1 \overline{) x^4 - 1}$

For Review

Multiply using special formulas.

31. $(4x - 1)(4x + 1)$

32. $(6x - 5)^2$

33. $(2a + 13)^2$

34. $(a^2 - 5)(a^2 + 5)$ **35.** $(y^3 - 2)^2$ **36.** $(2y^2 + 7)^2$

ANSWERS: **1.** $x^3 - 3x^2 + 9x + 1$ **2.** $2x^4 - x^2 + 4$ **3.** $5a^2 - 4a + 3$ **4.** $y^5 - y^3 + y^2$ **5.** $\frac{1}{2}x^3 - \frac{5}{2}x^2 + 1$
6. $x - 8$ **7.** $\frac{1}{3}a^{10} - a^4 + 3a^3 + 9a^2$ **8.** $\frac{1}{2}y^4 - \frac{3}{2}$ **9.** $4x^2 - 3x + 2$ **10.** $-40x^2 + 30x - 20$ **11.** $y + 2 + \frac{1}{y}$
12. $-4y^2 + 8y - \frac{2}{y}$ **13.** $5ab - 3b + 2$ **14.** $-2xy^2 + 3x$ **15.** $3xy - 2y^3 + 4$ **16.** $2xy - \frac{7}{xy}$ **17.** $x - 2$
18. $a + 2$ **19.** $y - 3$ **20.** $x - 2 - \frac{19}{x - 7}$ **21.** $y + 1$ **22.** $y - 4 - \frac{22}{y - 4}$ **23.** $3a - 8$ **24.** $-2y + 10$
25. $7x^2 - 4x - 5 - \frac{6}{4x + 6}$ **26.** $5x^3 - 4$ **27.** $2a^2 + 2a + 19 + \frac{148}{3a - 12}$ **28.** $y^4 - 2y^3 + 4y^2 - 8y + 16$
29. $3x + 10 - \frac{5x + 30}{3x^2 + 2}$ **30.** $x^2 + 1$ **31.** $16x^2 - 1$ **32.** $36x^2 - 60x + 25$ **33.** $4a^2 + 52a + 169$ **34.** $a^4 - 25$
35. $y^6 - 4y^3 + 4$ **36.** $4y^4 + 28y^2 + 49$

6.5 EXERCISES B

Divide.

1. $\dfrac{5x^4 - 25x^2 + 45x - 20}{5}$

2. $\dfrac{26x^3 - 16x^2 + 10}{-2}$

3. $(55a^6 - 22a^4 + 33a^2) \div (11a^2)$

4. $(-y^{12} + y^7 - y^5) \div (y^4)$

5. $\dfrac{-100x^7 + 50x^5 - 20x^3}{10x^2}$

6. $\dfrac{15x^3 - 30x^2 + 5x}{30x}$

7. $(35a^9 - 49a^7 - 14a^6 + 56a^5) \div (7a^4)$

8. $(-8y^4 - 6y^3 - 12y^2) \div (-3y^2)$

9. $\dfrac{54x^6 + 81x^4 - 36x^3}{9x^3}$

10. $\dfrac{54x^6 + 81x^4 - 36x^3}{0.9x^3}$

11. $\dfrac{y^4 - 6y^3 + y^2}{y^3}$

12. $\dfrac{-3y^6 + 12y^4 - 6y^2}{3y^4}$

13. $\dfrac{4a^2b^2 - 5a^2b + 7ab^2}{ab}$

14. $\dfrac{18x^3y^4 - 12x^4y^3}{3x^3y^3}$

15. $\dfrac{44x^6y^6 + 20x^4y^2 - 28x^2y^4}{4x^2y}$

16. $\dfrac{-5x^6y^4 + 10x^4y^6}{5x^6y^6}$

17. $(x^2 - 6x + 8) \div (x - 4)$

18. $(a^2 + 6a + 8) \div (a + 2)$

19. $(y^2 - 5y - 6) \div (y + 1)$

20. $(x^2 + 7x - 8) \div (x + 5)$

21. $(7y + 12 + y^2) \div (y + 4)$

22. $(3 - 5y - y^2) \div (3 - y)$

23. $(3a^2 + 13a - 30) \div (a + 6)$

24. $(7a - 12 + 12a^2) \div (4a - 3)$

25. $(2y^3 + y^2 - y + 1) \div (2y + 3)$

26. $(6x^3 + x^2 - 7x + 2) \div (3x - 1)$

27. $(a^4 - 1) \div (a^2 + 1)$

28. $(x^5 - 32) \div (x - 2)$

29. $(21x^4 - 7x^3 - 6x + 2) \div (3x - 1)$

30. $(2x^5 - x^3 + 16x^2 - 8) \div (2x^2 - 1)$

For Review

Multiply using special formulas.

31. $(3x - 5)(3x + 5)$
32. $(3x + 7)^2$
33. $(2x - 11)^2$
34. $(a^2 + 7)(a^2 - 7)$
35. $(y^3 + 5)^2$
36. $(3y^2 - 4)^2$

6.5 EXERCISES C

Divide.

1. $(2x^6 - 7x^3 - 30) \div (2x^3 + 5)$
2. $(6x^6 + x^3 - 12) \div (3x^3 - 4)$ [Answer: $2x^3 + 3$]
3. $(5x^5 + 21 - 26x + x^2 + 10x^4 - 11x^3) \div (x^2 + 2x - 3)$
4. $(-34x^4 - 5x^3 + 3x^5 + 6x^6 + 54x^2 - 56 + 7x) \div (2x^2 + x - 8)$ [Answer: $3x^4 - 5x^2 + 7$]

CHAPTER 6 SUMMARY

Key Words and Phrases for Review

6.1 polynomial
numerical coefficient
monomial
binomial
trinomial
polynomial in one variable
polynomial in several variables
degree of a term

degree of a polynomial
descending order
ascending order
collecting like terms

6.3 FOIL method
6.4 difference of squares
perfect square trinomial

Key Concepts

6.1 1. Only like terms can be collected. For example, $5x + 3 \neq 8x$ since $5x$ and 3 are *not* like terms. Similarly, $x^2 - 2x + 1$ has no like terms to combine.

2. Change all signs when removing parentheses preceded by a minus sign. For example, $3x - (2x^2 - 4x - 2) = 3x - 2x^2 + 4x + 2$.

6.2 When subtracting polynomials, be sure to change all signs in the polynomial to be subtracted.

6.3 Use the FOIL or column method to multiply polynomials, and write out all details.

6.4 (1) $(a + b)(a - b) = a^2 - b^2$
(2) $(a + b)^2 = a^2 + 2ab + b^2$ (not $a^2 + b^2$)
(3) $(a - b)^2 = a^2 - 2ab + b^2$ (not $a^2 - b^2$)

6.5 Write polynomials in descending order before dividing.

Review Exercises

6.1 *Determine whether the following are monomials, binomials, or trinomials.*

1. $4x + 1$
2. $3x^2$
3. $5y^4 + y^2$
4. $-7y^3 + y - 5$

Determine the degree of each polynomial.

5. $6x^4$

6. $15x^2 - 14x + 6x^7 - 4x^3$

7. $-x^3 + 14x^4 - x^{10} + 8x^2$

8. $2x^8 - 4x^4 - 16$

Collect like terms and write in descending order.

9. $5x + 2 - 6x$

10. $6y^2 - 2y + 8y^2 + 5 - 3y^2 + y$

11. $4a^3 - 6a + 7a^2 - 4a^3 + a - a^2$

12. $-7x^2 - 30x^4 + x^5 - 3x + 5x^5 - 4x + 22x^4 - 5$

Collect like terms.

13. $8xy + 5 - 4xy - 8$

14. $-a^2b^3 - a^2b^2 - 5a^2b^2 - ab^2 + 5a^2b^3$

Solve.

15. The profit in dollars when x refrigerators are sold is given by the expression $4x^3 - 140$. Find the profit when 5 refrigerators are sold.

6.2 **16.** Add $5x^2 - 4x + 5$ and $-7x^2 + 4x - 8$.

17. Subtract $-2x^2 - 1$ from $x^3 - 5x^2 + 14$.

Perform the indicated operations.

18. $(4x^3 - x^4 + 6x - 2) + (3x - x^5 + 2x^3 - 5x^4 + 2)$

19. $(16x - 2x^4 + 3x^2 - 5) - (-6 + x^4 - 3x^3 - 3x^2 + 2x)$

20. $(6x^2y^2 - 5xy) + (-4xy + 3) - (3x^2y^2 - 8xy + 5)$

POLYNOMIALS

Add.

21.
$$\begin{aligned} x^4 - 2x^3 + 7x^2 - 20 \\ -8x^4 - 5x^2 + 3x + 7 \\ 36x^4 - 18x^3 + 5x - 31 \end{aligned}$$

22.
$$\begin{aligned} 6.1x^3 - 0.2x^2 + 1.1x + 5 \\ -5.7x^3 + 2.3x - 7 \\ 1.1x^2 - 2.3x + 8 \end{aligned}$$

6.3
6.4 *Multiply.*

23. $(x + 8)(x - 10)$

24. $(2x + 1)(3x - 4)$

25. $(4a - 3)(4a + 3)$

26. $(4a - 3)^2$

27. $(4a + 3)^2$

28. $(5y - 6)(4y - 7)$

29. $\left(\dfrac{3}{4} + 2y^3\right)\left(\dfrac{3}{4} - 2y^3\right)$

30. $(4x - 9)(x + 2)$

31. $(x + 4)(2x^2 + x - 3)$

32.
$$\begin{aligned} 3x^2 - 2x + 1 \\ x^2 + x - 2 \end{aligned}$$

33.
$$\begin{aligned} 2x^2 + xy - 3y^2 \\ x + 4y \end{aligned}$$

34.
$$\begin{aligned} 6x^2y^2 - xy + 5 \\ 2xy - 3 \end{aligned}$$

35. $(7a - b)(4a + 3b)$

36. $(5a - 7b)^2$

37. $(4x - y)(4x + y)$

38. $(8x + y)^2$

6.5 *Divide.*

39. $\dfrac{25x^4 - 50x^3 + 10x^2}{5x^2}$

40. $\dfrac{x^{10} - x^8 + x^6}{-x^4}$

41. $\dfrac{6x^4y^4 - 2x^3y^3 + 10x^2y^2}{2x^2y^2}$

42. $\dfrac{-14a^4b^2 + 21a^2b^4 - 35ab^2}{-7a^2b^2}$

43. $(x^2 + 3x - 28) \div (x - 4)$

44. $(8x^2 + 10x + 3) \div (2x + 1)$

45. $2x + 3 \overline{)14x^4 + 27x^3 + 5x^2 - 16x - 15}$

46. $x - 5 \overline{)x^3 \qquad\qquad\qquad -100}$

ANSWERS: **1.** binomial **2.** monomial **3.** binomial **4.** trinomial **5.** 4 **6.** 7 **7.** 10 **8.** 8 **9.** $-x + 2$ **10.** $11y^2 - y + 5$ **11.** $6a^2 - 5a$ **12.** $6x^5 - 8x^4 - 7x^2 - 7x - 5$ **13.** $4xy - 3$ **14.** $4a^2b^3 - 6a^2b^2 - ab^2$ **15.** \$360 **16.** $-2x^2 - 3$ **17.** $x^3 - 3x^2 + 15$ **18.** $-x^5 - 6x^4 + 6x^3 + 9x$ **19.** $-3x^4 + 3x^3 + 6x^2 + 14x + 1$ **20.** $3x^2y^2 - xy - 2$ **21.** $29x^4 - 20x^3 + 2x^2 + 8x - 44$ **22.** $0.4x^3 + 0.9x^2 + 1.1x + 6$ **23.** $x^2 - 2x - 80$ **24.** $6x^2 - 5x - 4$ **25.** $16a^2 - 9$ **26.** $16a^2 - 24a + 9$ **27.** $16a^2 + 24a + 9$ **28.** $20y^2 - 59y + 42$ **29.** $\frac{9}{16} - 4y^6$ **30.** $4x^2 - x - 18$ **31.** $2x^3 + 9x^2 + x - 12$ **32.** $3x^4 + x^3 - 7x^2 + 5x - 2$ **33.** $2x^3 + 9x^2y + xy^2 - 12y^3$ **34.** $12x^3y^3 - 20x^2y^2 + 13xy - 15$ **35.** $28a^2 + 17ab - 3b^2$ **36.** $25a^2 - 70ab + 49b^2$ **37.** $16x^2 - y^2$ **38.** $64x^2 + 16xy + y^2$ **39.** $5x^2 - 10x + 2$ **40.** $-x^6 + x^4 - x^2$ **41.** $3x^2y^2 - xy + 5$ **42.** $2a^2 - 3b^2 + \frac{5}{a}$ **43.** $x + 7$ **44.** $4x + 3$ **45.** $7x^3 + 3x^2 - 2x - 5$ **46.** $x^2 + 5x + 25 + \frac{25}{x - 5}$

CHAPTER 6 TEST

1. Is $3x^2 + 5$ a monomial, binomial, or trinomial?

    1. _____

2. Give the degree of $x^3 - 7x^2 + 10x^4 - 8$.

    2. _____

3. Collect like terms and write in descending order.

    $3y^3 - 7y^5 + y^3 - 4y + 6 + y^5 - 2y$

    3. _____

4. Collect like terms.

    $3x^2y^2 - 4x^2 + 3x^2y - 7x^2y^2 + 2x^2y$

    4. _____

5. The cost in dollars of making x pairs of shoes is given by the expression $20x + 45$. Find the cost when 12 pairs of shoes are made.

    5. _____

Add.

6. $(3x + 2) + (-5x + 8)$

    6. _____

7. $(6x^3 + 5x^2 - 3x^4 + 2) + (-8x^2 + 4x^3 + 6x^4 + 5)$

    7. _____

8. $(3a^2b^2 - 2ab + 5) + (4a^2b^2 + 8ab - 3)$

    8. _____

Subtract.

9. $(-5x + 4) - (2x - 3)$

    9. _____

10. $(4x^2 - 6x^4 + 2) - (-2x^2 + x^3 - 2x^4 + 1)$

    10. _____

11. $(8a^2b - 3ab^2 - 2ab) - (4a^2b + ab^2 - ab)$

11. _____

Multiply.

12. $-4x^2(3x^2 - 4x + 1)$

12. _____

13. $(x + 5)(x - 8)$

13. _____

14. $(4a - 3)(2a + 7)$

14. _____

15. $(y + 8)(3y^2 - 2y + 1)$

15. _____

16. $(2x + y)(2x - y)$

16. _____

17. $(3x - 4y)^2$

17. _____

18. $(2x + 5)^2$

18. _____

19. $(3x + 2)(2x^2 - 3x + 4)$

19. _____

CHAPTER 6 TEST Continued

20. $(x^2 - 3)(x^2 + 3)$

20. _____

Divide.

21. $\dfrac{25x^4 - 10x^3 - 15x^2}{5x^2}$

21. _____

22. $(3x^3 - 11x^2 + 10x - 12) \div (x - 3)$

22. _____

7 FACTORING POLYNOMIALS

7.1 COMMON FACTORS AND GROUPING

STUDENT GUIDEPOSTS

1 Prime factors
2 Common factors
3 Greatest common factor (GCF)
4 Finding the GCF
5 Factoring by removing the GCF
6 Prime polynomials
7 Factoring by grouping

Factoring is the reverse of multiplying. Remember that in Chapter 1 we factored integers such as 28 by writing

$$28 = 4 \cdot 7 = 2 \cdot 2 \cdot 7 = 2^2 \cdot 7.$$

1 We say that 28 can be factored into 4 times 7 or that 4 and 7 are **factors** of 28. Further, we call 2 and 7 **prime factors** of 28. To factor $28x^3$ into prime factors we write

$$28x^3 = 2 \cdot 2 \cdot 7 \cdot x \cdot x \cdot x = 2^2 \cdot 7 \cdot x^3.$$

2 A **common factor** of two or more terms is a factor that occurs in each of the terms. For example, a common factor of 14 and 28 is 7 since

$$14 = 2 \cdot 7 \quad \text{and} \quad 28 = 2^2 \cdot 7.$$

3 Other common factors of 14 and 28 are 2 and 14. The **greatest common factor (GCF)** of two integers is the largest factor of both integers. Thus, 14 is the greatest common factor of 14 and 28. As shown in the next example, if we express integers as a product of primes, it is easier to determine the greatest common factor.

EXAMPLE 1

Find the greatest common factor of the integers.

(A) 35 and 75

$$35 = 5 \cdot 7 \quad \text{and} \quad 75 = 3 \cdot 5 \cdot 5$$

GCF: 5

Practice Exercise 1

Find the greatest common factor of the integers.

(A) 55 and 33

289

(B) 54 and 90

$$54 = 2 \cdot 3 \cdot 3 \cdot 3 \quad \text{and} \quad 90 = 2 \cdot 3 \cdot 3 \cdot 5$$

GCF: $2 \cdot 3 \cdot 3 = 18$ ◀◀

(B) 140 and 84

Answers: **(A)** 11 **(B)** 28

We can also find common factors of monomials such as $14x^3$ and $28x^2$. First we write the monomials in prime factored form.

$$14x^3 = 2 \cdot 7 \cdot x^3 = \boxed{2 \cdot 7 \cdot x^2} \cdot x$$
$$28x^2 = 2^2 \cdot 7 \cdot x^2 = 2 \cdot \boxed{2 \cdot 7 \cdot x^2}$$

— Factors common to both

Each shaded factor is a common factor of $14x^3$ and $28x^2$, and the shaded product, $2 \cdot 7 \cdot x^2 = 14x^2$, is the greatest common factor of the two monomials. We can use the following to find the greatest common factor of the terms of a polynomial.

To find the greatest common factor

1. Write each term as a product of prime factors, expressing repeated factors as powers.
2. Select the factors which are common to all terms and raise each factor to the *lowest* power to which it is raised in any one term.
3. The product of these factors is the greatest common factor (GCF).

EXAMPLE 2

Find the greatest common factor of the terms of each polynomial.

(A) $12x - 9 = 2^2 \cdot 3 \cdot x - 3^2$ Express each term as a product of primes

Since 3 is the only common factor of $2^2 \cdot 3 \cdot x$ and 3^2, and 3^1 is the lowest power of 3 that occurs, the GCF is $3^1 = 3$.

(B) $18x^2 + 30x = 2 \cdot 3^2 \cdot x^2 + 2 \cdot 3 \cdot 5 \cdot x$

The factors 2, 3, and x occur in both terms. The lowest power of each is 1. Thus, the GCF is $2 \cdot 3 \cdot x = 6x$.

(C) $3x^4 + 12x^2 = 3 \cdot x^4 + 2^2 \cdot 3 \cdot x^2$

Both 3 and x occur in each term. The lowest power of 3 is 1 and the lowest power of x is 2. Thus, the GCF is $3 \cdot x^2 = 3x^2$.

(D) $7y + 3 = 7 \cdot y + 3$

There are no common prime factors of the terms $7 \cdot y$ and 3. When this happens we say that the GCF is 1.

(E) $14x^4 - 28x^3 + 21x^2 = 2 \cdot 7 \cdot x^4 - 2^2 \cdot 7 \cdot x^3 + 3 \cdot 7 \cdot x^2$

The prime factors 7 and x are in each of the three terms. The lowest power of 7 is 1 and the lowest power of x is 2. The GCF is $7x^2$.

(F) $5x^2y^3 - 15xy^2 = 5 \cdot x^2 \cdot y^3 - 3 \cdot 5 \cdot x \cdot y^2$

Each term has the factors 5, x, and y. The factors of the GCF will be 5, x, and y^2. Thus, the GCF is $5xy^2$. ◀◀

Practice Exercise 2

Find the greatest common factor of the terms of each polynomial.

(A) $15a - 25$

(B) $24y^2 + 20y$

(C) $7w^5 + 21w^3$

(D) $9a + 8$

(E) $18y^5 - 27y^4 + 45y^3$

(F) $22u^3v^3 - 44u^3v + 77u^2v^4$

Answers: **(A)** 5 **(B)** $4y$ **(C)** $7w^3$
(D) 1 **(E)** $9y^3$ **(F)** $11u^2v$

7.1 COMMON FACTORS AND GROUPING

To factor by removing the greatest common factor
1. Write each term as a product of primes.
2. Find the greatest common factor and write each term in the form

    GCF · remaining factors.

3. Use the distributive property to remove the **GCF**. The factors left in each term are the **remaining factors.**
4. To check, multiply the GCF by the polynomial factor.

EXAMPLE 3

Factor by removing the greatest common factor.

(A) $12x - 9 = 2^2 \cdot 3 \cdot x - 3^2$ Prime factors
$ = 3 \cdot 2^2 \cdot x - 3 \cdot 3$ GCF is 3
$ = 3(2^2 \cdot x - 3)$ Remove the GCF using the distributive property
$ = 3(4x - 3)$

To check we multiply.

$$3(4x - 3) = 12x - 9$$

(B) $18x^2 + 30x = 2 \cdot 3^2 \cdot x^2 + 2 \cdot 3 \cdot 5 \cdot x$ Prime factors
$ = 2 \cdot 3 \cdot x \cdot 3 \cdot x + 2 \cdot 3 \cdot x \cdot 5$ GCF is $2 \cdot 3 \cdot x = 6x$
$ = 6x(3x + 5)$ Remove the GCF

To check we multiply.

$$6x(3x + 5) = 18x^2 + 30x$$

(C) $3x^4 + 12x^2 = 3 \cdot x^4 + 2^2 \cdot 3 \cdot x^2$ Prime factors
$ = 3 \cdot x^2 \cdot x^2 + 3 \cdot x^2 \cdot 2^2$ GCF is $3x^2$
$ = 3x^2(x^2 + 4)$ Remove the GCF

To check we multiply.

$$3x^2(x^2 + 4) = 3x^4 + 12x^2$$

(D) $7y + 3 = 7 \cdot y + 3$ Prime factors
$ = 7y + 3$ GCF is 1

When the greatest common factor is 1 we say that the polynomial cannot be factored using the distributive property. Polynomials like $7y + 3$ which cannot be factored are called **prime polynomials.**

(E) $14x^4 - 28x^3 + 21x^2 = 2 \cdot 7 \cdot x^4 - 2^2 \cdot 7 \cdot x^3 + 3 \cdot 7 \cdot x^2$
$ = 7x^2 \cdot 2x^2 - 7x^2 \cdot 4x + 7x^2 \cdot 3$ GCF is $7x^2$
$ = 7x^2(2x^2 - 4x + 3)$

(F) $5x^2y^3 - 15xy^2 = 5 \cdot x^2 \cdot y^3 - 3 \cdot 5 \cdot x \cdot y^2$
$ = 5xy^2 \cdot xy - 5xy^2 \cdot 3$ GCF is $5xy^2$
$ = 5xy^2(xy - 3)$

Practice Exercise 3

Factor by removing the greatest common factor.

(A) $15a - 25$

(B) $24y^2 + 20y$

(C) $7w^5 + 21w^3$

(D) $9a + 8$

(E) $18y^5 - 27y^4 + 45y^3$

(F) $22u^3v^3 - 44u^3v + 77u^2v^4$

Answers: (A) $5(3a - 5)$
(B) $4y(6y + 5)$ (C) $7w^3(w^2 + 3)$
(D) $9a + 8$ is a prime polynomial
(E) $9y^3(2y^2 - 3y + 5)$
(F) $11u^2v(2uv^2 - 4u + 7v^3)$

292 FACTORING POLYNOMIALS

▶▶ **CAUTION** ◀◀ When told to factor a polynomial such as $18x^2 + 30x$, a common error is to give

$$2 \cdot 3 \cdot 3 \cdot x \cdot x + 2 \cdot 3 \cdot 5 \cdot x$$

for the answer. Here the terms of the polynomial have been factored, but the polynomial itself has not. The correct factorization is given in Example 3(b). ◀◀

We may also factor out negative factors from a polynomial.

EXAMPLE 4

Factor $-6x^5 - 12x^4 - 15x$.

$$-6x^5 - 12x^4 - 15x = (-3x)2x^4 + (-3x)4x^3 + (-3x)5$$
$$= -3x(2x^4 + 4x^3 + 5)$$

In this example the greatest common factor is $3x$ and we could factor as $3x(-2x^4 - 4x^3 - 5)$. Either answer is considered correct. ◀◀

Practice Exercise 4

Factor $-10x^4 - 35x^3 - 40x^2$.

Answer: $-5x^2(2x^2 + 7x + 8)$

EXAMPLE 5

Factor $3x^2y^3 - xy$.

$$3x^2y^3 - xy = (xy) \cdot 3xy^2 - (xy) \cdot 1 \quad \text{GCF is } xy$$
$$= xy(3xy^2 - 1) \quad \text{Distributive property} \; \blacktriangleleft\blacktriangleleft$$

Practice Exercise 5

Factor $6u^4v^4 + u^2v^3$.

Answer: $u^2v^3(6u^2v + 1)$

▶▶ **CAUTION** ◀◀ When we factored xy out of $(xy) \cdot 1$ in Example 5, we were left with a 1 in the term. Do not leave this term out and write $xy(3xy^2)$ as the factors of $3x^2y^3 - xy$. If we multiply $xy(3xy^2)$ we get $3x^2y^3$, which is just the first term of $3x^2y^3 - xy$. ◀◀

In some cases a polynomial may contain symbols of grouping or we may introduce grouping symbols to aid in factoring. For example, the polynomial

$$3x(x + 2) + 5(x + 2)$$

is really a binomial made up of the two terms $3x(x + 2)$ and $5(x + 2)$. These terms have a greatest common factor of $(x + 2)$.

$$3x(x + 2) + 5(x + 2) = 3x \cdot (x + 2) + 5 \cdot (x + 2) \quad \text{GCF is } (x + 2)$$
$$= (3x + 5)(x + 2) \quad \text{Distributive property}$$

Here the $x + 2$ is considered as a single factor and is removed using the distributive property.

EXAMPLE 6

Factor $3x(a + b) + y(a + b)$.

$$3x(a + b) + y(a + b) = 3x \cdot (a + b) + y \cdot (a + b) \quad \text{GCF is } (a + b)$$
$$= (3x + y)(a + b) \quad \text{Distributive property} \; \blacktriangleleft\blacktriangleleft$$

Practice Exercise 6

Factor $2y(a - b) - x(a - b)$.

Answer: $(2y - x)(a - b)$

Notice that if the polynomial in Example 6 is multiplied out completely we obtain

$$3x(a + b) + y(a + b) = 3xa + 3xb + ya + yb.$$

7.1 COMMON FACTORS AND GROUPING

Suppose we are given the polynomial on the right to factor. We may do so by grouping the first two terms and the last two terms. Then we factor as in Example 6. This process is called **factoring by grouping** and is illustrated in the next example.

EXAMPLE 7

Factor each polynomial by grouping.

(A) $3xa + 3xb + ya + yb = (3xa + 3xb) + (ya + yb)$
$= 3x(a + b) + y(a + b)$
$= (3x + y)(a + b)$

(B) $5ay - 5ax + 3by - 3bx = (5ay - 5ax) + (3by - 3bx)$
$= 5a(y - x) + 3b(y - x)$
$= (5a + 3b)(y - x)$ ◀◀

Practice Exercise 7

Factor each polynomial by grouping.

(A) $2ay + by + 2ax + bx$

(B) $6xu - 5yu + 6xv - 5yv$

Answers: (A) $(2a + b)(y + x)$
(B) $(6x - 5y)(u + v)$

7.1 EXERCISES A

Find the greatest common factor.

1. 10, 15
2. 28, 42
3. 17, 23

4. $10x^2$, $15x$
5. $28y^3$, $42y$
6. $17y^2$, $23y^4$

7. 10, 15, 35
8. 28, 42, 12
9. 17, 23, 16

10. $10x^2$, $15x$, $35x^3$
11. $28y^3$, $42y^2$, $12y^5$
12. $17y^2$, $23y^4$, 16

13. $6x^3$, $12x^2$, $18x^2$
14. $36y^4$, $6y^3$, $42y^5$
15. $8y^2$, $9x$, $12y^5$

16. $5x^2y$, $40xy^2$
17. $16x^2y^2$, $12x^3y^4$, $8xy^5$
18. x^4y^3, x^5y^2, x^3

Find the greatest common factor of the terms of the polynomials.

19. $4x + 8$
20. $8y^2 - 16y$
21. $3a^3 - 9a^2$

22. $15x^3 - 25x^2$
23. $9y^4 + 18y^2$
24. $50a^{10} + 75a^8$

25. $4x^3 + 2x^2 - 6x$
26. $18y^5 - 24y^3 + 36y$
27. $3a(a + 2) + 5(a + 2)$

28. $6x^2y^2 - 4x^2y$
29. $5a^3b^3 - 15a^2b^2 + 10ab^2$
30. $x(a + b) + y(a + b)$

294 FACTORING POLYNOMIALS

Factor.

31. $3x + 9$

32. $21x - 14$

33. $24y - 6$

34. $4a^2 + 2a$

35. $18y^2 + 11y$

36. $23a^2 - 5$

37. $x^{10} - x^8 + x^6$

38. $6y^4 - 24y^2 + 12y$

39. $16a^5 + 48a^3 - 24$

40. $60x^3 + 50x^2 - 25x$

41. $-6y^{10} - 8y^8 - 4y^5$

42. $a^2 + 2a + 2$

43. $x^2y^2 + xy$

44. $6a^2b - 2ab^2$

45. $27x^3y^2 + 45xy^2$

46. $15x^3y^3 + 5x^2y^2 + 10xy$

47. $a^2(a + 2) + 3(a + 2)$

48. $x^2(a + b) + y^2(a + b)$

Factor by grouping.

49. $a^3 + 2a^2 + 3a + 6$

50. $x^2a + x^2b + y^2a + y^2b$

51. $a^2b - a^2 + 5b - 5$

52. $x^3 + 2x^2 - 7x - 14$

53. $a^3b^2 + a^3 + 2b^2 + 2$

54. $x^2y - 3x^2 + y - 3$

55. $a^4b + a^3 - ab^3 - b^2$

56. $-x^2y - x^2 - 3y - 3$

For Review

Use the FOIL method to multiply.

57. $(x + 3)(x + 7)$

58. $(x + 3)(x - 7)$

59. $(x - 3)(x + 7)$

60. $(x - 3)(x - 7)$

61. $(x + 4y)(x + 3y)$

62. $(x + 4y)(x - 3y)$

ANSWERS: **1.** 5 **2.** 14 **3.** 1 **4.** $5x$ **5.** $14y$ **6.** y^2 **7.** 5 **8.** 2 **9.** 1 **10.** $5x$ **11.** $2y^2$ **12.** 1 **13.** $6x^2$ **14.** $6y^3$ **15.** 1 **16.** $5xy$ **17.** $4xy^2$ **18.** x^3 **19.** 4 **20.** $8y$ **21.** $3a^2$ **22.** $5x^2$ **23.** $9y^2$ **24.** $25a^8$ **25.** $2x$ **26.** $6y$ **27.** $a + 2$ **28.** $2x^2y$ **29.** $5ab^2$ **30.** $a + b$ **31.** $3(x + 3)$ **32.** $7(3x - 2)$ **33.** $6(4y - 1)$ **34.** $2a(2a + 1)$ **35.** $y(18y + 11)$ **36.** cannot be factored **37.** $x^6(x^4 - x^2 + 1)$ **38.** $6y(y^3 - 4y + 2)$ **39.** $8(2a^5 + 6a^3 - 3)$ **40.** $5x(12x^2 + 10x - 5)$ **41.** $-2y^5(3y^5 + 4y^3 + 2)$ **42.** cannot be factored **43.** $xy(xy + 1)$ **44.** $2ab(3a - b)$ **45.** $9xy^2(3x^2 + 5)$ **46.** $5xy(3x^2y^2 + xy + 2)$ **47.** $(a + 2)(a^2 + 3)$ **48.** $(a + b)(x^2 + y^2)$ **49.** $(a + 2)(a^2 + 3)$ **50.** $(a + b)(x^2 + y^2)$ **51.** $(b - 1)(a^2 + 5)$ **52.** $(x + 2)(x^2 - 7)$ **53.** $(b^2 + 1)(a^3 + 2)$ **54.** $(y - 3)(x^2 + 1)$ **55.** $(ab + 1)(a^3 - b^2)$ **56.** $-(x^2 + 3)(y + 1)$ **57.** $x^2 + 10x + 21$ **58.** $x^2 - 4x - 21$ **59.** $x^2 + 4x - 21$ **60.** $x^2 - 10x + 21$ **61.** $x^2 + 7xy + 12y^2$ **62.** $x^2 + xy - 12y^2$

7.1 EXERCISES B

Find the greatest common factor.

1. 12, 18
2. 32, 48
3. 11, 29
4. $12x$, $18x^3$
5. $32y^4$, $48y^2$
6. $11y^3$, $29y^2$
7. 12, 18, 9
8. 32, 48, 64
9. 11, 29, 30
10. $12x^4$, $18x^2$, $9x^3$
11. $32y^3$, $48y^5$, $64y^4$
12. $11y^2$, $29y^3$, 30
13. $15x^2$, $20x^3$, $5x^4$
14. $22y^5$, $33y^4$, $44y^3$
15. $36y^4$, $15x^2$, $5xy$
16. $6x^2y^3$, $24x^2y$
17. $30x^5y^5$, $2x^2y^4$, $12x^4y^3$
18. x^8y^7, x^4y^6, y^5

Find the greatest common factor of the terms of the polynomials.

19. $9x + 3$
20. $22y^2 - 11y$
21. $15a^3 - 10a^2$
22. $16x^4 - 24x^2$
23. $14y^5 + 35y^4$
24. $100a^{16} + 200a^{14}$
25. $5x^3 + 15x^2 - 15x$
26. $48y^6 - 24y^5 + 12y^2$
27. $7a(a - 5) + 6(a - 5)$
28. $9x^3y^3 - 12xy^4$
29. $4a^3b^3 + 6a^2b^2 - 6ab^2$
30. $2a(x + y) + b(x + y)$

Factor.

31. $5y - 20$
32. $15x - 35$
33. $54y + 12$
34. $22y - 11$
35. $25x^4 - 15x^3$
36. $16a^2 - 7$
37. $x^{12} - x^6 + x^4$
38. $26y^5 - 13y^3 + 39y^2$
39. $8a^6 - 18a^3 + 12$
40. $90x^4 - 45x^3 + 180x^2$
41. $-5y^8 - 10y^6 - 15y^4$
42. $a^2 + a + 5$
43. $x^4y^4 - x^2y$
44. $11a^3b^3 - 22a^2b^2$
45. $24x^4y^3 + 36x^3y^4$
46. $28x^4y^4 + 14x^3y^3 + 35x^2y^2$
47. $2a^3(a - 5) + 5(a - 5)$
48. $a^2(x + y) + 2b(x + y)$

Factor by grouping.

49. $a^3 + 7a^2 + 2a + 14$
50. $a^2x + a^2y + b^2x + b^2y$
51. $a^3b - 2a^3 + 4b - 8$
52. $x^3 + 5x^2 - 2x - 10$
53. $a^5b^3 + 2a^5 + 6b^3 + 12$
54. $xy - 2y + x - 2$
55. $ab^2 + ab - 5b - 5$
56. $-x^3y^2 - 3x^3y - 7y - 21$

For Review

Use the FOIL method to multiply.

57. $(x + 2)(x + 3)$
58. $(x + 2)(x - 3)$
59. $(x - 2)(x + 3)$
60. $(x - 2)(x - 3)$
61. $(x - 2y)(x - 2y)$
62. $(x + 2y)(x + 2y)$

7.1 EXERCISES C

Factor by grouping. Do not collect like terms first.

1. $6x^2 + 21x - 10x - 35$
2. $10x^2 - 15xy - 2xy + 3y^2$
3. $a^2b^2c^2 - 5abc + 3abc - 15$
4. $6a^2b^2c - 15ab - 4abc^2 + 10c$
   [Answer: $(3ab - 2c)(2abc - 5)$]

7.2 FACTORING TRINOMIALS OF THE FORM $x^2 + bx + c$

STUDENT GUIDEPOSTS

1. Review of FOIL
2. Factoring $x^2 + bx + c$
3. Factoring $x^2 + bxy + cy^2$

1 To factor a trinomial into two binomials we must recognize the patterns involved in multiplying binomials. Consider the following multiplication.

$$(x + 3)(x + 7) = x \cdot x + x \cdot 7 + 3 \cdot x + 3 \cdot 7$$
$$= x^2 + 7x + 3x + 21$$
$$= x^2 + 10x + 21$$

Notice that x^2 is the product of the first terms, (F), 21 is the product of the last terms, (L), and $10x$ is the sum of the products (O) and (I).

If we are given the trinomial $x^2 + 10x + 21$ to factor, we reverse the multiplication steps above in order to find $(x + 3)(x + 7)$. That is, we must fill in the blanks in

$$x^2 + 10x + 21 = (x + \underline{})(x + \underline{}).$$

The numbers in the blanks must multiply to give 21 and must add to give 10.

$$x^2 + 10x + 21 = (x + \underline{3})(x + \underline{7})$$

In general, to factor $x^2 + bx + c$ we look for a pair of integers whose product is c and whose sum is b.

$$x^2 + bx + c = (x + \underline{})(x + \underline{})$$

2
To factor $x^2 + bx + c$
1. Write $x^2 + bx + c = (x + \underline{})(x + \underline{})$.
2. List all pairs of integers whose product is c.
3. Fill in the blanks with the pair from this list whose sum is b.

EXAMPLE 1

Factor $x^2 + 6x + 8$.   ($b = 6$ and $c = 8$)

$x^2 + 6x + 8 = (x + \underline{})(x + \underline{})$

Practice Exercise 1

Factor $x^2 + 9x + 18$.

7.2 FACTORING TRINOMIALS OF THE FORM $x^2 + bx + c$

Factors of $c = 8$	Sum of factors
1, 8	$1 + 8 = 9$
2, 4	$2 + 4 = 6$
$-1, -8$	$-1 + (-8) = -9$
$-2, -4$	$-2 + (-4) = -6$

Factors of $c = 18$	Sum of Factors
1, 18	$1 + 18 = 19$
2, 9	$2 + 9 = 11$
3, 6	$3 + 6 = 9$

Since $2 \cdot 4$ is 8, which equals c, and $2 + 4$ is 6, which equals b, we fill in the blanks with 2 and 4.

$$x^2 + 6x + 8 = (x + \underline{2})(x + \underline{4}).$$

To check we multiply.

$$(x + 2)(x + 4) = x^2 + 4x + 2x + 8 = x^2 + 6x + 8 \quad \blacktriangleleft\blacktriangleleft$$

Answer: $(x + 3)(x + 6)$

EXAMPLE 2

Factor each trinomial.

(A) $x^2 + 5x + 6 \quad (b = 5 \text{ and } c = 6)$

$x^2 + 5x + 6 = (x + \underline{})(x + \underline{})$

Factors of $c = 6$	Sum of factors
6, 1	$6 + 1 = 7$
2, 3	$2 + 3 = 5$
$-6, -1$	$-6 + (-1) = -7$
$-2, -3$	$-2 + (-3) = -5$

$x^2 + 5x + 6 = (x + 2)(x + 3) \quad 2 \cdot 3 = 6 \text{ and } 2 + 3 = 5$

Note that in our table since $c = 6 > 0$, the factors in each pair must have the same sign. Also, since $b = 5 > 0$, negative factors will not work.

(B) $x^2 + 5x + 6 \quad (b = -5 \text{ and } c = 6)$

$x^2 - 5x + 6 = (x + \underline{})(x + \underline{})$

Factors of $c = 6$	Sum of factors
6, 1	$6 + 1 = 7$
2, 3	$2 + 3 = 5$
$-6, -1$	$-6 + (-1) = -7$
$-2, -3$	$-2 + (-3) = -5$

$x^2 - 5x + 6 = (x - 2)(x - 3) \quad (-2)(-3) = 6 \text{ and } -2 + (-3) = -5$

Since $b = -5$, positive factors will not work.

(C) $x^2 + x - 6 \quad (b = 1 \text{ and } c = -6)$

$x^2 + x - 6 = (x + \underline{})(x + \underline{})$

Factors of $c = -6$	Sum of factors
6, -1	$6 + (-1) = 5$
-6, 1	$-6 + 1 = -5$
2, -3	$2 + (-3) = -1$
-2, 3	$-2 + 3 = 1$

$x^2 + x - 6 = (x - 2)(x + 3) \quad -2 \cdot 3 = -6 \text{ and } -2 + 3 = 1$

Since $c = -6 < 0$, the factors of c must have opposite signs.

Practice Exercise 2

Factor each trinomial.

(A) $x^2 + 10x + 21$

(B) $x^2 - 10x + 21$

(C) $x^2 + 4x - 21$

(D) $x^2 - x - 6$ $(b = -1$ and $c = -6)$

$x^2 - x - 6 = (x + \_\_)(x + \_\_)$

Factors of $c = -6$	Sum of factors
6, −1	$6 + (-1) = 5$
−6, 1	$-6 + 1 = -5$
2, −3	$-2 + (-3) = -1$
−2, 3	$-2 + 3 = 1$

$x^2 - x - 6 = (x + 2)(x - 3)$ $2(-3) = -6$ and $2 + (-3) = -1$ ◀◀

(D) $x^2 - 4x - 21$

Answers: **(A)** $(x + 3)(x + 7)$
(B) $(x - 3)(x - 7)$
(C) $(x - 3)(x + 7)$
(D) $(x + 3)(x - 7)$

---

3 To factor trinomials in two variables such as $x^2 + bxy + cy^2$, we use the same procedures as above. The only major difference occurs when writing the blanks in the binomials. For example, we write

$$x^2 + 5xy + 6y^2 = (x + \_\_y)(x + \_\_y).$$

The following table compares factoring trinomials in two variables with factoring similar trinomials in one variable.

One variable	Two variables
$x^2 + 5x + 6 = (x + 2)(x + 3)$	$x^2 + 5xy + 6y^2 = (x + 2y)(x + 3y)$
$x^2 - 5x + 6 = (x - 2)(x - 3)$	$x^2 - 5xy + 6y^2 = (x - 2y)(x - 3y)$
$x^2 + x - 6 = (x - 2)(x + 3)$	$x^2 + xy - 6y^2 = (x - 2y)(x + 3y)$
$x^2 - x - 6 = (x + 2)(x - 3)$	$x^2 - xy - 6y^2 = (x + 2y)(x - 3y)$

EXAMPLE 3

Factor each trinomial.

(A) $x^2 + 7xy + 10y^2$ $(b = 7$ and $c = 10)$

$x^2 + 7xy + 10y^2 = (x + \_\_y)(x + \_\_y)$

Factors of $c = 10$	Sum of factors
10, 1	$10 + 1 = 11$
5, 2	$5 + 2 = 7$
−10, −1	$-10 + (-1) = -11$
−5, −2	$-5 + (-2) = -7$

$x^2 + 7xy + 10y^2 = (x + 5y)(x + 2y)$ $5 \cdot 2 = 10$ and $5 + 2 = 7$

(B) $x^2 - xy - 12y^2$ $(b = -1$ and $c = -12)$

$x^2 - xy - 12y^2 = (x + \_\_y)(x + \_\_y)$

Factors of $c = -12$	Sum of factors
12, −1	$12 + (-1) = 11$
−12, 1	$-12 + 1 = -11$
6, −2	$6 + (-2) = 4$
−6, 2	$-6 + 2 = -4$
4, −3	$4 + (-3) = 1$
−4, 3	$-4 + 3 = -1$

$x^2 - xy - 12y^2 = (x - 4y)(x + 3y)$ $-4 \cdot 3 = -12$ and $-4 + 3 = -1$

Practice Exercise 3

Factor each trinomial.

(A) $x^2 + 7xy + 12y^2$

(B) $x^2 + xy - 12y^2$

With practice you will be able to find the factors in a problem like this without constructing the complete table. ◀◀

Answers: **(A)** $(x + 4y)(x + 3y)$
(B) $(x + 4y)(x - 3y)$

Some trinomials are not in the form $x^2 + bxy + cy^2$ but can be put into this form by removing a common factor.

EXAMPLE 4

Factor $3x^2 - 12xy + 12y^2$.

$3x^2 - 12xy + 12y^2 = 3 \cdot x^2 - 3 \cdot 4xy + 3 \cdot 4y^2$
$ = 3(x^2 - 4xy + 4y^2)$

We can now factor $x^2 - 4xy + 4y^2$ where $b = -4$ and $c = 4$.

$x^2 + 4xy + 4y^2 = (x + \underline{}y)(x + \underline{}y)$

Factors of $c = 4$	Sum of factors
4, 1	$4 + 1 = 5$
2, 2	$2 + 2 = 4$
$-4, -1$	$-4 + (-1) = -5$
$-2, -2$	$-2 + (-2) = -4$

$x^2 - 4xy + 4y^2 = (x - 2y)(x - 2y)$ $(-2)(-2) = 4$ and $-2 + (-2) = -4$

Thus,

$$3x^2 - 12xy + 12y^2 = 3(x - 2y)(x - 2y). \quad ◀◀$$

Practice Exercise 4

Factor $7x^2 + 21xy - 70y^2$.

Answer: $7(x + 5y)(x - 2y)$

▶▶ **CAUTION** ◀◀ Always include the common factor as a part of the answer in a factoring problem. Notice that the final answer in Example 4 includes the common factor 3. ◀◀

7.2 EXERCISES A

Factor and check by multiplying.

1. $x^2 + 4x + 3$

2. $x^2 + 2x - 3$

3. $x^2 - 2x - 3$

4. $x^2 - 4x + 3$

5. $u^2 - 12u + 35$

6. $u^2 - 2u - 35$

7. $u^2 + 12u + 35$
8. $u^2 + 2u - 35$
9. $y^2 + 10y + 21$

10. $y^2 + 5y - 24$
11. $x^2 - 12x + 27$
12. $x^2 + 4x - 45$

13. $y^2 - y - 56$
14. $x^2 - 2x - 63$
15. $x^2 - 2x - 120$

16. $x^2 + 4x - 77$
17. $x^2 + 4xy + 3y^2$
18. $x^2 + 2xy - 3y^2$

19. $x^2 - 2xy - 3y^2$
20. $u^2 - 9uv + 20v^2$
21. $x^2 - 4xy + 3y^2$

22. $u^2 + uv - 20v^2$
23. $u^2 + 9uv + 20v^2$
24. $u^2 - uv - 20v^2$

25. $x^2 + 13xy - 30y^2$
26. $x^2 + 11xy + 24y^2$
27. $u^2 - 8uv + 15v^2$

28. $u^2 - 3uv - 40v^2$

29. $x^2 - 10xy + 24y^2$

30. $x^2 + 11xy + 30y^2$

31. $u^2 + 12uv + 36v^2$

32. $u^2 - uv - 42v^2$

33. $x^2 + xy - 90y^2$

34. $x^2 - 4xy - 32y^2$

35. $u^2 - 22uv + 121v^2$

36. $u^2 + 18uv + 77v^2$

For Review

Find the GCF and factor.

37. $35x - 70$

38. $88y^3 - 33y^2$

39. $9a^3 + 24a^2 - 15a$

40. $6x^3y^2 - 12x^2y^3$

41. $28a^4b^4 - 14a^3b^3 - 21a^2b^2$

42. $2a(x + y) - 3b(x + y)$

Factor by grouping.

43. $x^3 - 6x^2 + 5x - 30$

44. $5ax^2 + 2bx^2 + 5ay^2 + 2by^2$

302 FACTORING POLYNOMIALS

Use the FOIL method to multiply.

45. $(2x + 1)(x + 3)$

46. $(3x + 2)(x + 1)$

47. $(3x - 5)(2x - 1)$

48. $(5x - 2)(2x + 1)$

49. $(2x + 5)(x - 2)$

50. $(2x + 1)(x - 3)$

ANSWERS: **1.** $(x + 1)(x + 3)$ **2.** $(x - 1)(x + 3)$ **3.** $(x + 1)(x - 3)$ **4.** $(x - 1)(x - 3)$ **5.** $(u - 5)(u - 7)$
6. $(u + 5)(u - 7)$ **7.** $(u + 5)(u + 7)$ **8.** $(u - 5)(u + 7)$ **9.** $(y + 3)(y + 7)$ **10.** $(y - 3)(y + 8)$ **11.** $(x - 3)(x - 9)$
12. $(x - 5)(x + 9)$ **13.** $(y + 7)(y - 8)$ **14.** $(x + 7)(x - 9)$ **15.** $(x + 10)(x - 12)$ **16.** $(x - 7)(x + 11)$
17. $(x + y)(x + 3y)$ **18.** $(x - y)(x + 3y)$ **19.** $(x + y)(x - 3y)$ **20.** $(u - 4v)(u - 5v)$ **21.** $(x - y)(x - 3y)$
22. $(u - 4v)(u + 5v)$ **23.** $(u + 4v)(u + 5v)$ **24.** $(u + 4v)(u - 5v)$ **25.** $(x - 2y)(x + 15y)$ **26.** $(x + 3y)(x + 8y)$
27. $(u - 3v)(u - 5v)$ **28.** $(u - 8v)(u + 5v)$ **29.** $(x - 4y)(x - 6y)$ **30.** $(x + 5y)(x + 6y)$ **31.** $(u + 6v)(u + 6v)$
32. $(u + 6v)(u - 7v)$ **33.** $(x - 9y)(x + 10y)$ **34.** $(x + 4y)(x - 8y)$ **35.** $(u - 11v)(u - 11v)$ **36.** $(u + 7v)(u + 11v)$
37. $35(x - 2)$ **38.** $11y^2(8y - 3)$ **39.** $3a(3a^2 + 8a - 5)$ **40.** $6x^2y^2(x - 2y)$ **41.** $7a^2b^2(4a^2b^2 - 2ab - 3)$
42. $(x + y)(2a - 3b)$ **43.** $(x - 6)(x^2 + 5)$ **44.** $(5a + 2b)(x^2 + y^2)$ **45.** $2x^2 + 7x + 3$ **46.** $3x^2 + 5x + 2$
47. $6x^2 - 13x + 5$ **48.** $10x^2 + x - 2$ **49.** $2x^2 + x - 10$ **50.** $2x^2 - 5x - 3$

7.2 EXERCISES B

Factor and check by multiplying.

1. $x^2 + 6x + 5$
2. $x^2 + 4x - 5$
3. $x^2 - 4x - 5$
4. $x^2 - 6x + 5$

5. $u^2 - 8u + 15$
6. $u^2 - 2u - 15$
7. $u^2 + 8u + 15$
8. $u^2 + 2u - 15$

9. $y^2 - 10y + 21$
10. $x^2 - 5x - 24$
11. $u^2 + 12u + 27$
12. $y^2 + y - 20$

13. $x^2 - 4x - 32$
14. $u^2 - 2u - 48$
15. $y^2 + 16y + 60$
16. $x^2 + 8x - 33$

17. $x^2 + 6xy + 5y^2$
18. $x^2 + 4xy - 5y^2$
19. $x^2 - 4xy - 5y^2$
20. $x^2 - 6xy + 5y^2$

21. $u^2 - 9uv + 18v^2$
22. $u^2 + 3uv - 18v^2$
23. $u^2 + 9uv + 18v^2$
24. $u^2 - 3uv + 18v^2$

25. $x^2 + 6xy - 27y^2$
26. $x^2 + 12xy + 20y^2$
27. $u^2 - 11uv + 28v^2$
28. $u^2 + uv - 30v^2$

29. $x^2 - 13xy + 40y^2$
30. $x^2 + 16xy + 63y^2$
31. $u^2 + 8uv + 16v^2$
32. $u^2 - 3uv - 40v^2$

33. $x^2 + 6xy - 72y^2$
34. $x^2 - 5xy - 50y^2$
35. $u^2 - 26uv + 169v^2$
36. $u^2 + 19uv + 88v^2$

For Review

Find the GCF and factor.

37. $18x + 36$

38. $35y^4 - 21y^2$

39. $16a^5 - 24a^4 - 32a^3$

40. $25x^4y^5 + 45x^3y^6$

41. $8a^5b^5 + 20a^3b^4 - 32a^2b^5$

42. $6u(x^2 + y^2) + 7(x^2 + y^2)$

Factor by grouping.

43. $x^3 + 3x^2 - 7x - 21$

44. $2a^2x - 2a^2y + b^2x - b^2y$

Use the FOIL method to multiply.

45. $(3x + 1)(x + 7)$

46. $(5x + 2)(x + 3)$

47. $(3x + 5)(2x - 1)$

48. $(2x - 1)(x + 10)$

49. $(5x - 3)(x + 6)$

50. $(2x - 5)(2x + 7)$

7.2 EXERCISES C

Factor.

1. $x^2 + \dfrac{1}{3}x - \dfrac{2}{9}$

2. $x^2 - 0.04x + 0.0003$

3. $(x + 2)^2 - 10(x + 2) + 16$
[Answer: $x(x - 6)$]

4. $x^4 + 5x^2 + 6$

7.3 FACTORING TRINOMIALS OF THE FORM $ax^2 + bx + c$

STUDENT GUIDEPOSTS

1 Factoring $ax^2 + bx + c$ for $a \neq 0$
2 Factoring $ax^2 + bxy + cy^2$
3 Factoring $ax^2 + bx + c$ by grouping

In this section we factor $ax^2 + bx + c$ where $a \neq 1$. As before, it is helpful to consider first a product such as the following.

$$(5x + 2)(2x + 3) = (5x)\cdot(2x) + (5x)\cdot(3) + (2)\cdot(2x) + (2)(3)$$
$$ = 10x^2 + 15x + 4x + 6$$
$$ = 10x^2 + 19x + 6$$

with labels F, O, I, L over the terms.

The $10x^2$ is the product of the first terms, Ⓕ, the 6 is the product of the last terms, Ⓛ, and $19x$ is the sum of the products Ⓞ and Ⓘ. With $a \neq 1$ we use the following trial and error method to factor.

To factor $ax^2 + bx + c$

1. Write $ax^2 + bx + c = (\_x + \_)(\_x + \_)$.
2. List all pairs of integers whose product is a and use these in the first blanks in each binomial factor.
3. List all pairs of integers whose product is c and use these in the second blanks.
4. Use trial and error to determine which pair gives Ⓞ + Ⓘ = b.

EXAMPLE 1

Factor $2x^2 + 7x + 3$.   ($a = 2$, $b = 7$, $c = 3$)

$$2x^2 + 7x + 3 = (\_x + \_)(\_x + \_)$$

with Factors of 3 on top and Factors of 2 on bottom.

Since all terms of the trinomial are positive, we need only list the positive factors of a and c. We list the factors of c in both orders as a reminder to try all possibilities.

Factors of $a = 2$ Factors of $c = 3$
2, 1   3, 1
                             1, 3

$2x^2 + 7x + 3 = (\_x + \_)(\_x + \_)$
$\qquad\qquad\quad = (2x + \_)(x + \_)$ The only factors of a are 2 and 1
$\qquad\qquad\quad \stackrel{?}{=} (2x + 3)(x + 1)$ Does not work because $2x + 3x = 5x \neq 7x$
$\qquad\qquad\quad \stackrel{?}{=} (2x + 1)(x + 3)$ This works because $6x + x = 7x$

$2x^2 + 7x + 3 = (2x + 1)(x + 3)$

To check we multiply.

$(2x + 1)(x + 3) = 2x^2 + 6x + x + 3 = 2x^2 + 7x + 3$ ◀◀

EXAMPLE 2

Factor $6x^2 - 13x + 5$.   ($a = 6$, $b = -13$, $c = 5$)

$$6x^2 - 13x + 5) = (\_x + \_)(\_x + \_)$$

with Factors of 5 on top and Factors of 6 on bottom.

Here the factors of $c = 5$ must both be negative in order to obtain the term $-13x$. As before, we list them in both orders.

Factors of $a = 6$ Factors of $c = 5$
6, 1 $-5, -1$ Try 6, 1 with both $-5, -1$ and $-1, -5$

3, 2 $-1, -5$ Try 3, 2 with both $-5, -1$ and $-1, -5$

$6x^2 - 13x + 5 = (\_x + \_)(\_x + \_)$
$\qquad\qquad\quad \stackrel{?}{=} (6x - 5)(x - 1)$ Does not work because $-6x - 5x = -11x \neq -13x$
$\qquad\qquad\quad \stackrel{?}{=} (6x - 1)(x - 5)$ Does not work because $-30x - x = -31x \neq -13x$
$\qquad\qquad\quad \stackrel{?}{=} (3x - 5)(2x - 1)$ This works because $-3x - 10x = -13x$

$6x^2 - 13x + 5 = (3x - 5)(2x - 1)$

Practice Exercise 1

Factor $3x^2 + 5x + 2$.

Factors of $a = 3$ Factors of $c = 2$
3, 1   2, 1
                               1, 2

$(\_x + \_)(\_x + \_)$ with Factors of 2 on top and Factors of 3 on bottom.

Answer: $(3x + 2)(x + 1)$

Practice Exercise 2

Factor $10x^2 + x - 2$.

To check we multiply.

$$(3x - 5)(2x - 1) = 6x^2 - 3x - 10x + 5 = 6x^2 - 13x + 5$$ ◀◀

Answer: $(5x - 2)(2x + 1)$

EXAMPLE 3

Factor $8y^2 - 10y - 7$.   ($a = 8$, $b = -10$, $c = -7$)

Practice Exercise 3

Factor $8y^2 + 10y - 7$.

The factors of $c = -7$ will have opposite signs.

Factors of $a = 8$	Factors of $c = -7$
8, 1	7, −1 and −7, 1
4, 2	1, −7 and −1, 7

With this many cases to try, we can minimize the number of trials. It appears that if 8 and 1 are used, the middle term will be too big if we have $8 \cdot 7y - 1 \cdot 1y = 55y$, and too small if we have $8 \cdot 1y - 7 \cdot 1y = y$. Thus, we try 4 and 2 as factors of 8.

$$8y^2 - 10y - 7 = (\_y + \_)(\_y + \_)$$
$$= (4y + \_)(2y + \_) \quad \text{Try 4, 2}$$
$$\stackrel{?}{=} (4y + 7)(2y - 1) \quad \text{Does not work}$$
$$\stackrel{?}{=} (4y - 7)(2y + 1) \quad \text{This works}$$
$$8y^2 - 10y - 7 = (4y - 7)(2y + 1)$$

Check this by multiplying. ◀◀

Answer: $(4y + 7)(2y - 1)$

EXAMPLE 4

Factor $-4x^2 - 2x + 20$.

Practice Exercise 4

Factor $-6x^2 + 15x + 9$.

First we factor out the common factor including -1 to make a positive. If a is always positive we only have to consider positive factors of a.

$$-4x^2 - 2x + 20 = -2(2x^2 + x - 10)$$

Now we factor $2x^2 + x - 10$, where $a = 2$, $b = 1$, and $c = -10$.

Factors of $a = 2$	Factors of $c = -10$
2, 1	10, −1 and −10, 1
	−1, 10 and 1, −10
	5, −2 and −5, 2
	−2, 5 and 2, −5

$$-4x^2 - 2x + 20 = -2(2x^2 + x - 10)$$
$$= -2(2x + \_)(x + \_)$$
$$\stackrel{?}{=} -2(2x + 10)(x - 1) \quad \text{Factors of 10 give too large a middle term}$$
$$\stackrel{?}{=} -2(2x + 5)(x - 2) \quad \text{This works}$$
$$-4x^2 - 2x + 20 = -2(2x + 5)(x - 2)$$

We must include the common factor -2 in our answer. ◀◀

Answer: $-3(2x + 1)(x - 3)$

306 FACTORING POLYNOMIALS

2 Factoring trinomials of the form $ax^2 + bxy + cy^2$ can be done by filling in the blanks as indicated below.

$$ax^2 + bxy + cy^2 = (\_x + \_y)(\_x + \_y)$$

EXAMPLE 5

Factor $3x^2 - 14xy + 8y^2$.   ($a = 3$, $b = -14$, $c = 8$)

Both factors of $c = 8$ must be negative since $b = -14$.

Factors of $a = 3$	Factors of $c = 8$
3, 1	$-8, -1$
	$-1, -8$
	$-4, -2$
	$-2, -4$

$3x^2 - 14xy + 8y^2 = (3x + \_y)(x + \_y)$
$\stackrel{?}{=} (3x - 8y)(x - y)$ Does not work
$\stackrel{?}{=} (3x - y)(x - 8y)$ Does not work
$\stackrel{?}{=} (3x - 4y)(x - 2y)$ Does not work
$\stackrel{?}{=} (3x - 2y)(x - 4y)$ This works

$3x^2 - 14xy + 8y^2 = (3x - 2y)(x - 4y)$ ◀

Practice Exercise 5

Factor $3a^2 + ab - 10b^2$.

Answer: $(3a - 5b)(a + 2b)$

EXAMPLE 6

Factor $10u^2 + 7uv - 3v^2$.   ($a = 10$, $b = 7$, $c = -3$)

Factors of $a = 10$	Factors of $c = -3$
10, 1	3, -1
5, 2	-3, 1
	1, -3
	-1, 3

$10u^2 + 7uv - 3v^2 = (\_u + \_v)(\_u + \_v)$
$\stackrel{?}{=} (10u + 3v)(u - v)$ Does not work since $-10uv + 3uv = -7uv$
$\stackrel{?}{=} (10u - 3v)(u + v)$ We know this would work from our first trial

$10u^2 + 7uv - 3v^2 = (10u - 3v)(u + v)$ ◀

Practice Exercise 6

Factor $10x^2 - 13xy + 3y^2$.

Answer: $(10x - 3y)(x - y)$

There is another method that can be used to factor trinomials. Consider the following procedure.

$2x^2 + 13x + 15 = 2x^2 + 10x + 3x + 15$ $13x = 10x + 3x$
$= (2x^2 + 10x) + (3x + 15)$ Group terms
$= 2x(x + 5) + 3(x + 5)$ Factor the groups
$= (2x + 3)(x + 5)$ Factor out the common factor $x + 5$

We have factored $2x^2 + 13x + 15$ by grouping the appropriate terms. But how do we decide to write $13x = 10x + 3x$? Notice that 10 and 3 are factors of $2 \cdot 15 = 30$. That is, 10 and 3 are factors of the product ac in $ax^2 + bx + c$.

> **To factor $ax^2 + bx + c$ by grouping**
> 1. Find the product ac.
> 2. List the factors of ac until a pair is found which add to give b.
> 3. Write bx as a sum using these factors as coefficients of x.
> 4. Factor the result by grouping.

EXAMPLE 7

Factor $3x^2 - 10x + 7$ by grouping.

The product $ac = 3 \cdot 7 = 21$.

Factors of $ac = 21$	Sum of factors
21, 1	$21 + 1 = 22$
$-21, -1$	$-21 + (-1) = -22$
7, 3	$7 + 3 = 10$
$-7, -3$	$-7 + (-3) = -10$

The factors -7 and -3 will work, so write $-10x = -7x - 3x$.

$3x^2 - 10x + 7 = 3x^2 - 7x - 3x + 7$
$= x(3x - 7) - 1 \cdot (3x - 7)$ Factor out x and -1
$= (x - 1)(3x - 7)$ The common factor is $3x - 7$ ◀◀

Practice Exercise 7

Factor $3x^2 + 10x + 7$ by grouping.

Answer: $(x + 1)(3x + 7)$

EXAMPLE 8

Factor $5x^2 + 6xy - 8y^2$ by grouping.

For two variables the procedure is the same.

$$ac = 5(-8) = -40.$$

We try factors whose sum seems most likely to equal 6.

Factors of $ac = -40$	Sum of factors
8, -5	$8 + (-5) = 3$
$-8, 5$	$-8 + 5 = -3$
$-10, 4$	$-10 + 4 = -6$
10, -4	$10 + (-4) = 6$

Write $6xy$ as $10xy - 4xy$.

$5x^2 + 6xy - 8y^2 = 5x^2 + 10xy - 4xy - 8y^2$
$= 5x(x + 2y) - 4y(x + 2y)$ Factor groups
$= (5x - 4y)(x + 2y)$ Common factor is $x + 2y$ ◀◀

Practice Exercise 8

Factor $5x^2 - 6xy - 8y^2$ by grouping.

Answer: $(5x + 4y)(x - 2y)$

7.3 EXERCISES A

Factor and check by multiplying.

1. $2x^2 + 7x + 5$
2. $2x^2 - 5x + 3$
3. $2u^2 - 5u - 3$

4. $2u^2 + 13u - 7$
5. $2y^2 + 13y - 24$
6. $2y^2 + 19y + 24$

7. $2y^2 - 19y + 24$
8. $2y^2 - 13y - 24$
9. $3z^2 - 14z - 5$

10. $6z^2 - 13z - 28$
11. $5x^2 - 33x + 40$
12. $-6x^2 - 39x - 54$
    [*Hint*: Don't forget the common factor.]

13. $7u^2 - 14u + 7$
14. $16u^2 - 16u + 4$
15. $y^2 - 9$ [*Hint*: Note that $b = 0$, $a = 1$, and $c = -9$.]

16. $3y^3 + 11y^2 + 10y$
17. $-45x^2 + 150x - 125$
18. $6x^2 - 3x + 21$

19. $6u^2 - 23u + 20$
20. $5u^2 + 7u - 24$
21. $2x^2 - 5xy + 3y^2$

22. $2x^2 + 7xy + 5y^2$
23. $2u^2 - 5uv - 3v^2$
24. $2u^2 + 13uv - 7v^2$

25. $3x^2 - 10xy + 3y^2$
26. $3x^2 + 13xy + 14y^2$
27. $5u^2 + 21uv + 4v^2$

28. $6u^2 + uv - v^2$
29. $-3x^2 + 18xy - 24y^2$
30. $x^2 + xy + y^2$

31. $4u^2 - v^2$
32. $3u^3v + 14u^2v^2 - 5uv^3$
33. $6x^2 - 35xy - 6y^2$

34. $6x^2 + 29xy + 30y^2$
35. $14x^2 + 30xy + 16y^2$
36. $3x^3y - 2x^2y^2 - xy^3$

For Review

Factor.

37. $x^2 + 17x + 72$
38. $x^2 - 17x + 72$
39. $y^2 - 2y - 80$

40. $x^2 + 6xy - 27y^2$
41. $x^2 - 18xy + 81y^2$
42. $-3x^2 + 9x + 120$

Find the following special products.

43. $(x + 3)(x - 3)$
44. $(x + 3)^2$
45. $(x - 3)^2$

46. $(4u + 5)^2$ **47.** $(4u - 5)^2$ **48.** $(4u + 5)(4u - 5)$

ANSWERS: **1.** $(2x + 5)(x + 1)$ **2.** $(2x - 3)(x - 1)$ **3.** $(2u + 1)(u - 3)$ **4.** $(2u - 1)(u + 7)$ **5.** $(2y - 3)(y + 8)$ **6.** $(2y + 3)(y + 8)$ **7.** $(2y - 3)(y - 8)$ **8.** $(2y + 3)(y - 8)$ **9.** $(3z + 1)(z - 5)$ **10.** $(2z - 7)(3z + 4)$ **11.** $(5x - 8)(x - 5)$ **12.** $(-3)(2x + 9)(x + 2)$ **13.** $7(u - 1)(u - 1)$ **14.** $4(2u - 1)(2u - 1)$ **15.** $(y + 3)(y - 3)$ **16.** $y(3y + 5)(y + 2)$ **17.** $(-5)(3x - 5)(3x - 5)$ **18.** $3(2x^2 - x + 7)$; the trinomial cannot be factored **19.** $(2u - 5)(3u - 4)$ **20.** $(5u - 8)(u + 3)$ **21.** $(2x - 3y)(x - y)$ **22.** $(2x + 5y)(x + y)$ **23.** $(2u + v)(u - 3v)$ **24.** $(2u - v)(u + 7v)$ **25.** $(3x - y)(x - 3y)$ **26.** $(3x + 7y)(x + 2y)$ **27.** $(5u + v)(u + 4v)$ **28.** $(3u - v)(2u + v)$ **29.** $(-3)(x - 4y)(x - 2y)$ **30.** cannot be factored **31.** $(2u + v)(2u - v)$ **32.** $uv(3u - v)(u + 5v)$ **33.** $(6x + y)(x - 6y)$ **34.** $(3x + 10y)(2x + 3y)$ **35.** $2(7x + 8y)(x + y)$ **36.** $xy(3x + y)(x - y)$ **37.** $(x + 8)(x + 9)$ **38.** $(x - 8)(x - 9)$ **39.** $(y + 8)(y - 10)$ **40.** $(x - 3y)(x + 9y)$ **41.** $(x - 9y)(x - 9y)$ **42.** $-3(x + 5)(x - 8)$ **43.** $x^2 - 9$ **44.** $x^2 + 6x + 9$ **45.** $x^2 - 6x + 9$ **46.** $16u^2 + 40u + 25$ **47.** $16u^2 - 40u + 25$ **48.** $16u^2 - 25$

7.3 EXERCISES B

Factor and check by multiplying.

1. $2x^2 + 5x + 3$
2. $2x^2 - 7x + 5$
3. $2u^2 + 3u - 14$
4. $2u^2 - 9u - 5$
5. $2u^2 + 9u - 5$
6. $2y^2 + 17y + 30$
7. $2x^2 - 15x + 25$
8. $2u^2 - 19u - 10$
9. $3y^2 + 10y - 8$
10. $6x^2 + x - 15$
11. $5u^2 - 33u + 18$
12. $-4y^2 - 22y - 28$
13. $5x^2 - 20x + 20$
14. $27u^2 + 18u + 3$
15. $y^2 - 25$
16. $2x^4 + 7x^3 + 3x^2$
17. $-16u^2 + 80u - 100$
18. $6y^2 + 36y + 36$
19. $20x^2 - 13x + 2$
20. $4u^2 - 4u - 35$
21. $2x^2 + 5xy + 3y^2$
22. $2x^2 - 7xy + 5y^2$
23. $2u^2 + 3uv - 14v^2$
24. $2u^2 - 9uv - 5v^2$
25. $3x^2 - 7xy + 2y^2$
26. $3x^2 + 22xy + 7y^2$
27. $5u^2 + 7uv + 2v^2$
28. $6u^2 - uv - 2v^2$
29. $-5x^2 + 25xy - 30y^2$
30. $x^2 - 2xy + 2y^2$
31. $9u^2 - v^2$
32. $3u^4 - 4u^3v - 4u^2v^2$
33. $8x^2 - 7xy - y^2$
34. $6x^2 + 25xy + 11y^2$
35. $14x^2 - 30xy + 16y^2$
36. $3x^3y + 2x^2y^2 - xy^3$

For Review

Factor.

37. $x^2 + 17x + 70$
38. $x^2 - 17x + 70$
39. $y^2 + 2y - 80$
40. $x^2 - 5xy - 66y^2$
41. $x^2 + 24xy + 144y^2$
42. $-5x^2 - 20x + 105$

Find the following special products.

43. $(y + 4)(y - 4)$
44. $(y + 4)^2$
45. $(y - 4)^2$
46. $(3u + 7)^2$
47. $(3u - 7)^2$
48. $(3u + 7)(3u - 7)$

7.3 EXERCISES C

Factor.

1. $2x^2 - \frac{1}{5}x - \frac{1}{25}$
2. $0.02x^2 - 0.9x + 4$

3. $2(3x - 1)^2 - (3x - 1) - 1$
[*Hint:* Substitute u for $3x - 1$.]

4. $3x^4 + 10x^2 - 8$

7.4 FACTORING PERFECT SQUARE TRINOMIALS AND DIFFERENCE OF SQUARES

STUDENT GUIDEPOSTS

1. Perfect square trinomials
2. Factoring perfect square trinomials
3. Factoring a difference of two squares
4. Summary of factoring techniques

Factoring such as

$$x^2 + 8x + 16 = (x + 4)(x + 4) = (x + 4)^2$$

can be done using the methods of the previous two sections. However, there are formulas that allow us to factor this type of trinomial more easily. We used these first in Section 6.4 to square binomials. Now we reverse the formulas to factor.

$$a^2 + 2ab + b^2 = (a + b)^2$$
$$a^2 - 2ab + b^2 = (a - b)^2$$

1 To factor using these formulas, we must check to see if a trinomial is a perfect square. To fit either formula, we see that a trinomial must contain two terms that are perfect squares. The remaining term must be either plus or minus twice the product of the numbers whose squares are the other terms. The trinomial $x^2 + 8x + 16$ fits these conditions since x^2 and 16 are both perfect squares and $8x = 2 \cdot x \cdot 4$. Such a trinomial is called a **perfect square trinomial** and we can use the first formula above to factor it.

$x^2 + 8x + 16 = x^2 + 8x + 4^2$	x^2 and 4^2 are perfect squares
$= x^2 + 2 \cdot x \cdot 4 + 4^2$	$x = a$ and $4 = b$
$= (x + 4)^2$	Use the formula

Notice that the trinomial $x^2 + 10x + 16$ is *not* a perfect square trinomial because of its middle term, even though both x^2 and 16 are perfect squares.

The middle term of a perfect square trinomial is either $2ab$ or $-2ab$. Therefore the numerical coefficient of the middle term must always be even. If a trinomial has an odd coefficient for its middle term, we know immediately it can not be a perfect square trinomial.

2
To factor perfect square trinomials

1. Determine if two terms are perfect squares and the remaining term is twice the product of the numbers whose squares are the other terms.
2. Use the formulas

$$a^2 + 2ab + b^2 = (a + b)^2$$
$$a^2 - 2ab + b^2 = (a - b)^2.$$

EXAMPLE 1

Factor the trinomials.

(A) $x^2 + 6x + 9 = x^2 + 6x + 3^2$ x^2 and 9 are perfect squares
$= x^2 + 2 \cdot x \cdot 3 + 3^2$ $x = a$ and $3 = b$
$= (x + 3)^2$ $(x + 3)^2$ instead of $(x - 3)^2$ since $6x$ is positive

(B) $x^2 - 6x + 9 = x^2 - 6x + 3^2$
$= x^2 - 2 \cdot x \cdot 3 + 3^2$ $x = a$ and $3 = b$
$= (x - 3)^2$ Note the minus sign in this case

(C) $16u^2 + 40u + 25 = (4u)^2 + 40u + 5^2$
$= (4u)^2 + 2 \cdot 4 \cdot u \cdot 5 + 5^2$ $4u = a$ and $5 = b$
$= (4u + 5)^2$

(D) $16u^2 - 40u + 25 = (4u)^2 - 2 \cdot 4 \cdot u \cdot 5 + 5^2$
$= (4u - 5)^2$ ◀◀

Practice Exercise 1

Factor the trinomials.

(A) $y^2 + 22y + 121$

(B) $y^2 - 22y + 121$

(C) $9u^2 + 42u + 49$

(D) $9u^2 - 42u + 49$

Answers: (A) $(y + 11)^2$
(B) $(y - 11)^2$ (C) $(3u + 7)^2$
(D) $(3u - 7)^2$

EXAMPLE 2

Factor the trinomials in two variables.

(A) $x^2 + 18xy + 81y^2 = x^2 + 18xy + (9y)^2$ x^2 and $(9y)^2$ are perfect squares
$= x^2 + 2 \cdot x \cdot 9y + (9y)^2$ $x = a$ and $9y = b$
$= (x + 9y)^2$ Use the formula

(B) $2u^2 - 40uv + 200v^2 = 2(u^2 - 20uv + 100v^2)$ Factor out common factor
$= 2(u^2 - 2 \cdot u \cdot 10v + (10v)^2)$ $u = a$ and $10v = b$
$= 2(u - 10v)^2$ ◀◀

Practice Exercise 2

Factor the trinomials in two variables.

(A) $x^2 - 18xy + 81y^2$

(B) $3u^2 + 60uv + 300v^2$

Answers: (A) $(x - 9y)^2$
(B) $3(u + 10v)^2$

Another special formula can be used whenever we are factoring the difference of two squares.

$$a^2 - b^2 = (a + b)(a - b)$$

Notice that we are factoring a binomial in this case.

To factor a difference of two squares
1. Determine if the binomial is the difference of two perfect squares.
2. Use the formulas
$$a^2 - b^2 = (a + b)(a - b).$$

▶▶ CAUTION ◀◀ $(a + b)(a - b) = a^2 - b^2 \neq (a - b)^2$
$= a^2 - 2ab + b^2$ ◀◀

EXAMPLE 3

Factor the binomials.

(A) $x^2 - 9 = x^2 - 3^2$ x^2 and 3^2 are perfect squares
$= (x + 3)(x - 3)$ $x = a$ and $3 = b$

Practice Exercise 3

Factor the binomials.

(A) $x^2 - 121$

(B) $16u^2 - 25 = (4u)^2 - 5^2$ $(4u)^2$ and 5^2 are perfect squares
$\qquad = (4u + 5)(4u - 5)$ $4u = a$ and $5 = b$
(C) $5x^3 - 500x = 5x(x^2 - 100)$ Factor out common factor
$\qquad = 5x(x^2 - 10^2)$
$\qquad = 5x(x + 10)(x - 10)$
(D) $y^6 - 49 = (y^3)^2 - 7^2$ $(y^3)^2$ and 7^2 are perfect squares
$\qquad = (y^3 + 7)(y^3 - 7)$ $y^3 = a$ and $7 = b$
(E) $x^2 + 9 = x^2 + 3^2$ Cannot be factored ◀◀

(B) $9u^2 - 49$
(C) $2y^2 - 288$
(D) $z^4 - 4$
(E) $w^2 + 25$

Answers: **(A)** $(x + 11)(x - 11)$
(B) $(3u + 7)(3u - 7)$
(C) $2(z + 12)(z - 12)$
(D) $(z^2 + 2)(z^2 - 2)$ **(E)** cannot be factored

In Example 3(E) we said that $x^2 + 9$ could not be factored. There is no formula for factoring $a^2 + b^2$.

EXAMPLE 4

Factor the binomials in two variables.

(A) $25x^2 - 9y^2 = (5x)^2 - (3y)^2$ $(5x)^2$ and $(3y)^2$ are perfect squares
$\qquad = (5x + 3y)(5x - 3y)$ $5x = a$ and $3y = b$
(B) $27u^3 - 48uv^2 = 3u(9u^2 - 16v^2)$ Factor out common factor
$\qquad = 3u[(3u)^2 - (4v)^2]$ $3u = a$ and $4v = b$
$\qquad = 3u(3u + 4v)(3u - 4v)$
(C) $x^4 - y^4 = (x^2)^2 - (y^2)^2$ $x^2 = a$ and $y^2 = b$
$\qquad = (x^2 + y^2)(x^2 - y^2)$ Now $x = a$ and $y = b$
$\qquad = (x^2 + y^2)(x + y)(x - y)$ Factor again ◀◀

Practice Exercise 4

Factor the binomials in two variables.

(A) $100a^2 - b^2$

(B) $50x^3 - 162x$

(C) $x^4 + y^4$

Answers: **(A)** $(10a + b)(10a - b)$
(B) $2x(5x + 9)(5x - 9)$ **(C)** cannot be factored

7.4 EXERCISES A

Factor using the special formulas.

1. $x^2 + 10x + 25$
2. $x^2 - 10x + 25$
3. $x^2 - 25$

4. $x^2 + 25$
5. $u^2 - 14u + 49$
6. $u^2 + 14u + 49$

7. $u^2 - 49$
8. $u^2 + 49$
9. $x^2 - 24x + 144$

10. $x^2 - 121$
11. $9u^2 + 6u + 1$
12. $9u^2 - 6u + 1$

13. $4y^2 - 12y + 9$
14. $4y^2 - 9$
15. $25x^2 + 20x + 4$

16. $4x^2 + 20x + 25$
17. $9u^2 - 25$
18. $9u^2 + 25$

19. $-12y^2 + 60y - 75$
20. $4y^2 - 36y + 81$
21. $9x^2 + 30x + 25$

22. $x^2 + x + 4$
23. $8u^3 - 8u$
24. $u^6 - 16$

25. $5y^3 - 100y^2 + 500y$
26. $7y^4 - 63$
27. $x^2 + 6xy + 9y^2$

28. $25x^2 - 10xy + y^2$
29. $64u^2 - 9v^2$
30. $u^4 - v^4$

Before we complete this exercise set we summarize the factoring techniques that we have learned.

1. Factor out any common factor, including -1 if a is negative in
$$ax^2 + bx + c.$$
2. To factor a binomial use the rule
$$a^2 - b^2 = (a + b)(a - b).$$
3. To factor a trinomial use the technique of Section 7.2 and Section 7.3, but also consider perfect squares using the rules
$$a^2 + 2ab + b^2 = (a + b)^2 \quad \text{and} \quad a^2 - 2ab + b^2 = (a - b)^2.$$
4. To factor a polynomial of more than three terms, try to factor by grouping.

Factor.

31. $x^2 + 9x + 18$
32. $x^2 - 12x + 36$
33. $u^2 - 14u + 48$

34. $u^2 + 10u + 16$
35. $4y^2 - 49$
36. $4y^2 - 28y + 49$

37. $-6x^2 + 486$ **38.** $2x^2 + 24x + 64$ **39.** $25u^2 - 20u + 4$

40. $16u^2 - 9$ **41.** $4y^2 - 16y + 15$ **42.** $40y^2 + 11y - 2$

43. $3x^9 - 147x^3$ **44.** $98x^6 - 18$ **45.** $x^2 - 9y^2$

46. $3x^2 - 19xy - 14y^2$ **47.** $36u^2 - 60uv + 25v^2$ **48.** $18u^2 - 98v^2$

49. $100x^5 - 60x^4y + 9x^3y^2$ **50.** $(x + y)^2 - 25$ [Hint: $x + y = a$, $5 = b$]

ANSWERS: **1.** $(x + 5)^2$ **2.** $(x - 5)^2$ **3.** $(x + 5)(x - 5)$ **4.** cannot be factored **5.** $(u - 7)^2$ **6.** $(u + 7)^2$
7. $(u + 7)(u - 7)$ **8.** cannot be factored **9.** $(x - 12)^2$ **10.** $(x + 11)(x - 11)$ **11.** $(3u + 1)^2$ **12.** $(3u - 1)^2$
13. $(2y - 3)^2$ **14.** $(2y + 3)(2y - 3)$ **15.** $(5x + 2)^2$ **16.** $(2x + 5)^2$ **17.** $(3u + 5)(3u - 5)$ **18.** cannot be factored
19. $(-3)(2y - 5)^2$ **20.** $(2y - 9)^2$ **21.** $(3x + 5)^2$ **22.** cannot be factored **23.** $8u(u + 1)(u - 1)$ **24.** $(u^3 + 4)(u^3 - 4)$
25. $5y(y - 10)^2$ **26.** $7(y^2 + 3)(y^2 - 3)$ **27.** $(x + 3y)^2$ **28.** $(5x - y)^2$ **29.** $(8u + 3v)(8u - 3v)$
30. $(u^2 + v^2)(u + v)(u - v)$ **31.** $(x + 3)(x + 6)$ **32.** $(x - 6)^2$ **33.** $(u - 6)(u - 8)$ **34.** $(u + 2)(u + 8)$
35. $(2y + 7)(2y - 7)$ **36.** $(2y - 7)^2$ **37.** $-6(x + 9)(x - 9)$ **38.** $2(x + 4)(x + 8)$ **39.** $(5u - 2)^2$ **40.** $(4u + 3)(4u - 3)$
41. $(2y - 3)(2y - 5)$ **42.** $(8y - 1)(5y + 2)$ **43.** $3x^3(x^3 + 7)(x^3 - 7)$ **44.** $2(7x^3 + 3)(7x^3 - 3)$ **45.** $(x + 3y)(x - 3y)$
46. $(3x + 2y)(x - 7y)$ **47.** $(6u - 5v)^2$ **48.** $2(3u + 7v)(3u - 7v)$ **49.** $x^3(10x - 3y)^2$ **50.** $(x + y + 5)(x + y - 5)$

7.4 EXERCISES B

Factor using the special formulas.

1. $x^2 + 12x + 36$ **2.** $x^2 - 12x + 36$ **3.** $x^2 - 36$ **4.** $x^2 + 36$

5. $u^2 - 16u + 64$ **6.** $u^2 + 16u + 64$ **7.** $u^2 - 64$ **8.** $u^2 + 64$

9. $x^2 - 22x + 121$ **10.** $x^2 - 169$ **11.** $4u^2 + 4u + 1$ **12.** $4u^2 - 4u + 1$

13. $9y^2 - 12y + 4$ **14.** $9y^2 - 4$ **15.** $36x^2 + 12x + 1$ **16.** $9x^2 + 30x + 25$

17. $4u^2 - 25$ **18.** $4u^2 + 25$ **19.** $-32y^2 + 48y - 18$ **20.** $9y^2 - 42y + 49$

21. $25x^2 + 30x + 9$ **22.** $x^2 + 2x + 4$ **23.** $4u^3 - 16u$ **24.** $u^6 - 81$

25. $-4y^2 - 64y - 256$ **26.** $y^4 - 81$ **27.** $x^2 + 8xy + 16y^2$ **28.** $36x^2 - 12xy + y^2$

29. $49u^2 - 16v^2$ **30.** $81u^4 - v^4$

Factor.

31. $x^2 + 12x + 35$ **32.** $x^2 + 20x + 100$ **33.** $u^2 - 22u + 121$ **34.** $u^2 - 14u + 45$

316 FACTORING POLYNOMIALS

35. $9y^2 - 64$
36. $-4y^2 + 400$
37. $9x^2 - 60x + 100$
38. $28x^2 - 41x + 15$
39. $-10u^2 - 21u + 10$
40. $18u^2 - 98$
41. $42y^2 - 3y - 9$
42. $16y^2 + 6y - 27$
43. $25x^2 - 90x + 81$
44. $x^2 - 3x - 180$
45. $x^2 - 25y^2$
46. $2x^2 - 52xy + 338y^2$
47. $49u^2 + 42uv + 9v^2$
48. $200u^2 - 242v^2$
49. $9x^4y^2 - 60x^3y^3 + 100x^2y^4$
50. $(x - y)^2 - 16$

7.4 EXERCISES C

Use the formulas,

$$x^3 + y^3 = (x + y)(x^2 - xy + y^2)$$
$$x^3 - y^3 = (x - y)(x^2 + xy + y^2),$$

to factor the following binomials.

1. $a^3 + 8$
   [Hint: $a^3 + 8 = a^3 + 2^3$]
2. $a^3 - 8$
3. $27b^3 - 8$
4. $27b^3 + 8$
5. $8a^3 - 125b^3$ [Answer: $(2a - 5b)(4a^2 + 10ab + 25b^2)$]
6. $8a^3 + 125b^3$

7.5 FACTORING TO SOLVE EQUATIONS

STUDENT GUIDEPOSTS

1 Zero-product rule
2 Quadratic equations

We now consider an important property of our number system that is used to solve certain types of equations. We know that any number times zero yields a zero product. For example,

$$4 \cdot 0 = 0 \quad \text{and} \quad 0 \cdot \frac{1}{2} = 0.$$

Moreover, if a product of two or more numbers is zero, then at least one of the factors must be zero. This property gives us the zero-product rule.

1 Zero-product rule

If an equation has a product of two or more factors on one side of the equation and 0 on the other, the solutions to the equation are found by setting each factor equal to 0. That is,

if $a \cdot b = 0$ then $a = 0$ or $b = 0$.

The zero-product rule can be used to solve for x in equations of the form

$$(x - 3)(x + 5) = 0.$$

That is, $x - 3$ and $x + 5$ are multiplied to give zero. The rule says that either $x - 3 = 0$ or $x + 5 = 0$. Thus, one solution of

$$(x - 3)(x + 5) = 0$$

is found by solving
$$x - 3 = 0$$
$$x = 3. \quad \text{One solution is 3}$$

Another solution comes by solving
$$x + 5 = 0$$
$$x = -5. \quad \text{Another solution is } -5$$

Thus, there are two solutions to the equation, 3 and -5. To check the solutions we substitute back into the original equations.

Check of 3	Check of -5
$(x - 3)(x + 5) = 0$	$(x - 3)(x + 5) = 0$
$(3 - 3)(3 + 5) \stackrel{?}{=} 0$	$(-5 - 3)(-5 + 5) \stackrel{?}{=} 0$
$0 \cdot 8 = 0$	$(-8) \cdot 0 = 0$
Thus 3 checks.	Thus -5 checks.

EXAMPLE 1

Solve $(2x + 1)(x - 1) = 0$.

$$2x + 1 = 0 \quad \text{or} \quad x - 1 = 0$$
$$2x + 1 - 1 = 0 - 1 \quad\quad x - 1 + 1 = 0 + 1$$
$$2x = -1 \quad\quad x = 1$$
$$\frac{1}{2} \cdot 2x = \frac{1}{2} \cdot (-1)$$
$$x = -\frac{1}{2}$$

The solutions are $-\frac{1}{2}$ and 1. ◀◀

Practice Exercise 1

Solve $(3x + 2)(x - 9) = 0$.

Answer: $-\frac{2}{3}$ and 9

EXAMPLE 2

Solve $x(3x + 1) = 0$.

$$x = 0 \quad \text{or} \quad 3x + 1 = 0$$
$$3x + 1 - 1 = 0 - 1$$
$$3x = -1$$
$$\frac{1}{3} \cdot 3x = \frac{1}{3}(-1)$$
$$x = -\frac{1}{3}$$

The solutions are 0 and $-\frac{1}{3}$. ◀◀

Practice Exercise 2

Solve $y(y + 5) = 0$.

Answer: 0 and -5

When one of the factors has only one term, say x, as in Example 2, one of the solutions will always be zero. Be sure that you give this solution, and do not divide both sides of the equation by x.

Suppose we solve $(x - 2)(x + 3) = 0$.

$$x - 2 = 0 \quad \text{or} \quad x + 3 = 0$$
$$x = 2 \quad\quad x = -3$$

318 FACTORING POLYNOMIALS

The solutions are 2 and -3. In this case, we could have started with $x^2 + x - 6 = 0$ and factored to get $(x - 2)(x + 3) = 0$. Then we would have been using the zero-product rule to solve the *quadratic equation* $x^2 + x - 6 = 0$. A **quadratic equation** has the form

$$ax^2 + bx + c = 0.$$

Thus, in this section we are solving quadratic equations, a subject that we will study in more detail in Chapter 10.

EXAMPLE 3

Solve $x^2 + 4x - 21 = 0$.

$(x + 7)(x - 3) = 0$ Factor the trinomial

$x + 7 = 0$ or $x - 3 = 0$ Use zero-product rule

$x = -7$ $x = 3$

The solutions are -7 and 3.

Practice Exercise 3

Solve $2x^2 - 11x - 40 = 0$.

Answer: 8 and $-\dfrac{5}{2}$

▶▶ **CAUTION** ◀◀ Using the zero-product rule, we can only solve equations in the form $a \cdot b = 0$. If $a \cdot b = 5$, we have no rule of *five* products. Nor can we apply this rule to $a + b = 0$ or $a - b = 0$. ◀◀

EXAMPLE 4

Solve $x(x - 4) = 5$.

The zero-product rule does not apply immediately because this product does *not* equal zero. When this happens, try to rewrite it as a product that does equal zero. We begin by clearing the parentheses.

$x^2 - 4x = 5$

$x^2 - 4x - 5 = 0$

$(x - 5)(x + 1) = 0$

$x - 5 = 0$ or $x + 1 = 0$

$x = 5$ $x = -1$

The solutions are 5 and -1.

Practice Exercise 4

Solve $x(x + 5) = 14$.

Answer: 2 and -7

▶▶ **CAUTION** ◀◀ Do not apply the zero-product rule unless the equation can be written as a product of factors equal to zero. ◀◀

EXAMPLE 5

Solve $(2x + 1) - (x + 5) = 0$.

The zero-product rule does not apply because the left side of the equation is a difference, *not* a product. To solve, we remove parentheses and then combine like terms.

$2x + 1 - x - 5 = 0$

$x - 4 = 0$

$x = 4$

The solution is 4.

Practice Exercise 5

Solve $(x + 2) - (3x + 8) = 0$.

Answer: -3

7.5 EXERCISES A

Solve.

1. $(x - 5)(x + 7) = 0$
2. $(x + 6)(x + 1) = 0$
3. $(y - 8)(y - 9) = 0$

4. $(y + 10)(y - 20) = 0$
5. $(u + 2)(u - 2) = 0$
6. $(u - 0.5)(u + 0.2) = 0$

7. $\left(x - \dfrac{1}{2}\right)\left(x + \dfrac{2}{5}\right) = 0$
8. $(x - 8)^2 = 0$
9. $(3y + 2)(y - 6) = 0$

10. $(y + 8)(y - 8) = 0$
11. $(5u + 1)(2u - 3) = 0$
12. $(8u - 5)(3u - 8) = 0$

13. $x(3x + 7) = 0$
14. $(2x - 5)(x + 12) = 0$
15. $y^2 - 5y + 6 = 0$

16. $y^2 - y - 6 = 0$
17. $u^2 - 8u + 7 = 0$
18. $u^2 + u - 20 = 0$

19. $x^2 + 10x + 25 = 0$
20. $x^2 - 13x + 40 = 0$
21. $y^2 - y - 42 = 0$

22. $y^2 - 16 = 0$
23. $4u^2 - 8u = 0$
24. $4u^2 + 12u + 9 = 0$

25. $9x^2 - 49 = 0$
26. $(3x - 5) - (x + 7) = 0$
27. $y(y + 3) = 10$

28. $y^2 - 2y = 35$
29. $u^2 = -2u + 35$
30. $16u^2 - 42u + 5 = 0$

For Review

Factor.

31. $x^2 - 4x - 21$

32. $x^2 + 14x + 49$

33. $25y^2 - 16$

34. $7y^2 - 70y + 175$

35. $2x^2 + 9xy - 5y^2$

36. $45x^2 - 80y^2$

ANSWERS: **1.** 5, −7   **2.** −6, −1   **3.** 8, 9   **4.** −10, 20   **5.** −2, 2   **6.** 0.5, −0.2   **7.** $\frac{1}{2}, -\frac{2}{5}$ **8.** 8   **9.** $-\frac{2}{3}, 6$ **10.** −8, 8   **11.** $-\frac{1}{5}, \frac{3}{2}$ **12.** $\frac{5}{8}, \frac{8}{3}$ **13.** 0, $-\frac{7}{3}$ **14.** $\frac{5}{2}, -12$ **15.** 2, 3   **16.** −2, 3   **17.** 1, 7   **18.** −5, 4   **19.** −5   **20.** 5, 8   **21.** −6, 7   **22.** −4, 4   **23.** 0, 2   **24.** $-\frac{3}{2}$ **25.** $\frac{7}{3}, -\frac{7}{3}$ **26.** 6   **27.** 2, −5   **28.** −5, 7   **29.** 5, −7   **30.** $\frac{1}{8}, \frac{5}{2}$ **31.** $(x+3)(x-7)$ **32.** $(u+7)^2$ **33.** $(5y+4)(5y-4)$ **34.** $7(y-5)^2$ **35.** $(2x-y)(x+5y)$ **36.** $5(3x+4y)(3x-4y)$

7.5 EXERCISES B

Solve.

1. $(x+2)(x-6) = 0$

2. $(x+8)(x+3) = 0$

3. $(y-5)(y-7) = 0$

4. $(y+8)(y+12) = 0$

5. $(u+6)(u-11) = 0$

6. $(u+4)(u-4) = 0$

7. $(x-0.7)(x+0.3) = 0$

8. $\left(x + \frac{1}{3}\right)\left(x + \frac{2}{3}\right) = 0$

9. $(2y-3)(y+8) = 0$

10. $(y-5)^2 = 0$

11. $(5u-6)(2u+1) = 0$

12. $(3u+14)(2u-5) = 0$

13. $x(4x+3) = 0$

14. $(2x-7)(2x+9) = 0$

15. $y^2 + 8y + 15 = 0$

16. $y^2 - 2y - 15 = 0$

17. $u^2 - 11u + 18 = 0$

18. $u^2 - 3u - 40 = 0$

19. $x^2 - 18x + 81 = 0$

20. $x^2 + 14x + 40 = 0$

21. $2y^2 - 13y + 6 = 0$

22. $49y^2 - 1 = 0$

23. $3u^2 - 27u = 0$

24. $9u^2 - 18u + 9 = 0$

25. $25y^2 - 16 = 0$

26. $(5x+1) - (x-7) = 0$

27. $y(y-8) = -16$

28. $-63 = y^2 - 16y$

29. $63 - u^2 = 2u$

30. $6u^2 - 29u - 5 = 0$

For Review

Factor.

31. $x^2 + 4x - 32$

32. $4x^2 - 12x + 9$

33. $81y^2 - 49$

34. $-6y^2 - 72y - 216$

35. $15x^2 + xy - 2y^2$

36. $40x^2 - 1210y^2$

7.5 EXERCISES C

Solve.

1. $(x + 3)(2x - 5)(x - 4) = 0$
2. $(x - 3)^2 - 3(x - 3) = 4$
3. $x^4 - 16 = 0$
4. $9(x + 2)^2 - 6(x + 2) = -1$
   $\left[\text{Answer: } -\dfrac{5}{3}\right]$
5. $16x^4 - 8x^2 + 1 = 0$
6. $5(x + 4)^2 = -9(x + 4) + 2$
   [*Hint:* Substitute u for $x + 4$.]

7.6 APPLICATIONS OF FACTORING

▶▶ STUDENT GUIDEPOSTS

1 Factoring to solve applied problems
2 Making a sketch in geometry problems

1 ▶▶ Many applied problems can be solved using the techniques of Section 7.5. We first consider several number problems.

EXAMPLE 1

If two times the square of a number minus three times the number is 27, find the number.

Let x = the number.

two times the square of a number minus three times the number is 27
$\qquad 2x^2 \qquad\qquad - \qquad 3x \qquad = \qquad 27$

$2x^2 - 3x = 27$
$2x^2 - 3x - 27 = 0$
$(2x - 9)(x + 3) = 0$
$2x - 9 = 0 \quad\text{or}\quad x + 3 = 0$
$2x = 9 \qquad\qquad x = -3$
$x = \dfrac{9}{2}$

The solutions are $\dfrac{9}{2}$ and -3. ◀◀

Practice Exercise 1

The product of 1 less than a number and 5 more than the same number is 16. Find the number.

Let x = the desired number

$x - 1 = 1$ less than the number

$x + 5 = 5$ more than the number

Solve the equation

$(x - 1)(x + 5) = \underline{}$.

Answer: 3 and -7

EXAMPLE 2

The product of two consecutive even integers is 440. Find the integers.

Let n = first integer,
$n + 2$ = next consecutive even integer,
$n(n + 2)$ = product of the two consecutive even integers.

$n(n + 2) = 440$ Product is 440
$n^2 + 2n = 440$ Remove parentheses
$n^2 + 2n - 440 = 0$ Subtract 440 from both sides
$(n + 22)(n - 20) = 0$ Factor

Practice Exercise 2

The product of two consecutive positive integers is 132. Find the integers.

Let n = first integer

$n + 1$ = next consecutive integer

$$n + 22 = 0 \quad \text{or} \quad n - 20 = 0$$
$$n = -22 \qquad\qquad n = 20$$
$$n + 2 = -20 \qquad n + 2 = 22$$

Thus, one solution is -22 and -20 while the other is 20 and 22. ◀◀

Answer: 11 and 12

 Some geometry problems may also require factoring in the solution. It is always helpful to make a sketch to aid in the solution of a geometry problem. Also, be sure that you answer the stated question and that your solution is reasonable.

EXAMPLE 3

If the area of a rectangle is 48 m² and the length is three times the width, find the dimensions of the rectangle.

Make a sketch, as in Figure 7.1.

Let x = width of rectangle in meters,

$3x$ = length of rectangle in meters.

For a rectangle,

width · length = area.

$$x \cdot 3x = 48$$
$$3x^2 = 48$$
$$3x^2 - 48 = 0$$
$$3(x^2 - 16) = 0 \quad \text{Divide both sides by 3.}$$
$$\text{Remember } \frac{0}{3} = 0$$
$$x^2 - 16 = 0$$
$$(x - 4)(x + 4) = 0$$
$$x - 4 = 0 \quad \text{or} \quad x + 4 = 0$$
$$x = 4 \qquad\qquad x = -4$$

We can rule out $x = -4$ since we are looking for the width of a rectangle

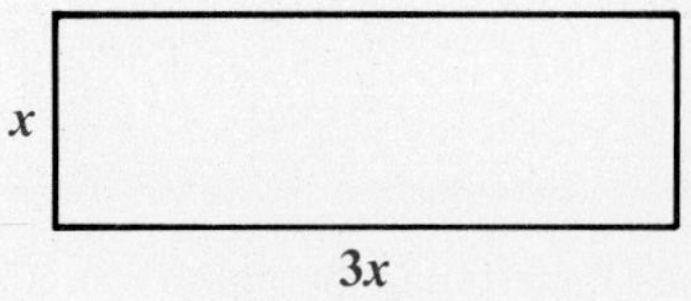

Figure 7.1

Thus, since $x = 4$ and $3x = 12$, the rectangle is 4 m by 12 m. ◀◀

Practice Exercise 3

A triangle has area 60 ft², and its base is 2 ft longer than its height. Find the length of each.

Answer: base: 12 ft; height: 10 ft

EXAMPLE 4

A clothing store owner finds that her daily profit on jeans is given by $P = n^2 - 6n - 20$, where n is the number of jeans sold. How many jeans must be sold for a profit of $260?

Since $P = \$260$, we must solve the following equation.

$$P = n^2 - 6n - 20 = 260 \quad \text{Profit is to be \$260}$$
$$n^2 - 6n - 280 = 0 \quad \text{Set equal to zero}$$
$$(n + 14)(n - 20) = 0 \quad \text{Factor}$$
$$n + 14 = 0 \quad \text{or} \quad n - 20 = 0$$
$$n = -14 \qquad\qquad n = 20$$

Practice Exercise 4

A chemical reaction is described by the equation $C = 2n^2 - 7n + 1$, where n is always positive. Find n when C is 16.

Since n is the number of jeans sold, the -14 answer is discarded. Thus, 20 jeans must be sold to make \$260. To check we evaluate P for $n = 20$.

$$P = n^2 - 6n - 20$$
$$= (20)^2 - 6(20) - 20 \quad \text{Substitute 20 for } n$$
$$= 400 - 120 - 20$$
$$= 260 \quad \text{This checks} \blacktriangleleft\blacktriangleleft$$

Answer: 5

7.6 EXERCISES A

Solve.

1. The product of 5 more than a number and 3 less than the number is zero. Find the number.

   Let $\quad x =$ the number,
   $x + 5 = 5$ more than the number,
   $x - 3 = 3$ less than the number.

2. If the square of a number plus seven times the number is 44, what is the number?

3. The product of a number and 4 less than the number is 21. Find the number.

4. The length of a rectangle is 5 cm more than the width. If the area is 24 cm², find the dimensions.

5. The product of two consecutive integers is 240. Find the integers.

6. If the square of a number is increased by 3 times the number, the result is 70. Find the number.

7. The product of two consecutive odd integers is 143. Find the integers.

8. The sum of the squares of two consecutive even positive integers is 100. Find the integers.

324 FACTORING POLYNOMIALS

9. The area of a rectangle is 40 ft² and the length is 3 feet more than the width. Find the dimensions of the rectangle.

10. The area of a square is numerically 4 less than the perimeter. Find the length of the side in cm.

11. The area of a triangle is 98 cm². If the base is four times the height, determine the base and height.

12. The area of a triangle is 30 in². If the base is 4 inches more than the height, determine the base and height.

The profit on a small appliance is given by the equation $P = n^2 - 3n - 60$ where n is the number of appliances sold per day. Use this equation to do Exercises 13–16.

13. What is the profit when 30 appliances are sold?

14. What is the profit when 5 appliances are sold?

15. How many appliances were sold on a day when the profit was $120?

16. How many appliances were sold on a day when there was a $20 loss?

The number of ways of choosing two people from an organization to serve on a committee is given by $N = \frac{1}{2}n(n - 1)$, where n is the number of people in the organization. Use this equation to do Exercises 17–20.

17. How many committees of two can be formed from a group with 7 members?

18. How many committees of two can be formed from a club with 10 members?

19. If 10 different committees of two can be formed from the members of a club, how many members are in the club?

20. If 28 different committees of two can be formed from the members of a sorority, how many women are in the sorority?

For Review

Solve.

21. $(2x - 5)(x + 10) = 0$

22. $9x^2 + 6x + 1 = 0$

23. $50y^2 - 32 = 0$

24. $3y^2 = 13y + 10$

ANSWERS: **1.** $-5, 3$ **2.** $-11, 4$ **3.** $-3, 7$ **4.** 3 cm, 8 cm **5.** $-16, -15$ or $15, 16$ **6.** 7 or -10 **7.** $-13, -11$; 11, 13 **8.** 6, 8 **9.** 5 ft by 8 ft **10.** 2 cm **11.** 28 cm, 7 cm **12.** 10 in, 6 in **13.** $750 **14.** $-$50 (loss of $50) **15.** 15 **16.** 8 **17.** 21 **18.** 45 **19.** 5 **20.** 8 **21.** $\frac{5}{2}, -10$ **22.** $-\frac{1}{3}$ **23.** $-\frac{4}{5}, \frac{4}{5}$ **24.** $-\frac{2}{3}, 5$

7.6 EXERCISES B

Solve.

1. The product of 4 more than a number and 6 less than the number is 0. Find the number.

2. The square of a number, plus 15, is the same as 8 times the number. Find the number.

3. The product of 4 more than a number and 6 less than the number is 24. Find the number.

4. The length of a rectangle is 10 m more than the width. If the area is 144 m², find the dimensions.

5. If the length of the side of a square is increased by 3 inches, the area is 49 in². Find the length of the side of the original square.

6. The product of two consecutive integers is 132. Find the integers.

7. The product of two consecutive even integers is 224. Find the integers.

8. The sum of the squares of two consecutive odd positive integers is 202. Find the integers.

9. The area of a rectangle is 84 cm² and the length is 5 cm more than the width. Find the dimensions of the rectangle.

10. The area of a square is numerically 12 more than the perimeter. Find the length of a side in ft.

11. The area of a triangle is 150 cm². If the base is three times the height, determine the base and height.

12. The area of a triangle is 56 in². If the base is 6 inches less than the height, determine the base and height.

The profit on one type of shoe is given by the equation $P = n^2 - 5n - 200$, where n is the number of shoes sold per day. Use this equation to do Exercises 13–16.

13. What is the profit if 20 pairs of shoes are sold?

14. What is the profit if 10 pairs of shoes are sold?

15. How many pairs of shoes were sold on a day when the profit was $550?

16. How many pairs of shoes were sold on a day when there was a loss of $50?

326 FACTORING POLYNOMIALS

In a basketball league containing n teams, the number of different ways that the league champion and second place team can be chosen is given by $N = n(n - 1)$. Use this equation to do Exercises 17–20.

17. How many different first and second place finishers are possible in a league with 8 members?

18. How many different first and second place finishers are possible in the NBA, which has 23 teams?

19. If it is known that there are 240 different ways for two teams to finish first and second in a league, how many teams are in the league?

20. If it has been determined that there are 90 different ways for two teams to finish first and second in a conference, how many teams are in the conference?

For Review

Solve.

21. $(3x + 2)(x - 8) = 0$ **22.** $x^2 - 14x + 49 = 0$ **23.** $-45y^2 + 245 = 0$ **24.** $5y^2 = 18y + 8$

7.6 EXERCISES C

Solve.

1. A chemical reaction is described by the equation $C = 2n^2 - 7n + 1$. Find n when $C = 10$ if n is a positive integer.   [Answer: no solution]

2. A banker uses the equation $M = (n + 2)(n - 7)$. Find n when $M = -20$.   [*Hint:* Do not set each of the given factors equal to -20.]

CHAPTER 7 SUMMARY

Key Words and Phrases for Review

7.1 factor
prime factor
greatest common factor (GCF)
prime polynomial

7.4 factor by grouping
perfect square trinomial
difference of squares

7.5 zero-product rule

Key Concepts

7.1 Always factor completely. For example, factor $20x + 30$ as $10(2x + 3)$, not $5(4x + 6)$.

7.2 To factor $x^2 + bx + c = (x + \_\_)(x + \_\_)$, list all pairs of integers whose product is c and fill the blanks with the pair from this list whose sum is b.

7.3 To factor $ax^2 + bx + c = (\_\_x + \_\_)(\_\_x + \_\_)$, list all pairs of integers whose product is a for the first blanks in each binomial factor. Do the same for c using these in the second blanks. By trial and error, select the correct pairs that give the middle term bx.

7.4 (1) $a^2 - b^2 = (a + b)(a - b)$, not $(a - b)^2$

(2) $a^2 + 2ab + b^2 = (a + b)^2$

(3) $a^2 - 2ab + b^2 = (a - b)^2$

7.5 1. The zero-product rule,

if $a \cdot b = 0$, then $a = 0$ or $b = 0$,

only applies when one side of the equation is zero. For example, if $a \cdot b = 7$, we *cannot* conclude that $a = 7$ or $b = 7$.

2. Do not use the zero-product rule on a zero-sum or zero-difference equation. For example, $(2x + 1) - (x + 5) = 0$ is *not* a zero-product equation. To solve it, clear parentheses and combine terms. *Do not* set $2x + 1$ and $x + 5$ equal to zero.

Review Exercises

7.1 *Factor by removing the greatest common factor.*

1. $8x + 2$
2. $3y^4 - 9y^2$
3. $6a^2 - 1$
4. $50x^3 - 40x^2 + 20x$
5. $6x^4y^2 + 12x^2y^4$
6. $a^2(a + b) + 5(a + b)$

Factor by grouping.

7. $x^3 - 3x^2 + 2x - 6$
8. $x^2y^2 + x^2 + 6y^2 + 6$

7.2
7.3
7.4 *Factor.*

9. $x^2 + 3x - 10$
10. $9y^2 + 6y + 1$
11. $x^2 + 7x + 10$
12. $y^2 - 8y + 7$
13. $4x^2 - 9$
14. $4y^2 - 12y + 9$
15. $x^2 - 12x - 45$
16. $y^2 + 12y - 45$
17. $x^2 + 10x + 16$
18. $2y^2 - 19y + 42$
19. $5x^2 + 12x - 9$
20. $2y^2 - 50$
21. $x^4 - 25$
22. $y^4 - 81$
23. $6x^2 + 13x - 5$
24. $6y^2 - 13y - 63$
25. $x^2 - 10xy + 25y^2$
26. $x^2 - 13xy + 42y^2$

FACTORING POLYNOMIALS

27. $81x^2 - 64y^2$

28. $81x^2 + 64y^2$

29. $5x^2 - 90xy + 405y^2$

30. $6x^2 + 13xy - 28y^2$

7.5 *Solve.*

31. $(2x - 1)(12x + 5) = 0$

32. $81y^2 - 9 = 0$

33. $4x^2 + 5x + 1 = 0$

34. $y^2 - 4y + 4 = 0$

35. $12x^2 + 6x = 0$

36. $y(y - 5) = 84$

7.6

37. The product of a number and 5 less than the number is -6. Find the number.

38. The sum of the squares of two consecutive even integers is 452. Find the integers.

39. The length of a rectangle is five times the width. If the area is 180 cm², find the dimensions.

40. The profit on the sale of sport coats is given by the equation $P = n^2 - 6n - 27$, where n is the number of coats sold per day.
  (A) What is the profit when 12 coats are sold?
  (B) How many coats were sold on a day when the profit was $160?

ANSWERS: **1.** $2(4x + 1)$ **2.** $3y^2(y^2 - 3)$ **3.** cannot be factored   **4.** $10x(5x^2 - 4x + 2)$ **5.** $6x^2y^2(x^2 + 2y^2)$
6. $(a + b)(a^2 + 5)$ **7.** $(x - 3)(x^2 + 2)$ **8.** $(y^2 + 1)(x^2 + 6)$ **9.** $(x + 5)(x - 2)$ **10.** $(3y + 1)^2$ **11.** $(x + 5)(x + 2)$
12. $(y - 7)(y - 1)$ **13.** $(2x + 3)(2x - 3)$ **14.** $(2y - 3)^2$ **15.** $(x + 3)(x - 15)$ **16.** $(y - 3)(y + 15)$
17. $(x + 2)(x + 8)$ **18.** $(2y - 7)(y - 6)$ **19.** $(5x - 3)(x + 3)$ **20.** $2(y + 5)(y - 5)$ **21.** $(x^2 + 5)(x^2 - 5)$
22. $(y^2 + 9)(y + 3)(y - 3)$ **23.** $(3x - 1)(2x + 5)$ **24.** $(3y + 7)(2y - 9)$ **25.** $(x - 5y)^2$ **26.** $(x - 6y)(x - 7y)$
27. $(9x + 8y)(9x - 8y)$ **28.** cannot be factored   **29.** $5(x - 9y)^2$ **30.** $(3x - 4y)(2x + 7y)$ **31.** $\frac{1}{2}, -\frac{5}{12}$ **32.** $-\frac{1}{3}, \frac{1}{3}$
33. $-\frac{1}{4}, -1$ **34.** 2   **35.** $0, -\frac{1}{2}$ **36.** $-7, 12$ **37.** $2, 3$ **38.** $-16, -14; 14, 16$ **39.** 6 cm by 30 cm
40. (A) $45   **(B)** 17

CHAPTER 7 TEST

Factor by removing the greatest common factor.

1. $20x + 12$

2. $35y - 7$

3. $24x^4 - 12x^3 + 18x^2$

4. $5x^3y^3 + 40x^3y^2$

5. Factor $7x^2y^2 - y^2 + 14x^2 - 2$ by grouping.

Factor.

6. $x^2 + 15x + 56$

7. $x^2 - 16x + 64$

8. $3y^2 - 75$

9. $x^2 - 8x - 20$

1. _____

2. _____

3. _____

4. _____

5. _____

6. _____

7. _____

8. _____

9. _____

10. $2x^2 + 13x - 24$

11. $3x^2 + 16xy + 5y^2$

Solve.

12. $(4y - 5)(2y + 3) = 0$

13. $x^2 - x - 30 = 0$

14. The product of a number and 3 more than the number is 4. Find the number.

15. The area of a triangle is 12 cm². If the base is 2 cm less than the altitude, find the base and altitude.

16. The profit on the sale of one type of dress is given by the equation $P = n^2 - 3n - 8$, where n is the number of dresses sold per day.
    (A) What is the profit when 20 dresses are sold?

    (B) How many dresses were sold on a day when the profit was $100?

10. _____

11. _____

12. _____

13. _____

14. _____

15. _____

16. (A) _____

16. (B) _____

8 RATIONAL EXPRESSIONS

8.1 BASIC CONCEPTS OF ALGEBRAIC FRACTIONS

STUDENT GUIDEPOSTS

1. Algebraic fractions
2. Rational expressions
3. Omitting values that make the denominator zero
4. Equivalent fractions
5. Reducing to lowest terms
6. Canceling factors when the quotient is −1

1 In Chapter 1 we saw that a fraction is a quotient of integers. In this chapter we define an **algebraic fraction** as a fraction containing a variable in either the numerator or denominator (or both). The following are algebraic fractions:

$$\frac{x}{x+1}, \quad \frac{y^2+1}{y-1}, \quad \frac{\sqrt{a}+5}{2}, \quad \frac{7}{(x-2)(x+1)}, \quad \frac{x+y}{x-y}.$$

2 If an algebraic fraction is the quotient of two polynomials it is called a **rational expression**. Of the algebraic fractions listed above only $\frac{\sqrt{a}+5}{2}$ is *not* a rational expression ($\sqrt{a}+5$ is not a polynomial because $\sqrt{a}$ is not raised to a whole number power). The algebraic fractions that are considered in this chapter will be rational expressions.

When working with rational expressions, we must be concerned with problems of division by zero, which is undefined. Any value of the variable that makes the denominator zero must be omitted from consideration.

3 **To find the values which must be omitted in a rational expression**
1. Set the denominator equal to zero.
2. Solve this equation; any solution must be omitted.

EXAMPLE 1

Find the value of the variable that must be omitted in each rational expression.

(A) $\dfrac{2x}{x+7}$

Set the denominator equal to zero and solve.
$$x + 7 = 0$$
$$x = -7$$

Check: $\dfrac{2(-7)}{-7+7} = \dfrac{-14}{0}$

Since division by zero is not defined, -7 must be omitted.

(B) $\dfrac{(a-1)(a+1)}{2}$

Set the denominator equal to zero and solve.
$$2 = 0$$

Since there is no solution to this equation, there are no values to omit.

(C) $\dfrac{3y}{(y-1)(y+4)}$

We must solve $(y-1)(y+4) = 0$, so we use the zero-product rule.

$$y - 1 = 0 \quad \text{or} \quad y + 4 = 0$$
$$y = 1 \qquad\qquad y = -4$$

The values to omit are 1 and -4.

(D) $\dfrac{x^2 + 1}{x^2 + 5x + 6}$

We must solve $x^2 + 5x + 6 = 0$. First factor and then use the zero-product rule.

$$x^2 + 5x + 6 = 0$$
$$(x + 2)(x + 3) = 0$$
$$x + 2 = 0 \quad \text{or} \quad x + 3 = 0$$
$$x = -2 \qquad\qquad x = -3$$

The values to omit are -2 and -3. ◀◀

Practice Exercise 1

Find the values of the variable that must be omitted.

(A) $\dfrac{a}{a-2}$

(B) $\dfrac{x^2 + x - 3}{7}$

(C) $\dfrac{4w}{(w+1)(w-7)}$

(D) $\dfrac{y^2}{3y^2 - 11y - 4}$

Answers: (A) 2 (B) There are none. (C) -1 and 7 (D) 4 and $-\dfrac{1}{3}$

In Chapter 1 we saw that multiplying or dividing the numerator and denominator of a fraction by the same nonzero number gives us an equivalent fraction.

$$\dfrac{4}{6} = \dfrac{4 \cdot 3}{6 \cdot 3} = \dfrac{12}{18} \qquad \dfrac{4}{6} \text{ is equivalent to } \dfrac{12}{18}$$

$$\dfrac{4}{6} = \dfrac{4 \div 2}{6 \div 2} = \dfrac{2}{3} \qquad \dfrac{4}{6} \text{ is equivalent to } \dfrac{2}{3}$$

We can generalize the idea of equivalent fractions to include algebraic fractions.

334 RATIONAL EXPRESSIONS

4 If the numerator and denominator of an algebraic fraction are multiplied or divided by the same nonzero expression or number, the resulting algebraic fraction is **equivalent** to the original.

EXAMPLE 2

Are the given fractions equivalent?

(A) $\dfrac{3}{x+2}$ and $\dfrac{3(x+1)}{(x+2)(x+1)}$

$$\dfrac{3 \cdot (x+1)}{(x+2) \cdot (x+1)} = \dfrac{3(x+1)}{(x+2)(x+1)} \qquad \text{Numerator and denominator are both multiplied by } x+1$$

Also, $\dfrac{3(x+1) \div (x+1)}{(x+2)(x+1) \div (x+1)} = \dfrac{\frac{3(x+1)}{(x+1)}}{\frac{(x+2)(x+1)}{(x+1)}} = \dfrac{3}{x+2}.$

Thus, the fractions are equivalent. We often abbreviate the division process by simply crossing out or **canceling** the common factors. For example,

$$\dfrac{3(x+1)}{(x+2)(x+1)} = \dfrac{3}{x+2}.$$

(B) $\dfrac{3}{x+2}$ and $\dfrac{7}{x+1}$

These are *not* equivalent since there is no expression that one can be multiplied by, in both numerator and denominator, to give the other. ◀◀

Practice Exercise 2

Are the given fractions equivalent?

(A) $\dfrac{a(a+2)}{(a-3)(a+2)}$ and $\dfrac{a}{a-3}$

(B) $\dfrac{5}{y+1}$ and $\dfrac{5}{y-1}$

Answers: **(A)** yes **(B)** no

5 A fraction is **reduced to lowest terms** when 1 (or −1) is the only number or expression that divides both numerator and denominator. For example, we reduce $\frac{30}{12}$ to lowest terms as follows.

$$\dfrac{30}{12} = \dfrac{2 \cdot 3 \cdot 5}{2 \cdot 2 \cdot 3} = \dfrac{5}{2} \qquad \text{Factor completely and cancel common factors}$$

▶▶ **CAUTION** ◀◀ Cancel *factors* only, never cancel terms. The only expressions that can be canceled are those that are multiplied (never added or subtracted) by every other expression in the numerator or denominator. For example,

$$\dfrac{2 \cdot 4}{2} = \dfrac{2 \cdot 4}{2} \qquad \text{but} \qquad \dfrac{2+4}{2} \neq \dfrac{2+4}{2}. \quad \text{◀◀}$$

We can generalize this procedure to reduce rational expressions.

To reduce a rational expression to lowest terms

1. Factor numerator and denominator completely.
2. Divide (cancel) all common factors.
3. Multiply the remaining factors.

8.1 BASIC CONCEPTS OF ALGEBRAIC FRACTIONS

EXAMPLE 3

Reduce each fraction to lowest terms.

(A) $\dfrac{12x^2}{18x^3}$

$\dfrac{12x^2}{18x^3} = \dfrac{\cancel{2} \cdot 2 \cdot \cancel{3} \cdot \cancel{x} \cdot \cancel{x}}{\cancel{2} \cdot \cancel{3} \cdot 3 \cdot \cancel{x} \cdot \cancel{x} \cdot x} = \dfrac{2}{3x}$ Factor completely and cancel common factors

(B) $\dfrac{2x^2 + x}{3x^3 + 3x}$

$\dfrac{2x^2 + x}{3x^3 + 3x} = \dfrac{\cancel{x}(2x + 1)}{3\cancel{x}(x^2 + 1)} = \dfrac{2x + 1}{3(x^2 + 1)}$ ◀◀

Remember that canceling is dividing; when an expression is canceled, the quotient is 1. For example,

$$\dfrac{x}{2x^2} = \dfrac{\cancel{x}}{2 \cdot \cancel{x} \cdot x} = \dfrac{1}{2x}.$$

EXAMPLE 4

Reduce each fraction to lowest terms.

(A) $\dfrac{7a + 14}{7}$

$\dfrac{7a + 14}{7} = \dfrac{\cancel{7}(a + 2)}{\cancel{7}} = \dfrac{a + 2}{1} = a + 2$

(B) $\dfrac{2 + 2y}{y}$

$\dfrac{2 + 2y}{y} = \dfrac{2(1 + y)}{y}$ Cannot be reduced

(C) $\dfrac{2y}{4y^2 + 20y}$

$\dfrac{2y}{4y^2 + 20y} = \dfrac{2y}{4y(y + 5)} = \dfrac{\cancel{2} \cdot \cancel{y}}{\cancel{2} \cdot 2 \cdot \cancel{y} \cdot (y + 5)} = \dfrac{1}{2(y + 5)}$

(D) $\dfrac{x^2 + 3x + 2}{x^2 - 4}$

$\dfrac{x^2 + 3x + 2}{x^2 - 4} = \dfrac{\cancel{(x + 2)}(x + 1)}{\cancel{(x + 2)}(x - 2)} = \dfrac{x + 1}{x - 2}$

Notice that we can *not* cancel the x in the answer, because x appears as a term, not a factor, in both the numerator and denominator. ◀◀

Fractions of the form $\dfrac{a - b}{b - a}$ often occur in our work. For example, suppose we consider $\dfrac{5 - x}{x - 5}$ and reduce it to lowest terms.

Practice Exercise 3

Reduce each fraction to lowest terms.

(A) $\dfrac{15y^6}{35y^2}$

(B) $\dfrac{a^2 - 3a}{5a^3 + 10a}$

Answers: (A) $\dfrac{3y^4}{7}$ (B) $\dfrac{a - 3}{5(a^2 + 2)}$

Practice Exercise 4

Reduce each fraction to lowest terms.

(A) $\dfrac{3}{3y - 6}$

(B) $\dfrac{-5x}{x - 5}$

(C) $\dfrac{3w^2}{15w^3 + 90w^2}$

(D) $\dfrac{y^2 + 2y - 15}{y^2 - 2y - 35}$

Answers: (A) $\dfrac{1}{y - 2}$ (B) Cannot be reduced. (C) $\dfrac{1}{5(w + 6)}$ (D) $\dfrac{y - 3}{y - 7}$

$$\frac{5-x}{x-5} = \frac{(-1)(-5+x)}{(x-5)}$$ Factor -1 from each term in the numerator

$$= \frac{(-1)(x-5)}{(x-5)}$$ $-5 + x = x - 5$ by the commutative law

$$= -1$$

In general, we can show that

$$\frac{a-b}{b-a} = -1.$$

Whenever a fraction of this type results, simply replace it with -1.

▶▶ **CAUTION** ◀◀ Remember that $a + b = b + a$ so that $\frac{a+b}{b+a} = 1$, not -1. ◀◀

EXAMPLE 5

Reduce $\frac{x^2 - xy}{3y - 3x}$ to lowest terms.

$$\frac{x^2 - xy}{3y - 3x} = \frac{x(x-y)}{3(y-x)} = \frac{x}{3} \cdot \frac{x-y}{y-x}$$

$$= \frac{x}{3} \cdot (-1) \qquad \frac{x-y}{y-x} = -1$$

$$= -\frac{x}{3} \quad ◀◀$$

Practice Exercise 5

Reduce $\frac{a^2b - a^3}{5a - 5b}$ to lowest terms.

Answer: $-\frac{a^2}{5}$

8.1 EXERCISES A

Find the values of the variable that must be omitted.

1. $\dfrac{3}{x+1}$

2. $\dfrac{a}{a-7}$

3. $\dfrac{y+1}{y(y+3)}$

4. $\dfrac{z+2}{2(z+4)}$

5. $\dfrac{x+2}{(x-1)(x+4)}$

6. $\dfrac{a-3}{(a-1)(a+1)}$

7. $\dfrac{2x+7}{x^2 + 2x + 1}$

8. $\dfrac{3y-4}{y^2 - 9}$

9. $\dfrac{(x-1)(x+1)}{7}$

Are the given fractions equivalent? Explain.

10. $\dfrac{2}{7}, \dfrac{6}{21}$

11. $\dfrac{3}{y}, \dfrac{-3}{-y}$

12. $\dfrac{2x}{4}, \dfrac{x}{4}$

13. $\dfrac{a^2}{3}, \dfrac{-2a^2}{-6}$

14. $\dfrac{x+1}{2}, \dfrac{3(x+1)}{6}$

15. $\dfrac{x+1}{9}, \dfrac{(x+1)(x-1)}{9(x-1)}$

16. $\dfrac{2}{x^2}, \dfrac{2+x}{x^2+x}$

17. $\dfrac{x-1}{3x-1}, \dfrac{x}{3x}$

18. $\dfrac{z-2}{z^2-4}, \dfrac{1}{z+2}$

Reduce to lowest terms.

19. $\dfrac{21}{30}$

20. $\dfrac{2y^2}{10y^3}$

21. $\dfrac{4x^3y}{2xy^2}$

22. $\dfrac{22ab^5}{11ab^2}$

23. $\dfrac{3x^2z}{21x^6z^2}$

24. $\dfrac{a(a+1)}{5(a+1)}$

25. $\dfrac{(x+2)(2x-3)}{(x+2)(2x+3)}$

26. $\dfrac{z+9}{z^2+9}$

27. $\dfrac{a+3}{a^2-9}$

28. $\dfrac{w^2+7w}{w^2-49}$

29. $\dfrac{x^2-3x-10}{x^2-6x+5}$

30. $\dfrac{2a(a+5)}{10a+2a^2}$

31. $\dfrac{x-7}{7-x}$

32. $\dfrac{x^2-9}{3-x}$

33. $\dfrac{y+4}{y-4}$

338 RATIONAL EXPRESSIONS

34. $\dfrac{x+y}{x^2-y^2}$

35. $\dfrac{a^2+2ab+b^2}{a+b}$

36. $\dfrac{x^2-7xy+6y^2}{x^2-4xy-12y^2}$

ANSWERS: **1.** -1 **2.** 7 **3.** $0, -3$ **4.** -4 **5.** $1, -4$ **6.** $1, -1$ **7.** -1 **8.** $3, -3$ **9.** none **10.** yes
11. yes **12.** no **13.** yes **14.** yes **15.** yes **16.** no **17.** no **18.** yes **19.** $\dfrac{7}{10}$ **20.** $\dfrac{1}{5y}$ **21.** $\dfrac{2x^2}{y}$ **22.** $2b^3$
23. $\dfrac{1}{7x^4z}$ **24.** $\dfrac{a}{5}$ **25.** $\dfrac{2x-3}{2x+3}$ **26.** $\dfrac{z+9}{z^2+9}$ **27.** $\dfrac{1}{a-3}$ **28.** $\dfrac{w}{w-7}$ **29.** $\dfrac{x+2}{x-1}$ **30.** 1 **31.** -1
32. $-(x+3)$ **33.** $\dfrac{y+4}{y-4}$ **34.** $\dfrac{1}{x-y}$ **35.** $a+b$ **36.** $\dfrac{x-y}{x+2y}$

8.1 EXERCISES B

Find the values of the variable that must be omitted.

1. $\dfrac{5}{x-6}$

2. $\dfrac{x^2+8}{x+5}$

3. $\dfrac{y-3}{y(y-4)}$

4. $\dfrac{z-8}{2(z+10)}$

5. $\dfrac{x+1}{(x-2)(x+6)}$

6. $\dfrac{a+5}{(a-3)(a+3)}$

7. $\dfrac{3x-1}{x^2-2x+1}$

8. $\dfrac{3y+7}{y^2-16}$

9. $\dfrac{z+5}{z^2+9}$

Are the given fractions equivalent? Explain.

10. $\dfrac{3}{5}, \dfrac{6}{10}$

11. $\dfrac{5}{y}, \dfrac{-5}{-y}$

12. $\dfrac{3x}{6}, \dfrac{x}{6}$

13. $\dfrac{a^3}{4}, \dfrac{-2a^3}{-8}$

14. $\dfrac{x+5}{5}, \dfrac{2(x+5)}{10}$

15. $\dfrac{x+3}{8}, \dfrac{(x+3)(x-4)}{8(x-4)}$

16. $\dfrac{3}{x^3}, \dfrac{3-x}{x^3-x}$

17. $\dfrac{z+3}{z^2-9}, \dfrac{1}{z-3}$

18. $\dfrac{x^2+2}{3x^2+2}, \dfrac{x^2}{3x^2}$

Reduce to lowest terms.

19. $\dfrac{22}{55}$

20. $\dfrac{45x^6}{15x^2}$

21. $\dfrac{9a^4b}{3ab^3}$

22. $\dfrac{25xy^3}{5xy^4}$

23. $\dfrac{6ay}{18a^2y^3}$

24. $\dfrac{a^2(a-5)}{3(a-5)}$

25. $\dfrac{(2x+1)(x+5)}{(2x+1)(x-5)}$

26. $\dfrac{z+16}{z^2+16}$

27. $\dfrac{a+4}{a^2-16}$

28. $\dfrac{x^3+5x^2}{x(x^2-25)}$

29. $\dfrac{x^2-3x-10}{x^2-8x+15}$

30. $\dfrac{3a^2(a+4)}{9a+6a^2}$

31. $\dfrac{8-x}{x-8}$

32. $\dfrac{16-x^2}{x-4}$

33. $\dfrac{2y+3}{2y-3}$

34. $\dfrac{x - y}{x^2 - y^2}$

35. $\dfrac{a^2 + 7ab + 10b^2}{a + 5b}$

36. $\dfrac{x^2 - 10xy + 25y^2}{x^2 - 4xy - 5y^2}$

8.1 EXERCISES C

Reduce to lowest terms. The factoring formulas in Exercises C of Section 7.4 are used in these exercises.

1. $\dfrac{5x^2 - 25xy + 30y^2}{10x^2 - 40y^2}$

2. $\dfrac{27x^3 - y^3}{18x^2 + 6xy + 2y^2}$

   $\left[\text{Answer: } \dfrac{3x - y}{2}\right]$

3. $\dfrac{x^4 + 8xy^3}{x^4 - 16y^4}$

8.2 MULTIPLICATION AND DIVISION OF FRACTIONS

STUDENT GUIDEPOSTS

1. Multiplying fractions
2. Reciprocal of a fraction
3. Dividing fractions

In Chapter 1 we saw that the product of two or more fractions is equal to the product of all numerators divided by the product of all denominators. The resulting fraction should be reduced to lowest terms. For example,

$$\dfrac{3}{4} \cdot \dfrac{2}{9} = \dfrac{3 \cdot 2}{4 \cdot 9} = \dfrac{6}{36}.$$

To reduce $\dfrac{6}{36}$ to lowest terms, factor the numerator and denominator.

$$\dfrac{6}{36} = \dfrac{\cancel{2} \cdot \cancel{3}}{\cancel{2} \cdot \cancel{3} \cdot 2 \cdot 3} = \dfrac{1}{6}$$

An easier way to find this product is to factor numerators and denominators, divide out common factors, and then multiply. The resulting fraction is the product.

$$\dfrac{3}{4} \cdot \dfrac{2}{9} = \dfrac{3}{2 \cdot 2} \cdot \dfrac{2}{3 \cdot 3} = \dfrac{\cancel{3} \cdot \cancel{2}}{\cancel{2} \cdot 2 \cdot \cancel{3} \cdot 3} = \dfrac{1}{2 \cdot 3} = \dfrac{1}{6}$$

We can generalize this procedure to multiply algebraic fractions.

To multiply fractions

1. Factor all numerators and denominators completely.
2. Place the product in factored form of all numerators over the product in factored form of all denominators.
3. Divide common factors before multiplying the remaining numerator factors and multiplying the remaining denominator factors.

EXAMPLE 1

Multiply.

$$\frac{x}{6} \cdot \frac{3}{2x^2} = \frac{x}{2 \cdot 3} \cdot \frac{3}{2 \cdot x \cdot x} \quad \text{Factor}$$

$$= \frac{\cancel{3} \cdot \cancel{x}}{2 \cdot \cancel{3} \cdot 2 \cdot \cancel{x} \cdot x} \quad \text{Indicate product and cancel}$$

$$= \frac{1}{2 \cdot 2 \cdot x} = \frac{1}{4x} \quad \text{Multiply} \blacktriangleleft\blacktriangleleft$$

Practice Exercise 1

Multiply. $\dfrac{y^2}{15} \cdot \dfrac{5}{y}$

Answer: $\dfrac{y}{3}$

EXAMPLE 2

Multiply.

$$\frac{a^2 - 9}{a^2 - 4a + 4} \cdot \frac{a - 2}{a + 3}$$

$$= \frac{(a-3)(a+3)}{(a-2)(a-2)} \cdot \frac{(a-2)}{(a+3)} \quad \text{Factor}$$

$$= \frac{(a-3)\cancel{(a+3)}\cancel{(a-2)}}{(a-2)\cancel{(a-2)}\cancel{(a+3)}} \quad \text{Cancel common factors}$$

$$= \frac{(a-3)}{(a-2)} \blacktriangleleft\blacktriangleleft$$

Practice Exercise 2

Multiply. $\dfrac{x^2 - 2x}{x^2 + 11x + 28} \cdot \dfrac{x + 7}{x - 2}$

Answer: $\dfrac{x}{x+4}$

EXAMPLE 3

Multiply.

$$\frac{x^2 - y^2}{xy} \cdot \frac{xy - x^2}{x^2 - 2xy + y^2}$$

$$= \frac{(x-y)(x+y)}{xy} \cdot \frac{x(y-x)}{(x-y)(x-y)} \quad \text{Factor}$$

$$= \frac{\cancel{(x-y)}(x+y) \cdot \cancel{x} \cdot (y-x)}{\cancel{x} \cdot y \cdot \cancel{(x-y)}(x-y)} \quad \text{Cancel common factors}$$

$$= \frac{x+y}{y} \cdot \frac{(y-x)}{(x-y)}$$

$$= \frac{x+y}{y} \cdot (-1) \qquad \frac{y-x}{x-y} = -1$$

$$= -\frac{x+y}{y} \blacktriangleleft\blacktriangleleft$$

Practice Exercise 3

Multiply.

$$\frac{a^2 - ab}{2b^2 - ab - a^2} \cdot \frac{2a^2 + 3ab + b^2}{2a^2 + ab}$$

Answer: $-\dfrac{a+b}{2b+a}$

When we divide one arithmetic fraction by another, for example,

$$\frac{2}{3} \div \frac{3}{4} \quad \text{or} \quad \frac{\frac{2}{3}}{\frac{3}{4}},$$

2 ▶▶ the fraction $\frac{3}{4}$ is called the **divisor.** The **quotient** of the two fractions (the resulting fraction after the division is performed) is obtained by taking the **reciprocal** of the divisor $\frac{3}{4}$, which is $\frac{4}{3}$, and multiplying by $\frac{2}{3}$.

$$\frac{2}{3} \div \boxed{\frac{3}{4}} = \frac{2}{3} \cdot \boxed{\frac{4}{3}} = \frac{8}{9}$$

8.2 MULTIPLICATION AND DIVISION OF FRACTIONS

Exactly the same procedure applies when we divide algebraic fractions.

> **Division of algebraic fractions**
> To divide one algebraic fraction by another, find the reciprocal of the divisor and multiply.

▶▶ **CAUTION** ◀◀ The divisor is always to the right of the ÷ sign or below the fraction bar. No change is made to the numerator, which is to the left of the ÷ sign or above the fraction bar. ◀

The following table lists several algebraic fractions and their reciprocals.

Fraction	Reciprocal	
$\dfrac{2}{x}$	$\dfrac{x}{2}$	
$\dfrac{x+1}{(x-5)^2}$	$\dfrac{(x-5)^2}{x+1}$	
5	$\dfrac{1}{5}$	5 can be thought of as $\dfrac{5}{1}$
x	$\dfrac{1}{x}$	x is $\dfrac{x}{1}$
$\dfrac{1}{3}$	3	3 is $\dfrac{3}{1}$
$x + y$	$\dfrac{1}{x+y}$	Not $\dfrac{1}{x} + \dfrac{1}{y}$; $\dfrac{1}{2+3} = \dfrac{1}{5} \neq \dfrac{1}{2} + \dfrac{1}{3} = \dfrac{5}{6}$

What number has no reciprocal? Why?

EXAMPLE 4

Divide.

(A) $\dfrac{x}{x+1} \div \dfrac{x}{2} = \dfrac{x}{x+1} \cdot \dfrac{2}{x} = \dfrac{\cancel{x} \cdot 2}{(x+1) \cdot \cancel{x}} = \dfrac{2}{x+1}$

(B) $\dfrac{y+3}{y} \div y = \dfrac{y+3}{y} \cdot \dfrac{1}{y} = \dfrac{y+3}{y^2}$

(C) $\dfrac{a^2 - 4a + 4}{a^2 - 9} \div \dfrac{a-2}{a+3} = \dfrac{a^2 - 4a + 4}{a^2 - 9} \cdot \dfrac{a+3}{a-2}$

$= \dfrac{(a-2)(a-2)}{(a-3)(a+3)} \cdot \dfrac{a+3}{a-2}$

$= \dfrac{(\cancel{a-2})(a-2)(\cancel{a+3})}{(a-3)(\cancel{a+3})(\cancel{a-2})} = \dfrac{a-2}{a-3}$ ◀◀

Practice Exercise 4

Divide.

(A) $\dfrac{y-5}{y} \div \dfrac{5}{y}$

(B) $\dfrac{b}{a+b} \div b$

(C) $\dfrac{y^2 + 2y - 15}{y^2 - 25} \div \dfrac{y-3}{y-5}$

Answers: (A) $\dfrac{y-5}{5}$

(B) $\dfrac{1}{a+b}$ (C) 1

EXAMPLE 5

Divide.

$\dfrac{x^2 - 3xy + 2y^2}{x^2 - 4y^2} \div (x^2 + xy - 2y^2)$

Practice Exercise 5

Divide.

$\dfrac{a^2 + 8a - 33}{a^2 - 2a - 3} \div (a^2 + 12a + 11)$

$$= \frac{x^2 - 3xy + 2y^2}{x^2 - 4y^2} \cdot \frac{1}{x^2 + xy - 2y^2} \qquad \text{Take reciprocal and multiply}$$

$$= \frac{(x - y)(x - 2y)}{(x + 2y)(x - 2y)} \cdot \frac{1}{(x - y)(x + 2y)} \qquad \text{Factor}$$

$$= \frac{\cancel{(x - y)}\cancel{(x - 2y)}}{(x + 2y)\cancel{(x - 2y)}\cancel{(x - y)}(x + 2y)} \qquad \text{Cancel}$$

$$= \frac{1}{(x + 2y)^2} \qquad\qquad\qquad\qquad\qquad\qquad \text{Answer: } \frac{1}{(a + 1)^2}$$

8.2 EXERCISES A

Multiply.

1. $\dfrac{1}{x^2} \cdot \dfrac{x}{3}$

2. $\dfrac{3a}{4} \cdot \dfrac{6}{9a^2}$

3. $\dfrac{2y}{5} \cdot \dfrac{25}{4y^3}$

4. $\dfrac{2x}{(x + 1)^2} \cdot \dfrac{x + 1}{4x^2}$

5. $\dfrac{x^2 - 36}{x^2 - 6x} \cdot \dfrac{x}{x + 6}$

6. $\dfrac{z^2}{z + 2} \cdot \dfrac{z^2 - 4}{z^2 - 2z}$

7. $\dfrac{5x + 5}{x - 2} \cdot \dfrac{x^2 - 4x + 4}{x^2 - 1}$

8. $\dfrac{y^2 - 3y - 10}{y^2 - 4y + 4} \cdot \dfrac{y - 2}{y - 5}$

9. $\dfrac{x^2 + 2x + 1}{9x^2} \cdot \dfrac{3x^3}{x^2 - 1}$

10. $\dfrac{z^2 - z - 20}{z^2 + 7z + 12} \cdot \dfrac{z + 3}{z^2 - 25}$

11. $\dfrac{y^2 + 6y + 5}{7y^2 - 63} \cdot \dfrac{7y + 21}{(y + 5)^2}$

12. $\dfrac{16 - x^2}{5x - 1} \cdot \dfrac{5x^2 - x}{16 - 8x + x^2}$

13. $\dfrac{a^2 - 4}{a^2 - 4a + 4} \cdot \dfrac{a^2 - 9a + 14}{a^3 + 2a^2}$

14. $\dfrac{2x^2 - 5xy + 3y^2}{x^2 - y^2} \cdot (x^2 + 2xy + y^2)$

15. $\dfrac{x^2 - y^2}{(x + y)^2} \cdot \dfrac{x + y}{x - y}$

16. $\dfrac{x^2 - 7xy + 6y^2}{x^2 - xy - 2y^2} \cdot \dfrac{x + y}{x - 6y}$

Find the reciprocal.

17. $\dfrac{3}{7}$ 18. $\dfrac{1}{2}$ 19. 5 20. $\dfrac{1}{x + 1}$ 21. $\dfrac{2x - 3}{x + 5}$ 22. $z + 2$

Divide.

23. $\dfrac{x}{x + 2} \div \dfrac{x}{3}$

24. $\dfrac{2a}{a + 1} \div \dfrac{5a}{a + 1}$

25. $\dfrac{2(a + 3)}{7} \div \dfrac{4(a + 3)}{21}$

26. $\dfrac{3(x^2 + 5)}{4(x^2 - 3)} \div \dfrac{x^2 + 5}{4(x^2 - 3)}$

27. $\dfrac{5y^4}{y^2 - 1} \div \dfrac{5y^3}{y^2 + 2y + 1}$

28. $\dfrac{6a^2}{a^2 - 25} \div \dfrac{3a^3}{5a^2 - 25a}$

29. $(x + 6) \div \dfrac{x^2 - 36}{x^2 - 6x}$

30. $\dfrac{z^2 - 4z + 4}{z^2 - 1} \div \dfrac{z - 2}{5z + 5}$

31. $\dfrac{y^2 - y - 20}{y^2 + 7y + 12} \div \dfrac{y^2 - 25}{y + 3}$

32. $\dfrac{a^2 + 10a + 21}{a^2 - 2a - 15} \div \dfrac{a + 7}{a^2 - 25}$

33. $\dfrac{y^3 - 64y}{2y^2 + 16y} \div \dfrac{y^2 - 9y + 8}{y^2 + 4y - 5}$

34. $\dfrac{x^2 + 16x + 64}{2x^2 - 128} \div \dfrac{3x^2 + 30x + 48}{x^2 - 6x - 16}$

35. $\dfrac{x^2 - 5xy + 6y^2}{x^2 - 4y^2} \div (x^2 - 2xy - 3y^2)$

36. $\dfrac{x + 2y}{x^2 - y^2} \div \dfrac{x + 2y}{x + y}$

37. $\dfrac{x^2 - 81y^2}{x^2 + 18xy + 81y^2} \div \dfrac{x - 9y}{x + 9y}$

38. $\dfrac{a^2 - y^2}{a^2 - ay} \cdot \dfrac{2a^2 + ay}{a^2 - 4y^2} \div \dfrac{a + y}{a + 2y}$

For Review

Find the values of the variable that must be omitted.

39. $\dfrac{y - 1}{y^2 + 2y}$

40. $\dfrac{a}{a^2 + 5}$

41. $\dfrac{(x + 1)(x - 5)}{3x}$

Are the given fractions equivalent?

42. $\dfrac{x + 5}{2x^2 + 5}, \dfrac{x}{2x^2}$

43. $\dfrac{x}{3}, \dfrac{x^2 + 2x}{3(x + 2)}$

44. $\dfrac{1}{x + 1}, \dfrac{x + 1}{1}$

45. Change $\dfrac{x}{2}$ to an equivalent fraction by multiplying its numerator and denominator by each of the following.

(A) 5 (B) x (C) x^2

(D) $x+1$ (E) $2x-1$ (F) x^2-1

46. Burford reduced the fraction $\dfrac{1+2x}{1+5x}$ in the following way:

$$\dfrac{1+2x}{1+5x} \div \dfrac{\cancel{x}+2\cancel{x}}{\cancel{x}+5\cancel{x}} = \dfrac{2}{5}$$

What is wrong with Burford's work?

ANSWERS: **1.** $\dfrac{1}{3x}$ **2.** $\dfrac{1}{2a}$ **3.** $\dfrac{5}{2y^2}$ **4.** $\dfrac{1}{2x(x+1)}$ **5.** 1 **6.** z **7.** $\dfrac{5(x-2)}{x-1}$ **8.** $\dfrac{y+2}{y-2}$ **9.** $\dfrac{x(x+1)}{3(x-1)}$
10. $\dfrac{1}{z+5}$ **11.** $\dfrac{y+1}{(y-3)(y+5)}$ **12.** $\dfrac{x(4+x)}{4-x}$ **13.** $\dfrac{a-7}{a^2}$ **14.** $(2x-3y)(x+y)$ **15.** 1 **16.** $\dfrac{x-y}{x-2y}$ **17.** $\dfrac{7}{3}$
18. 2 **19.** $\dfrac{1}{5}$ **20.** $x+1$ **21.** $\dfrac{x+5}{2x-3}$ **22.** $\dfrac{1}{z+2}$ **23.** $\dfrac{3}{x+2}$ **24.** $\dfrac{2}{5}$ **25.** $\dfrac{3}{2}$ **26.** 3 **27.** $\dfrac{y(y+1)}{y-1}$
28. $\dfrac{10}{a+5}$ **29.** x **30.** $\dfrac{5(z-2)}{z-1}$ **31.** $\dfrac{1}{y+5}$ **32.** $a+5$ **33.** $\dfrac{y+5}{2}$ **34.** $\dfrac{1}{6}$ **35.** $\dfrac{1}{(x+2y)(x+y)}$
36. $\dfrac{1}{x-y}$ **37.** 1 **38.** $\dfrac{2a+y}{a-2y}$ **39.** 0, −2 **40.** none **41.** 0 **42.** no **43.** yes **44.** no **45.** (A) $\dfrac{5x}{10}$ (B) $\dfrac{x^2}{2x}$
(C) $\dfrac{x^3}{2x^2}$ (D) $\dfrac{x(x+1)}{2(x+1)}$ (E) $\dfrac{x(2x-1)}{2(2x-1)}$ (F) $\dfrac{x(x^2-1)}{2(x^2-1)}$ **46.** Burford is canceling terms not factors. It is clear that the two fractions are not equal when x is replaced with a number.

8.2 EXERCISES B

Multiply.

1. $\dfrac{2}{x^3} \cdot \dfrac{x^2}{6}$ **2.** $\dfrac{5a^2}{8} \cdot \dfrac{4a}{10a^3}$ **3.** $\dfrac{7y^3}{3} \cdot \dfrac{15}{5y}$

4. $\dfrac{3x^2}{x+5} \cdot \dfrac{(x+5)^2}{9x}$ **5.** $\dfrac{x^2-9}{x^2+3x} \cdot \dfrac{x}{x+9}$ **6.** $\dfrac{z^3}{(z+3)} \cdot \dfrac{z^2-9}{z^2-3z}$

7. $\dfrac{4x+4}{4x-4} \cdot \dfrac{x^2-2x+1}{x^2-1}$ **8.** $\dfrac{y^2-5y+6}{y^2-9} \cdot \dfrac{y+2}{y-2}$ **9.** $\dfrac{12x^4}{x^2+8x+16} \cdot \dfrac{x^2-16}{3x^2}$

10. $\dfrac{z^2+z-20}{z^2-7z+12} \cdot \dfrac{z-3}{z^2-25}$ **11.** $\dfrac{y^2+8y+7}{5y^2-125} \cdot \dfrac{5y-25}{(y+1)^2}$ **12.** $\dfrac{9-x^2}{3x-2} \cdot \dfrac{3x^2-2x}{9-6x+x^2}$

13. $\dfrac{a^2-25}{a^2-10a+25} \cdot \dfrac{a^2-8a+15}{a^3-3a^2}$ **14.** $\dfrac{2x^2+xy-10y^2}{x^2-4y^2} \cdot (x^2+4xy+4y^2)$

15. $\dfrac{x^2-2xy+y^2}{x+y} \cdot \dfrac{(x+y)^2}{x-y}$ **16.** $\dfrac{x^2-7xy+10y^2}{x^2-3xy-10y^2} \cdot \dfrac{x+2y}{x-2y}$

Find the reciprocal.

17. 6 **18.** $\dfrac{6}{y}$ **19.** $7a$ **20.** $\dfrac{x}{x+6}$ **21.** $\dfrac{3x-2}{x-2}$ **22.** $5a-1$

Divide.

23. $\dfrac{x+3}{x} \div \dfrac{4}{x}$

24. $\dfrac{a+5}{3a} \div \dfrac{a+5}{9a}$

25. $\dfrac{6(a+4)}{5} \div \dfrac{4(a+4)}{35}$

26. $\dfrac{8(x^2+x+1)}{6(x^2-2)} \div \dfrac{x^2+x+1}{3(x^2-2)}$

27. $\dfrac{y^2+4y+4}{8y^3} \div \dfrac{y^2-4}{2y^2}$

28. $\dfrac{a^2-16}{5a^3} \div \dfrac{3a^2-12a}{20a}$

29. $\dfrac{z^2+4z+4}{z^2-4} \div \dfrac{z+2}{3z-6}$

30. $\dfrac{x^2-49}{x^2+7x} \div (x-7)$

31. $\dfrac{y^2+y-20}{y^2-7y+12} \div \dfrac{y^2-25}{y-3}$

32. $\dfrac{a^2-10a+21}{a^2+2a-15} \div 5a-35$

33. $\dfrac{x^2-16x+64}{3x^2-192} \div \dfrac{3x^2-30x+48}{x^2+6x-16}$

34. $\dfrac{5y^2+10y}{2y^3-8y} \div \dfrac{y^2+5y-6}{y^2-3y+2}$

35. $\dfrac{x^2-y^2}{x-3y} \div \dfrac{x-y}{x-3y}$

36. $\dfrac{x^2+8xy+15y^2}{x^2-25y^2} \div (x^2+2xy-3y^2)$

37. $\dfrac{9x^2-y^2}{9x^2-6xy+y^2} \div \dfrac{3x+y}{3x-y}$

38. $\dfrac{u^2-4v^2}{u^2+uv-2v^2} \cdot \dfrac{4u^2-4uv-3v^2}{2u^2-3uv-2v^2} \div \dfrac{3u+v}{u-v}$

For Review

Find the values of the variable that must be omitted.

39. $\dfrac{x-3}{x^2-8x}$

40. $\dfrac{a^2+1}{a^2-9}$

41. $\dfrac{y^2+3y}{6}$

Are the given fractions equivalent?

42. $\dfrac{6x^3-7}{7x-7},\ \dfrac{6x^3}{7x}$

43. $\dfrac{x^2-5x}{3x-15},\ \dfrac{x^2}{3x}$

44. $\dfrac{2x+y}{1},\ \dfrac{1}{2x+y}$

45. Change $\dfrac{x^2}{5}$ to an equivalent fraction by multiplying its numerator and denominator by each of the following.

(A) 7

(B) x^3

(C) $x-2$

(D) $x+3$

(E) $5x+1$

(F) x^3+2x

46. What is wrong with the following?

$$\dfrac{3-2y}{3+5y} = \dfrac{\cancel{3}-2\cancel{y}}{\cancel{3}+5\cancel{y}} = -\dfrac{2}{5}$$

8.2 EXERCISES C

Perform the indicated operations.

1. $\dfrac{x^2-5x-6}{x+1} \div \dfrac{x^2-12x+36}{x^2-1} \cdot \dfrac{x}{x-1}$

$\left[\text{Answer: } \dfrac{x(x+1)}{x-6}\right]$

2. $\dfrac{y^2-3y-4}{y^2-1} \div \dfrac{y+3}{y^2-9} \cdot \dfrac{y-1}{y^2-6y+9}$

3. $\dfrac{a^2-b^2}{2a+b} \div \left[\dfrac{a-b}{2a^2+3ab+b^2} \cdot \dfrac{a-b}{a+b}\right]$

$\left[\text{Answer: } \dfrac{(a+b)^3}{a-b}\right]$

4. $\dfrac{a+2b}{a^2-4b^2} \div \left[\dfrac{a-b}{a^2-3ab+2b^2} \cdot \dfrac{a+2b}{a-2b}\right]$

8.3 ADDITION AND SUBTRACTION OF LIKE FRACTIONS

STUDENT GUIDEPOSTS
1. Like (unlike) fractions
2. Adding and subtracting like fractions
3. Changing the sign of the denominator of a fraction

1 Fractions are called **like fractions** if they have the same denominators. **Unlike fractions** have different denominators.

2
> **To add or subtract like fractions**
> 1. Add or subtract the numerators to find the numerator of the answer.
> 2. Use the common denominator as the denominator of the answer.
> 3. Reduce the sum or difference to lowest terms.

We can use the rule to add or subtract both arithmetic and algebraic fractions.

EXAMPLE 1
Add.

(A) $\dfrac{3}{7} + \dfrac{2}{7} = \dfrac{3+2}{7} = \dfrac{5}{7}$

(B) $\dfrac{2+x}{x} + \dfrac{x^2+1}{x} = \dfrac{2+x+x^2+1}{x} = \dfrac{x^2+x+3}{x}$ ◀◀

Always reduce the answer to lowest terms.

Practice Exercise 1
Add.

(A) $\dfrac{1}{11} + \dfrac{5}{11}$

(B) $\dfrac{y-1}{y} + \dfrac{3+y^2}{y}$

Answers: (A) $\dfrac{6}{11}$ (B) $\dfrac{y^2+y+2}{y}$

EXAMPLE 2
Subtract.

(A) $\dfrac{1}{4} - \dfrac{3}{4} = \dfrac{1-3}{4} = \dfrac{-2}{4} = \dfrac{-\cancel{2}}{\cancel{2}\cdot 2} = -\dfrac{1}{2}$

(B) $\dfrac{a^2}{a-1} - \dfrac{1}{a-1} = \dfrac{a^2-1}{a-1} = \dfrac{\cancel{(a-1)}(a+1)}{\cancel{(a-1)}} = a+1$ ◀◀

Practice Exercise 2
Subtract.

(A) $\dfrac{5}{12} - \dfrac{1}{12}$

(B) $\dfrac{x^2}{x+y} - \dfrac{y^2}{x+y}$

Answers: (A) $\dfrac{1}{3}$ (B) $x-y$

▶▶ **CAUTION** ◀◀ When subtracting, it is important to enclose the numerators in parentheses. This helps to prevent sign errors. Study the example below and be sure to observe the step where $-(x-3)$ becomes $-x+3$. ◀◀

EXAMPLE 3

Subtract.

$$\frac{2x+1}{x-5} - \frac{x-3}{x-5} = \frac{(2x+1) - (x-3)}{x-5} \quad \text{Use parentheses}$$

$$= \frac{2x + 1 - x + 3}{x - 5} \quad \text{Change signs}$$

$$= \frac{x+4}{x-5} \blacktriangleleft$$

Practice Exercise 3

Subtract. $\dfrac{y^2 - 3}{y - 1} - \dfrac{-y - 1}{y - 1}$

Answer: $y + 2$

3 ▶▶ Some fractions have denominators that are negatives of each other. These can be made into like fractions by changing the sign of both the numerator and the denominator of one of the fractions. This can be done by multiplying the fraction by 1 in the form $\dfrac{(-1)}{(-1)}$.

> To change the sign of the denominator of a fraction, multiply both numerator and denominator by -1.

EXAMPLE 4

Perform the indicated operations.

(A) $\dfrac{x}{5} + \dfrac{2x+1}{-5} = \dfrac{x}{5} + \dfrac{(-1)(2x+1)}{(-1)(-5)} = \dfrac{x}{5} + \dfrac{(-2x-1)}{5}$

$$= \frac{x + (-2x - 1)}{5} = \frac{-x - 1}{5}$$

(B) $\dfrac{2x}{x-3} - \dfrac{x+3}{3-x}$

$= \dfrac{2x}{x-3} - \dfrac{(-1)(x+3)}{(-1)(3-x)} \quad$ $x - 3$ and $3 - x$ are negatives of each other

$= \dfrac{2x}{x-3} - \dfrac{(-x-3)}{-3+x}$

$= \dfrac{2x}{x-3} - \dfrac{(-x-3)}{x-3}$

$= \dfrac{2x - (-x - 3)}{x - 3}$

$= \dfrac{2x + x + 3}{x - 3} \quad \text{Watch all signs}$

$= \dfrac{3x + 3}{x - 3} \blacktriangleleft$

Practice Exercise 4

Perform the indicated operations.

(A) $\dfrac{a+3}{2} + \dfrac{3a-1}{-2}$

(B) $\dfrac{-3u}{u-v} - \dfrac{3u+1}{v-u}$

Answers: (A) $2 - a$ (B) $\dfrac{1}{u-v}$

8.3 EXERCISES A

Perform the indicated operations.

1. $\dfrac{3}{5} + \dfrac{7}{5}$

2. $\dfrac{x}{6} - \dfrac{3}{6}$

3. $\dfrac{2}{x} + \dfrac{7}{x}$

4. $\dfrac{2x+1}{x-7} + \dfrac{-1-2x}{x-7}$

5. $\dfrac{2y}{y+2} - \dfrac{y+1}{y+2}$

6. $\dfrac{z-1}{z+1} - \dfrac{2z-1}{z+1}$

7. $\dfrac{a+1}{a+1} + \dfrac{a^2+1}{a+1}$

8. $\dfrac{7x^2}{3x^2-1} - \dfrac{4x^2}{3x^2-1}$

9. $\dfrac{a}{(a+1)^2} + \dfrac{1}{(a+1)^2}$

10. $\dfrac{z}{2} + \dfrac{3z-1}{-2}$

11. $\dfrac{4}{a} + \dfrac{2a-3}{-a}$

12. $\dfrac{2x}{-3} + \dfrac{3x+7}{3}$

13. $\dfrac{7}{2z} - \dfrac{3z+1}{-2z}$

14. $\dfrac{2a}{a-1} + \dfrac{3a}{1-a}$

15. $\dfrac{3x}{1-x} - \dfrac{x+1}{x-1}$

16. $\dfrac{z}{(z-1)(z+1)} - \dfrac{1}{(z-1)(z+1)}$

17. $\dfrac{3}{a^2+2a+1} - \dfrac{2-a}{a^2+2a+1}$

18. $\dfrac{x}{6x-12} - \dfrac{4}{3(4-2x)}$

19. $\dfrac{2z - 3}{z^2 + 3z - 4} - \dfrac{z - 7}{z^2 + 3z - 4}$

20. $\dfrac{2x}{x^2 + x - 6} + \dfrac{x - 3}{6 - x - x^2}$

21. $\dfrac{x^2 - 15}{x^2 + 4x + 3} - \dfrac{2x}{x^2 + 4x + 3}$

22. $\dfrac{y}{x} + \dfrac{2y - 1}{x}$

23. $\dfrac{3a}{a + b} - \dfrac{2a - b}{a + b}$

24. $\dfrac{x + y}{x - y} - \dfrac{x + y}{y - x}$

25. $\dfrac{5x}{x + 1} + \dfrac{2x - 1}{x + 1} - \dfrac{3x}{x + 1}$

26. $\dfrac{x + 2y}{2x - y} - \dfrac{3x - 2y}{2x - y} - \dfrac{4y}{2x - y}$

For Review

Perform the indicated operations.

27. $\dfrac{3x - 6}{2x} \cdot \dfrac{24x^2}{3(x^2 - 4x + 4)}$

28. $\dfrac{a^2 - 9}{4a + 12} \div \dfrac{a - 3}{6}$

29. $\dfrac{y^2 + 10y + 21}{y^2 - 2y - 15} \div (y^2 + 2y - 35)$

30. $\dfrac{x^2 - 5xy + 6y^2}{x^2 + 3xy - 10y^2} \cdot \dfrac{x^2 + 5xy}{x^2 - 3xy}$

ANSWERS: **1.** 2 **2.** $\dfrac{x-3}{6}$ **3.** $\dfrac{9}{x}$ **4.** 0 **5.** $\dfrac{y-1}{y+2}$ **6.** $\dfrac{-z}{z+1}$ **7.** $\dfrac{a^2+a+2}{a+1}$ **8.** $\dfrac{3x^2}{3x^2-1}$ **9.** $\dfrac{1}{a+1}$
10. $\dfrac{-2z+1}{2}$ **11.** $\dfrac{7-2a}{a}$ **12.** $\dfrac{x+7}{3}$ **13.** $\dfrac{3z+8}{2z}$ **14.** $\dfrac{-a}{a-1}$ **15.** $\dfrac{4x+1}{1-x}$ **16.** $\dfrac{1}{z+1}$ **17.** $\dfrac{1}{a+1}$
18. $\dfrac{x+4}{6(x-2)}$ **19.** $\dfrac{1}{z-1}$ **20.** $\dfrac{1}{x-2}$ **21.** $\dfrac{x-5}{x+1}$ **22.** $\dfrac{3y-1}{x}$ **23.** 1 **24.** $\dfrac{2(x+y)}{x-y}$ **25.** $\dfrac{4x-1}{x+1}$
26. $\dfrac{-2x}{2x-y}$ **27.** $\dfrac{12x}{x-2}$ **28.** $\dfrac{3}{2}$ **29.** $\dfrac{1}{(y-5)^2}$ **30.** 1

8.3 EXERCISES B

Perform the indicated operations.

1. $\dfrac{x}{3} + \dfrac{7}{3}$

2. $\dfrac{4}{5} - \dfrac{9}{5}$

3. $\dfrac{5}{3x} - \dfrac{8}{3x}$

4. $\dfrac{6}{x+2} + \dfrac{10}{x+2}$

5. $\dfrac{3x-2}{x-4} + \dfrac{2-3x}{x-4}$

6. $\dfrac{8y}{y-5} - \dfrac{4y+3}{y-5}$

7. $\dfrac{a+2}{a+2} + \dfrac{3a-2}{a+2}$

8. $\dfrac{6x^2}{5x^2+2} - \dfrac{9x^2}{5x^2+2}$

9. $\dfrac{a^2}{a^2+1} - \dfrac{-1}{a^2+1}$

10. $\dfrac{z}{3} + \dfrac{2z-3}{-3}$

11. $\dfrac{9}{a} + \dfrac{5a-2}{-a}$

12. $\dfrac{3x}{-5} + \dfrac{4x-7}{5}$

13. $\dfrac{10}{5z} - \dfrac{5z+4}{-5z}$

14. $\dfrac{5a}{a-4} + \dfrac{7a}{4-a}$

15. $\dfrac{4x}{7-x} - \dfrac{2x+5}{x-7}$

16. $\dfrac{2z}{(2z+1)(2z-1)} - \dfrac{1}{(2z+1)(2z-1)}$

17. $\dfrac{5}{a^2+4a+4} - \dfrac{3-a}{a^2+4a+4}$

18. $\dfrac{2x}{5(x-2)} - \dfrac{3}{10-5x}$

19. $\dfrac{3z+2}{z^2+4z-12} - \dfrac{2z-4}{z^2+4z-12}$

20. $\dfrac{2x}{x^2-2x-15} + \dfrac{x+5}{15+2x-x^2}$

21. $\dfrac{x^2-8}{x^2-4x+3} - \dfrac{-7x}{x^2-4x+3}$

22. $\dfrac{x+1}{y} - \dfrac{x-2}{y}$

23. $\dfrac{a-2b}{a-b} - \dfrac{a+2b}{a-b}$

24. $\dfrac{x-y}{2x-y} - \dfrac{x-y}{y-2x}$

25. $\dfrac{3x-1}{x+2} + \dfrac{4x-2}{x+2} - \dfrac{5x-3}{x+2}$

26. $\dfrac{x+3y}{x-2y} - \dfrac{4x+y}{x-2y} - \dfrac{2y}{x-2y}$

For Review

Perform the indicated operations.

27. $\dfrac{5x+5}{x-2} \cdot \dfrac{x^2-4x+4}{x^2-1}$

28. $\dfrac{a^2-16}{5a-20} \div \dfrac{a+4}{10}$

29. $\dfrac{y^2-11y+28}{y^2-2y-35} \div (y^2+y-20)$

30. $\dfrac{2x^2+9xy+4y^2}{x^2+xy-12y^2} \cdot \dfrac{3xy-x^2}{2x^2+xy}$

8.3 EXERCISES C

Perform the indicated operations.

1. $\dfrac{2x}{x-5} - \dfrac{2}{5-x} + \dfrac{3x}{5-x}$ $\left[\text{Answer: } \dfrac{2-x}{x-5}\right]$

2. $\dfrac{5y}{x^2-y^2} + \dfrac{5y}{y^2-x^2} - \dfrac{x-y}{y^2-x^2}$

3. $\dfrac{a+5}{(a-2)(a-3)} - \dfrac{2a-1}{(2-a)(3-a)} + \dfrac{3a-2}{(a-2)(3-a)}$

4. $\dfrac{a+b}{2a-b} - \dfrac{2b-a}{b-2a} - \dfrac{a-b}{2a-b} + \dfrac{3b-a}{b-2a}$ $\left[\text{Answer: } \dfrac{b}{2a-b}\right]$

8.4 ADDITION AND SUBTRACTION OF UNLIKE FRACTIONS

STUDENT GUIDEPOSTS

1 Least common denominator (LCD)
2 Adding and subtracting fractions

Before we can add or subtract fractions that have different denominators, we must convert them to equivalent fractions having the same denominators. Recall that to add $\frac{2}{15}$ and $\frac{1}{6}$ we must first find a common denominator. One such denominator is $15 \cdot 6 = 90$. Since

$$\frac{2}{15} = \frac{2 \cdot 6}{15 \cdot 6} = \frac{12}{90} \quad \text{and} \quad \frac{1}{6} = \frac{1 \cdot 15}{6 \cdot 15} = \frac{15}{90},$$

we could add as follows.

$$\frac{2}{15} + \frac{1}{6} = \frac{12}{90} + \frac{15}{90} = \frac{12+15}{90} = \frac{27}{90} = \frac{\cancel{9} \cdot 3}{\cancel{9} \cdot 10} = \frac{3}{10}$$

Often it is wiser to try to find a common denominator that is smaller than the product of denominators. If we use the *least common denominator*, this shortens the computation required and decreases the effort needed to reduce the final sum or difference to lowest terms. In the above example, we could have used 30 as the least common denominator.

$$\frac{2}{15} + \frac{1}{6} = \frac{2 \cdot 2}{15 \cdot 2} + \frac{1 \cdot 5}{6 \cdot 5} = \frac{4}{30} + \frac{5}{30} = \frac{9}{30} = \frac{3}{10}$$

1 The same remarks apply to algebraic fractions. We will be concerned with finding the **least common denominator (LCD)** of all fractions involved.

To find the LCD of two or more fractions

1. Factor the denominators completely.
2. Put each factor in the LCD as many times as it appears in the denominator where it is found the greatest number of times.

8.4 ADDITION AND SUBTRACTION OF UNLIKE FRACTIONS

EXAMPLE 1
Find the LCD of the given fractions.

(A) $\dfrac{7}{90}$ and $\dfrac{5}{24}$

$$90 = 2 \cdot \overset{1}{3} \cdot \overset{2}{3} \cdot \overset{1}{5} \quad \text{and} \quad 24 = \overset{3}{2 \cdot 2 \cdot 2} \cdot \overset{1}{3}.$$

The LCD must consist of three 2's, two 3's, and one 5.

$$\text{LCD} = 2 \cdot 2 \cdot 2 \cdot 3 \cdot 3 \cdot 5 = 360$$

(B) $\dfrac{3}{x}$ and $\dfrac{2}{x+1}$

Since x and $x+1$ are already completely factored, and since there are no common factors in the two denominators, the LCD $= x(x+1)$.

(C) $\dfrac{3}{2y}$ and $\dfrac{y+1}{y^2}$

The denominators are essentially factored already. We see that the LCD has one 2 and two y's. Thus, the LCD is $2y^2$.

(D) $\dfrac{2x-3}{x^2-25}$ and $\dfrac{7x}{2x-10}$

$$x^2 - 25 = (x-5)(x+5) \quad \text{and} \quad 2x - 10 = 2(x-5)$$

We need one $(x-5)$, one $(x+5)$, and one 2 for the LCD. That is, the LCD $= 2(x-5)(x+5)$.

(E) $\dfrac{y}{y^2-4}$ and $\dfrac{4}{y^2-4y+4}$

$$y^2 - 4 = (y-2)(y+2) \quad \text{and} \quad y^2 - 4y + 4 = (y-2)(y-2)$$

The LCD must consist of two $(y-2)$'s and one $(y+2)$. Thus, the LCD $= (y-2)^2(y+2)$. ◀◀

Practice Exercise 1
Find the LCD of the given fractions.

(A) $\dfrac{5}{12}$ and $\dfrac{11}{126}$

(B) $\dfrac{5}{a+1}$ and $\dfrac{1}{a-1}$

(C) $\dfrac{7}{x^2}$ and $\dfrac{x+11}{5x}$

(D) $\dfrac{1-y}{7y+21}$ and $\dfrac{4y}{y^2-9}$

(E) $\dfrac{w}{w^2+2w+1}$ and $\dfrac{3}{w^2-5w-6}$

Answers: **(A)** 252
(B) $(a+1)(a-1)$ **(C)** $5x^2$
(D) $7(y+3)(y-3)$
(E) $(w+1)^2(w-6)$

To add or subtract unlike fractions, we first find their LCD and then transform each fraction into an equivalent fraction having the LCD as its denominator. We illustrate this method in the following numerical example.

$$\dfrac{2}{15} + \dfrac{3}{35}$$

$$= \dfrac{2}{3 \cdot 5} + \dfrac{3}{5 \cdot 7} \qquad \text{Since } 15 = 3 \cdot 5 \text{ and } 35 = 5 \cdot 7, \text{ the LCD is } 3 \cdot 5 \cdot 7 = 105$$

$$= \dfrac{2 \cdot (7)}{3 \cdot 5 \cdot (7)} + \dfrac{3 \cdot (3)}{5 \cdot 7 \cdot (3)} \qquad \text{Multiply numerator and denominator of } \dfrac{2}{15} \text{ by 7 and numerator and denominator of } \dfrac{3}{35} \text{ by 3 so the denominators will be the same}$$

$$= \dfrac{2 \cdot 7 + 3 \cdot 3}{3 \cdot 5 \cdot 7} \qquad \text{Since denominators are now the same, add numerators and place the sum over the LCD}$$

$$= \dfrac{14 + 9}{3 \cdot 5 \cdot 7} \qquad \text{Leave the denominator in factored form and simplify the numerator}$$

$$= \dfrac{23}{3 \cdot 5 \cdot 7} = \dfrac{23}{105} \qquad \text{Since 23 has no factor of 3, 5, or 7, the resulting fraction is in lowest terms}$$

354 RATIONAL EXPRESSIONS

Exactly the same procedure applies to algebraic fractions.

> **To add or subtract two (or more) algebraic fractions**
> 1. Rewrite the indicated sum or difference with all denominators expressed in factored form.
> 2. Determine the LCD.
> 3. Multiply the numerator and denominator of each fraction by all factors present in the LCD but missing in the denominator of the particular fraction.
> 4. Indicate the sum or difference of all numerators, using parentheses if applicable, and place the result over the LCD.
> 5. Simplify and factor (if possible) the resulting numerator and divide any factors common to the LCD.

EXAMPLE 2

Add.

$$\frac{5}{6x} + \frac{3}{10x^2}$$

$$= \frac{5}{2 \cdot 3 \cdot x} + \frac{3}{2 \cdot 5 \cdot x \cdot x}$$
Factor denominators; LCD $= 2 \cdot 3 \cdot 5 \cdot x \cdot x$

$$= \frac{5 \cdot (5 \cdot x)}{2 \cdot 3 \cdot x \cdot (5 \cdot x)} + \frac{3 \cdot (3)}{2 \cdot 5 \cdot x \cdot x \cdot (3)}$$
Supply missing factors

$$= \frac{5 \cdot 5 \cdot x + 3 \cdot 3}{2 \cdot 3 \cdot 5 \cdot x \cdot x}$$
Add numerators over LCD

$$= \frac{25x + 9}{30x^2}$$
No common factors exist, so this is in lowest terms ◂◂

Practice Exercise 2

Add.

$$\frac{1}{12y^3} + \frac{7}{4y}$$

Answer: $\dfrac{1 + 21y^2}{12y^3}$

EXAMPLE 3

Subtract.

$$\frac{2}{y + 2} - \frac{2}{y + 3}$$
Denominators already factored; LCD $= (y + 2)(y + 3)$

$$= \frac{2(y + 3)}{(y + 2)(y + 3)} - \frac{2(y + 2)}{(y + 2)(y + 3)}$$
Supply missing factors.

$$= \frac{2(y + 3) - 2(y + 2)}{(y + 2)(y + 3)}$$
Subtract numerators over LCD

$$= \frac{2y + 6 - 2y - 4}{(y + 2)(y + 3)}$$
Watch signs

$$= \frac{2}{(y + 2)(y + 3)}$$
No common factors exist, so this is in lowest terms ◂◂

Practice Exercise 3

Subtract.

$$\frac{3}{a - 1} - \frac{1}{a + 3}$$

Answer: $\dfrac{2(a + 5)}{(a - 1)(a + 3)}$

8.4 ADDITION AND SUBTRACTION OF UNLIKE FRACTIONS

EXAMPLE 4
Add.

$$\frac{5}{x^2 - 4} + \frac{7}{x^2 + 2x}$$

$$= \frac{5}{(x-2)(x+2)} + \frac{7}{x(x+2)} \quad \text{Factor denominators;}$$
$$\text{LCD} = x(x-2)(x+2)$$

$$= \frac{5\,(x)}{(x-2)(x+2)\,(x)} + \frac{7\,(x-2)}{x(x+2)\,(x-2)} \quad \text{Supply missing factors}$$

$$= \frac{5x + 7(x-2)}{x(x+2)(x-2)} \quad \text{Add numerators}$$

$$= \frac{5x + 7x - 14}{x(x+2)(x-2)} \quad \text{Clear parentheses}$$

$$= \frac{12x - 14}{x(x+2)(x-2)}$$

$$= \frac{2(6x-7)}{x(x+2)(x-2)} \quad \text{No common factors exist, so this is in lowest terms} \blacktriangleleft$$

Practice Exercise 4
Add.

$$\frac{3}{y^2 - 5y} + \frac{-2}{y^2 - 3y - 10}$$

Answer: $\dfrac{y + 6}{y(y-5)(y+2)}$

EXAMPLE 5
Add.

$$\frac{5}{x^2 + x - 6} + \frac{3x}{x^2 - 4x + 4}$$

$$= \frac{5}{(x-2)(x+3)} + \frac{3x}{(x-2)(x-2)} \quad \text{LCD} = (x+3)(x-2)^2$$

$$= \frac{5\,(x-2)}{(x-2)(x+3)\,(x-2)} + \frac{3x\,(x+3)}{(x-2)(x-2)\,(x+3)}$$

$$= \frac{5(x-2) + 3x(x+3)}{(x+3)(x-2)^2}$$

$$= \frac{5x - 10 + 3x^2 + 9x}{(x+3)(x-2)^2}$$

$$= \frac{3x^2 + 14x - 10}{(x+3)(x-2)^2} \blacktriangleleft$$

Practice Exercise 5
Add.

$$\frac{y}{y^2 + 2y - 15} + \frac{1 - y}{y^2 - 6y + 9}$$

Answer: $\dfrac{5 - 7y}{(y-3)^2(y+5)}$

The procedures are the same even if there are two variables and more than two terms.

EXAMPLE 6
Perform the indicated operations.

$$\frac{2x}{x+y} - \frac{y}{x-y} - \frac{-4xy}{x^2 - y^2}$$

$$= \frac{2x}{x+y} - \frac{y}{x-y} - \frac{-4xy}{(x+y)(x-y)} \quad \text{LCD} = (x+y)(x-y)$$

Practice Exercise 6
Perform the indicated operations.

$$\frac{2a}{a + 2b} - \frac{b}{a - 2b} + \frac{2b^2 + ab}{a^2 - 4b^2}$$

$$= \frac{2x(x-y)}{(x+y)(x-y)} - \frac{y(x+y)}{(x-y)(x+y)} - \frac{-4xy}{(x+y)(x-y)} \quad \text{Supply missing factors}$$

$$= \frac{2x(x-y) - y(x+y) + 4xy}{(x+y)(x-y)}$$

$$= \frac{2x^2 - 2xy - xy - y^2 + 4xy}{(x+y)(x-y)}$$

$$= \frac{2x^2 + xy - y^2}{(x+y)(x-y)} \quad \text{This will simplify}$$

$$= \frac{(2x-y)(x+y)}{(x+y)(x-y)} \quad \text{Cancel } (x+y)$$

$$= \frac{2x-y}{x-y} \blacktriangleleft$$

Answer: $\dfrac{2a}{a+2b}$

8.4 EXERCISES A

Find the LCD of the given fractions.

1. $\dfrac{1}{20}$ and $\dfrac{7}{30}$

2. $\dfrac{2}{39}$ and $\dfrac{4}{35}$

3. $\dfrac{y+1}{y^2}$ and $\dfrac{1}{3y}$

4. $\dfrac{a+1}{9a^2}$ and $\dfrac{5}{12a^3}$

5. $\dfrac{2}{3x}$ and $\dfrac{x+2}{3x+3}$

6. $\dfrac{7}{y+1}$ and $\dfrac{y}{y-1}$

7. $\dfrac{3}{2z+4}$ and $\dfrac{-5}{3z+6}$

8. $\dfrac{2a+1}{a^2-25}$ and $\dfrac{3}{a+5}$

9. $\dfrac{3x+2}{3x+6}$ and $\dfrac{x}{x^2-4}$

10. $\dfrac{1}{x^2+2x+1}$ and $\dfrac{2}{x^2-1}$

11. $\dfrac{3}{z^2-9}$ and $\dfrac{2}{z^2+z-6}$

12. $\dfrac{2}{a^2-9}$ and $\dfrac{2}{a^2+2a-3}$

13. $\dfrac{1}{x^2 - y^2}$ and $\dfrac{3xy}{x - y}$

14. $\dfrac{5x}{x + y}$, $\dfrac{7y}{2x + 2y}$, and $\dfrac{2xy}{3x + 3y}$

15. $\dfrac{16}{5x}$, $\dfrac{5}{3y}$, and $\dfrac{xy}{2x - y}$

Perform the indicated operations.

16. $\dfrac{4}{21} + \dfrac{5}{14}$

17. $\dfrac{5}{12} - \dfrac{11}{30}$

18. $\dfrac{2}{9x} + \dfrac{4}{15x^2}$

19. $\dfrac{a - 5}{a} - \dfrac{3a - 1}{4a}$

20. $\dfrac{5}{y + 5} - \dfrac{3}{y - 5}$

21. $\dfrac{3a}{a + 2} + \dfrac{a}{a - 2}$

22. $\dfrac{2}{3x + 21} - \dfrac{3}{5x + 35}$

23. $\dfrac{4y}{y^2 - 36} - \dfrac{4}{y + 6}$

24. $\dfrac{8}{7 - a} - \dfrac{8a}{49 - a^2}$

25. $\dfrac{3}{2x^2 - 2x} - \dfrac{5}{2x - 2}$

26. $\dfrac{-8}{y^2 - 4} - \dfrac{4}{y + 2}$

27. $\dfrac{8}{a^2 - 16} + \dfrac{1}{a + 4}$

28. $\dfrac{3z + 2}{3z + 6} + \dfrac{z - 2}{z^2 - 4}$

29. $\dfrac{2}{y^2 - 1} + \dfrac{1}{y^2 + 2y - 3}$

30. $\dfrac{2}{z^2 - 9} - \dfrac{2}{z^2 + 2z - 3}$

31. $\dfrac{3}{a^2 - a - 12} - \dfrac{2}{a^2 - 9}$

32. $\dfrac{5}{x^2 - 4x + 3} + \dfrac{7}{x^2 + x - 2}$

33. $\dfrac{y + 1}{y + 4} + \dfrac{4 - y^2}{y^2 - 16}$

34. $\dfrac{2}{a-1} + \dfrac{a}{a^2-1}$ **35.** $\dfrac{3}{x+1} + \dfrac{5}{x-1} - \dfrac{10}{x^2-1}$ **36.** $\dfrac{5x}{x+y} + \dfrac{6y-2x}{x-y}$

37. $\dfrac{2y}{y+5} - \dfrac{3y}{y-2} - \dfrac{2y^2}{y^2+3y-10}$ **38.** $\dfrac{1}{x(x-y)} - \dfrac{2}{y(x+y)}$

For Review

Perform the indicated operations.

39. $\dfrac{3}{z^2+2z+1} - \dfrac{2-z}{z^2+2z+1}$ **40.** $\dfrac{2}{2x-1} - \dfrac{3}{1-2x}$ **41.** $\dfrac{2a-1}{a-5} - \dfrac{6-a}{5-a}$

42. $\dfrac{3-y}{y-7} + \dfrac{2y-5}{7-y}$ **43.** $\dfrac{2x}{3x-2} - \dfrac{5x}{3x-2} - \dfrac{-2}{3x-2}$ **44.** $\dfrac{x^2}{x-y} + \dfrac{x^2}{x-y} - \dfrac{y^2}{y-x}$

ANSWERS: **1.** 60   **2.** 1365   **3.** $3y^2$ **4.** $36a^3$ **5.** $3x(x+1)$ **6.** $(y+1)(y-1)$ **7.** $6(z+2)$ **8.** $(a+5)(a-5)$
9. $3(x+2)(x-2)$ **10.** $(x-1)(x+1)^2$ **11.** $(z-3)(z+3)(z-2)$ **12.** $(a-3)(a+3)(a-1)$ **13.** $(x+y)(x-y)$
14. $6(x+y)$ **15.** $15xy(2x-y)$ **16.** $\dfrac{23}{42}$ **17.** $\dfrac{1}{20}$ **18.** $\dfrac{2(5x+6)}{45x^2}$ **19.** $\dfrac{a-19}{4a}$ **20.** $\dfrac{2(y-20)}{(y+5)(y-5)}$
21. $\dfrac{4a(a-1)}{(a+2)(a-2)}$ **22.** $\dfrac{1}{15(x+7)}$ **23.** $\dfrac{24}{(y-6)(y+6)}$ **24.** $\dfrac{56}{(7-a)(7+a)}$ **25.** $\dfrac{3-5x}{2x(x-1)}$
26. $\dfrac{-4y}{(y-2)(y+2)}$ **27.** $\dfrac{1}{a-4}$ **28.** $\dfrac{3z+5}{3(z+2)}$ **29.** $\dfrac{3y+7}{(y-1)(y+1)(y+3)}$ **30.** $\dfrac{4}{(z-3)(z+3)(z-1)}$
31. $\dfrac{a-1}{(a-4)(a+3)(a-3)}$ **32.** $\dfrac{12x-11}{(x-3)(x-1)(x+2)}$ **33.** $\dfrac{-3y}{(y-4)(y+4)}$ **34.** $\dfrac{3a+2}{(a-1)(a+1)}$ **35.** $\dfrac{8}{x+1}$
36. $\dfrac{3x^2-xy+6y^2}{(x+y)(x-y)}$ **37.** $\dfrac{-y(3y+19)}{(y+5)(y-2)}$ **38.** $\dfrac{y^2+3xy-2x^2}{xy(x+y)(x-y)}$ **39.** $\dfrac{1}{z+1}$ **40.** $\dfrac{5}{2x-1}$ **41.** $\dfrac{a+5}{a-5}$
42. $\dfrac{8-3y}{y-7}$ **43.** -1 **44.** $\dfrac{2x^2+y^2}{x-y}$

8.4 EXERCISES B

Find the LCD of the given fractions.

1. $\dfrac{1}{40}$ and $\dfrac{1}{30}$
2. $\dfrac{4}{39}$ and $\dfrac{2}{15}$
3. $\dfrac{3}{x}$ and $\dfrac{8}{7x}$
4. $\dfrac{a+1}{6a^2}$ and $\dfrac{7}{9a^3}$
5. $\dfrac{3}{5x}$ and $\dfrac{x+5}{5x+5}$
6. $\dfrac{3}{x+1}$ and $\dfrac{x}{x-1}$
7. $\dfrac{-8}{3z-6}$ and $\dfrac{2}{7z-14}$
8. $\dfrac{2z+3}{z^2-9}$ and $\dfrac{12}{z+3}$
9. $\dfrac{4x-1}{2x+6}$ and $\dfrac{5x+1}{x^2-9}$
10. $\dfrac{4}{y^2-1}$ and $\dfrac{y}{y^2-2y+1}$
11. $\dfrac{a}{a^2-4}$ and $\dfrac{2a+5}{a^2-a-6}$
12. $\dfrac{6y^2}{y^2-25}$ and $\dfrac{3y}{y^2+y-20}$
13. $\dfrac{1}{4x^2-y^2}$ and $\dfrac{x^2}{2x+y}$
14. $\dfrac{2}{4x-4y}$, $\dfrac{3x}{2x-2y}$, and $\dfrac{3xy}{5x-5y}$
15. $\dfrac{21y}{8x}$, $\dfrac{15x}{7y}$, and $\dfrac{x^2y^2}{x+3y}$

Perform the indicated operations.

16. $\dfrac{2}{15} + \dfrac{3}{25}$
17. $\dfrac{11}{20} - \dfrac{8}{35}$
18. $\dfrac{3}{4y} + \dfrac{5}{8y^2}$
19. $\dfrac{a+8}{2a} - \dfrac{5a-3}{3a}$
20. $\dfrac{6}{x-6} - \dfrac{4}{x+6}$
21. $\dfrac{5y}{y+4} + \dfrac{2y}{y-4}$
22. $\dfrac{7}{4x+8} - \dfrac{5}{7x+14}$
23. $\dfrac{4}{z+2} + \dfrac{8}{z^2-4}$
24. $\dfrac{14}{49-a^2} - \dfrac{7}{7-a}$
25. $\dfrac{5}{3x^2-3x} - \dfrac{7}{3x-3}$
26. $\dfrac{8}{y^2-16} + \dfrac{1}{y+4}$
27. $\dfrac{10}{x^2-25} - \dfrac{3}{x-5}$
28. $\dfrac{2z-3}{2z-4} + \dfrac{z+2}{z^2-4}$
29. $\dfrac{2}{a^2-a-2} + \dfrac{a}{a^2-1}$
30. $\dfrac{y}{y^2-1} - \dfrac{y+2}{y^2+y-2}$
31. $\dfrac{4}{x^2-9} - \dfrac{4}{x^2-2x-3}$
32. $\dfrac{8}{x^2+4x+3} + \dfrac{3}{x^2-x-2}$
33. $\dfrac{y-3}{y+5} + \dfrac{9-y^2}{y^2-25}$
34. $\dfrac{4}{a-2} + \dfrac{2a}{a^2-4}$
35. $\dfrac{2}{x+5} + \dfrac{3}{x-5} - \dfrac{7}{x^2-25}$
36. $\dfrac{2y}{x-y} + \dfrac{3x-y}{x+y}$
37. $\dfrac{3y}{y+4} - \dfrac{2y}{y-3} - \dfrac{y^2}{y^2+y-12}$
38. $\dfrac{3}{x(x+y)} - \dfrac{2}{y(x-y)}$

For Review

Perform the indicated operations.

39. $\dfrac{z}{z^2+2z+1} + \dfrac{2+z}{z^2+2z+1}$
40. $\dfrac{3x}{x-1} - \dfrac{-2x}{1-x}$
41. $\dfrac{y+2}{y-2} - \dfrac{y^2-2}{2-y}$
42. $\dfrac{2x^2}{2x^2-1} + \dfrac{x^2}{1-2x^2}$
43. $\dfrac{6x}{2x-3} - \dfrac{8x}{2x-3} - \dfrac{5x}{2x-3}$
44. $\dfrac{2x^2}{2x-y} + \dfrac{3y^2}{2x-y} - \dfrac{x^2+y^2}{y-2x}$

8.4 EXERCISES C

Perform the indicated operations.

1. $\dfrac{2}{x^2 - y^2} - \dfrac{3}{x^2 + 2xy + y^2} - \dfrac{3}{x^2 - 2xy + y^2}$
$\left[\text{Answer: } \dfrac{-4(x^2 + 2y^2)}{(x + y)^2(x - y)^2}\right]$

2. $\dfrac{3}{x^2 - 7xy + 10y^2} + \dfrac{2}{x^2 + 2xy - 8y^2} - \dfrac{2}{x^2 - xy - 20y^2}$

3. $\dfrac{1}{x + 5y} \div \dfrac{1}{x - 5y} - \dfrac{1}{x + 5y} \cdot \dfrac{-6xy}{x - y}$

4. $\dfrac{2x - y}{x^2 - y^2} \div \dfrac{2x - y}{x + y} - \dfrac{x + y}{(x - y)^2} \div \dfrac{x + y}{2y}$
$\left[\text{Answer: } \dfrac{x - 3y}{(x - y)^2}\right]$

8.5 SOLVING FRACTIONAL EQUATIONS

STUDENT GUIDEPOSTS

1. Fractional equations
2. Solving fractional equations
3. Age and number problems resulting in fractional equations
4. Work problems

1 An equation involving one or more algebraic fractions is called a **fractional equation**. Once all fractions have been eliminated (we call this **clearing the fractions**), the resulting equation can be solved using the same techniques we learned in Chapter 3.

2
> **To solve a fractional equation**
> 1. Find the LCD of all fractions in the equation.
> 2. Multiply both sides of the equation by the LCD, making sure that *all* terms are multiplied. Once simplified, the resulting equation will be free of fractions.
> 3. Solve this equation.
> 4. Check your solutions to be certain your answers do not make one of the denominators zero.

EXAMPLE 1

Solve $\dfrac{2}{5} + \dfrac{x}{15} = \dfrac{1}{3}$.

$15\left(\dfrac{2}{5} + \dfrac{x}{15}\right) = 15 \cdot \dfrac{1}{3}$ Multiply both sides by the LCD, 15

Practice Exercise 1

Solve $\dfrac{x}{24} + \dfrac{3}{8} = \dfrac{2}{3}$.

$$15 \cdot \frac{2}{5} + 15 \cdot \frac{x}{15} = 5 \qquad \text{Clear the parentheses}$$

$$6 + x = 5$$

$$6 - 6 + x = 5 - 6 \qquad \text{Subtract 6 from both sides}$$

$$x = -1$$

Check: $\dfrac{2}{5} + \dfrac{-1}{15} \stackrel{?}{=} \dfrac{1}{3}$

$$\frac{2 \cdot 3}{5 \cdot 3} + \frac{-1}{15} \stackrel{?}{=} \frac{1 \cdot 5}{3 \cdot 5}$$

$$\frac{6}{15} + \frac{-1}{15} \stackrel{?}{=} \frac{5}{15}$$

$$\frac{5}{15} = \frac{5}{15}$$

The solution is -1. ◀◀

Answer: 7

Notice that we are using the LCD in this section to clear the fractions and not to write them as equivalent fractions with the same denominator.

EXAMPLE 2

Solve $\dfrac{1}{y} = \dfrac{1}{8 - y}$.

$$y(8-y)\left(\frac{1}{y}\right) = y(8-y)\left(\frac{1}{8-y}\right) \qquad \begin{array}{l}\text{Multiply both sides by the} \\ \text{LCD, } y(8-y)\end{array}$$

$$8 - y = y$$

$$8 = 2y \qquad \text{Add } y \text{ to both sides}$$

$$4 = y$$

Check: $\dfrac{1}{4} \stackrel{?}{=} \dfrac{1}{8-4}$

$$\frac{1}{4} = \frac{1}{4}$$

The solution is 4. ◀◀

Practice Exercise 2

Solve $\dfrac{2}{x+3} = \dfrac{5}{x-1}$.

Answer: $-\dfrac{17}{3}$

Recall that two numerical fractions are equal if their cross products are equal. That is,

$$\frac{a}{b} = \frac{c}{d} \quad \text{if} \quad ad = bc.$$

The same is true for algebraic fractions. In some fractional equations, such as the one in Example 2, we can save a step and proceed immediately to the **cross-product equation** $ad = bc$. Thus, to solve

$$\frac{1}{y} = \frac{1}{8-y}$$

362 RATIONAL EXPRESSIONS

we can solve the cross-product equation

$$(1)(8 - y) = (y)(1).$$

Remember that this only works when the equation involves one fraction that is equal to another fraction.

EXAMPLE 3

Solve $\dfrac{3x}{x - 3} - 3 = \dfrac{1}{x}$.

$$x(x - 3)\left[\dfrac{3x}{x - 3} - 3\right] = x(x - 3)\left[\dfrac{1}{x}\right] \quad \text{The LCD} = x(x - 3)$$

$$x(3x) - 3\,x(x - 3) = x - 3 \quad \text{Be sure to multiply 3 by LCD}$$

$$3x^2 - 3x^2 + 9x = x - 3$$

$$8x = -3$$

$$x = -\dfrac{3}{8}$$

Substituting $-\dfrac{3}{8}$ into the original equation requires much calculation. We might note that the only two numbers that make the denominator of either fraction in the equation zero are 0 and 3, neither of which is the answer we obtained, $-\dfrac{3}{8}$. After checking our work for mistakes, we conclude that the solution is $-\dfrac{3}{8}$. ◀◀

Practice Exercise 3

Solve $\dfrac{2x}{x - 4} - 2 = \dfrac{-24}{x}$.

Answer: 3

EXAMPLE 4

Solve $\dfrac{x}{x + 4} - \dfrac{4}{x - 4} = \dfrac{x^2 + 16}{x^2 - 16}$.

The LCD $= (x - 4)(x + 4)$.

$$(x - 4)(x + 4)\left(\dfrac{x}{x + 4} - \dfrac{4}{x - 4}\right) = (x - 4)(x + 4)\left(\dfrac{x^2 + 16}{x^2 - 16}\right)$$

$$x(x - 4) - 4(x + 4) = x^2 + 16$$

$$x^2 - 4x - 4x - 16 = x^2 + 16$$

$$-8x - 16 = 16$$

$$-8x = 32$$

$$x = -4$$

Check: $\dfrac{-4}{-4 + 4} - \dfrac{4}{-4 - 4} \stackrel{?}{=} \dfrac{(-4)^2 + 16}{(-4)^2 - 16}$.

But $\dfrac{-4}{-4 + 4} = \dfrac{-4}{0}$, which is undefined. Thus, -4 *does not check*. There are no solutions. ◀◀

Practice Exercise 4

Solve $\dfrac{2}{y - 3} - \dfrac{12}{y^2 - 9} = \dfrac{3}{y + 3}$.

Answer: No solution (notice that 3 makes the denominators of the fractions on the left side equal to zero).

▶▶ **CAUTION** ◀◀ Be sure to check your answers in the *original* equation. Example 4 shows that sometimes a possible solution may in fact have to be omitted because it makes one or more denominators equal to zero. ◀◀

When the fractions are cleared in some fractional equations, the resulting equation must be solved by the factoring method that we studied in Chapter 7.

EXAMPLE 5

Solve $\dfrac{2}{x+2} + \dfrac{3}{x-2} = 1$.

$(x+2)(x-2)\left[\dfrac{2}{x+2} + \dfrac{3}{x-2}\right] = (x+2)(x-2)$ LCD $= (x+2)(x-2)$

$\qquad 2(x-2) + 3(x+2) = x^2 - 4$

$\qquad 2x - 4 + 3x + 6 = x^2 - 4$

$\qquad\qquad\qquad 5x + 2 = x^2 - 4$

$\qquad\qquad -x^2 + 5x + 6 = 0$

$\qquad\qquad\ \ x^2 - 5x - 6 = 0$ Multiply by -1

$\qquad\qquad (x-6)(x+1) = 0$ Factor

$\quad x - 6 = 0 \quad$ or $\quad x + 1 = 0$ Zero-product Rule

$\qquad x = 6 \qquad\qquad\ x = -1$

Check: $\dfrac{2}{6+2} + \dfrac{3}{6-2} \stackrel{?}{=} 1 \qquad \dfrac{2}{-1+2} + \dfrac{3}{-1-2} \stackrel{?}{=} 1$

$\qquad\qquad \dfrac{2}{8} + \dfrac{3}{4} \stackrel{?}{=} 1 \qquad\qquad \dfrac{2}{1} + \dfrac{3}{-3} \stackrel{?}{=} 1$

$\qquad\qquad \dfrac{1}{4} + \dfrac{3}{4} = 1 \qquad\qquad 2 - 1 = 1$

The solutions are 6 and -1.

Practice Exercise 5

Solve $\dfrac{y}{y-1} + \dfrac{2}{y+1} = \dfrac{8}{y^2-1}$.

Answer: 2 and -5

In Chapter 3 we considered age and number problems that resulted in simple equations. We now consider similar problems that result in fractional equations.

EXAMPLE 6

The sum of Bill's and Bob's ages is 45 years, and Bill's age divided by Bob's age is $\frac{5}{4}$. How old is each?

Let $\quad x =$ Bill's age,

$\quad 45 - x =$ Bob's age.   The sum of x and $45 - x$ is 45

We must solve the following equation.

$\dfrac{x}{45-x} = \dfrac{5}{4}$ The LCD $= 4(45-x)$

$4(45-x)\left(\dfrac{x}{45-x}\right) = 4(45-x)\left(\dfrac{5}{4}\right)$ Multiply by the LCD

$\qquad 4x = (45-x)5$

$\qquad 4x = 225 - 5x$

$\qquad 9x = 225$

$\qquad\ x = 25$ Bill's age is 25

Since $45 - x = 45 - 25 = 20$, Bill is 25 years old and Bob is 20 years old.

Practice Exercise 6

Terry's age divided by Pam's age is $\dfrac{7}{5}$, and the sum of their ages is 24. How old is each?

Let $\quad x =$ Terry's age

$\quad 24 - x =$ Pam's age

Answer: Terry is 14 and Pam is 10.

EXAMPLE 7

The reciprocal of 2 more than a number is three times the reciprocal of the number. Find the number.

Let x = the number,

$x + 2$ = 2 more than the number,

$\dfrac{1}{x+2}$ = the reciprocal of 2 more than the number,

$\dfrac{1}{x}$ = the reciprocal of the number,

$3 \cdot \dfrac{1}{x} = \dfrac{3}{x}$ = three times the reciprocal of the number.

The equation we must solve is $\dfrac{1}{x+2} = \dfrac{3}{x}$.

$$x(x+2)\left(\dfrac{1}{x+2}\right) = x(x+2)\left(\dfrac{3}{x}\right) \quad \text{The LCD} = x(x+2)$$

$$x = (x+2)3$$
$$x = 3x + 6$$
$$-2x = 6$$
$$x = -3$$

The number is -3. ◀◀

Practice Exercise 7

The numerator of a fraction is 3 more than the denominator, and the value of the fraction is $\dfrac{4}{3}$. Find the fraction.

Let x = denominator of fraction

$x + 3$ = numerator of fraction

$\dfrac{x+3}{x}$ = the fraction

The equation to solve is

$$\dfrac{x+3}{x} = \underline{}.$$

Answer: $\dfrac{12}{9}$

   Consider the following problem: Bob can do a job in 2 hours and Pete can do the same job in 4 hours. How long would it take them to do the job if they worked together?

To solve this type of problem, usually called a *work problem,* three important principles must be kept in mind:

1. The time required to do a job when the individuals work together must be less than the time required for the fastest worker to complete the job alone. Thus, the time together is *not* the average of the two times (which would be 3 hours in this case). Since Bob can do the job alone in 2 hr, with help the time must clearly be less than 2 hr.

2. If a job can be done in t hr, in 1 hour $\dfrac{1}{t}$ of the job would be completed. For example, since Bob can do the job in 2 hours, in 1 hour he would do $\frac{1}{2}$ the job. Similarly, Pete would do $\frac{1}{4}$ the job in 1 hr since he does it all in 4 hr.

3. The amount of work done by Bob in 1 hr added to the amount of work done by Pete in 1 hr equals the amount of work done together in 1 hr. Thus, if t is the time required to complete the job working together,

(amount Bob does in 1 hr) + (amount Pete does in 1 hr)
= (amount done together in 1 hr)

translates to the equation

$$\dfrac{1}{2} + \dfrac{1}{4} = \dfrac{1}{t}.$$

We see that a work problem translates to a fractional equation that may result in either a first-degree equation or a second-degree (quadratic) equation. In our example, we obtain a first-degree equation after we multiply both sides by the LCD, $4t$. We will work only with first-degree equations in this section.

$$4t\left(\frac{1}{2} + \frac{1}{4}\right) = 4t\left(\frac{1}{t}\right)$$

$$4t \cdot \frac{1}{2} + 4t \cdot \frac{1}{4} = 4$$

$$2t + t = 4$$

$$3t = 4$$

$$t = \frac{4}{3}$$

Thus, it would take $\frac{4}{3}$ hr (1 hr 20 min) for Bob and Pete to do the job working together.

EXAMPLE 8

Jan can wash the dishes in 20 minutes and her mother can wash them in 15 minutes. How long would it take them to wash the dishes if they worked together?

20 = number of minutes for Jan to do the job

$\frac{1}{20}$ = amount done by Jan in 1 minute

15 = number of minutes for Jan's mother to do the job

$\frac{1}{15}$ = amount done by Jan's mother in 1 minute

Let t = time required to do the job if Jan and her mother work together

Then $\frac{1}{t}$ = amount done in 1 minute when they work together

Thus, we have the following equation.

(amount by Jan) + (amount by mother) = (amount together)

$$\frac{1}{20} + \frac{1}{15} = \frac{1}{t}$$

$$60t\left(\frac{1}{20} + \frac{1}{15}\right) = 60t\left(\frac{1}{t}\right) \quad \text{Multiply by the LCD, } 60t$$

$$60t \cdot \frac{1}{20} + 60t \cdot \frac{1}{15} = 60t \cdot \frac{1}{t}$$

$$3t + 4t = 60$$

$$7t = 60$$

$$t = \frac{60}{7} = 8\frac{4}{7}$$

Working together, it would take Jan and her mother $8\frac{4}{7}$ minutes to wash the dishes. Is this reasonable in view of principle (1)? ◀◀

Practice Exercise 8

Barbara can do a job in 3 days and Paula can do the same job in 8 days. How long would it take them to do the job if they work together?

Let t = number of days to do job together

$\frac{1}{t}$ = amount done together in 1 day

$\frac{1}{3}$ = amount done by Barbara in 1 day

$\frac{1}{8}$ = _____

The equation to solve is

$$\frac{1}{3} + \frac{1}{8} = \text{\_\_\_\_\_}.$$

Answer: $\frac{24}{11}$ days

A simple variation of this kind of work problem is to request the time for one individual when given the time for the other and the time working together.

EXAMPLE 9

When pipe A and pipe B are turned on together, it takes 5 hours to fill a tank. If pipe A can fill the tank in 7 hours, how long would it take pipe B?

5 = number of hrs to do the job together

$\dfrac{1}{5}$ = amount done together in 1 hr

7 = number of hrs for pipe A to do the job

$\dfrac{1}{7}$ = amount done by pipe A in 1 hr

Let t = number of hrs for pipe B to do the job

$\dfrac{1}{t}$ = amount done by pipe B in 1 hr

We have the following equation.

(amount by A) + (amount by B) = (amount together)

$$\frac{1}{7} + \frac{1}{t} = \frac{1}{5}$$

$$35t\left(\frac{1}{7} + \frac{1}{t}\right) = 35t\left(\frac{1}{5}\right) \quad \text{The LCD is } 35t$$

$$35t\left(\frac{1}{7}\right) + 35t\left(\frac{1}{t}\right) = 35t \cdot \frac{1}{5}$$

$$5t + 35 = 7t$$

$$35 = 2t$$

$$\frac{35}{2} = t$$

Pipe B would take $\frac{35}{2}$ hours or $17\frac{1}{2}$ hours to fill the tank alone. Does this make sense in view of principle (1)? ◀◀

Practice Exercise 9

Working together, Phil Mortensen and his son Brad can build a barn in 45 days. Working by himself, Phil can build the barn in 60 days. How long would it take Brad working alone?

Answer: 180 days

8.5 EXERCISES A

Solve.

1. $\dfrac{3x + 5}{6} = \dfrac{4 + 3x}{5}$

2. $\dfrac{1}{x} - \dfrac{2}{x} = 6$

3. $\dfrac{2}{y + 1} = \dfrac{3}{y}$

4. $\dfrac{1}{z-3} + \dfrac{3}{z-5} = 0$

5. $\dfrac{1}{2x-3} - \dfrac{3}{4x-15} = 0$

6. $\dfrac{a-1}{a+1} = \dfrac{a-3}{a+2}$

7. $\dfrac{2y+1}{3y-2} = \dfrac{2y-3}{3y+2}$

8. $\dfrac{z}{z+1} - 1 = \dfrac{1}{z}$

9. $\dfrac{2x}{x+2} - 2 = \dfrac{1}{x}$

10. $\dfrac{1}{y+2} + \dfrac{1}{y-2} = \dfrac{1}{y^2-4}$

11. $\dfrac{4}{z-3} + \dfrac{2z}{z^2-9} = \dfrac{1}{z+3}$

12. $\dfrac{3}{x+3} + \dfrac{1}{x-3} = \dfrac{10}{x^2-9}$

13. $\dfrac{y}{y+3} - \dfrac{3}{y-3} = \dfrac{y^2+9}{y^2-9}$

14. $\dfrac{10}{x^2-25} + 1 = \dfrac{x}{x+5}$

15. $\dfrac{y}{y-4} - 1 = \dfrac{8}{y^2-16}$

16. $\dfrac{3}{x-3} + \dfrac{2}{x+3} = -1$

17. $\dfrac{4}{y-4} + \dfrac{6}{y+4} = -1$

18. $\dfrac{2z}{z-2} - 1 = \dfrac{1}{z^2 - 4}$

Solve. (Some problems have been started.)

19. Find two numbers whose sum is 42 and whose quotient is $\tfrac{3}{4}$.

    Let $\quad x$ = one number,
    $\quad\quad 42 - x$ = second number. [The sum of the two is $x + (42 - x) = 42$.]

20. Two thirds of a certain number is one more than five eighths of the number. Find the number.

21. The denominator of a certain fraction is 2 greater than the numerator, and the value of the fraction is $\tfrac{4}{5}$. Find the fraction.

22. If Joe is three fourths as old as Jeff and the difference in their ages is 6 years, how old is each?

    Let x = Jeff's age,
    $\dfrac{3}{4}x$ = Joe's age.

23. Find the number which when added to both the numerator and denominator of $\tfrac{5}{7}$ will produce a fraction equivalent to $\tfrac{4}{5}$.

24. The reciprocal of 3 less than a number is four times the reciprocal of the number. Find the number.

    Let x = the desired number,
    $x - 3$ = 3 less than the number,
    $\dfrac{1}{x - 3}$ = the reciprocal of 3 less than the number.

25. The numerator of a fraction is 4 less than its denominator. If both the numerator and denominator are increased by 1, the value of the fraction is $\tfrac{2}{3}$. Find the fraction.

26. Mary's age is 4 years more than Ruth's age, and the quotient of their ages is $\tfrac{7}{5}$. How old is each?

27. The denominator of a fraction exceeds the numerator by 6, and the value of the fraction is $\frac{5}{6}$. Find the fraction.

   Let x = numerator of fraction,
   $x + 6$ = denominator of fraction.

28. The reciprocal of 4 less than a number is three times the reciprocal of the number. Find the number.

29. If Burford can mow a lawn in 7 hours and Hilda can mow the same lawn in 3 hours, how long would it take them to mow it if they worked together?

   7 = number of hrs for Burford to do the job
   $\frac{1}{7}$ = amount done by Burford in 1 hr
   3 = number of hrs for Hilda to do the job
   $\frac{1}{3}$ = amount done by Hilda in 1 hr
   Let t = time to do the job if they work together
   $\frac{1}{t}$ = amount done in 1 hr when they work together
   $\frac{1}{7} + \frac{1}{3} = \frac{1}{t}$

30. Bob can paint a house alone in 12 days. When he and his father work together it takes only 9 days. How long would it take his father to paint the house if he worked alone?

31. A 6-inch pipe can fill a reservoir in 3 weeks. When this pipe and a second pipe are both turned on, it takes only 2 weeks to fill the reservoir. How long would it take the second pipe to fill it by itself?

32. If Irv takes 4 times longer to repair a car than Max, and together they can repair it in 8 hours, how long does it take each working alone?

33. If a 1-inch pipe takes 40 minutes to drain a pond and a 2-inch pipe takes 10 minutes to drain the same pond, how long would it take them to drain it together?

34. If Art can frame a shed in 4 days and together Art and Clyde can frame the same shed in 3 days, how long would it take Clyde to frame it working by himself?

35. An older machine requires twice as much time to do a job as a modern one. Together they accomplish the job in 8 minutes. How long would it take each to do the job alone?

36. Susan can make 100 widgets in 3 hours. It takes Harry 5 hours to produce 100 widgets. How long would it take them to produce 100 widgets if they worked together?

For Review

Perform the indicated operations.

37. $\dfrac{3}{5-x} - \dfrac{2-x}{x-5}$

38. $\dfrac{2}{x-3} + \dfrac{5}{x+2}$

39. $\dfrac{-5y}{y^2-25} - \dfrac{y}{y+5}$

40. $\dfrac{5}{y^2-25} + \dfrac{5}{y^2-3y-10}$

41. $\dfrac{3}{x^2-x-2} - \dfrac{2}{x^2+6x+5}$

42. $\dfrac{x+y}{x-y} - \dfrac{x-y}{x+y}$

ANSWERS: **1.** $\dfrac{1}{3}$ **2.** $-\dfrac{1}{6}$ **3.** -3 **4.** $\dfrac{7}{2}$ **5.** -3 **6.** $-\dfrac{1}{3}$ **7.** $\dfrac{1}{5}$ **8.** $-\dfrac{1}{2}$ **9.** $-\dfrac{2}{5}$ **10.** $\dfrac{1}{2}$ **11.** no solution **12.** 4 **13.** no solution **14.** 3 **15.** -2 **16.** 1, -6 **17.** 2, -12 **18.** $-1, -3$ **19.** 18, 24 **20.** 24 **21.** $\dfrac{8}{10}$ **22.** Joe is 18, Jeff is 24 **23.** 3 **24.** 4 **25.** $\dfrac{7}{11}$ **26.** Mary is 14, Ruth is 10 **27.** $\dfrac{30}{36}$ **28.** 6 **29.** $\dfrac{21}{10}$ hr **30.** 36 days **31.** 6 weeks **32.** 10 hr, 40 hr **33.** 8 min **34.** 12 days **35.** 12 min, 24 min **36.** $\dfrac{15}{8}$ hr **37.** 1 **38.** $\dfrac{7x-11}{(x-3)(x+2)}$ **39.** $\dfrac{-y^2}{y^2-25}$ **40.** $\dfrac{5(2y+7)}{(y+5)(y-5)(y+2)}$ **41.** $\dfrac{x+19}{(x-2)(x+1)(x+5)}$ **42.** $\dfrac{4xy}{(x+y)(x-y)}$

8.5 EXERCISES B

Solve.

1. $\dfrac{5x-2}{3} = \dfrac{2+7x}{4}$

2. $\dfrac{3}{x} - \dfrac{4}{x} = 3$

3. $\dfrac{8}{a+4} = \dfrac{7}{a-5}$

4. $\dfrac{2}{z-4} + \dfrac{3}{z-6} = 0$

5. $\dfrac{2}{3x-1} - \dfrac{1}{4x-5} = 0$

6. $\dfrac{x-3}{x-2} = \dfrac{x-2}{x+3}$

7. $\dfrac{2y-3}{4y-1} = \dfrac{y-6}{2y-1}$

8. $\dfrac{2z}{z+3} - 2 = \dfrac{3}{z}$

9. $\dfrac{x+1}{x} = 1 + \dfrac{1}{x+1}$

10. $\dfrac{1}{y+3} + \dfrac{2}{y-3} = \dfrac{3}{y^2-9}$
11. $\dfrac{5}{z+2} + \dfrac{3z}{z^2-4} = \dfrac{2}{z-2}$
12. $\dfrac{-2}{x+5} + \dfrac{1}{x-5} = \dfrac{4}{x^2-25}$

13. $\dfrac{y}{y+5} - \dfrac{5}{y-5} = \dfrac{y^2+25}{y^2-25}$
14. $\dfrac{64}{x^2-16} + 2 = \dfrac{2x}{x-4}$
15. $\dfrac{y}{y+6} - 1 = \dfrac{6}{y^2-36}$

16. $\dfrac{6}{x+5} + \dfrac{1}{x-5} = 1$
17. $\dfrac{1}{y-6} - \dfrac{6}{y+6} = -1$
18. $\dfrac{3z}{z-3} - 2 = \dfrac{10}{z^2-9}$

Solve.

19. Find two numbers whose sum is 32 and whose quotient is $\tfrac{3}{5}$.

20. Three fourths of a number is three more than three eighths of the number. Find the number.

21. The denominator of a fraction is 6 greater than the numerator, and the value of the fraction is $\tfrac{7}{9}$. Find the fraction.

22. Art is five eighths as old as Toni and the difference in their ages is 21 years. How old is each?

23. Find the number which when added to both the numerator and denominator of $\tfrac{3}{11}$ will produce a fraction equivalent to $\tfrac{1}{2}$.

24. The reciprocal of 6 less than a number is four times the reciprocal of the number. Find the number.

25. The numerator of a fraction is 5 less than the denominator. If both the numerator and the denominator are decreased by 2, the value of the fraction is $\tfrac{1}{6}$. Find the fraction.

26. Cindy is 4 years younger than Becky and the quotient of their ages is $\tfrac{7}{9}$. How old is each?

27. The numerator of a fraction exceeds the denominator by 10, and the value of the fraction is $\tfrac{7}{5}$. Find the fraction.

28. The reciprocal of 6 more than a number is four times the reciprocal of the number. Find the number.

29. Alfonse can paint a garage in 20 hours and Heidi can do the same job in 10 hours. How long would it take them if they worked together?

30. Walter can dig a well in 8 days. When he and Sean work together they can dig the same size well in 6 days. How long would it take Sean to do the job working alone?

31. If Andy takes 3 times as long to put new brakes on a car as Hortense does, and together they can do the job in 4 hours, how long does it take each working alone?

32. There are two drains to a water tank. If one is used, the tank is drained in 24 minutes. The other requires 32 minutes. How long will it take if both are opened?

33. If Nancy can process an order of 1000 appliances in 20 minutes and together she and Hans can process the order in 12 minutes, how long will it take Hans to process the order by himself?

34. Using old equipment it requires 5 times as long to make 100 sportcoats as with new equipment. If together the equipment can accomplish the job in $7\tfrac{1}{2}$ hours, how long would it take each working alone?

35. Charles can make 200 dolls in 3 weeks and Wilma can make 200 dolls in $2\tfrac{1}{2}$ weeks. How long will it take to make 200 dolls if they work together?

36. Bertha can clean the stable in 4 hours. If Bertha and Jose work together they can do the same job in 1 hour. How long would Jose take to do the job working alone?

For Review

Perform the indicated operations.

37. $\dfrac{2}{9-x} - \dfrac{3-x}{x-9}$

38. $\dfrac{8}{z-7} + \dfrac{2}{z+1}$

372 RATIONAL EXPRESSIONS

39. $\dfrac{-9y}{y^2 - 16} - \dfrac{5y}{y - 4}$

40. $\dfrac{3}{y^2 - 6y + 8} + \dfrac{7}{y^2 - 16}$

41. $\dfrac{5}{x^2 + x - 12} - \dfrac{3}{x^2 + 10x + 24}$

42. $\dfrac{x - y}{x + y} - \dfrac{x + y}{x - y}$

8.5 EXERCISES C

Solve.

1. $\dfrac{3}{x^2 - 6x + 5} - \dfrac{2}{x^2 - 25} = \dfrac{4}{x^2 + 4x - 5}$

2. $\dfrac{2x}{x^2 - 3x - 4} - \dfrac{x - 1}{x^2 + 2x + 1} = \dfrac{x + 4}{x^2 + 5x + 4}$
[Answer: 0]

3. A man walks a distance of 12 mi at a rate 8 mph slower than the rate he rides a bicycle for a distance of 24 mi. If the total time of the trip is 5 hr, how fast does he walk? How fast does he ride?

4. A swimming pool can be filled by an inlet pipe in 6 hr. It can be drained by an outlet in 9 hr. If the inlet and outlet are both opened, how long will it take to fill the empty pool?   [Answer: 18 hr]

8.6 RATIO, PROPORTION, AND VARIATION

STUDENT GUIDEPOSTS

1 Ratios
2 Proportions
3 Variation
4 Direct Variation

1 The **ratio** of one number x to another number y is the quotient

$$x \div y \text{ or } \dfrac{x}{y}.$$

We sometimes express the ratio $\dfrac{x}{y}$ using the notation $x:y$, which is read "the ratio of x to y." Ratios occur in applications such as percent, rate of speed, gas mileage, unit cost, and number comparisons. Consider the following examples.

Applications	Ratio involved
5% sales tax	$\dfrac{\$5}{\$100} = \$5 \text{ per } \$100$
200 miles in 4 hours	$\dfrac{200 \text{ mi}}{4 \text{ hr}} = 50 \dfrac{\text{mi}}{\text{hr}} = 50 \text{ mi per hr}$
200 miles on 10 gallons of gas	$\dfrac{200 \text{ mi}}{10 \text{ gal}} = 20 \dfrac{\text{mi}}{\text{gal}} = 20 \text{ mi per gal}$
$12 for 2 kg of meat	$\dfrac{\$12}{2 \text{ kg}} = \$6 \text{ per kg}$
30 children and 6 adults	$\dfrac{30 \text{ children}}{6 \text{ adults}} = 5 \text{ children per adult}$

An equation that states that two ratios are equal is called a **proportion.** For example,

$$\frac{2}{3} = \frac{4}{6} \quad \text{and} \quad \frac{x}{3} = \frac{3}{9}$$

are proportions. When a variable occurs in a proportion, as in

$$\frac{x}{x+2} = \frac{3}{4},$$

the proportion is a fractional equation. Equations like this were solved in Section 8.5 by forming the cross-product equation

$$x \cdot 4 = 3 \cdot (x+2).$$

Proportions can be used to solve many problems. Consider the following: If 3 hours are required to travel 150 miles, how many hours would be required to travel 250 miles? The simple proportion

$$\begin{array}{c}\text{first time} \rightarrow \\ \text{first distance} \rightarrow\end{array} \frac{3}{150} = \frac{x}{250}, \begin{array}{c}\leftarrow \text{second time} \\ \leftarrow \text{second distance}\end{array}$$

where x is the number of hours required to travel 250 miles, completely describes this problem. That is, the ratio of the first time to the first distance must equal the ratio of the second time to the second distance. Forming the cross-product equation, we obtain

$$\begin{pmatrix}\text{first}\\\text{time}\end{pmatrix} \cdot \begin{pmatrix}\text{second}\\\text{distance}\end{pmatrix} = \begin{pmatrix}\text{first}\\\text{distance}\end{pmatrix} \cdot \begin{pmatrix}\text{second}\\\text{time}\end{pmatrix}$$
$$\downarrow \quad \downarrow \quad \quad \downarrow \quad \downarrow$$
$$3 \cdot 250 = 150 \cdot x$$

or

$$5 = x.$$

Thus, it would take 5 hours to go 250 miles.

EXAMPLE 1

If a family drinks 5 gallons of milk in two weeks, how many gallons of milk will be consumed in a year (52 weeks)?

Let x = the number of gallons of milk consumed in 52 weeks. The proportion is

$$\begin{array}{c}\text{gallons consumed} \rightarrow \\ \text{in 2 weeks}\end{array} \frac{5}{2} = \frac{x}{52}. \begin{array}{c}\leftarrow \text{gallons consumed} \\ \text{in 52 weeks}\end{array}$$

Form the cross-product equation and solve.

$$5 \cdot 52 = 2x$$
$$130 = x$$

The family drinks 130 gallons of milk in a year.

Practice Exercise 1

A sample of 235 television sets contained 4 that were defective. How many defective sets would you expect to find in a sample of 1645?

Answer: 28

EXAMPLE 2

If a secretary spends 37 minutes typing 3 pages of a report, how long will it take her to type the remaining 10 pages?

Let x be the number of minutes required to complete the typing. We set up the following proportion.

Practice Exercise 2

If it costs $18.00 to operate a freezer for 3 months, how much would it cost to operate the freezer for one year?

374 RATIONAL EXPRESSIONS

$$\text{minutes to type 3 pages} \rightarrow \frac{3}{37} = \frac{10}{x} \leftarrow \text{minutes to type 10 pages}$$

$$3x = 370$$

$$x = 123\frac{1}{3}.$$

It will take another $123\frac{1}{3}$ minutes or 2 hours, 3 minutes, and 20 seconds. ◀◀

Answer: $72.00

---

3 ▶▶ Another way of looking at proportion problems is in terms of **variation.** For example, the distance that a car travels varies (or changes) as we vary (or change) the speed. Also, the time to cook a roast will vary as we change the temperature.

4 ▶▶
> When two variables have a constant ratio k,
>
> $$\frac{y}{x} = k \quad \text{or} \quad y = kx,$$
>
> we say that y **varies directly as** x and we call k the **constant of variation.**

One common example of this is the fact that the ratio of the circumference to the diameter of a circle is π for any circle.

$$c = kd = \pi d \qquad k = \pi$$

In a variation problem if we know the value of y for one value of x, we can determine k. With k known we can determine y for any x.

EXAMPLE 3

Find the equation of variation if y varies directly as x and we know that $y = 30$ when $x = 6$. Also, find y when $x = 10$.

First write the equation for y varies directly as x. Then substitute $y = 30$ and $x = 6$ to find k.

$$y = kx \qquad y \text{ varies directly as } x$$
$$30 = k \cdot 6 \qquad y = 30 \text{ and } x = 6$$
$$\frac{30}{6} = k \qquad k \text{ is a ratio}$$
$$5 = k$$

Thus, the equation of variation is

$$y = 5x.$$

To find y when $x = 10$, substitute 10 for x.

$$y = 5x \qquad \text{Equation of variation}$$
$$y = 5(10) \qquad \text{Substitute 10 for } x$$
$$y = 50$$

Thus, y is 50 when x is 10. ◀◀

Practice Exercise 3

Find the equation of variation if u varies directly as v and we know that $u = 4$ when $v = 3$. Use this equation to find u when $v = 45$.

Answer: $u = \frac{4}{3}v$; 60

EXAMPLE 4

The cost c of steak varies directly as the weight w of the steak purchased. If \$36 is required for 8 kg of steak, how much will 30 kg of a steak cost?

First we write the equation for c varies directly as w. Then substitute in $c = \$36$ and $w = 8$ kg to find k.

$$c = kw \quad \text{c varies directly as w}$$
$$36 = k \cdot 8 \quad \text{$c = \$36$ and $w = 8$ kg}$$
$$4.5 = k \quad \frac{36}{8} = 4.5$$

The equation of variation that we can use for any weight of meat is

$$c = 4.5\,w.$$

To find the cost of 30 kg of steak, substitute 30 for w.

$$c = 4.5\,w$$
$$c = 4.5\,(30)$$
$$= 135$$

Thus, 30 kg (about 66 lb) of steak would cost \$135. ◂◂

Practice Exercise 4

The weight m of an object on the moon varies directly as its weight e on earth. If a man weighing 150 lb on earth weighs 24 lb on the moon, how much will a module that weighs 9500 lb on earth weigh on the moon?

Answer: 1520 lb

8.6 EXERCISES A

Indicate as a ratio and simplify.

1. 320 mi in 8 hr

2. \$10 for 4 lb of meat

3. 20 children for 2 adults

4. 420 mi on 12 gal of gas

5. 200 trees on 4 acres

6. 45°F in 30 min

Solve the proportions.

7. $\dfrac{x}{2} = \dfrac{9}{27}$

8. $\dfrac{y+8}{y-2} = \dfrac{7}{3}$

9. $\dfrac{w-5}{6} = \dfrac{w+6}{2}$

10. $\dfrac{5}{x-7} = \dfrac{4}{x+4}$

11. $\dfrac{y-3}{y+2} = \dfrac{y-2}{y+3}$

12. $\dfrac{z+5}{z-2} = \dfrac{z-7}{z-3}$

RATIONAL EXPRESSIONS

Solve. (Some problems have been started.)

13. In an election the winning candidate won by a 5 to 3 margin. If she received 1025 votes, how many votes did the losing candidate receive?

Let x = number of votes for losing candidate

$$\frac{1025}{x} = \frac{5}{3}$$

14. If a boat uses 7 gallons of gas to go 51 miles, how many gallons would be needed to go 255 miles?

Let x = number of gallons needed to go 255 miles

$$\frac{7}{51} = \frac{x}{255}$$

15. If 50 feet of wire weigh 15 pounds, what will 80 feet of the same wire weigh?

Let x = weight of 80 feet of wire

16. If $\frac{3}{4}$ inch on a map represents 10 miles, how many miles will be represented by 9 inches?

Let x = number of miles represented by 9 inches

17. A sample of 184 tires contained 6 that were defective. How many defective tires would you expect in a sample of 1288?

18. If 2 bricks weigh 9 pounds, how much will 558 bricks weigh?

19. In an election, the successful candidate won by a 3 to 2 margin. If he received 324 votes, how many did the losing candidate receive?

20. If a car uses 8 gallons of gas for a trip of 132 miles, how much gas will be used for a trip of 550 miles?

21. The reciprocal of 4 more than a number is equal to five times the reciprocal of the number. Find the number.

22. If Sal is 7 years older than Ev and the ratio of their ages is equal to $\frac{2}{3}$, how old is each?

23. On a map 2 centimeters represent 14 kilometers. How many centimeters are used to represent 35 kilometers?

24. In a sample of 212 light bulbs, 3 were defective. How many defective bulbs would you expect in a shipment of 3240?

25. Pat was charged a commission of $65.00 on the purchase of 400 shares of stock. What commission would be charged on 700 shares of the same stock?

26. To determine the approximate number of fish in a lake, the State Game and Fish Department caught 100 fish, tagged their fins, and returned them to the water. After a period of time they caught 70 fish and discovered that 14 were tagged. Approximately how many fish are in the lake?

Solve using direct variation.

27. If y varies directly as x, and $y = 10$ when $x = 15$, find y when $x = 42$.

28. If u varies directly as w, and $u = 24$ when $w = 6$, find u when $w = 20$.

29. If y varies directly as x, and $y = 12$ when $x = 9$, find y when $x = 6$.

30. The cost c of peaches varies directly as the total weight w of the peaches. If 12 pounds of peaches cost $4.80, what would 66 pounds cost?

31. The number d of defective tires in a shipment varies directly as the number n of tires in the shipment. If there were 10 defective tires in a shipment of 2200, how many defective tires would you expect in a shipment of 8800 tires?

32. The number g of gallons of gas required varies directly as the number n of miles traveled. If 25 gallons are required to travel 650 miles, how many gallons will be required to travel 3900 miles?

33. The size of Maria's paycheck varies directly as the number of hours she works. If she is paid $334.00 for working 40 hours each week, how much would she be paid for working 2080 hours (1 year)?

34. The amount of money a man spends on recreation varies directly as his total income. If he spends $1920 on recreation when earning $16,000, how much would he spend if his income is increased by $5,000?

For Review

Solve.

35. $\dfrac{3}{x+1} - \dfrac{2}{x+1} = 5$

36. $\dfrac{3}{2y+1} - \dfrac{1}{2y-1} = 0$

37. $\dfrac{x+1}{x-1} - \dfrac{x-1}{x+1} = 0$

38. $\dfrac{3x}{x-3} - 3 = \dfrac{1}{x}$

39. $\dfrac{1}{y-5} + \dfrac{1}{y+5} = \dfrac{1}{y^2 - 25}$

40. $\dfrac{2x}{x-3} - 1 = \dfrac{16}{x^2 - 9}$

41. Phillip can paint a wall in 3 hours and Elizabeth can paint the same size wall in 2 hours. How long would it take to paint the wall if they worked together?

42. Steve can type a manuscript 3 times as fast as Randy, and working together they can do the job in 5 hours. How long would it take each working alone?

ANSWERS: **1.** 40 mph **2.** $2.50 per lb **3.** 10 children per adult **4.** 35 mpg **5.** 50 trees per acre **6.** 1.5°F per min **7.** $\dfrac{2}{3}$ **8.** $\dfrac{19}{2}$ **9.** $-\dfrac{23}{2}$ **10.** -48 **11.** no solution **12.** $\dfrac{29}{11}$ **13.** 615 votes **14.** 35 gal **15.** 24 lb **16.** 120 mi **17.** 42 **18.** 2511 lb **19.** 216 votes **20.** $33\dfrac{1}{3}$ gal **21.** -5 **22.** Ev is 14, Sal is 21 **23.** 5 cm **24.** approximately 46 **25.** $113.75 **26.** 500 fish **27.** 28 **28.** 80 **29.** 8 **30.** $26.40 **31.** 40 **32.** 150 gal **33.** $17,368 **34.** $2520 **35.** $-\dfrac{4}{5}$ **36.** 1 **37.** 0 **38.** $-\dfrac{3}{8}$ **39.** $\dfrac{1}{2}$ **40.** $1, -7$ **41.** $\dfrac{6}{5}$ hr **42.** Steve $\dfrac{20}{3}$ hr; Randy 20 hr

8.6 EXERCISES B

Indicate as a ratio and simplify.

1. 480 mi in 6 hr
2. $6 for 15 lb of apples
3. 60 children for 4 adults
4. 2232 mi on 72 gal of gas
5. 600 bushels from 12 acres
6. 320 gal in 10 min

Solve the proportions.

7. $\dfrac{x}{18} = \dfrac{12}{20}$

8. $\dfrac{y-3}{y-7} = \dfrac{5}{8}$

9. $\dfrac{w+8}{5} = \dfrac{w-4}{3}$

10. $\dfrac{6}{x+10} = \dfrac{5}{x-10}$

11. $\dfrac{y+8}{y-4} = \dfrac{y+1}{y-6}$

12. $\dfrac{z-1}{z-2} = \dfrac{z-3}{z-4}$

Solve.

13. In an election the winning candidate won by a 6 to 5 margin. If the loser received 1520 votes, how many did the winner receive?

14. If a boat uses 12 gallons of gas to go 82 miles, how many gallons would be required to go 123 miles?

15. If 20 meters of hose weighs 6 kilograms, what will 320 meters of the same hose weigh?

16. If 3 centimeters on a map represent 14 kilometers, how many kilometers are represented by 36 centimeters?

17. A sample of 350 bolts contained 5 that were defective. How many defective bolts should be expected in a shipment of 9870 bolts?

18. If 5 blocks weigh 14 pounds, how much will 255 blocks weigh?

19. In an election the losing candidate lost by a 3 to 4 margin. If he received 5280 votes, how many votes did the winner receive?

20. A boat uses 12 gallons of gas in 4 hours. Under the same conditions how many gallons will be used in 10 hours?

21. The reciprocal of 5 less than a number is equal to the reciprocal of twice the number. Find the number.

22. If Walter is 3 years older than Molly and the ratio of their ages is equal to $\frac{5}{6}$, how old is each?

23. On a map 5 inches represent 24 miles. How many miles are represented by 2 inches?

24. In a sample of 280 toasters, 8 were defective. How many defective toasters should be expected in a warehouse containing 42,700 toasters?

25. Marvin received a commission of $1200 on the sale of $150,000 worth of life insurance. If the same person had bought $250,000 worth of insurance, what commission would Marvin have earned?

26. To determine the number of antelope on a game preserve, a ranger caught 50 antelope and tagged their ears. Some time later he captured 18 antelope and discovered that 5 were tagged. Approximately how many antelope are on the preserve?

Solve using direct variation.

27. If y varies directly as x, and $y = 5$ when $x = 35$, find y when $x = 10$.

28. If u varies directly as w, and $u = 52$ when $w = 13$, find u when $w = 22$.

29. If y varies directly as x, and $y = 7$ when $x = 3$, find y when $x = 39$.

30. The cost c of carrots varies directly as the total weight w. If 5 kg of carrots cost $3.20, what would 80 kg of carrots cost?

380 RATIONAL EXPRESSIONS

31. The number d of defective bolts in a shipment varies directly as the number n of bolts in the shipment. If there were 16 defective bolts in a shipment of 4800, how many would be expected in a shipment of 8400?

32. The number g of gallons of gas required varies directly as the number n of miles traveled. If 16 gallons are required to travel 840 miles, how many gallons will be required to travel 2800 miles?

33. The amount of tax paid on real estate varies directly as its assessed value. If the Austins paid $320.00 in taxes on a home valued at $90,000, how much tax will the Wilsons pay on their home valued at $65,000?

34. The amount of money a family spends on food varies direclty as its income. If the Carpenters spend $3780 on food and have an income of $18,000, how much will the Woods spend if their income is $31,000?

For Review

Solve.

35. $\dfrac{5}{x-2} - \dfrac{1}{x-2} = 8$

36. $\dfrac{10}{3y-2} + \dfrac{3}{3y+1} = 0$

37. $\dfrac{x-2}{x+2} - \dfrac{x+2}{x-2} = 0$

38. $\dfrac{2x}{x+5} - 2 = \dfrac{1}{x}$

39. $\dfrac{1}{y+6} + \dfrac{1}{y-6} = \dfrac{4}{y^2-36}$

40. $\dfrac{3x}{x+4} - 2 = \dfrac{12}{x^2-16}$

41. Grady can build a log cabin in 5 weeks and Smitty can build the same type of cabin in 4 weeks. How long would it take to build a cabin if they worked together?

42. Lynda can process an order twice as fast as Maria, and working together they can do the job in 3 hours. How long would it take each working alone?

8.6 EXERCISES C

Solve.

1. In a sample of 120 tires, 4 were defective. Because of manufacturing problems it is estimated that the defective rate will increase by 50% on the next shipment of 480 tires. How many of the new shipment would you predict to be defective.   [Answer: 24]

2. A light-bulb manufacturer detected 12 defective bulbs in a sample of 840. After buying new equipment he was able to decrease the number of defective bulbs by 40 percent. How many defective bulbs should now be expected in a shipment of 1400.

3. If z varies directly as x and inversely as y, show that x varies directly as y and z. $\left[\text{Hint: } z \text{ varies inversely as } y \text{ means } z = \dfrac{k}{y}\right]$

4. The volume V of a gas varies directly as the temperature T and inversely as the pressure P. If the volume is 50 ft^3 when the temperature is 300° and the pressure is 30 pounds per square foot, find V when $T =$ 400° and $P =$ 20 lb/ft^2.   [Answer: 100 ft^3]

8.7 SIMPLIFYING COMPLEX FRACTIONS

STUDENT GUIDEPOSTS

1 Complex fractions
2 Simplifying complex fractions

1 A **complex fraction** is a fraction that contains at least one other fraction within it. The following are complex fractions.

$$\dfrac{\frac{2}{3}}{5}, \quad \dfrac{\frac{1}{2}}{\frac{4}{5}}, \quad \dfrac{\frac{1}{x}}{3}, \quad \dfrac{1 + \frac{1}{x}}{2}, \quad \dfrac{1 + \frac{1}{a}}{1 - \frac{1}{a}}$$

8.7 SIMPLIFYING COMPLEX FRACTIONS

2 A complex fraction is **simplified** when all its component fractions have been eliminated and a simple fraction obtained.

> **To simplify a complex fraction**
> 1. Find the LCD of all fractions in the complex fraction.
> 2. Multiply the numerator and denominator by the LCD and simplify if necessary.

EXAMPLE 1

Simplify $\dfrac{1 + \frac{1}{2}}{1 - \frac{1}{3}}$.

The denominator in $1 + \frac{1}{2}$ is 2, and the denominator in $1 - \frac{1}{3}$ is 3. Thus, the LCD = $2 \cdot 3 = 6$.

$$\dfrac{1 + \frac{1}{2}}{1 - \frac{1}{3}} = \dfrac{\left(1 + \frac{1}{2}\right) \cdot 6}{\left(1 - \frac{1}{3}\right) \cdot 6}$$
Multiply numerator and denominator by LCD = 6

$$= \dfrac{1 \cdot 6 + \frac{1}{2} \cdot 6}{1 \cdot 6 - \frac{1}{3} \cdot 6}$$
Multiply all terms by 6

$$= \dfrac{6 + 3}{6 - 2} = \dfrac{9}{4}$$ ◀◀

Practice Exercise 1

Simplify $\dfrac{2 + \frac{3}{4}}{3 - \frac{1}{6}}$.

Answer: $\dfrac{33}{34}$

EXAMPLE 2

Simplify $\dfrac{1 + \frac{1}{x}}{2}$.

The denominator in $1 + \dfrac{1}{x}$ is x and the denominator in $2 = \frac{2}{1}$ is 1. Thus, the LCD = $x \cdot 1 = x$.

$$\dfrac{1 + \frac{1}{x}}{2} = \dfrac{\left(1 + \frac{1}{x}\right) \cdot x}{2 \cdot x}$$
Multiply numerator and denominator by the LCD = x

$$= \dfrac{1 \cdot x + \frac{1}{x} \cdot x}{2x}$$
Multiply all terms by x

$$= \dfrac{x + 1}{2x}$$ ◀◀

Practice Exercise 2

Simplify $\dfrac{2 + \frac{2}{y}}{2y}$.

Answer: $\dfrac{y + 1}{y^2}$

382 RATIONAL EXPRESSIONS

EXAMPLE 3

Simplify $\dfrac{1 + \dfrac{2}{a}}{1 - \dfrac{4}{a^2}}$.

The denominator in $1 + \dfrac{2}{a}$ is a, and in $1 - \dfrac{4}{a^2}$ is a^2. Thus, the LCD = a^2.

$$\dfrac{1 + \dfrac{2}{a}}{1 - \dfrac{4}{a^2}} = \dfrac{\left(1 + \dfrac{2}{a}\right) \cdot a^2}{\left(1 - \dfrac{4}{a^2}\right) \cdot a^2} \quad \text{Multiply by LCD} = a^2$$

$$= \dfrac{1 \cdot a^2 + \dfrac{2}{a} \cdot a^2}{1 \cdot a^2 - \dfrac{4}{a^2} \cdot a^2} \quad \text{Multiply all terms by } a^2$$

$$= \dfrac{a^2 + 2a}{a^2 - 4}$$

$$= \dfrac{a(a+2)}{(a+2)(a-2)} \quad \text{Factor and cancel}$$

$$= \dfrac{a}{a - 2} \;\;\blacktriangleleft\!\blacktriangleleft$$

Practice Exercise 3

Simplify $\dfrac{\dfrac{1}{x} + 3}{\dfrac{1}{x^2} - 9}$.

Answer: $\dfrac{x}{1 - 3x}$

EXAMPLE 4

Simplify $\dfrac{1 + \dfrac{2}{x - 2}}{\dfrac{2}{x + 2} - 1}$.

The denominator in $1 + \dfrac{2}{x - 2}$ is $x - 2$ and the denominator in $\dfrac{2}{x + 2} - 1$ is $x + 2$. Thus, the LCD $= (x - 2)(x + 2)$.

$$\dfrac{1 + \dfrac{2}{x - 2}}{\dfrac{2}{x + 2} - 1} = \dfrac{\left(1 + \dfrac{2}{x - 2}\right)(x - 2)(x + 2)}{\left(\dfrac{2}{x + 2} - 1\right)(x - 2)(x + 2)} \quad \text{Multiply by LCD}$$

$$= \dfrac{1 \cdot (x - 2)(x + 2) + \dfrac{2}{x - 2}(x - 2)(x + 2)}{\dfrac{2}{x + 2}(x - 2)(x + 2) - 1 \cdot (x - 2)(x + 2)}$$

$$= \dfrac{(x^2 - 4) + 2(x + 2)}{2(x - 2) - (x^2 - 4)}$$

$$= \dfrac{x^2 - 4 + 2x + 4}{2x - 4 - x^2 + 4} \quad \text{Watch signs when removing parentheses}$$

Practice Exercise 4

Simplify $\dfrac{\dfrac{1}{y + 2} - 4}{4 + \dfrac{15}{y - 2}}$.

$$= \frac{x^2 + 2x}{-x^2 + 2x}$$

$$= \frac{\cancel{x}(x+2)}{\cancel{x}(-x+2)} = \frac{x+2}{2-x} \quad \text{Factor and simplify} \blacktriangleleft \qquad \text{Answer: } \frac{2-y}{y+2}$$

8.7 EXERCISES A

Simplify.

1. $\dfrac{\frac{2}{3}}{\frac{1}{3}}$

2. $\dfrac{3 + \frac{2}{3}}{1 - \frac{1}{3}}$

3. $\dfrac{\frac{1}{3} + \frac{1}{5}}{\frac{2}{3} - \frac{3}{5}}$

4. $\dfrac{\frac{2}{x}}{\frac{3}{x}}$

5. $\dfrac{6}{1 + \frac{1}{y}}$

6. $\dfrac{1 - \frac{1}{a}}{3}$

7. $\dfrac{\frac{1}{x} + 3}{\frac{1}{x} - 3}$

8. $\dfrac{y - 1}{y - \frac{1}{y}}$

9. $\dfrac{1 - \frac{4}{y}}{\frac{4-y}{y}}$

10. $\dfrac{\frac{2}{a} + a}{\frac{a}{2} + a}$

11. $\dfrac{\frac{1}{2x} - 1}{\frac{1}{x} - 2}$

12. $\dfrac{\frac{2}{y} + 3}{2 - \frac{3}{y}}$

13. $\dfrac{\frac{1}{x} + 1}{\frac{1}{x} - 1}$

14. $\dfrac{4 - \frac{1}{a^2}}{2 - \frac{1}{a}}$

15. $\dfrac{z - 2 - \frac{3}{z}}{1 + \frac{1}{z}}$

384 RATIONAL EXPRESSIONS

16. $\dfrac{1 - \dfrac{2}{y} - \dfrac{3}{y^2}}{1 + \dfrac{1}{y}}$

17. $\dfrac{x - 3 + \dfrac{2}{x}}{x - 4 + \dfrac{3}{x}}$

18. $\dfrac{\dfrac{3}{a-3} + 1}{\dfrac{3}{a+3} - 1}$

For Review

Solve.

19. $\dfrac{x + 10}{x - 8} = \dfrac{2}{3}$

20. $\dfrac{y + 3}{5} = \dfrac{y - 7}{3}$

21. $\dfrac{z + 1}{z - 5} = \dfrac{z + 2}{z - 6}$

22. A boat can go 38 miles on 6 gallons of gas. How far can it go on 21 gallons of gas?

23. If y varies directly as x, and $y = 42$ when $x = 18$, find y when $x = 3$.

24. The cost c of fish varies directly as the total weight w of the fish. If 8 pounds of fish cost $18.60, what would 60 pounds of fish cost?

25. A recipe requires 2.5 cups of flour to feed 6 people. How many cups would be required to feed 18 people?

ANSWERS: 1. 2   2. $\dfrac{11}{2}$ 3. 8   4. $\dfrac{2}{3}$ 5. $\dfrac{6y}{y+1}$ 6. $\dfrac{a-1}{3a}$ 7. $\dfrac{1+3x}{1-3x}$ 8. $\dfrac{y}{y+1}$ 9. -1 10. $\dfrac{4 + 2a^2}{3a^2}$
11. $\dfrac{1}{2}$ 12. $\dfrac{2 + 3y}{2y - 3}$ 13. $\dfrac{1 + x}{1 - x}$ 14. $\dfrac{2a + 1}{a}$ 15. $z - 3$ 16. $\dfrac{y - 3}{y}$ 17. $\dfrac{x - 2}{x - 3}$ 18. $\dfrac{a + 3}{3 - a}$ 19. -46
20. 22   21. 2   22. 133 mi   23. 7   24. $139.50   25. 7.5 cups

8.7 EXERCISES B

Simplify.

1. $\dfrac{\dfrac{3}{5}}{\dfrac{2}{5}}$

2. $\dfrac{7 + \dfrac{1}{2}}{2 - \dfrac{1}{2}}$

3. $\dfrac{\dfrac{1}{6} + \dfrac{1}{8}}{\dfrac{5}{6} - \dfrac{1}{8}}$

4. $\dfrac{\dfrac{4}{x}}{\dfrac{2}{x}}$

5. $\dfrac{10}{1-\dfrac{1}{y}}$

6. $\dfrac{2+\dfrac{1}{a}}{7}$

7. $\dfrac{\dfrac{1}{x}+1}{\dfrac{1}{x}-1}$

8. $\dfrac{2-\dfrac{1}{y}}{\dfrac{2}{y}}$

9. $\dfrac{a-\dfrac{1}{a}}{1-\dfrac{1}{a}}$

10. $\dfrac{\dfrac{3}{a}+a}{\dfrac{a}{3}+a}$

11. $\dfrac{\dfrac{1}{3y}-2}{\dfrac{1}{y}+2}$

12. $\dfrac{\dfrac{1}{a}+\dfrac{2}{a}}{\dfrac{3}{a}+\dfrac{4}{a}}$

13. $\dfrac{\dfrac{2}{x-2}+1}{\dfrac{2}{x+2}-1}$

14. $\dfrac{\dfrac{1}{y^2}-4}{\dfrac{1}{y}-2}$

15. $\dfrac{\dfrac{3}{z}+z+4}{1+\dfrac{3}{z}}$

16. $\dfrac{1-\dfrac{2}{x}-\dfrac{3}{x^2}}{\dfrac{1}{x}+1}$

17. $\dfrac{\dfrac{5}{y+5}-1}{\dfrac{5}{y-5}+1}$

18. $\dfrac{z-5+\dfrac{6}{z}}{z-1-\dfrac{2}{z}}$

For Review

Solve.

19. $\dfrac{x-3}{x+4}=\dfrac{7}{8}$

20. $\dfrac{2y+8}{6}=\dfrac{y-9}{9}$

21. $\dfrac{z-3}{z-4}=\dfrac{z-6}{z-5}$

22. In a shipment of 200 tires 6 were defective. How many defective tires would you expect in a shipment of 9000 tires?

23. If y varies directly as x, and $y = 14$ when $x = 5$, find y when $x = 20$.

24. The distance d traveled varies directly as the rate r of travel. If 280 miles are traveled at a rate of 40 mph, what distance could be traveled at 60 mph?

25. If 1 quart contains 0.95 liters, how many quarts are in 190 liters?

8.7 EXERCISES C

Simplify.

1. $\dfrac{\dfrac{a+b}{a-b}-\dfrac{a-b}{a+b}}{\dfrac{a}{a-b}+\dfrac{b}{a+b}}$

2. $\dfrac{\dfrac{1}{xy}+\dfrac{1}{yz}+\dfrac{1}{xz}}{\dfrac{x+y+z}{xyz}}$

[Answer: 1]

3. $a-\dfrac{a}{1-\dfrac{a}{1-a}}$

$\left[\text{Answer: }\dfrac{-a^2}{1-2a}\right]$

4. An important formula in physics gives V in terms of S_1 and S_2 by

$$V = \frac{3}{\dfrac{1}{S_1} + \dfrac{1}{S_2}}.$$

Express V as a simple fraction and use the result to find V when S_1 is 2 and S_2 is 5.

CHAPTER 8 SUMMARY

Key Words and Phrases for Review

8.1 algebraic fraction
rational expression
equivalent fractions
canceling
reduce to lowest terms

8.2 divisor
quotient
reciprocal

8.3 like fractions
unlike fractions

8.4 least common denominator (LCD)

8.5 fractional equation
clearing fractions
cross-product equation

8.6 ratio
proportion
variation
varies directly as
constant of variation

8.7 complex fraction
simplified fraction

Key Concepts

8.1
1. Always be aware of values that must be omitted when working with algebraic fractions. For example, 4 cannot replace x in $\dfrac{3}{x-4}$.

2. When reducing fractions, cancel or divide factors only, never cancel terms.

3. When reducing fractions by canceling common factors, do not make the numerator 0 when it is actually 1. For example,

$$\frac{\cancel{(x+1)}}{3\cancel{(x+1)}} = \frac{1}{3}, \quad \text{not } \frac{0}{3}.$$

8.2 When multiplying or dividing fractions, factor numerators and denominators and cancel common factors. Do not find the LCD.

8.3
8.4
1. Find the LCD of all fractions when adding or subtracting.

2. When subtracting fractions, use parentheses to avoid sign errors. For example,

$$\frac{2x+1}{x-5} - \frac{x-3}{x-5} =$$

$$\frac{2x+1-(x-3)}{x-5} =$$

$$\frac{2x+1-x+3}{x-5} = \frac{x+4}{x-5}.$$

3. Do not cancel terms when adding or subtracting algebraic fractions. For example,

$$\frac{3(x-2)}{(x+2)(x-2)} - \frac{7(x+2)}{(x-2)(x+2)}$$

$$= \frac{3(x-2) - 7(x+2)}{(x+2)(x-2)}$$

$$= \frac{3x - 6 - 7x - 14}{(x+2)(x-2)},$$

$$not \quad \frac{3\cancel{(x-2)} - 7\cancel{(x+2)}}{\cancel{(x+2)}\cancel{(x-2)}}.$$

8.5
1. To solve a fractional equation, multiply both sides by the LCD of all fractions.

2. Check all answers in the *original* fractional equation. Watch for division by zero.

8.6
1. Write out solutions to word problems in detail.

2. Solve an equation by finding the cross-product equation only when the equation is a proportion. For example,
$$\frac{1}{x} = \frac{2}{x-2}$$
can be solved by finding the cross-product equation, but
$$\frac{1}{x} - \frac{2}{x-2} = 5$$
cannot.

8.7 Multiply each term of a complex fraction by the LCD of all fractions. For example,
$$\frac{1+\frac{1}{x}}{1-\frac{1}{x}} = \frac{\left(1+\frac{1}{x}\right)\cdot x}{\left(1-\frac{1}{x}\right)\cdot x} = \frac{1\cdot x + \frac{1}{x}\cdot x}{1\cdot x - \frac{1}{x}\cdot x} = \frac{x+1}{x-1}.$$

Review Exercises

8.1 *Find the values of the variable that must be omitted.*

1. $\dfrac{x+2}{x(x-1)}$

2. $\dfrac{a}{a^2+1}$

3. $\dfrac{y^2}{(y+1)(y-5)}$

Are the given fractions equivalent?

4. $\dfrac{x-2}{5x-2}, \dfrac{x}{5x}$

5. $\dfrac{a-3}{a^2-9}, \dfrac{1}{a+3}$

6. $\dfrac{x+y}{x}, \dfrac{x}{x+y}$

Reduce to lowest terms.

7. $\dfrac{x^2+2x}{x}$

8. $\dfrac{x^2-9}{x+3}$

9. $\dfrac{x^2+2xy+y^2}{x+y}$

8.2 *Perform the indicated operations.*

10. $\dfrac{x}{x+6} \cdot \dfrac{x^2-36}{x^2-6x}$

11. $\dfrac{y^2-y-2}{y^2-2y-3} \cdot \dfrac{y^2-3y}{y+2}$

12. $\dfrac{x^2-4y^2}{x+2y} \cdot \dfrac{x+y}{x-2y}$

13. $\dfrac{a+7}{a-7} \div \dfrac{a^2+7}{a^2-49}$ 14. $\dfrac{4y^4}{y^2-1} \div \dfrac{2y^3}{y^2-2y+1}$ 15. $\dfrac{9x^2-y^2}{x+y} \div \dfrac{3x+y}{x^2-y^2}$

8.3
8.4 *Perform the indicated operations.*

16. $\dfrac{3x}{x-2} + \dfrac{2x-1}{2-x}$ 17. $\dfrac{2x}{x^2-x} + \dfrac{x}{x^2-1}$ 18. $\dfrac{2}{x^2-4x+3} + \dfrac{3}{x^2+x-2}$

19. $\dfrac{x+1}{x-3} - \dfrac{x-1}{3-x}$ 20. $\dfrac{2}{y^2-y-12} - \dfrac{1}{y^2-9}$ 21. $\dfrac{a}{a^2-1} + \dfrac{a+2}{a^2+a-2}$

22. $\dfrac{3}{x-2} + \dfrac{x+2}{x-2} - \dfrac{2x}{2-x}$ 23. $\dfrac{6x}{x^2-9} - \dfrac{2}{x-3} - \dfrac{5}{x+3}$ 24. $\dfrac{x}{x+y} + \dfrac{y}{x-y}$

8.5 *Solve.*

25. $\dfrac{2}{z+5} = \dfrac{3}{z}$ 26. $\dfrac{3}{x+2} + \dfrac{2}{x-2} = \dfrac{3}{x^2-4}$ 27. $\dfrac{x}{x-1} + 1 = \dfrac{x^2+1}{x^2-1}$

CHAPTER 8 TEST

1. What are the values of the variable that must be omitted in $\dfrac{5}{x(x+5)}$?

2. Are the fractions $\dfrac{y}{5}$ and $\dfrac{6y}{30}$ equivalent?

3. Reduce to lowest terms. $\dfrac{a+2}{a^2+2a}$

4. Multiply. $\dfrac{x^2-16}{7x^2} \cdot \dfrac{35x}{x-4}$

5. Divide. $\dfrac{y}{y^2+y-12} \div \dfrac{y^2+3y}{y^2+7y+12}$

6. Add. $\dfrac{3}{a-b} + \dfrac{2}{b-a}$

7. Add. $\dfrac{x^2-6x}{x^2-4} + \dfrac{x}{x-2}$

8. Subtract. $\dfrac{4}{y^2-1} - \dfrac{3}{y^2-y-2}$

1. _____

2. _____

3. _____

4. _____

5. _____

6. _____

7. _____

8. _____

CHAPTER 8 TEST Continued

9. Solve. $\dfrac{x}{x+1} = \dfrac{x^2+1}{x^2-1} + \dfrac{3}{x-1}$

9. _____

10. The denominator of a fraction is 3 more than the numerator. If 1 is subtracted from both the numerator and denominator, the result has value $\frac{3}{4}$. Find the fraction.

10. _____

11. A car required 16 gallons of gas to go 352 miles. How many gallons would be required to go 814 miles?

11. _____

12. The cost c of ground beef varies directly as the total weight w of the meat purchased. If 5 lb of ground beef cost $5.75, what would 12 lb cost?

12. _____

13. Ralph can do a job in 3 days and Harry can do the same job in 5 days. How long would it take them to do the same job if they work together?

13. _____

14. Simplify. $\dfrac{\dfrac{1}{y} - 1}{\dfrac{1}{y} - y}$

14. _____

RADICALS

9.1 ROOTS AND RADICALS

STUDENT GUIDEPOSTS
1 Basic terminology
2 Simplifying $\sqrt{x^2}$ when $x \geq 0$

1 In Section 2.7 we introduced square roots to discuss irrational numbers. In general, if a and x are real numbers such that

$$x = a^2,$$

then a is a **square root** of x. Thus, since $9 = 3^2$, 3 is a square root of 9. Also, since $9 = (-3)^2$, -3 is also a square root of 9. Every *positive* real number x has two square roots since if a is a square root of x then

$$x = a^2 \quad \text{and} \quad x = (-a)^2.$$

Thus, both a and $-a$ are square roots of x. Remember that we called the positive square root the **principal square root** and indicated it by using a **radical,** $\sqrt{}$. Thus, if we write $\sqrt{9}$, we call 9 the **radicand.** The principal square root of 9 is $\sqrt{9}$ or 3. If we want to indicate the negative square root of 9, we place a minus sign in front of the radical. For example, $-\sqrt{9} = -3$. If we write $\pm\sqrt{9}$ we mean both 3 and -3. Therefore, $\pm\sqrt{9}$ represents two numbers, not one.

Although every positive number has two square roots, 0 has only one square root, namely 0, since $-0 = 0$. What about negative numbers? Does -25 have a square root? We could try 5 and -5, but they are square roots of 25. In fact, if we square any real number a, we have

$$a^2 \geq 0.$$

Since no real number squared can be negative, negative numbers cannot have real number square roots. Thus, -25 does not have a square root, that is, $\sqrt{-25}$ is not a real number.

EXAMPLE 1

Evaluate the following radicals.

(A) $\sqrt{100} = 10$

(B) $-\sqrt{100} = -10$

(C) $\pm\sqrt{100} = \pm 10$ (two numbers, 10 and -10)

(D) $\sqrt{-100}$ is not a real number

(E) $\sqrt{0} = 0$

(F) $\sqrt{1} = 1$

(G) $\sqrt{5^2} = \sqrt{25} = 5$

(H) $\sqrt{(-5)^2} = \sqrt{25} = 5$

(I) $\sqrt{-5^2} = \sqrt{-25}$ is not a real number

Note the difference between $(-5)^2$ and -5^2. ◀◀

Practice Exercise 1

Evaluate the following radicals.

(A) $\sqrt{169}$

(B) $-\sqrt{169}$

(C) $\pm\sqrt{169}$

(D) $\sqrt{-169}$

(E) $-\sqrt{0}$

(F) $-\sqrt{1}$

(G) $\sqrt{9^2}$

(H) $\sqrt{(-9)^2}$

(I) $\sqrt{-9^2}$

Answers: (A) 13 (B) -13
(C) ± 13 (D) not a real number
(E) 0 (F) -1 (G) 9 (H) 9
(I) not a real number

In Example 1 we saw that $\sqrt{5^2} = 5$ and $\sqrt{(-5)^2} = 5$. Thus, if x is a nonnegative number, $\sqrt{x^2} = x$. For example, $\sqrt{5^2} = 5$. But if x is a negative number, $\sqrt{x^2} = -x$. For example, $\sqrt{(-5)^2} = -(-5) = 5$. This means that when we do not know if x is positive or negative, in order to obtain the principal (positive) root, we should write

$$\sqrt{x^2} = |x| \quad \text{for any real number } x,$$

where $|x|$ is the absolute value of x. To avoid the problem of working with absolute value, we will assume for our work in this chapter that *all variables under radicals represent nonnegative numbers*. Under this assumption, $|x|$ will equal x and we have

$$\sqrt{x^2} = x \quad \text{for } x \geq 0.$$

We will also assume that *all algebraic expressions under radicals are nonnegative numbers*. Thus, for $x - 7 \geq 0$,

$$\sqrt{(x-7)^2} = x - 7.$$

EXAMPLE 2

Evaluate the following radicals assuming all variables and algebraic expressions are nonnegative.

(A) $\sqrt{49} = \sqrt{7^2} = 7$

(B) $\sqrt{y^2} = y$

(C) $\sqrt{25x^2} = \sqrt{5^2 x^2} = \sqrt{(5x)^2} = 5x \quad a^n b^n = (ab)^n$

(D) $\sqrt{x^2 + 6x + 9} = \sqrt{(x+3)^2} = x + 3$

(E) $-\sqrt{81} = -\sqrt{9^2} = -9$

(F) $-\sqrt{u^2} = -u$ ◀◀

Practice Exercise 2

Evaluate the following radicals assuming all variables and algebraic expressions are nonnegative.

(A) $\sqrt{36}$

(B) $\sqrt{a^2}$

(C) $\sqrt{36a^2}$

(D) $\sqrt{x^2 - 6x + 9}$

(E) $-\sqrt{196}$

(F) $-\sqrt{w^2}$

Answers: (A) 6 (B) a (C) $6a$
(D) $x - 3$ (E) -14 (F) $-w$

9.1 EXERCISES A

Indicate whether each number has 0, 1, or 2 square roots.

1. 25   **2.** −25   **3.** 0   **4.** 175

Evaluate each of the square roots assuming that all variables and algebraic expressions under the radical are nonnegative.

5. $\sqrt{6^2}$ **6.** $\sqrt{(-6)^2}$ **7.** $\sqrt{-6^2}$ **8.** $\sqrt{121}$

9. $-\sqrt{121}$ **10.** $\pm\sqrt{121}$ **11.** $\sqrt{\dfrac{9}{4}}$ **12.** $\pm\sqrt{\dfrac{9}{4}}$

13. $\sqrt{\dfrac{144}{25}}$ **14.** $\sqrt{\dfrac{1000}{10}}$ **15.** $\sqrt{\dfrac{50}{2}}$ **16.** $\sqrt{-\dfrac{49}{9}}$

17. $\sqrt{(x+1)^2}$ **18.** $\sqrt{(x-1)^2}$ **19.** $\sqrt{49a^2}$ **20.** $\sqrt{x^2y^2}$

21. $\sqrt{-x^2y^2}$ **22.** $-\sqrt{x^2y^2}$ **23.** $\sqrt{x^2+10x+25}$ **24.** $\sqrt{y^2-8y+16}$

In a particular manufacturing plant, the productivity, p, is related to the work force, w, by the equation $p = \sqrt{w}$.

25. Find p when $w = 25$.   **26.** Find p when $w = 8^2$.

27. Find p when $w = 0$.   **28.** Find w when $p = 4$.

29. Find w when $p = 12$.   **30.** Find p when $w = -4$.

ANSWERS: **1.** 2   **2.** 0   **3.** 1   **4.** 2   **5.** 6   **6.** 6   **7.** meaningless   **8.** 11   **9.** −11   **10.** ±11 (11 and −11)   **11.** $\dfrac{3}{2}$ **12.** $\pm\dfrac{3}{2}\left(\dfrac{3}{2} \text{ and } -\dfrac{3}{2}\right)$ **13.** $\dfrac{12}{5}$ **14.** 10   **15.** 5   **16.** meaningless   **17.** $x+1$ **18.** $x-1$ **19.** $7a$ **20.** xy **21.** meaningless   **22.** $-xy$ **23.** $x+5$ **24.** $y-4$ **25.** 5   **26.** 8   **27.** 0   **28.** 16   **29.** 144   **30.** no value for p when w is negative

396 RADICALS

9.1 EXERCISES B

Indicate whether each number has 0, 1, or 2 square roots.

1. 81   **2.** -81 **3.** 19   **4.** -0

Evaluate each of the square roots assuming that all variables and algebraic expressions under the radical are nonnegative.

5. $\sqrt{49}$ **6.** $\sqrt{(-7)^2}$ **7.** $\sqrt{-7^2}$ **8.** $\sqrt{144}$

9. $-\sqrt{144}$ **10.** $\pm\sqrt{144}$ **11.** $\pm\sqrt{\dfrac{16}{9}}$ **12.** $\sqrt{\dfrac{121}{16}}$

13. $\sqrt{\dfrac{75}{3}}$ **14.** $\pm\sqrt{\dfrac{64}{25}}$ **15.** $\sqrt{-0}$ **16.** $\sqrt{-\dfrac{9}{16}}$

17. $\sqrt{(y+2)^2}$ **18.** $\sqrt{(y-2)^2}$ **19.** $\sqrt{36x^2}$ **20.** $\sqrt{u^2v^2}$

21. $\sqrt{-u^2v^2}$ **22.** $-\sqrt{u^2v^2}$ **23.** $\sqrt{x^2+12x+36}$ **24.** $\sqrt{y^2-14y+49}$

In a particular retail operation, the cost of production c, is related to the number of items sold, n, by the equation $c = \sqrt{n}$.

25. Find c when $n = 49$.   **26.** Find c when $n = 0$.   **27.** Find c when $n = 3^2$.

28. Find n when $c = 6$.   **29.** Find n when $c = 11$.   **30.** Find c when $n = -1$.

9.1 EXERCISES C

Use the definition of cube root, $\sqrt[3]{a^3} = a$, to do the following exercises.

1. $\sqrt[3]{5^3}$ **2.** $\sqrt[3]{27}$ **3.** $\sqrt[3]{8x^3}$ [Answer: $2x$]   **4.** $\sqrt[3]{(2x+1)^6}$

9.2 SIMPLIFYING RADICALS

▶▶ STUDENT GUIDEPOSTS

1 Simplifying square roots of products   3 Rationalizing a denominator
2 Simplifying square roots of quotients   4 Simplified radicals

We know that $\sqrt{9} = 3$, but what if we are asked to simplify $\sqrt{18}$? Table 1 in the Appendix or a calculator gives only an approximation of $\sqrt{18}$. However, since $18 = 9 \cdot 2$, we can write

$$\sqrt{18} = \sqrt{9 \cdot 2} = \sqrt{9} \cdot \sqrt{2} = 3\sqrt{2}.$$

This is an example of the following rule.

1 ▶▶ **Simplifying rule 1**

If $a \geq 0$ and $b \geq 0$, then

$$\sqrt{ab} = \sqrt{a}\sqrt{b}.$$

The square root of a product is equal to the product of square roots.

9.2 SIMPLIFYING RADICALS

To apply this rule, look for perfect square factors of the radicand. Recall that a number is a perfect square when it is the square of an integer. For example, in $\sqrt{18} = \sqrt{9 \cdot 2} = \sqrt{9}\sqrt{2} = 3\sqrt{2}$, 9 is a perfect square factor of 18 and $\sqrt{9}$ can be written in simpler form as 3. When there are no perfect square factors, the rule does not help us. For example, it does no good in trying to simplify $\sqrt{6} = \sqrt{2 \cdot 3}$ since neither 2 nor 3 are perfect squares. In fact $\sqrt{6}$ is in its simplest form.

EXAMPLE 1

Simplify the radicals.

(A) $\sqrt{27} = \sqrt{9 \cdot 3}$ 9 is a perfect square
$= \sqrt{9}\sqrt{3}$ $\sqrt{ab} = \sqrt{a}\sqrt{b}$
$= 3\sqrt{3}$ $\sqrt{9} = \sqrt{3^2} = 3$

(B) $\sqrt{52} = \sqrt{4 \cdot 13}$ 4 is a perfect square
$= \sqrt{4}\sqrt{13}$ $\sqrt{ab} = \sqrt{a}\sqrt{b}$
$= 2\sqrt{13}$ $\sqrt{4} = \sqrt{2^2} = 2$

(C) $\sqrt{15} = \sqrt{3 \cdot 5}$ Cannot be simplified since neither 3 nor 5 is a perfect square
$= \sqrt{15}$

(D) $\sqrt{3^4} = \sqrt{(3^2)^2}$ $3^4 = 3^{2 \cdot 2} = (3^2)^2$, which is a perfect square
$= 3^2 = 9$ $\sqrt{(3^2)^2} = 3^2$

(E) $2\sqrt{3^6} = 2\sqrt{(3^3)^2}$ $3^6 = 3^{3 \cdot 2} = (3^3)^2$
$= 2 \cdot 3^3 = 2 \cdot 27 = 54$

(F) $\sqrt{3^7} = \sqrt{3^6 \cdot 3}$ $3^7 = 3^{6+1} = 3^6 \cdot 3^1 = 3^6 \cdot 3$
$= \sqrt{(3^3)^2}\sqrt{3}$
$= 3^3\sqrt{3} = 27\sqrt{3}$ ◀◀

Practice Exercise 1

Simplify the radicals.

(A) $\sqrt{45}$

(B) $\sqrt{44}$

(C) $\sqrt{55}$

(D) $\sqrt{2^4}$

(E) $3\sqrt{2^6}$

(F) $\sqrt{2^9}$

Answers: (A) $3\sqrt{5}$ (B) $2\sqrt{11}$
(C) cannot be simplified (D) 4
(E) 24 (F) $16\sqrt{2}$

As Example 1 illustrates, to simplify radicals we must recall the rules of exponents such as

$$a^{m+n} = a^m a^n \text{ and } a^{m \cdot n} = (a^m)^n.$$

Notice in Example 1(D) and (E) when the exponent under the radical is even, the radicand is a perfect square. When the exponent under the radical is odd, we rewrite the radicand as a product with a perfect square factor. For example, in Example 1 (F), we rewrite $3^7 = 3^6 \cdot 3$ and then simplify further. This process also works with variables. For example,

$\sqrt{x^7}\ \sqrt{x^6 \cdot x}$ x^6 is a perfect square
$= \sqrt{x^6}\sqrt{x}$ $\sqrt{ab} = \sqrt{a}\sqrt{b}$
$= \sqrt{(x^3)^2}\sqrt{x}$ $x^6 = x^{3 \cdot 2} = (x^3)^2$
$= x^3\sqrt{x}.$

Remember that all our variables are positive.

EXAMPLE 2

Simplify the radicals.

(A) $\sqrt{x^4} = \sqrt{(x^2)^2}$ $(x^2)^2$ is a perfect square
$= x^2$

Practice Exercise 2

Simplify the radicals.

(A) $\sqrt{y^6}$

398 RADICALS

(B) $\sqrt{x^5} = \sqrt{x^4 \cdot x}$ $x^5 = x^{4+1} = x^4 \cdot x$
$\phantom{\sqrt{x^5}} = \sqrt{x^4}\sqrt{x}$ $\sqrt{ab} = \sqrt{a}\sqrt{b}$
$\phantom{\sqrt{x^5}} = \sqrt{(x^2)^2}\sqrt{x}$
$\phantom{\sqrt{x^5}} = x^2\sqrt{x}$

(C) $\sqrt{27a^3} = \sqrt{9 \cdot 3 \cdot a^2 \cdot a}$ 9 and a^2 are perfect squares
$\phantom{\sqrt{27a^3}} = \sqrt{9 \cdot a^2 \cdot 3a}$ $3a$ will be left under the radical
$\phantom{\sqrt{27a^3}} = \sqrt{9} \cdot \sqrt{a^2} \cdot \sqrt{3a}$ The simplifying rule expanded to three factors, $\sqrt{abc} = \sqrt{a}\sqrt{b}\sqrt{c}$
$\phantom{\sqrt{27a^3}} = 3 \cdot a \cdot \sqrt{3a}$
$\phantom{\sqrt{27a^3}} = 3a\sqrt{3a}$

(D) $\sqrt{x^4y^7} = \sqrt{x^4y^6 \cdot y}$ x^4 and y^6 are perfect squares
$\phantom{\sqrt{x^4y^7}} = \sqrt{x^4}\sqrt{y^6}\sqrt{y}$ $\sqrt{abc} = \sqrt{a}\sqrt{b}\sqrt{c}$
$\phantom{\sqrt{x^4y^7}} = \sqrt{(x^2)^2}\sqrt{(y^3)^2}\sqrt{y}$
$\phantom{\sqrt{x^4y^7}} = x^2y^3\sqrt{y}$

(E) $\sqrt{288ab^2} = \sqrt{2 \cdot 144 \cdot a \cdot b^2}$ 144 and b^2 are perfect squares
$\phantom{\sqrt{288ab^2}} = \sqrt{144 \cdot b^2 \cdot 2a}$
$\phantom{\sqrt{288ab^2}} = \sqrt{144}\sqrt{b^2}\sqrt{2a}$
$\phantom{\sqrt{288ab^2}} = 12b\sqrt{2a}$ ◀◀

(B) $\sqrt{y^7}$

(C) $\sqrt{45x^2}$

(D) $\sqrt{a^5b^8}$

(E) $\sqrt{162x^3b^4}$

Answers: (A) y^3 (B) $y^3\sqrt{y}$
(C) $3x\sqrt{5}$ (D) $a^2b^4\sqrt{a}$
(E) $9xb^2\sqrt{2x}$

We can use the simplifying rule above to simplify

$$\sqrt{\frac{9}{4}} = \sqrt{\left(\frac{3}{2}\right)^2} = \frac{3}{2}.$$

This can also be simplified as

$$\sqrt{\frac{9}{4}} = \frac{\sqrt{9}}{\sqrt{4}} = \frac{3}{2}.$$

This is an example of the following rule which is used for radical expressions involving fractions.

2 **Simplifying rule 2**

If $a \geq 0$ and $b > 0$, then

$$\sqrt{\frac{a}{b}} = \frac{\sqrt{a}}{\sqrt{b}}.$$

The square root of a quotient is equal to the quotient of the square roots.

EXAMPLE 3

Simplify the radicals.

(A) $\sqrt{\dfrac{27}{4}} = \dfrac{\sqrt{27}}{\sqrt{4}}$ $\sqrt{\dfrac{a}{b}} = \dfrac{\sqrt{a}}{\sqrt{b}}$

$\phantom{(A) \sqrt{\dfrac{27}{4}}} = \dfrac{\sqrt{9 \cdot 3}}{2}$ $\sqrt{4} = 2$

Practice Exercise 3

Simplify the radicals.

(A) $\sqrt{\dfrac{8}{9}}$

$$= \frac{\sqrt{9}\sqrt{3}}{2} = \frac{3\sqrt{3}}{2} \qquad \sqrt{ab} = \sqrt{a}\sqrt{b}$$

(B) $\sqrt{\dfrac{4x^3}{y^2}} = \dfrac{\sqrt{4x^3}}{\sqrt{y^2}} \qquad \sqrt{\dfrac{a}{b}} = \dfrac{\sqrt{a}}{\sqrt{b}}$

$$= \frac{\sqrt{4 \cdot x^2 \cdot x}}{y} \qquad \sqrt{y^2} = y$$

$$= \frac{\sqrt{4}\sqrt{x^2}\sqrt{x}}{y} \qquad \sqrt{abc} = \sqrt{a}\sqrt{b}\sqrt{c}$$

$$= \frac{2x\sqrt{x}}{y}$$

(C) $\sqrt{\dfrac{48x^3y^2}{3xy}} = \sqrt{\dfrac{\cancel{3} \cdot 16 \cdot x^3 \cdot y^2}{\cancel{3}xy}}$ Simplify under the radical first

$\dfrac{x^3}{x} = x^{3-1} = x^2$ and

$= \sqrt{16x^2y}$

$\dfrac{y^2}{y} = y^{2-1} = y$

$= \sqrt{16}\sqrt{x^2}\sqrt{y}$

$= 4x\sqrt{y}$ ◀◀

(B) $\sqrt{\dfrac{25a^5}{w^2}}$

(C) $\sqrt{\dfrac{32a^4b^3}{2a^2b^2}}$

Answers: (A) $\dfrac{2\sqrt{2}}{3}$

(B) $\dfrac{5a^2\sqrt{a}}{w}$ (C) $4a\sqrt{b}$

Notice that in Example 3(C) we used

$$\frac{a^m}{a^n} = a^{m-n}.$$

Consider the following problem.

$$\sqrt{\frac{9}{7}} = \frac{\sqrt{9}}{\sqrt{7}} = \frac{3}{\sqrt{7}} \qquad \sqrt{7} \text{ is in the denominator}$$

When a square root is left in the denominator, the radical expression is not considered completely simplified. The process of removing radicals from the denominator (making the denominator a rational number) is called **rationalizing the denominator.** This is done by making the expression under the radical in the denominator a perfect square.

In order to rationalize denominators we need the following property. Since $\sqrt{ab} = \sqrt{a}\sqrt{b}$, if we let $a = b$, then

Thus, $\sqrt{7}\sqrt{7} = 7$ and $\sqrt{x^3}\sqrt{x^3} = x^3$.

EXAMPLE 4

Rationalize the denominators.

(A) $\dfrac{3}{\sqrt{7}} = \dfrac{3\sqrt{7}}{\sqrt{7}\sqrt{7}}$ Multiply numerator and denominator by $\sqrt{7}$

$= \dfrac{3\sqrt{7}}{7}$ Since $\sqrt{7}\sqrt{7} = 7$, the denominator is now rational

Practice Exercise 4

Rationalize the denominators.

(A) $\dfrac{5}{\sqrt{11}}$

(B) $\sqrt{\dfrac{25x^2}{y}} = \dfrac{\sqrt{25x^2}}{\sqrt{y}}$ $\sqrt{\dfrac{a}{b}} = \dfrac{\sqrt{a}}{\sqrt{b}}$ **(B)** $\sqrt{\dfrac{49u^4}{w}}$

$= \dfrac{\sqrt{25}\sqrt{x^2}}{\sqrt{y}}$

$= \dfrac{5x}{\sqrt{y}}$

(C) $\sqrt{\dfrac{32x^7y^9}{3y^4}}$

$= \dfrac{5x\sqrt{y}}{\sqrt{y}\sqrt{y}} = \dfrac{5x\sqrt{y}}{y}$ The denominator is rationalized

(C) $\sqrt{\dfrac{25x^2y}{yz^2}} = \sqrt{\dfrac{25x^2\cancel{y}}{\cancel{y}z^2}}$ Simplify under the radical first

$= \dfrac{\sqrt{25x^2}}{\sqrt{z^2}}$ $\sqrt{\dfrac{a}{b}} = \dfrac{\sqrt{a}}{\sqrt{b}}$

Answers: **(A)** $\dfrac{5\sqrt{11}}{11}$

$= \dfrac{5x}{z}$ Do not need to rationalize ◀◀

(B) $\dfrac{7u^2\sqrt{w}}{w}$ **(C)** $\dfrac{4x^3y^2\sqrt{6xy}}{3}$

In summary,

A radical is considered simplified
1. When there are no perfect square factors of the radicand.
2. When there are no fractions under the radical.
3. When there are no radicals in the denominator.

9.2 EXERCISES A

Simplify each of the radicals. Assume that all variables are positive.

1. $\sqrt{75}$
2. $\sqrt{10}$
3. $\sqrt{32}$
4. $3\sqrt{200}$

5. $\sqrt{147}$
6. $5\sqrt{36}$
7. $\sqrt{5^4}$
8. $\sqrt{625}$

9. $\sqrt{7^3}$
10. $\sqrt{5^5}$
11. $\sqrt{75y^2}$
12. $\sqrt{25x^3}$

13. $\sqrt{75y^3}$
14. $\sqrt{3^2x^3}$
15. $\sqrt{3^3y^3}$
16. $\sqrt{147x^5}$

17. $\sqrt{9x^2y^2}$
18. $\sqrt{27x^3y^2}$
19. $\sqrt{48x^2y^3}$
20. $\sqrt{75x^9y^5}$

9.2 SIMPLIFYING RADICALS

21. $\sqrt{\dfrac{75}{49}}$ 22. $\sqrt{\dfrac{8}{121}}$ 23. $\sqrt{\dfrac{50}{32}}$ 24. $\sqrt{\dfrac{25x^2}{y^2}}$

25. $\sqrt{\dfrac{16x^4}{y^4}}$ 26. $\sqrt{\dfrac{75x^2}{y^2}}$ 27. $\sqrt{\dfrac{75x^2y^4}{3}}$ 28. $\sqrt{\dfrac{x^3y^3}{49}}$

29. $\dfrac{5}{\sqrt{3}}$ 30. $\dfrac{8}{\sqrt{7}}$ 31. $\sqrt{\dfrac{9}{5}}$ 32. $\sqrt{\dfrac{45}{5}}$

33. $\sqrt{\dfrac{75x^2y^3}{yz^2}}$ 34. $\sqrt{\dfrac{36y^2}{x}}$ 35. $\sqrt{\dfrac{25x^3}{y^3z^4}}$ 36. $\sqrt{\dfrac{48x^2y^2}{3x^4y^4}}$

In a wholesale operation, it has been estimated that the cost, c, is related to the number of items sold, n, by the equation $c = \sqrt{n}$.

37. Find c if $n = 81$.   38. Find c if $n = 8$.   39. Find c if $n = 0$.

40. Find c if $n = 75$.   41. Find c if $n = 2^4$.   42. Find c if $n = 2^5$.

For Review

Evaluate.

43. $-\sqrt{121}$ 44. $\sqrt{-121}$ 45. $\sqrt{-0}$

46. $\sqrt{\dfrac{49}{25}}$ 47. $\sqrt{(x-7)^2}$ 48. $\sqrt{x^2 - 10x + 25}$

ANSWERS: 1. $5\sqrt{3}$ 2. $\sqrt{10}$ 3. $4\sqrt{2}$ 4. $30\sqrt{2}$ 5. $7\sqrt{3}$ 6. 30   7. 25   8. 25   9. $7\sqrt{7}$ 10. $25\sqrt{5}$
11. $5y\sqrt{3}$ 12. $5x\sqrt{x}$ 13. $5y\sqrt{3y}$ 14. $3x\sqrt{x}$ 15. $3y\sqrt{3y}$ 16. $7x^2\sqrt{3x}$ 17. $3xy$ 18. $3xy\sqrt{3x}$ 19. $4xy\sqrt{3y}$
20. $5x^4y^2\sqrt{3xy}$ 21. $\dfrac{5\sqrt{3}}{7}$ 22. $\dfrac{2\sqrt{2}}{11}$ 23. $\dfrac{5}{4}$ 24. $\dfrac{5x}{y}$ 25. $\dfrac{4x^2}{y^2}$ 26. $\dfrac{5x\sqrt{3}}{y}$ 27. $5xy^2$ 28. $\dfrac{xy\sqrt{xy}}{7}$
29. $\dfrac{5\sqrt{3}}{3}$ 30. $\dfrac{8\sqrt{7}}{7}$ 31. $\dfrac{3\sqrt{5}}{5}$ 32. 3   33. $\dfrac{5xy\sqrt{3}}{z}$ 34. $\dfrac{6y\sqrt{x}}{x}$ 35. $\dfrac{5x\sqrt{xy}}{y^2z^2}$ 36. $\dfrac{4}{xy}$ 37. 9   38. $2\sqrt{2}$
39. 0   40. $5\sqrt{3}$ 41. 4   42. $4\sqrt{2}$ 43. -11 44. meaningless   45. 0   46. $\dfrac{7}{5}$ 47. $x - 7$ 48. $x - 5$

9.2 EXERCISES B

Simplify each of the radicals. Assume that all variables are positive.

1. $\sqrt{98}$
2. $\sqrt{35}$
3. $\sqrt{50}$
4. $2\sqrt{300}$
5. $\sqrt{108}$
6. $4\sqrt{64}$
7. $\sqrt{3^4}$
8. $\sqrt{81}$
9. $\sqrt{5^3}$
10. $\sqrt{7^5}$
11. $\sqrt{98y^2}$
12. $\sqrt{49x^3}$
13. $\sqrt{98y^3}$
14. $\sqrt{5^2 x^3}$
15. $\sqrt{5^3 x^3}$
16. $\sqrt{125x^5}$
17. $\sqrt{16x^2 y^2}$
18. $\sqrt{32x^3 y^2}$
19. $\sqrt{27x^2 y^3}$
20. $\sqrt{12x^7 y^9}$
21. $\sqrt{\dfrac{32}{81}}$
22. $\sqrt{\dfrac{75}{144}}$
23. $\sqrt{\dfrac{45}{20}}$
24. $\sqrt{\dfrac{49x^2}{y^2}}$
25. $\sqrt{\dfrac{81x^4}{y^4}}$
26. $\sqrt{\dfrac{147x^2}{y^2}}$
27. $\sqrt{\dfrac{x^5 y^5}{25}}$
28. $\sqrt{\dfrac{147x^4 y^7}{3}}$
29. $\dfrac{3}{\sqrt{5}}$
30. $\dfrac{12}{\sqrt{11}}$
31. $\sqrt{\dfrac{16}{3}}$
32. $\sqrt{\dfrac{63}{7}}$
33. $\sqrt{\dfrac{4x^4}{y}}$
34. $\sqrt{\dfrac{147x^3 y^2}{xz^2}}$
35. $\sqrt{\dfrac{50x^3 y^3}{2x^5 y^5}}$
36. $\sqrt{\dfrac{16x^5}{y^5 z^2}}$

In a manufacturing process, two quantities p and r are related by the equation $p = \sqrt{r}$.

37. Find p when $r = 64$.
38. Find p when $r = 27$.
39. Find p when $r = 0$.
40. Find p when $r = 32$.
41. Find p when $r = 3^4$.
42. Find p when $r = 3^5$.

For Review

Evaluate.

43. $-\sqrt{144}$
44. $\sqrt{-144}$
45. $\sqrt{0}$
46. $\sqrt{\dfrac{81}{4}}$
47. $\sqrt{(x-3)^2}$
48. $\sqrt{x^2 + 10x + 25}$

9.2 EXERCISES C

Simplify each of the radicals. Assume all variables are positive.

1. $\sqrt{\dfrac{27x^9 y^7}{2x^6}}$
2. $\sqrt{\dfrac{243x^5 y^4}{8xyz^3}}$
3. $\sqrt{\dfrac{75xy}{9x^5 y^6 z^7}}$

$\left[\text{Answer: } \dfrac{5\sqrt{3yz}}{3x^2 y^3 z^4}\right]$

Use $\sqrt[3]{a} = a$ in the following exercises.

4. $\sqrt[3]{y^5}$
5. $\sqrt[3]{\dfrac{81x^5 y^4}{8z^3}}$
6. $\sqrt[3]{\dfrac{24x^4}{y}}$

$\left[\text{Answer: } \dfrac{2x\sqrt[3]{3xy^2}}{y}\right]$

9.3 MULTIPLICATION AND DIVISION OF RADICALS

STUDENT GUIDEPOSTS
1. Multiplication rule
2. Division rule

In the previous section we simplified radicals in the following manner.
$$\sqrt{4 \cdot 9} = \sqrt{4}\sqrt{9} = 2 \cdot 3 = 6$$

Consider

$$\begin{aligned}\sqrt{2}\sqrt{18} &= \sqrt{2 \cdot 18} && \text{Simplifying rule 1 used in the reverse order} \\ &= \sqrt{2 \cdot 2 \cdot 9} \\ &= \sqrt{4 \cdot 9} && \text{4 and 9 are perfect squares} \\ &= \sqrt{4}\sqrt{9} && \sqrt{ab} = \sqrt{a}\sqrt{b} \\ &= 2 \cdot 3 = 6.\end{aligned}$$

Here we used simplifying rule 1 from Section 9.2 in reverse order. This gives us the rule for multiplying radicals.

1 **Multiplication rule for radicals**
If $a \geq 0$ and $b \geq 0$, then
$$\sqrt{a}\sqrt{b} = \sqrt{ab}.$$
The product of square roots is equal to the square root of the product.

EXAMPLE 1

Multiply and then simplify.

(A) $\sqrt{7}\sqrt{7} = \sqrt{7 \cdot 7}$ $\sqrt{a}\sqrt{b} = \sqrt{ab}$
$= \sqrt{7^2} = 7$ We did this in Section 9.2

(B) $\sqrt{5}\sqrt{20} = \sqrt{5 \cdot 20}$ $\sqrt{a}\sqrt{b} = \sqrt{ab}$
$= \sqrt{5 \cdot 5 \cdot 4}$
$= \sqrt{25 \cdot 4}$ 25 and 4 are perfect squares
$= \sqrt{25}\sqrt{4}$ $\sqrt{ab} = \sqrt{a}\sqrt{b}$
$= 5 \cdot 2 = 10$

(C) $\sqrt{6}\sqrt{75} = \sqrt{6 \cdot 75}$ $\sqrt{a}\sqrt{b} = \sqrt{ab}$
$= \sqrt{2 \cdot 3 \cdot 3 \cdot 25}$ Factor to find perfect squares
$= \sqrt{9 \cdot 25 \cdot 2}$ 2 will be left under the radical
$= \sqrt{9}\sqrt{25}\sqrt{2}$
$= 3 \cdot 5 \cdot \sqrt{2} = 15\sqrt{2}$

(D) $\sqrt{5}\sqrt{7y} = \sqrt{5 \cdot 7 \cdot y}$ $\sqrt{a}\sqrt{b} = \sqrt{ab}$
$= \sqrt{35y}$ This is considered a simpler form than $\sqrt{5}\sqrt{7y}$

Practice Exercise 1

Multiply and then simplify.

(A) $\sqrt{11}\sqrt{11}$

(B) $\sqrt{2}\sqrt{18}$

(C) $\sqrt{48}\sqrt{6}$

(D) $\sqrt{2x}\sqrt{13}$

(E) $\sqrt{3x}\sqrt{75x} = \sqrt{3x \cdot 75x}$
$= \sqrt{3 \cdot 3 \cdot 25 \cdot x^2}$ $\sqrt{a}\sqrt{b} = \sqrt{ab}$
$= \sqrt{9}\sqrt{25}\sqrt{x^2}$ Look for perfect squares
$= 3 \cdot 5 \cdot x = 15x$ ◀◀   9, 25, and x^2 are perfect squares

(E) $\sqrt{24a}\sqrt{6a}$

Answers: **(A)** 11
(B) 6   **(C)** $12\sqrt{2}$ **(D)** $\sqrt{26x}$
(E) $12a$

EXAMPLE 2

Multiply and then simplify.

(A) $\sqrt{5x^3}\sqrt{3x^3}\sqrt{5x} = \sqrt{5x^3 \cdot 3x^3 \cdot 5x}$ $\sqrt{a}\sqrt{b}\sqrt{c} = \sqrt{abc}$
$= \sqrt{5 \cdot 5 \cdot x^3 \cdot x^3 \cdot 3x}$ Look for perfect squares
$= \sqrt{5^2 \cdot (x^3)^2 \cdot 3x}$ 5^2 and $(x^3)^2$ are perfect squares
$= \sqrt{5^2}\sqrt{(x^3)^2}\sqrt{3x}$
$= 5x^3\sqrt{3x}$

(B) $\sqrt{2xy}\sqrt{2x^3y^3} = \sqrt{2xy \cdot 2x^3y^3}$ $\sqrt{a}\sqrt{b} = \sqrt{ab}$
$= \sqrt{2^2x^4y^4}$
$= \sqrt{2^2}\sqrt{(x^2)^2}\sqrt{(y^2)^2}$ $x^4 = (x^2)^2$ and $y^4 = (y^2)^2$
$= 2x^2y^2$

(C) $7\sqrt{6ab}\sqrt{3a} = 7\sqrt{6ab \cdot 3a}$ Multiplication rule
$= 7\sqrt{2 \cdot 3 \cdot 3 \cdot a^2 \cdot b}$ 3^2 and a^2 are perfect squares
$= 7\sqrt{3^2a^2 \cdot 2b}$
$= 7\sqrt{3^2}\sqrt{a^2}\sqrt{2b}$
$= 7 \cdot 3 \cdot a \cdot \sqrt{2b}$
$= 21a\sqrt{2b}$ ◀◀

Practice Exercise 2

Multiply and then simplify.

(A) $\sqrt{7w^3}\sqrt{2w}\sqrt{7w^5}$

(B) $\sqrt{3ab^3}\sqrt{3ab}$

(C) $5\sqrt{14xy}\sqrt{7y}$

Answers: **(A)** $7w^4\sqrt{2w}$ **(B)** $3ab^2$
(C) $35y\sqrt{2x}$

In Section 9.2 we used simplifying rule 2 to simplify problems such as

$$\sqrt{\frac{9}{4}} = \frac{\sqrt{9}}{\sqrt{4}} = \frac{3}{2}.$$

Using the rule in reverse order gives

$$\frac{\sqrt{45}}{\sqrt{20}} = \sqrt{\frac{45}{20}} = \sqrt{\frac{\cancel{5} \cdot 9}{\cancel{5} \cdot 4}} = \sqrt{\frac{9}{4}} = \frac{\sqrt{9}}{\sqrt{4}} = \frac{3}{2}.$$

This gives us the method for dividing radicals. It is helpful when we can cancel to find perfect squares.

2 ▶▶ **Division rule for radicals**

If $a \geq 0$ and $b > 0$, then

$$\frac{\sqrt{a}}{\sqrt{b}} = \sqrt{\frac{a}{b}}.$$

The quotient of square roots is equal to the square root of the quotient.

EXAMPLE 3

Divide and then simplify.

(A) $\dfrac{\sqrt{75}}{\sqrt{12}} = \sqrt{\dfrac{75}{12}}$ $\quad \dfrac{\sqrt{a}}{\sqrt{b}} = \sqrt{\dfrac{a}{b}}$

$= \sqrt{\dfrac{\cancel{3} \cdot 25}{\cancel{3} \cdot 4}}$ Look for perfect squares and common factors

$= \sqrt{\dfrac{25}{4}}$ 25 and 4 are perfect squares

$= \dfrac{\sqrt{25}}{\sqrt{4}} = \dfrac{5}{2}$ $\quad \sqrt{\dfrac{a}{b}} = \dfrac{\sqrt{a}}{\sqrt{b}}$

(B) $\dfrac{\sqrt{15x^3}}{\sqrt{3x}} = \sqrt{\dfrac{15x^3}{3x}}$ $\quad \dfrac{\sqrt{a}}{\sqrt{b}} = \sqrt{\dfrac{a}{b}}$

$= \sqrt{\dfrac{\cancel{3} \cdot 5 \cdot \cancel{x} \cdot x^2}{\cancel{3} \cdot \cancel{x}}}$ 3 and x are common factors and x^2 is a perfect square

$= \sqrt{5x^2} = x\sqrt{5}$

(C) $\dfrac{\sqrt{12y^5}}{\sqrt{75y^9}} = \sqrt{\dfrac{12y^5}{75y^9}}$ Division rule

$= \sqrt{\dfrac{\cancel{3} \cdot 4 \cdot \cancel{y^5}}{\cancel{3} \cdot 25 \cdot \cancel{y^5} \cdot y^4}}$ 3 and y^5 are common factors since $y^9 = y^{5+4} = y^5 \cdot y^4$

$= \sqrt{\dfrac{4}{25y^4}}$

$= \dfrac{\sqrt{4}}{\sqrt{25}\sqrt{y^4}} = \dfrac{2}{5y^2}$ ◀◀

Practice Exercise 3

Divide and then simplify.

(A) $\dfrac{\sqrt{72}}{\sqrt{50}}$

(B) $\dfrac{\sqrt{21a^5}}{\sqrt{7a}}$

(C) $\dfrac{\sqrt{20a}}{\sqrt{405a^3}}$

Answers: (A) $\dfrac{6}{5}$ (B) $a^2\sqrt{3}$ (C) $\dfrac{2}{9a}$

EXAMPLE 4

Divide and then simplify.

(A) $\dfrac{\sqrt{3xy^3}}{\sqrt{4x^3y}} = \sqrt{\dfrac{3xy^3}{4x^3y}}$ Division rule

$= \sqrt{\dfrac{3 \cdot \cancel{x} \cdot y^2 \cdot \cancel{y}}{4 \cdot \cancel{x} \cdot x^2 \cdot \cancel{y}}}$ x and y are common factors.

$= \sqrt{\dfrac{3y^2}{4x^2}}$

$= \dfrac{\sqrt{y^2}\sqrt{3}}{\sqrt{4}\sqrt{x^2}} = \dfrac{y\sqrt{3}}{2x}$

(B) $\dfrac{2\sqrt{3x}}{\sqrt{12x^3}} = 2\sqrt{\dfrac{3x}{12x^3}}$ Division rule

$= 2\sqrt{\dfrac{\cancel{3} \cdot \cancel{x}}{\cancel{3} \cdot 4 \cdot x^2 \cdot \cancel{x}}}$ 3 and x are common factors

$= 2\sqrt{\dfrac{1}{4x^2}}$ 1 is left in the numerator

$= \dfrac{2\sqrt{1}}{\sqrt{4}\sqrt{x^2}}$

$= \dfrac{\cancel{2} \cdot 1}{\cancel{2}x} = \dfrac{1}{x}$ 2 is a common factor and $\sqrt{1} = 1$

Practice Exercise 4

Divide and then simplify.

(A) $\dfrac{\sqrt{7ab^3}}{\sqrt{9a^5b}}$

(B) $\dfrac{5\sqrt{2w^4}}{\sqrt{50w^6}}$

406 RADICALS

(C) $\dfrac{\sqrt{2x^2}}{\sqrt{6y}} = \sqrt{\dfrac{2x^2}{6y}}$ Division rule

$= \sqrt{\dfrac{x^2}{3y}}$ 2 is a common factor

$= \dfrac{\sqrt{x^2}}{\sqrt{3y}}$

$= \dfrac{x}{\sqrt{3y}}$

$= \dfrac{x\sqrt{3y}}{\sqrt{3y}\sqrt{3y}}$ Rationalize the denominator

$= \dfrac{x\sqrt{3y}}{3y}$ ◀◀

(C) $\dfrac{\sqrt{3u^4}}{\sqrt{15v}}$

Answers: (A) $\dfrac{b\sqrt{7}}{3a^2}$ (B) $\dfrac{1}{w}$
(C) $\dfrac{u^2\sqrt{5v}}{5v}$

9.3 EXERCISES A

Multiply and simplify.

1. $\sqrt{2}\,\sqrt{50}$
2. $\sqrt{3}\,\sqrt{75}$
3. $\sqrt{15}\,\sqrt{5}$
4. $\sqrt{18}\,\sqrt{98}$

5. $\sqrt{6}\,\sqrt{7}$
6. $\sqrt{6}\,\sqrt{30}$
7. $\sqrt{20}\,\sqrt{48}$
8. $2\sqrt{24}\,\sqrt{42}$

9. $\sqrt{98}\,\sqrt{75}$
10. $\sqrt{2}\,\sqrt{8x}$
11. $\sqrt{27x}\,\sqrt{3x}$
12. $\sqrt{15y}\,\sqrt{5}$

13. $\sqrt{15y}\,\sqrt{20y}$
14. $\sqrt{3a^2}\,\sqrt{12a}$
15. $\sqrt{3x^3}\,\sqrt{3x}$
16. $\sqrt{2xy}\,\sqrt{8xy}$

17. $\sqrt{6x^2y}\,\sqrt{2xy^2}$
18. $\sqrt{7x^3y^3}\,\sqrt{42xy^3}$
19. $\sqrt{5x}\,\sqrt{15xy}\,\sqrt{3y}$
20. $\sqrt{2xy}\,\sqrt{6xy}\,\sqrt{9xy}$

Divide and simplify.

21. $\dfrac{\sqrt{50}}{\sqrt{2}}$
22. $\dfrac{\sqrt{18}}{\sqrt{8}}$
23. $\dfrac{\sqrt{98}}{\sqrt{18}}$
24. $\dfrac{\sqrt{6}}{\sqrt{50}}$

25. $\dfrac{\sqrt{54}}{\sqrt{50}}$ 26. $\dfrac{5\sqrt{15}}{\sqrt{75}}$ 27. $\dfrac{\sqrt{4}}{\sqrt{3}}$ 28. $\dfrac{\sqrt{20}}{\sqrt{35}}$

29. $\dfrac{\sqrt{10}}{\sqrt{12}}$ 30. $\dfrac{\sqrt{50x^2}}{\sqrt{2x^2}}$ 31. $\dfrac{\sqrt{3y}}{\sqrt{4y}}$ 32. $\dfrac{\sqrt{27a^3}}{\sqrt{3}}$

33. $\dfrac{\sqrt{27x^2}}{\sqrt{3}}$ 34. $\dfrac{\sqrt{4y}}{\sqrt{y^3}}$ 35. $\dfrac{\sqrt{9x^2y}}{\sqrt{4y}}$ 36. $\dfrac{\sqrt{50x^3y^3}}{\sqrt{2xy}}$

37. $\dfrac{5\sqrt{2xy}}{\sqrt{25x}}$ 38. $\dfrac{\sqrt{18x}}{\sqrt{2y}}$ 39. $\dfrac{\sqrt{25}}{\sqrt{x}}$ 40. $\dfrac{\sqrt{2y^2}}{\sqrt{5x}}$

In an environmental study, three quantities m, p, and q have been found to be related by the equation $m = \sqrt{p}\sqrt{q}$.

41. Find m if p = 4 and q = 9.

42. Find m if p = 25 and q = 8.

43. Find m if p = 8 and q = 18.

44. Find m if p = 0 and q = 5.

45. Find m if p = 4 and q = −4.

46. Find m if p = 6 and q = 15.

408 RADICALS

For Review

Simplify.

47. $\sqrt{288x^2}$
48. $\sqrt{147x^2y}$
49. $\sqrt{\dfrac{98x^2}{25}}$
50. $\sqrt{\dfrac{36x^2}{5y}}$

ANSWERS: 1. 10 2. 15 3. $5\sqrt{3}$ 4. 42 5. $\sqrt{42}$ 6. $6\sqrt{5}$ 7. $8\sqrt{15}$ 8. $24\sqrt{7}$ 9. $35\sqrt{6}$ 10. $4\sqrt{x}$
11. $9x$ 12. $5\sqrt{3y}$ 13. $10y\sqrt{3}$ 14. $6a\sqrt{a}$ 15. $3x^2$ 16. $4xy$ 17. $2xy\sqrt{3xy}$ 18. $7x^2y^3\sqrt{6}$ 19. $15xy$
20. $6xy\sqrt{3xy}$ 21. 5 22. $\dfrac{3}{2}$ 23. $\dfrac{7}{3}$ 24. $\dfrac{\sqrt{3}}{5}$ 25. $\dfrac{3\sqrt{3}}{5}$ 26. $\sqrt{5}$ 27. $\dfrac{2\sqrt{3}}{3}$ 28. $\dfrac{2\sqrt{7}}{7}$ 29. $\dfrac{\sqrt{30}}{6}$
30. 5 31. $\dfrac{\sqrt{3}}{2}$ 32. $3a\sqrt{a}$ 33. $3x$ 34. $\dfrac{2}{y}$ 35. $\dfrac{3x}{2}$ 36. $5xy$ 37. $\sqrt{2y}$ 38. $\dfrac{3\sqrt{xy}}{y}$ 39. $\dfrac{5\sqrt{x}}{x}$
40. $\dfrac{y\sqrt{10x}}{5x}$ 41. 6 42. $10\sqrt{2}$ 43. 12 44. 0 45. no value for m if q is negative 46. $3\sqrt{10}$ 47. $12x\sqrt{2}$
48. $7x\sqrt{3y}$ 49. $\dfrac{7x\sqrt{2}}{5}$ 50. $\dfrac{6x\sqrt{5y}}{5y}$

9.3 EXERCISES B

Multiply and simplify.

1. $\sqrt{5}\,\sqrt{45}$
2. $\sqrt{7}\,\sqrt{28}$
3. $\sqrt{6}\,\sqrt{2}$
4. $\sqrt{20}\,\sqrt{45}$
5. $\sqrt{7}\,\sqrt{11}$
6. $\sqrt{10}\,\sqrt{18}$
7. $\sqrt{18}\,\sqrt{28}$
8. $3\sqrt{35}\,\sqrt{45}$
9. $\sqrt{72}\,\sqrt{48}$
10. $\sqrt{5}\,\sqrt{20x}$
11. $\sqrt{8x}\,\sqrt{2x}$
12. $\sqrt{12y}\,\sqrt{4}$
13. $\sqrt{14y}\,\sqrt{50y}$
14. $\sqrt{5a^2}\,\sqrt{20a}$
15. $\sqrt{5x}\,\sqrt{5x^3}$
16. $\sqrt{3xy}\,\sqrt{27xy}$
17. $\sqrt{10x^2y}\,\sqrt{2xy^2}$
18. $\sqrt{5x^3y^3}\,\sqrt{35xy^3}$
19. $\sqrt{2x}\,\sqrt{14xy}\,\sqrt{7y}$
20. $\sqrt{3xy}\,\sqrt{15xy}\,\sqrt{xy}$

Divide and simplify.

21. $\dfrac{\sqrt{3}}{\sqrt{48}}$
22. $\dfrac{\sqrt{27}}{\sqrt{12}}$
23. $\dfrac{\sqrt{125}}{\sqrt{45}}$
24. $\dfrac{\sqrt{14}}{\sqrt{72}}$
25. $\dfrac{\sqrt{150}}{\sqrt{8}}$
26. $\dfrac{6\sqrt{10}}{\sqrt{72}}$
27. $\dfrac{\sqrt{9}}{\sqrt{2}}$
28. $\dfrac{\sqrt{50}}{\sqrt{14}}$
29. $\dfrac{\sqrt{5}}{\sqrt{28}}$
30. $\dfrac{\sqrt{98x^2}}{\sqrt{2x^2}}$
31. $\dfrac{\sqrt{5y}}{\sqrt{9y}}$
32. $\dfrac{\sqrt{45x^2}}{\sqrt{5}}$
33. $\dfrac{\sqrt{8a^3}}{\sqrt{2}}$
34. $\dfrac{\sqrt{9y}}{\sqrt{y^3}}$
35. $\dfrac{\sqrt{25xy^2}}{\sqrt{9x}}$
36. $\dfrac{\sqrt{98x^3y^3}}{\sqrt{2xy}}$
37. $\dfrac{7\sqrt{3xy}}{\sqrt{49y}}$
38. $\dfrac{\sqrt{16}}{\sqrt{x}}$
39. $\dfrac{\sqrt{75y}}{\sqrt{3x}}$
40. $\dfrac{\sqrt{3x^2}}{\sqrt{7y}}$

During the course of a scientific experiment, it was determined that three quantities v, w, and d are related by the equation $v = \dfrac{\sqrt{w}}{\sqrt{d}}$.

41. Find v if $w = 4$ and $d = 25$.
42. Find v if $w = 8$ and $d = 16$.
43. Find v if $w = 8$ and $d = 50$.
44. Find v if $w = 0$ and $d = 5$.
45. Find v if $w = -9$ and $d = 7$.
46. Find v if $w = 21$ and $d = 14$.

For Review

Simplify.

47. $\sqrt{242x^2}$
48. $\sqrt{75xy^2}$
49. $\sqrt{\dfrac{147x^2}{16}}$
50. $\sqrt{\dfrac{25x^2}{6y}}$

9.3 EXERCISES C

Multiply or divide and simplify.

1. $\sqrt{3x^3y^{-3}}\,\sqrt{6xy^{-1}}$
2. $\dfrac{\sqrt{125x^{-6}y^{-3}}}{\sqrt{45x^4y^{-6}}}$
3. $\dfrac{\sqrt{16x^2y^5z^{-1}}}{\sqrt{6x^3y^{-4}z^{-3}}}$ $\left[\text{Answer: } \dfrac{2y^4z\sqrt{6xy}}{3x}\right]$

Use $\sqrt[3]{a^3} = a$ in the following exercises.

4. $\sqrt[3]{81x^4y^6}\,\sqrt[3]{18x^5y^5}$
5. $\dfrac{\sqrt[3]{9x^4y^7}}{\sqrt[3]{24xy^2}}$
6. $\dfrac{\sqrt[3]{40x^6y}}{\sqrt[3]{25xy^5}}$ $\left[\text{Answer: } \dfrac{2x\sqrt[3]{25x^2y^2}}{5y^2}\right]$

9.4 ADDITION AND SUBTRACTION OF RADICALS

STUDENT GUIDEPOSTS

1 Like radicals
2 Unlike radicals
3 Adding and subtracting radicals

The rules for multiplication and division of radicals,

$$\sqrt{a}\,\sqrt{b} = \sqrt{ab} \quad \text{and} \quad \dfrac{\sqrt{a}}{\sqrt{b}} = \sqrt{\dfrac{a}{b}},$$

do not have counterparts relative to addition and subtraction. For example,

$$5 = \sqrt{25} = \sqrt{9 + 16} \neq \sqrt{9} + \sqrt{16} = 3 + 4 = 7, \quad 5 \neq 7$$

so that in general,

$$\sqrt{a + b} \neq \sqrt{a} + \sqrt{b}.$$

Also, since

$$4 = \sqrt{16} = \sqrt{25 - 9} \neq \sqrt{25} - \sqrt{9} = 5 - 3 = 2, \quad 4 \neq 2$$

in general,

$$\sqrt{a - b} \neq \sqrt{a} - \sqrt{b}.$$

1 We may, however, use the distributive law to add or subtract *like radicals*. **Like radicals** have the same radicand. Thus, the terms

$$\sqrt{11},\ 3\sqrt{11},\ -5\sqrt{11}$$

410 RADICALS

involve like radicals. Also, terms such as

$$\sqrt{x}, \ -7\sqrt{x}, \ 100\sqrt{x}$$

contain like radicals. However,

$$\sqrt{11}, \ \sqrt{x}, \ 3\sqrt{y}$$

2 do not involve like radicals; they are **unlike radicals.**

We add or subtract like radicals just as we collect like terms of polynomials. Recall,

$$3x + 5x = (3 + 5)x = 8x.$$

We collect like radicals in the same way.

$$3\sqrt{x} + 5\sqrt{x} = (3 + 5)\sqrt{x} = 8\sqrt{x}$$

EXAMPLE 1

Add or subtract.

(A) $9\sqrt{5} + 4\sqrt{5} = (9 + 4)\sqrt{5}$ Distributive law, $ac + bc = (a + b)c$
$\phantom{(A) \ 9\sqrt{5} + 4\sqrt{5}} = 13\sqrt{5}$

(B) $9\sqrt{5} - 4\sqrt{5} = (9 - 4)\sqrt{5}$ Distributive law, $ac - bc = (a - b)c$
$\phantom{(B) \ 9\sqrt{5} - 4\sqrt{5}} = 5\sqrt{5}$

(C) $3\sqrt{xy} - 10\sqrt{xy} = (3 - 10)\sqrt{xy}$ Distributive law
$\phantom{(C) \ 3\sqrt{xy} - 10\sqrt{xy}} = -7\sqrt{xy}$

(D) $7\sqrt{x} + 5\sqrt{y}$ cannot be simplified since $\sqrt{x}$ and $\sqrt{y}$ are not like radicals.

Practice Exercise 1

Add or subtract.

(A) $2\sqrt{11} + 5\sqrt{11}$
(B) $2\sqrt{11} - 5\sqrt{11}$
(C) $6\sqrt{u} - 5\sqrt{u}$
(D) $8\sqrt{a} - 2\sqrt{w}$

Answers: (A) $7\sqrt{11}$ (B) $-3\sqrt{11}$ (C) $\sqrt{u}$ (D) Cannot be simplified since $\sqrt{a}$ and $\sqrt{w}$ are not like radicals.

Some unlike radicals can be changed into like radicals if we simplify. For example, $\sqrt{2}$ and $\sqrt{8}$ are not like radicals since they have different radicands. However,

$$\sqrt{8} = \sqrt{4 \cdot 2} = \sqrt{4}\sqrt{2} = 2\sqrt{2}.$$

Now the terms $\sqrt{2}$ and $2\sqrt{2}$ do contain like radicals.

3 **To add or subtract unlike radicals**
1. Simplify each radical as much as possible.
2. Use the distributive law to collect any like radicals.

EXAMPLE 2

Add or subtract.

(A) $3\sqrt{2} + 5\sqrt{8} = 3\sqrt{2} + 5\sqrt{4 \cdot 2}$ 4 is a perfect square
$\phantom{(A) \ 3\sqrt{2} + 5\sqrt{8}} = 3\sqrt{2} + 5\sqrt{4}\sqrt{2}$ $\sqrt{ab} = \sqrt{a}\sqrt{b}$
$\phantom{(A) \ 3\sqrt{2} + 5\sqrt{8}} = 3\sqrt{2} + 5 \cdot 2\sqrt{2}$ $\sqrt{4} = 2$
$\phantom{(A) \ 3\sqrt{2} + 5\sqrt{8}} = 3\sqrt{2} + 10\sqrt{2}$ Terms now contain like radicals
$\phantom{(A) \ 3\sqrt{2} + 5\sqrt{8}} = (3 + 10)\sqrt{2} = 13\sqrt{2}$ Distributive law

Practice Exercise 2

Add or subtract.

(A) $5\sqrt{5} + 3\sqrt{20}$

9.4 ADDITION AND SUBTRACTION OF RADICALS

(B) $\sqrt{12} + \sqrt{75} = \sqrt{4 \cdot 3} + \sqrt{25 \cdot 3}$ 4 and 25 are perfect squares

$= \sqrt{4}\sqrt{3} + \sqrt{25}\sqrt{3}$ $\sqrt{ab} = \sqrt{a}\sqrt{b}$

$= 2\sqrt{3} + 5\sqrt{3}$ $\sqrt{4} = 2$ and $\sqrt{25} = 5$

$= (2 + 5)\sqrt{3} = 7\sqrt{3}$ Distributive law

(C) $5\sqrt{12} - 7\sqrt{27} = 5\sqrt{4 \cdot 3} - 7\sqrt{9 \cdot 3}$ 4 and 9 are perfect squares

$= 5\sqrt{4}\sqrt{3} - 7\sqrt{9}\sqrt{3}$ $\sqrt{ab} = \sqrt{a}\sqrt{b}$

$= 5 \cdot 2 \cdot \sqrt{3} - 7 \cdot 3 \cdot \sqrt{3}$

$= 10\sqrt{3} - 21\sqrt{3}$ Terms contain like radicals

$= (10 - 21)\sqrt{3} = -11\sqrt{3}$ Distributive law

(D) $2\sqrt{25y} - 4\sqrt{36y} = 2\sqrt{25}\sqrt{y} - 4\sqrt{36}\sqrt{y}$ 25 and 36 are perfect squares

$= 2 \cdot 5\sqrt{y} - 4 \cdot 6\sqrt{y}$

$= 10\sqrt{y} - 24\sqrt{y}$

$= (10 - 24)\sqrt{y} = -14\sqrt{y}$ Distributive law ◀◀

(B) $7\sqrt{32} + 3\sqrt{18}$

(C) $6\sqrt{8} - \sqrt{50}$

(D) $8\sqrt{4x} - 3\sqrt{49x}$

Answers: **(A)** $11\sqrt{5}$ **(B)** $37\sqrt{2}$
(C) $7\sqrt{2}$ **(D)** $-5\sqrt{x}$

EXAMPLE 3

Perform the indicated operations.

(A) $4\sqrt{18} + 3\sqrt{27} - 6\sqrt{12}$

$= 4\sqrt{9 \cdot 2} + 3\sqrt{9 \cdot 3} - 6\sqrt{4 \cdot 3}$

$= 4\sqrt{9}\sqrt{2} + 3\sqrt{9}\sqrt{3} - 6\sqrt{4}\sqrt{3}$

$= 4 \cdot 3\sqrt{2} + 3 \cdot 3\sqrt{3} - 6 \cdot 2\sqrt{3}$

$= 12\sqrt{2} + 9\sqrt{3} - 12\sqrt{3}$ Only the last two terms are like terms

$= 12\sqrt{2} + (9 - 12)\sqrt{3}$ Distributive law

$= 12\sqrt{2} - 3\sqrt{3}$ As simple as possible

(B) $3\sqrt{x} - 2\sqrt{4x} - 5\sqrt{9x} = 3\sqrt{x} - 2\sqrt{4}\sqrt{x} - 5\sqrt{9}\sqrt{x}$

$= 3\sqrt{x} - 2 \cdot 2\sqrt{x} - 5 \cdot 3\sqrt{x}$

$= 3\sqrt{x} - 4\sqrt{x} - 15\sqrt{x}$

$= (3 - 4 - 15)\sqrt{x} = -16\sqrt{x}$ ◀◀

Practice Exercise 3

Perform the indicated operations.

(A) $3\sqrt{75} - \sqrt{18} + 2\sqrt{27}$

(B) $8\sqrt{w} - 3\sqrt{25w} + 5\sqrt{4w}$

Answers: **(A)** $21\sqrt{3} - 3\sqrt{2}$
(B) $3\sqrt{w}$

In order to obtain like terms, it may be necessary to rationalize denominators.

EXAMPLE 4

Add or subtract.

(A) $3\sqrt{2} + \dfrac{5}{\sqrt{2}} = 3\sqrt{2} + \dfrac{5\sqrt{2}}{\sqrt{2}\sqrt{2}}$ Rationalize denominator

$= 3\sqrt{2} + \dfrac{5\sqrt{2}}{2}$ $\sqrt{2}\sqrt{2} = 2$

Practice Exercise 4

Add or subtract.

(A) $5\sqrt{3} + \dfrac{1}{\sqrt{3}}$

$$= 3\sqrt{2} + \frac{5}{2}\sqrt{2} \qquad \frac{5\sqrt{2}}{2} = \frac{5 \cdot \sqrt{2}}{2 \cdot 1}$$
$$= \frac{5}{2} \cdot \frac{\sqrt{2}}{1} = \frac{5}{2}\sqrt{2}$$
$$= \left(3 + \frac{5}{2}\right)\sqrt{2} \qquad \text{Distributive law}$$
$$= \left(\frac{6}{2} + \frac{5}{2}\right)\sqrt{2} = \frac{11}{2}\sqrt{2} = \frac{11\sqrt{2}}{2}$$

(B) $\dfrac{6}{\sqrt{5}} - 4\sqrt{5} = \dfrac{6\sqrt{5}}{\sqrt{5}\sqrt{5}} - 4\sqrt{5}$ Rationalize the denominator **(B)** $\dfrac{9}{\sqrt{7}} - 6\sqrt{7}$

$$= \frac{6\sqrt{5}}{5} - 4\sqrt{5}$$
$$= \frac{6}{5}\sqrt{5} - 4\sqrt{5} \qquad \frac{6\sqrt{5}}{5} = \frac{6 \cdot \sqrt{5}}{5 \cdot 1} = \frac{6}{5} \cdot \frac{\sqrt{5}}{1}$$
$$= \frac{6}{5}\sqrt{5}$$
$$= \left(\frac{6}{5} - 4\right)\sqrt{5} \qquad \text{Distributive law}$$
$$= \left(\frac{6}{5} - \frac{20}{5}\right)\sqrt{5}$$
$$= -\frac{14}{5}\sqrt{5} = -\frac{14\sqrt{5}}{5} \;\blacktriangleleft$$

Answers: **(A)** $\dfrac{16\sqrt{3}}{3}$
(B) $-\dfrac{33\sqrt{7}}{7}$

9.4 EXERCISES A

Add or subtract.

1. $9\sqrt{2} + \sqrt{2}$

2. $-4\sqrt{5} + 3\sqrt{5}$

3. $-18\sqrt{xy} - 7\sqrt{xy}$

4. $3\sqrt{2} + 2\sqrt{3}$

5. $5\sqrt{12} - 3\sqrt{12}$

6. $\sqrt{50} + \sqrt{98}$

7. $\sqrt{18} - \sqrt{8}$

8. $6\sqrt{2} + 3\sqrt{8}$

9. $5\sqrt{3} - \sqrt{27}$

10. $4\sqrt{50} + 7\sqrt{18}$

11. $3\sqrt{18} - 5\sqrt{12}$

12. $-18\sqrt{7} - 2\sqrt{28}$

13. $15\sqrt{45} + 4\sqrt{20}$

14. $\sqrt{147} - 2\sqrt{75}$

15. $6\sqrt{44} + 2\sqrt{99}$

16. $\sqrt{121} - \sqrt{144}$

17. $2\sqrt{75} - 3\sqrt{125}$

18. $2\sqrt{4x} + 7\sqrt{9x}$

19. $-3\sqrt{25y} - 2\sqrt{9y}$

20. $8\sqrt{50y} + 2\sqrt{18y}$

21. $x\sqrt{y^3} + y\sqrt{x^2 y}$

Perform the indicated operations.

22. $\sqrt{18} + \sqrt{50} + \sqrt{72}$

23. $\sqrt{27} - \sqrt{75} - \sqrt{108}$

24. $5\sqrt{20} + 2\sqrt{45} - 9\sqrt{80}$

25. $3\sqrt{75} + 2\sqrt{12} - 5\sqrt{48}$

26. $-3\sqrt{24} - 5\sqrt{150} - 4\sqrt{54}$

27. $4\sqrt{99} - 7\sqrt{44} + 5\sqrt{52}$

28. $4\sqrt{98} - 8\sqrt{72} + 5\sqrt{32}$

29. $4\sqrt{300} - 2\sqrt{500} + 4\sqrt{125}$

30. $\sqrt{4x} + \sqrt{16x} - \sqrt{25x}$

31. $3\sqrt{5x} + 2\sqrt{20x} - 8\sqrt{45x}$

32. $9\sqrt{100a^2b^2} - 4\sqrt{36a^2b^2} - 10\sqrt{a^2b^2}$

33. $2\sqrt{25ab^3} - b\sqrt{36ab} - 5\sqrt{49ab^3}$

Rationalize the denominator and then add or subtract.

34. $\sqrt{3} + \dfrac{1}{\sqrt{3}}$

35. $2\sqrt{5} - \dfrac{3}{\sqrt{5}}$

36. $\dfrac{7}{\sqrt{2}} + 6\sqrt{2}$

37. $3\sqrt{7} - \dfrac{1}{\sqrt{7}}$

38. $3\sqrt{7} - \sqrt{\dfrac{1}{7}}$

39. $-3\sqrt{5} - \dfrac{9}{\sqrt{5}}$

40. $\dfrac{\sqrt{2}}{\sqrt{3}} + 2\sqrt{6}$

41. $\dfrac{\sqrt{2}}{\sqrt{10}} - \sqrt{5}$

42. $\sqrt{\dfrac{2}{3}} - 2\sqrt{6}$

9.4 ADDITION AND SUBTRACTION OF RADICALS

A structural architect often works with the formula $F = \sqrt{f} + 2\sqrt{g}$, relating the three quantities, F, f, and g.

43. Find F when $f = 8$ and $g = 18$.

44. Find F when $f = 60$ and $g = 15$.

45. Find F when $f = 0$ and $g = 7$.

46. Find F when $f = -4$ and $g = 9$.

For Review

Simplify.

47. $3\sqrt{10}\sqrt{40}$

48. $\sqrt{3a^2}\sqrt{15a}$

49. $\dfrac{\sqrt{150x}}{\sqrt{3x}}$

50. $\dfrac{\sqrt{12x^2}}{\sqrt{5y}}$

ANSWERS: **1.** $10\sqrt{2}$ **2.** $-\sqrt{5}$ **3.** $-25\sqrt{xy}$ **4.** $3\sqrt{2} + 2\sqrt{3}$ (cannot be simplified) **5.** $4\sqrt{3}$ **6.** $12\sqrt{2}$
7. $\sqrt{2}$ **8.** $12\sqrt{2}$ **9.** $2\sqrt{3}$ **10.** $41\sqrt{2}$ **11.** $9\sqrt{2} - 10\sqrt{3}$ **12.** $-22\sqrt{7}$ **13.** $53\sqrt{5}$ **14.** $-3\sqrt{3}$ **15.** $18\sqrt{11}$
16. -1 **17.** $10\sqrt{3} - 15\sqrt{5}$ **18.** $25\sqrt{x}$ **19.** $-21\sqrt{y}$ **20.** $46\sqrt{2y}$ **21.** $2xy\sqrt{y}$ **22.** $14\sqrt{2}$ **23.** $-8\sqrt{3}$
24. $-20\sqrt{5}$ **25.** $-\sqrt{3}$ **26.** $-43\sqrt{6}$ **27.** $-2\sqrt{11} + 10\sqrt{13}$ **28.** 0 **29.** $40\sqrt{3}$ **30.** $\sqrt{x}$ **31.** $-17\sqrt{5x}$
32. $56ab$ **33.** $-31b\sqrt{ab}$ **34.** $\dfrac{4\sqrt{3}}{3}$ **35.** $\dfrac{7\sqrt{5}}{5}$ **36.** $\dfrac{19\sqrt{2}}{2}$ **37.** $\dfrac{20\sqrt{7}}{7}$ **38.** $\dfrac{20\sqrt{7}}{7}$ **39.** $\dfrac{-24\sqrt{5}}{5}$
40. $\dfrac{7\sqrt{6}}{3}$ **41.** $\dfrac{-4\sqrt{5}}{5}$ **42.** $\dfrac{-5\sqrt{6}}{3}$ **43.** $8\sqrt{2}$ **44.** $4\sqrt{15}$ **45.** $2\sqrt{7}$ **46.** no value for F when f is negative
47. 60 **48.** $3a\sqrt{5a}$ **49.** $5\sqrt{2}$ **50.** $\dfrac{2x\sqrt{15y}}{5y}$

9.4 EXERCISES B

Add or subtract.

1. $5\sqrt{3} - 9\sqrt{3}$

2. $-5\sqrt{7} - 9\sqrt{7}$

3. $-9\sqrt{xy} - 3\sqrt{xy}$

4. $5\sqrt{5} + 3\sqrt{3}$

5. $2\sqrt{27} - 4\sqrt{27}$

6. $\sqrt{48} + \sqrt{75}$

7. $\sqrt{45} - \sqrt{20}$

8. $7\sqrt{3} + 2\sqrt{12}$

9. $4\sqrt{2} - \sqrt{50}$

10. $8\sqrt{32} + 3\sqrt{50}$

11. $6\sqrt{27} - 4\sqrt{18}$

12. $-8\sqrt{5} + 2\sqrt{45}$

13. $4\sqrt{28} + 8\sqrt{63}$

14. $3\sqrt{75} - 5\sqrt{147}$

15. $-8\sqrt{99} + 8\sqrt{44}$

16. $\sqrt{169} - \sqrt{121}$

17. $5\sqrt{45} - 2\sqrt{48}$

18. $4\sqrt{25x} + 2\sqrt{16x}$

19. $-4\sqrt{4y} - 9\sqrt{25y}$

20. $6\sqrt{8y} + 5\sqrt{32y}$

21. $y\sqrt{x^3} + x\sqrt{xy^2}$

22. $\sqrt{8} + \sqrt{72} + \sqrt{98}$

23. $\sqrt{12} - \sqrt{27} - \sqrt{147}$

24. $7\sqrt{48} - 4\sqrt{75} - 6\sqrt{12}$

25. $3\sqrt{125} + 4\sqrt{80} - 6\sqrt{5}$

26. $-2\sqrt{6} - 2\sqrt{54} - 5\sqrt{150}$

27. $5\sqrt{28} - 2\sqrt{63} + 4\sqrt{99}$

28. $5\sqrt{147} - 2\sqrt{108} + 7\sqrt{48}$

29. $7\sqrt{700} - 4\sqrt{175} + 8\sqrt{44}$

30. $\sqrt{25x} + \sqrt{36x} - \sqrt{49x}$

31. $5\sqrt{3x} + 6\sqrt{27x} - 7\sqrt{75x}$

32. $5\sqrt{25a^2b^2} - 3\sqrt{16a^2b^2} - 20\sqrt{a^2b^2}$

33. $2a\sqrt{16ab} + \sqrt{4a^3b} - 3\sqrt{25a^3b}$

416 RADICALS

Rationalize the denominator and then add or subtract.

34. $\sqrt{2} + \dfrac{1}{\sqrt{2}}$
35. $3\sqrt{5} - \dfrac{4}{\sqrt{5}}$
36. $\dfrac{8}{\sqrt{3}} + 2\sqrt{3}$

37. $4\sqrt{5} - \dfrac{1}{\sqrt{5}}$
38. $4\sqrt{5} - \sqrt{\dfrac{1}{5}}$
39. $-4\sqrt{7} - \dfrac{3}{\sqrt{7}}$

40. $\dfrac{\sqrt{3}}{\sqrt{2}} + 5\sqrt{6}$
41. $\sqrt{\dfrac{3}{2}} - 5\sqrt{6}$
42. $\dfrac{\sqrt{5}}{\sqrt{10}} - \sqrt{2}$

An environmental scientist frequently uses the formula $P = \sqrt{n} - 3\sqrt{p}$, *relating the three quantities P, n, and p.*

43. Find P when $n = 49$ and $p = 45$.
44. Find P when $n = 52$ and $p = 117$.
45. Find P when $n = -9$ and $p = 14$.
46. Find P when $n = 11$ and $p = 0$.

For Review

Simplify.

47. $5\sqrt{48}\,\sqrt{3}$
48. $\sqrt{8a^2}\,\sqrt{10a}$
49. $\dfrac{\sqrt{150x}}{\sqrt{2x}}$
50. $\dfrac{\sqrt{18y^2}}{\sqrt{7x}}$

9.4 EXERCISES C

Add or subtract.

1. $\sqrt{12x^4 - 4x^2y^2} - \sqrt{27x^4 - 9x^2y^2}$
2. $\sqrt{2x^2 - 20xy + 50y^2} + \sqrt{8x^2 - 80xy + 200y^2}$
   [Answer: $3(x - 5y)\sqrt{2}$]
3. $3\sqrt[3]{40x^4y^4} - 2x\sqrt[3]{5xy^4}$ [Hint: $\sqrt[3]{a^3} = a$]
4. $-6ab\sqrt[3]{27a^5b^3} + 10\sqrt[3]{27a^8b^6}$
   [Answer: $12a^2b^2\sqrt[3]{a^2}$]

9.5 SUMMARY OF TECHNIQUES AND RATIONALIZING DENOMINATORS

STUDENT GUIDEPOSTS

1 Simplifying radical expressions
2 Conjugates
3 Rationalizing binomial denominators

Below is a summary of the simplifying techniques from previous sections.

1 **To simplify a radical expression**

1. Combine all like radicals using the distributive laws.
2. When indicated use the rules of multiplication and division,

$$\sqrt{a}\,\sqrt{b} = \sqrt{ab} \quad \text{and} \quad \dfrac{\sqrt{a}}{\sqrt{b}} = \sqrt{\dfrac{a}{b}}.$$

3. Remove all perfect squares from under the radicals.
4. Rationalize all denominators.

9.5 SUMMARY OF TECHNIQUES AND RATIONALIZING DENOMINATORS

The following table illustrates the various techniques.

Simplification	Type
$3\sqrt{2} - \sqrt{2} = 2\sqrt{2}$	Combining like radicals
$\sqrt{2}\sqrt{3} = \sqrt{6}$	Using multiplication rule
$\dfrac{\sqrt{12}}{\sqrt{3}} = \sqrt{\dfrac{12}{3}} = \sqrt{4} = 2$	Using division rule and removing perfect squares
$\sqrt{9x^2y} = 3x\sqrt{y}$	Removing perfect squares
$\dfrac{3}{\sqrt{2}} = \dfrac{3\sqrt{2}}{\sqrt{2}\sqrt{2}} = \dfrac{3\sqrt{2}}{2}$	Rationalizing the denominator

We will use these methods in the following examples. The distributive law is used first to multiply a radical by a sum of radicals.

EXAMPLE 1
Multiply.

(A) $\sqrt{5}(\sqrt{15} + \sqrt{5}) = \sqrt{5}\sqrt{15} + \sqrt{5}\sqrt{5}$ Distributive law
$= \sqrt{5 \cdot 15} + \sqrt{5 \cdot 5}$ $\sqrt{a}\sqrt{b} = \sqrt{ab}$
$= \sqrt{5 \cdot 5 \cdot 3} + \sqrt{5 \cdot 5}$ Factor
$= \sqrt{5^2}\sqrt{3} + \sqrt{5^2}$ 5^2 is a perfect square
$= 5\sqrt{3} + 5$

(B) $\sqrt{3}(\sqrt{27} - \sqrt{12}) = \sqrt{3}\sqrt{27} - \sqrt{3}\sqrt{12}$ Distributive law
$= \sqrt{3 \cdot 3^3} - \sqrt{3 \cdot 3 \cdot 4}$ $\sqrt{a}\sqrt{b} = \sqrt{ab}$
$= \sqrt{3^4} - \sqrt{3^2 \cdot 2^2}$ $3^4, 3^2,$ and 2^2 are perfect squares
$= 3^2 - 3 \cdot 2$
$= 9 - 6 = 3$

(C) $\sqrt{7}\left(5\sqrt{7} - \dfrac{8}{\sqrt{7}}\right) = \sqrt{7}(5\sqrt{7}) - \sqrt{7}\left(\dfrac{8}{\sqrt{7}}\right)$ Distributive law
$= 5(\sqrt{7})^2 - \dfrac{\cancel{\sqrt{7}}}{\cancel{\sqrt{7}}} \cdot 8$ Commutative and associative laws
$= 5 \cdot 7 - 1 \cdot 8$ $(\sqrt{7})^2 = \sqrt{7}\sqrt{7} = 7$
$= 35 - 8 = 27$ ◀◀

Practice Exercise 1
Multiply.

(A) $\sqrt{2}(\sqrt{6} + 2\sqrt{2})$

(B) $\sqrt{5}(\sqrt{125} - \sqrt{20})$

(C) $\sqrt{3}\left(2\sqrt{3} - \dfrac{1}{\sqrt{3}}\right)$

Answers: (A) $2\sqrt{3} + 4$ (B) 15 (C) 5

In Chapter 6 we used the FOIL method to multiply binomials (F—first terms, O—outside terms, I—inside terms, and L—last terms).

$(x + 2y)(2x - y) = 2x \cdot x - x \cdot y + 2x \cdot 2y - 2y \cdot y$

418 RADICALS

The same rule can be used to multiply binomial radical expressions.

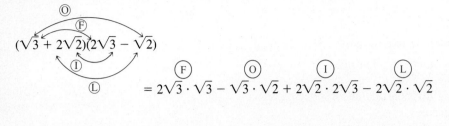

We multiplied the first terms (F), the outside terms (O), the inside terms (I), and the last terms (L). The simplification is completed by using the multiplication rule and collecting like terms.

EXAMPLE 2

Multiply.

(A) $(\sqrt{3} - \sqrt{2})(\sqrt{3} + 5\sqrt{2})$

$= \sqrt{3}\sqrt{3} + \sqrt{3}(5\sqrt{2}) - \sqrt{2}\sqrt{3} - \sqrt{2}(5\sqrt{2})$
$= 3 + 5\sqrt{6} - \sqrt{6} - 5(2)$
$= 3 - 10 + (5 - 1)\sqrt{6}$
$= -7 + 4\sqrt{6}$

(B) $(\sqrt{3} - \sqrt{2})(\sqrt{3} + \sqrt{2})$
$= \sqrt{3}\sqrt{3} + \sqrt{3}\sqrt{2} - \sqrt{2}\sqrt{3} - \sqrt{2}\sqrt{2}$
$= 3 + \sqrt{6} - \sqrt{6} - 2$
$= 3 - 2$
$= 1$ ◀◀

Practice Exercise 2

Multiply.

(A) $(\sqrt{7} - 2\sqrt{3})(\sqrt{7} + \sqrt{3})$

(B) $(\sqrt{11} - 2\sqrt{2})(\sqrt{11} + 2\sqrt{2})$

Answers: **(A)** $1 - \sqrt{21}$ **(B)** 3

Notice that in Example 2(B) the product is the rational number 1. This occurred because the expressions multiplied fit the rule

$$(a - b)(a + b) = a^2 - b^2.$$

Thus,

$$(\sqrt{3} - \sqrt{2})(\sqrt{3} + \sqrt{2}) = (\sqrt{3})^2 - (\sqrt{2})^2 \quad \sqrt{3} = a \text{ and } \sqrt{2} = b$$
$$= 3 - 2 = 1.$$

This observation leads to a way to rationalize the denominator of a fraction such as $\dfrac{1}{\sqrt{3} - \sqrt{2}}$. We can rationalize the denominator by multiplying both the numerator and denominator by $\sqrt{3} + \sqrt{2}$. This is the same as multiplying the fraction by 1, written in the form

$$\dfrac{\sqrt{3} + \sqrt{2}}{\sqrt{3} + \sqrt{2}}.$$

$$\dfrac{1}{\sqrt{3} - \sqrt{2}} = \dfrac{1(\sqrt{3} + \sqrt{2})}{(\sqrt{3} - \sqrt{2})(\sqrt{3} + \sqrt{2})}$$

9.5 SUMMARY OF TECHNIQUES AND RATIONALIZING DENOMINATORS

$$= \frac{\sqrt{3} + \sqrt{2}}{(\sqrt{3})^2 - (\sqrt{2})^2}$$
$$= \frac{\sqrt{3} + \sqrt{2}}{3 - 2}$$
$$= \frac{\sqrt{3} + \sqrt{2}}{1} = \sqrt{3} + \sqrt{2}$$

2 ▶▶ Of course if $\sqrt{3} + \sqrt{2}$ were in the denominator, we would multiply both numerator and denominator by $\sqrt{3} - \sqrt{2}$. The binomials $\sqrt{3} + \sqrt{2}$ and $\sqrt{3} - \sqrt{2}$ are called **conjugates** as are binomials of the form $\sqrt{3} + 2$ and $\sqrt{3} - 2$.

3 ▶▶ **To rationalize a binomial denominator**

Multiply both numerator and denominator of the fraction by the conjugate of the denominator and simplify.

EXAMPLE 3

Rationalize the denominators.

(A) $\dfrac{2}{\sqrt{7} + \sqrt{5}}$

$= \dfrac{2(\sqrt{7} - \sqrt{5})}{(\sqrt{7} + \sqrt{5})(\sqrt{7} - \sqrt{5})}$ Multiply numerator and denominator by $\sqrt{7} - \sqrt{5}$, the conjugate of $\sqrt{7} + \sqrt{5}$

$= \dfrac{2(\sqrt{7} - \sqrt{5})}{(\sqrt{7})^2 - (\sqrt{5})^2}$ $(a + b)(a - b) = a^2 - b^2$

$= \dfrac{2(\sqrt{7} - \sqrt{5})}{7 - 5}$ $(\sqrt{7})^2 = 7$ and $(\sqrt{5})^2 = 5$

$= \dfrac{\cancel{2}(\sqrt{7} - \sqrt{5})}{\cancel{2}}$

$= \sqrt{7} - \sqrt{5}$

(B) $\dfrac{\sqrt{3}}{\sqrt{3} - 2} = \dfrac{\sqrt{3}(\sqrt{3} + 2)}{(\sqrt{3} - 2)(\sqrt{3} + 2)}$ Multiply numerator and denominator by $\sqrt{3} + 2$, the conjugate of $\sqrt{3} - 2$

$= \dfrac{\sqrt{3}(\sqrt{3} + 2)}{(\sqrt{3})^2 - (2)^2}$ $(a - b)(a + b) = a^2 - b^2$

$= \dfrac{\sqrt{3}(\sqrt{3} + 2)}{3 - 4}$ $(\sqrt{3})^2 = 3$ and $(2)^2 = 4$

$= \dfrac{\sqrt{3}(\sqrt{3} + 2)}{-1}$

$= -\sqrt{3}(\sqrt{3} + 2)$ $\dfrac{a}{-1} = -a$

$= -\sqrt{3} \cdot \sqrt{3} - \sqrt{3} \cdot 2$ Distributive law

$= -3 - 2\sqrt{3}$ ◀◀

Practice Exercise 3

Rationalize the denominators.

(A) $\dfrac{4}{\sqrt{7} - \sqrt{3}}$

(B) $\dfrac{\sqrt{5}}{\sqrt{5} + 2}$

Answers: (A) $\sqrt{7} + \sqrt{3}$
(B) $5 - 2\sqrt{5}$

9.5 EXERCISES A

Refer to the summary of techniques and simplify.

1. $\sqrt{20} - \sqrt{45}$
2. $\sqrt{2}\sqrt{32}$
3. $\sqrt{3x}\sqrt{12x}$

4. $\dfrac{\sqrt{18xy^2}}{\sqrt{2x}}$
5. $\dfrac{\sqrt{25x^2}}{\sqrt{3y}}$
6. $\dfrac{3}{\sqrt{2}} + \dfrac{1}{\sqrt{2}}$

7. $\sqrt{147} - \sqrt{108}$
8. $\sqrt{3}\sqrt{12} + \sqrt{5}\sqrt{20}$
9. $\sqrt{2}\sqrt{10} - \sqrt{15}\sqrt{3}$

10. $\dfrac{\sqrt{2}}{\sqrt{50}} - \dfrac{\sqrt{3}}{\sqrt{75}}$
11. $\dfrac{\sqrt{98}}{\sqrt{2}} + \dfrac{\sqrt{125}}{\sqrt{5}}$
12. $\dfrac{\sqrt{3}}{\sqrt{2}} - \dfrac{\sqrt{5}}{\sqrt{2}}$

13. $\sqrt{3x}\sqrt{6x} + \sqrt{5x}\sqrt{10x}$
14. $\sqrt{x}\sqrt{y} + \dfrac{\sqrt{xy^2}}{\sqrt{y}}$
15. $\sqrt{5x} - \dfrac{\sqrt{x}}{\sqrt{5}}$

16. $6\sqrt{2}\sqrt{6} + 2\sqrt{5}\sqrt{15} - 3\sqrt{12}\sqrt{25}$
17. $\sqrt{x^3}\sqrt{y} - 3x\sqrt{x}\sqrt{y} + \dfrac{4x\sqrt{xy^2}}{\sqrt{y}}$

Multiply and simplify.

18. $\sqrt{3}(\sqrt{3} + \sqrt{2})$
19. $\sqrt{5}(\sqrt{125} - \sqrt{45})$

20. $\sqrt{2}(\sqrt{27} - \sqrt{12})$

21. $\sqrt{7}(\sqrt{5} - \sqrt{125})$

22. $\sqrt{5}(\sqrt{5} + 1)$

23. $\sqrt{6}(\sqrt{2} - 1)$

24. $\sqrt{3}\left(\sqrt{3} - \dfrac{1}{\sqrt{3}}\right)$

25. $\sqrt{7}\left(\dfrac{1}{\sqrt{7}} + \sqrt{7}\right)$

26. $\sqrt{15}\left(\dfrac{\sqrt{3}}{\sqrt{5}} + \dfrac{1}{\sqrt{15}}\right)$

27. $\sqrt{8}\left(\sqrt{2} - \dfrac{1}{\sqrt{8}}\right)$

28. $\sqrt{2}\left(\sqrt{6} + \dfrac{\sqrt{2}}{\sqrt{3}}\right)$

29. $\sqrt{5}\left(\sqrt{10} - \dfrac{\sqrt{5}}{\sqrt{2}}\right)$

30. $(\sqrt{5} + 1)(\sqrt{5} + 1)$

31. $(\sqrt{5} + 1)(\sqrt{5} - 1)$

32. $(\sqrt{2} - \sqrt{5})(\sqrt{2} - \sqrt{5})$

33. $(\sqrt{2} - \sqrt{5})(\sqrt{2} + \sqrt{5})$

34. $(2\sqrt{3} + 1)(\sqrt{3} + 1)$

35. $(\sqrt{2} - 2)(\sqrt{2} + 1)$

36. $(2\sqrt{5} + \sqrt{7})(2\sqrt{5} - \sqrt{7})$

37. $(3\sqrt{3} - \sqrt{2})(2\sqrt{3} + \sqrt{2})$

Rationalize the denominator.

38. $\dfrac{1}{\sqrt{2} - 1}$

39. $\dfrac{2}{\sqrt{5} - \sqrt{3}}$

40. $\dfrac{2}{\sqrt{7} + \sqrt{5}}$

41. $\dfrac{3}{\sqrt{5} - 2}$

42. $\dfrac{6}{\sqrt{5} + 2}$

43. $\dfrac{-8}{\sqrt{3} - \sqrt{7}}$

44. $\dfrac{2\sqrt{3}}{\sqrt{3} - 1}$

45. $\dfrac{\sqrt{2}}{\sqrt{2} + 1}$

46. $\dfrac{2\sqrt{5}}{\sqrt{5} - \sqrt{3}}$

47. $\dfrac{x - 1}{\sqrt{x} - 1}$

For Review

Add or subtract.

48. $2\sqrt{75} - 6\sqrt{48} + 9\sqrt{12}$

49. $6\sqrt{20x} - 2\sqrt{45x} - \sqrt{5x}$

50. An economist might use the formula $E = 3\sqrt{c} + \sqrt{e}$, relating the three quantities E, c, and e. Find E when $c = 98$ and $e = 50$.

ANSWERS: **1.** $-\sqrt{5}$ **2.** 8 **3.** $6x$ **4.** $3y$ **5.** $\dfrac{5x\sqrt{3y}}{3y}$ **6.** $2\sqrt{2}$ **7.** $\sqrt{3}$ **8.** 16 **9.** $-\sqrt{5}$ **10.** 0 **11.** 12 **12.** $\dfrac{\sqrt{6} - \sqrt{10}}{2}$ **13.** $8x\sqrt{2}$ **14.** $2\sqrt{xy}$ **15.** $\dfrac{4\sqrt{5x}}{5}$ **16.** $-8\sqrt{3}$ **17.** $2x\sqrt{xy}$ **18.** $3 + \sqrt{6}$ **19.** 10 **20.** $\sqrt{6}$ **21.** $-4\sqrt{35}$ **22.** $5 + \sqrt{5}$ **23.** $2\sqrt{3} - \sqrt{6}$ **24.** 2 **25.** 8 **26.** 4 **27.** 3 **28.** $\dfrac{8\sqrt{3}}{3}$ **29.** $\dfrac{5\sqrt{2}}{2}$ **30.** $6 + 2\sqrt{5}$ **31.** 4 **32.** $7 - 2\sqrt{10}$ **33.** -3 **34.** $7 + 3\sqrt{3}$ **35.** $-\sqrt{2}$ **36.** 13 **37.** $16 + \sqrt{6}$ **38.** $\sqrt{2} + 1$ **39.** $\sqrt{5} + \sqrt{3}$ **40.** $\sqrt{7} - \sqrt{5}$ **41.** $3\sqrt{5} + 6$ **42.** $6\sqrt{5} - 12$ **43.** $2\sqrt{3} + 2\sqrt{7}$ **44.** $3 + \sqrt{3}$ **45.** $2 - \sqrt{2}$ **46.** $5 + \sqrt{15}$ **47.** $\sqrt{x} + 1$ **48.** $4\sqrt{3}$ **49.** $5\sqrt{5x}$ **50.** $26\sqrt{2}$

9.5 EXERCISES B

Refer to the summary of techniques and simplify.

1. $\sqrt{98} + \sqrt{50}$

2. $\sqrt{5y}\,\sqrt{75y}$

3. $\dfrac{\sqrt{108}}{\sqrt{3}}$

4. $\dfrac{\sqrt{50x^2y}}{\sqrt{2y}}$

5. $\dfrac{\sqrt{4y^2}}{\sqrt{3x}}$

6. $\dfrac{5}{\sqrt{3}} + \dfrac{1}{\sqrt{3}}$

7. $2\sqrt{175} - 3\sqrt{28}$

8. $\sqrt{2}\,\sqrt{8} + \sqrt{3}\,\sqrt{27}$

9. $\sqrt{5}\,\sqrt{10} - \sqrt{6}\,\sqrt{3}$

10. $\dfrac{\sqrt{3}}{\sqrt{75}} + \dfrac{\sqrt{5}}{\sqrt{125}}$

11. $\dfrac{\sqrt{108}}{\sqrt{3}} + \dfrac{\sqrt{147}}{\sqrt{3}}$

12. $\dfrac{\sqrt{2}}{\sqrt{5}} - \dfrac{\sqrt{3}}{\sqrt{5}}$

13. $\sqrt{2x}\,\sqrt{6x} + \sqrt{15x}\,\sqrt{5x}$

14. $\dfrac{\sqrt{x^2y}}{\sqrt{x}} - \sqrt{x}\,\sqrt{y}$

15. $\dfrac{\sqrt{x}}{\sqrt{7}} + \sqrt{7x}$

16. $3\sqrt{5}\sqrt{15} + 5\sqrt{27} - 6\sqrt{6}\sqrt{8}$

17. $\dfrac{3y\sqrt{x^2y}}{\sqrt{y}} + 3x\sqrt{y^2} - \dfrac{5x\sqrt{y^3}}{\sqrt{y}}$

Multiply and simplify.

18. $\sqrt{2}(\sqrt{3} + \sqrt{2})$
19. $\sqrt{3}(\sqrt{27} - \sqrt{12})$
20. $\sqrt{5}(\sqrt{7} - \sqrt{28})$
21. $\sqrt{5}(\sqrt{12} - \sqrt{48})$
22. $\sqrt{7}(1 - \sqrt{7})$
23. $\sqrt{10}(\sqrt{5} + 1)$
24. $\sqrt{5}\left(\sqrt{5} - \dfrac{1}{\sqrt{5}}\right)$
25. $\sqrt{11}\left(\sqrt{11} + \dfrac{1}{\sqrt{11}}\right)$
26. $\sqrt{18}\left(\sqrt{2} - \dfrac{1}{\sqrt{18}}\right)$
27. $\sqrt{21}\left(\dfrac{\sqrt{3}}{\sqrt{7}} + \dfrac{1}{\sqrt{21}}\right)$
28. $\sqrt{3}\left(\sqrt{6} + \dfrac{\sqrt{3}}{\sqrt{2}}\right)$
29. $\sqrt{2}\left(\sqrt{10} - \dfrac{\sqrt{2}}{\sqrt{5}}\right)$
30. $(\sqrt{7} + 1)(\sqrt{7} + 1)$
31. $(\sqrt{7} + 1)(\sqrt{7} - 1)$
32. $(\sqrt{3} - \sqrt{7})(\sqrt{3} - \sqrt{7})$
33. $(\sqrt{3} - \sqrt{7})(\sqrt{3} + \sqrt{7})$
34. $(3\sqrt{5} + 1)(\sqrt{5} + 1)$
35. $(\sqrt{3} - 3)(\sqrt{3} + 1)$
36. $(2\sqrt{2} - \sqrt{3})(3\sqrt{2} + \sqrt{3})$
37. $(2\sqrt{3} + \sqrt{5})(2\sqrt{3} - \sqrt{5})$

Rationalize the denominator.

38. $\dfrac{2}{\sqrt{3} - 1}$
39. $\dfrac{3}{\sqrt{5} - \sqrt{2}}$
40. $\dfrac{4}{\sqrt{7} + \sqrt{3}}$
41. $\dfrac{-2}{\sqrt{7} - 3}$
42. $\dfrac{6}{\sqrt{5} - \sqrt{7}}$
43. $\dfrac{9}{\sqrt{7} + 2}$
44. $\dfrac{2\sqrt{3}}{\sqrt{3} + 1}$
45. $\dfrac{4\sqrt{5}}{\sqrt{5} - 1}$
46. $\dfrac{3\sqrt{5}}{\sqrt{5} - \sqrt{2}}$
47. $\dfrac{1 - x}{1 - \sqrt{x}}$

For Review

Add or subtract.

48. $5\sqrt{125} + 3\sqrt{5} - 8\sqrt{45}$
49. $7\sqrt{12x} - 3\sqrt{48x} - 2\sqrt{75x}$

50. A meteorologist may use the formula $W = 5\sqrt{f} - \sqrt{g}$, relating the three quantities $W, f,$ and g. Find W when $f = 72$ and $g = 8$.

9.5 EXERCISES C

Multiply and simplify.

1. $(\sqrt{ax} - \sqrt{by})(\sqrt{ax} + \sqrt{by})$
2. $(\sqrt{x} - \sqrt{y})(x + \sqrt{xy} + y)$
   [Answer: $x\sqrt{x} - y\sqrt{y}$]

Rationalize the denominator and simplify.

3. $\dfrac{x\sqrt{x} - y\sqrt{x}}{\sqrt{x} - \sqrt{y}}$
4. $\dfrac{x^2}{\sqrt{4 - x^2}} + \sqrt{4 - x^2}$

$\left[\text{Answer: } \dfrac{4\sqrt{4 - x^2}}{4 - x^2}\right]$

9.6 SOLVING RADICAL EQUATIONS

STUDENT GUIDEPOSTS

1. Rule of squaring
2. Solving equations involving radicals

To solve equations containing one or more radical expressions, use the following rule.

Rule of squaring

If $a = b$, then $a^2 = b^2$.

For example, if $\sqrt{x} = 5$, then the rule of squaring gives $(\sqrt{x})^2 = 5^2$ or $x = 25$.

When this rule is applied to an equation involving a variable, the resulting equation may have more solutions than the original. For example, 2 is the only solution to the equation $x = 2$, but the equation formed by squaring both sides, $x^2 = 4$, has two solutions, 2 and -2. Thus, when the rule of squaring is used, any possible solutions must always be checked in the original equation.

To solve an equation involving radicals

1. If only one radical is present, isolate this radical on one side of the equation and proceed to 3.
2. If two radicals are present, isolate one of the radicals on one side of the equation.
3. Square both sides to obtain an equation without the isolated radical expression(s).
4. Solve the resulting equation.
5. Check all possible solutions in the original equation.

EXAMPLE 1

Solve.

$\sqrt{3x} - 5 = 7$

$\sqrt{3x} = 12$ Add 5 to both sides to isolate radical on the left

$(\sqrt{3x})^2 = (12)^2$ Use the rule of squaring

$3x = 144$ Squaring

$x = 48$ Divide both sides by 3

Check: $\sqrt{3(48)} - 5 \stackrel{?}{=} 7$

$\sqrt{144} - 5 \stackrel{?}{=} 7$

$12 - 5 \stackrel{?}{=} 7$

$7 = 7$

The solution is 48.

Practice Exercise 1

Solve. $\sqrt{5y} + 2 = 12$

Answer: 20

EXAMPLE 2

Solve.

$$\sqrt{x+3} + 4 = 11$$
$$\sqrt{x+3} = 11 - 4 \quad \text{Isolate the radical}$$
$$\sqrt{x+3} = 7$$
$$(\sqrt{x+3})^2 = 7^2 \quad \text{Rule of squaring}$$
$$x + 3 = 49$$
$$x = 46$$

Check: $\sqrt{46+3} + 4 \stackrel{?}{=} 11$

$$\sqrt{49} + 4 \stackrel{?}{=} 11$$
$$7 + 4 \stackrel{?}{=} 11$$
$$11 = 11$$

The solution is 46. ◀◀

Practice Exercise 2

Solve. $\sqrt{z-2} - 1 = 3$

Answer: 18

EXAMPLE 3

Solve.

$$3\sqrt{2x-5} - \sqrt{x+23} = 0$$
$$3\sqrt{2x-5} = \sqrt{x+23} \quad \text{Isolate the radicals}$$
$$(3\sqrt{2x-5})^2 = (\sqrt{x+23})^2$$
$$9(2x-5) = x + 23 \quad \text{Be careful to square the 3 on the left side}$$
$$18x - 45 = x + 23$$
$$17x = 68$$
$$x = 4$$

Check: $3\sqrt{2 \cdot 4 - 5} - \sqrt{4+23} \stackrel{?}{=} 0$

$$3\sqrt{3} - \sqrt{27} \stackrel{?}{=} 0$$
$$3\sqrt{3} - 3\sqrt{3} = 0$$

The solution is 4. ◀◀

Practice Exercise 3

Solve. $2\sqrt{3y+1} - \sqrt{y+15} = 0$

Answer: 1

EXAMPLE 4

Solve.

$$\sqrt{x^2 - 5} - x + 5 = 0$$
$$\sqrt{x^2 - 5} = x - 5 \quad \text{Isolate the radical}$$
$$(\sqrt{x^2 - 5})^2 = (x-5)^2 \quad \text{Do } not \text{ square the right side as } x^2 - 25$$
$$x^2 - 5 = x^2 - 10x + 25$$
$$-5 = -10x + 25 \quad \text{Subtract } x^2 \text{ from both sides}$$
$$-30 = -10x$$
$$x = \frac{-30}{-10} = 3$$

Practice Exercise 4

Solve. $\sqrt{y^2 + 9} - y + 1 = 0$

Check: $\sqrt{3^2 - 5} - 3 + 5 \stackrel{?}{=} 0$

$\sqrt{9 - 5} - 3 + 5 \stackrel{?}{=} 0$

$\sqrt{4} - 3 + 5 \stackrel{?}{=} 0$

$2 - 3 + 5 \stackrel{?}{=} 0 \qquad \sqrt{4}$ is 2 not -2

$4 \neq 0 \qquad$ 3 does not check

The equation has no solution. ◀◀

Answer: no solution (-4 does not check)

▶▶ CAUTION ◀◀ Remember that the radical represents the principal (nonnegative) root. Thus, in Example 4, do not be tempted to replace $\sqrt{4}$ with -2 in the check. ◀

Sometimes the equation that results when the radical is eliminated must be solved by factoring and using the zero-product rule. This is illustrated in Example 5.

EXAMPLE 5

Solve.

$\sqrt{2x + 2} - x + 3 = 0$

$\sqrt{2x + 2} = x - 3 \qquad$ Isolate the radical

$(\sqrt{2x + 2})^2 = (x - 3)^2$

$2x + 2 = x^2 - 6x + 9 \qquad (a - b)^2 = a^2 - 2ab + b^2$

$0 = x^2 - 8x + 7 \qquad$ Subtract $2x$ and 2

$0 = (x - 1)(x - 7) \qquad$ Factor

$x - 1 = 0 \quad$ or $\quad x - 7 = 0 \qquad$ Zero-product rule

$x = 1 \qquad\qquad x = 7$

Check: $\sqrt{2(1) + 2} - 1 + 3 \stackrel{?}{=} 0 \qquad \sqrt{2(7) + 2} - 7 + 3 \stackrel{?}{=} 0$

$\sqrt{4} - 1 + 3 \stackrel{?}{=} 0 \qquad \sqrt{16} - 7 + 3 \stackrel{?}{=} 0$

$2 - 1 + 3 \stackrel{?}{=} 0 \qquad 4 - 7 + 3 \stackrel{?}{=} 0$

$4 \neq 0 \qquad\qquad 0 = 0$

1 does not check $\qquad$ 7 does check

The only solution is 7. ◀◀

Practice Exercise 5

Solve. $\sqrt{17 - 4x} - x - 1 = 0$

Answer: 2

9.6 EXERCISES A

Solve the following equations.

1. $\sqrt{x} = 4$

2. $3\sqrt{2x} - 9 = 0$

3. $\sqrt{4y} - 6 = 2$

4. $\sqrt{y+2} = 6$

5. $\sqrt{4x-3} = 5$

6. $3\sqrt{a-3} + 6 = 0$

7. $4\sqrt{2a-1} - 2 = 0$

8. $3\sqrt{x} = \sqrt{x+16}$

9. $\sqrt{y+3} = \sqrt{y+5}$

10. $3\sqrt{y+3} = \sqrt{y+35}$

11. $\sqrt{3a+2} - \sqrt{a+8} = 0$

12. $\sqrt{5a-3} - \sqrt{2a+1} = 0$

13. $3\sqrt{x-1} - \sqrt{x+31} = 0$

14. $\sqrt{2x+1} = \sqrt{x+10}$

15. $3\sqrt{y+7} = 4\sqrt{y}$

16. $2\sqrt{2x+5} - 3\sqrt{3x-2} = 0$

17. $\sqrt{a^2-5} + a - 5 = 0$

18. $\sqrt{a^2+2} - a - 2 = 0$

19. $-\sqrt{x^2-12} + x + 6 = 0$

20. $\sqrt{x^2+3} + x = 6$

21. $\sqrt{x^2 + 16} - x + 8 = 0$

22. $\sqrt{x^2 - 8} + 3 = x$

23. $\sqrt{x + 2} = x$

24. $\sqrt{x + 1} = 1 - x$

25. $\sqrt{x - 3} - x + 5 = 0$

26. $\sqrt{x + 4} - x + 2 = 0$

27. $\sqrt{x + 15} - x - 3 = 0$

28. $\sqrt{x + 10} + x - 2 = 0$

For Review

Simplify.

29. $\sqrt{5}(\sqrt{15} - \sqrt{12})$

30. $(\sqrt{3} + \sqrt{5})(2\sqrt{3} - \sqrt{5})$

31. $\dfrac{4}{\sqrt{11} - \sqrt{7}}$

32. $\dfrac{3\sqrt{5}}{\sqrt{5} - \sqrt{2}}$

ANSWERS: **1.** 16 **2.** $\dfrac{9}{2}$ **3.** 16 **4.** 34 **5.** 7 **6.** no solution **7.** $\dfrac{5}{8}$ **8.** 2 **9.** no solution **10.** 1 **11.** 3
12. $\dfrac{4}{3}$ **13.** 5 **14.** 9 **15.** 9 **16.** 2 **17.** 3 **18.** $-\dfrac{1}{2}$ **19.** -4 **20.** $\dfrac{11}{4}$ **21.** no solution **22.** $\dfrac{17}{6}$ **23.** 2
24. 0 **25.** 7 **26.** 5 **27.** 1 **28.** -1 **29.** $5\sqrt{3} - 2\sqrt{15}$ **30.** $1 + \sqrt{15}$ **31.** $\sqrt{11} + \sqrt{7}$ **32.** $5 + \sqrt{10}$

9.6 EXERCISES B

Solve the following equations.

1. $\sqrt{x} = 9$
2. $5\sqrt{3x} - 10 = 0$
3. $\sqrt{9y} - 6 = 9$
4. $\sqrt{y + 6} = 2$
5. $\sqrt{8x + 2} = 2$
6. $5\sqrt{a - 2} + 10 = 0$
7. $3\sqrt{3a + 1} - 6 = 0$
8. $5\sqrt{x} = \sqrt{x + 3}$
9. $\sqrt{2y + 1} = \sqrt{2y + 3}$
10. $3\sqrt{y - 2} = \sqrt{y + 12}$
11. $\sqrt{5a + 3} - \sqrt{a + 4} = 0$
12. $\sqrt{7a - 10} - \sqrt{4a - 5} = 0$
13. $2\sqrt{x - 5} - \sqrt{x + 22} = 0$
14. $\sqrt{3x + 5} = \sqrt{x + 15}$
15. $3\sqrt{y + 2} = 5\sqrt{y}$
16. $3\sqrt{2x - 4} - 2\sqrt{3x + 3} = 0$
17. $\sqrt{a^2 + 5} - a + 5 = 0$
18. $\sqrt{a^2 - 8} - a - 4 = 0$
19. $-\sqrt{x^2 - 20} + x + 10 = 0$
20. $\sqrt{x^2 + 7} - x + 7 = 0$
21. $\sqrt{x^2 - 15} + x = 5$
22. $\sqrt{x^2 + 3} - 3 = x$
23. $\sqrt{x + 6} = x$
24. $\sqrt{x + 1} = x - 1$
25. $\sqrt{x + 3} - x + 3 = 0$
26. $\sqrt{x + 6} - x - 4 = 0$
27. $\sqrt{x + 21} + x + 1 = 0$
28. $\sqrt{x + 5} + x - 1 = 0$

For Review

Simplify.

29. $\sqrt{7}(\sqrt{14} - \sqrt{21})$
30. $(\sqrt{5} - \sqrt{3})(\sqrt{5} - 2\sqrt{3})$
31. $\dfrac{4}{\sqrt{11} + \sqrt{7}}$
32. $\dfrac{4\sqrt{7}}{\sqrt{7} - \sqrt{5}}$

9.6 EXERCISES C

Solve.

1. $\sqrt{x + 5} - \sqrt{x - 3} = 2$
   [*Hint:* You will need to square twice.]
2. $\sqrt{x - 3} + \sqrt{x + 2} = 5$

9.7 APPLICATIONS USING RADICALS

STUDENT GUIDEPOSTS

1 Right triangles
2 Pythagorean theorem

There are several types of problems that require a radical for their description or solution. We first use our knowledge of radicals to solve right triangle problems.

1 Remember that a **right triangle** is a triangle with a 90° (right) angle. The sides of the triangle that form the 90° angle are called the **legs** and the side opposite the 90° angle is called the **hypotenuse**. We will always label the legs a and b and the hypotenuse will be labeled c, as in Figure 9.1. The next well-known rule is named after the Greek mathematician Pythagoras.

9.7 APPLICATIONS USING RADICALS

Pythagorean theorem

In a right triangle with legs a and b and hypotenuse c,
$$a^2 + b^2 = c^2.$$

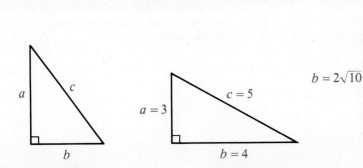

Figure 9.1   Figure 9.2   Figure 9.3

EXAMPLE 1

If the legs of a right triangle are 3 and 4, find the hypotenuse. See Figure 9.2.

$$\begin{aligned} c^2 &= a^2 + b^2 & \text{Pythagorean theorem} \\ &= 3^2 + 4^2 & a = 3 \text{ and } b = 4 \\ &= 9 + 16 \\ &= 25 \end{aligned}$$

Thus, we must solve $c^2 = 25$.

$$\begin{aligned} c^2 &= 25 \\ c^2 - 25 &= 0 & \text{Subtract 25} \\ c^2 - 5^2 &= 0 & 25 = 5^2 \\ (c + 5)(c - 5) &= 0 & \text{Factor using } a^2 - b^2 \\ & & = (a + b)(a - b) \\ c + 5 = 0 \quad &\text{or} \quad c - 5 = 0 \\ c = -5 \quad & \quad c = 5 \end{aligned}$$

Since we want the length of the hypotenuse of a triangle, the answer -5 is discarded. Thus, $c = 5$. Notice that if we take the principal square root of both sides, we obtain the desired answer.

$$\begin{aligned} c^2 &= 25 \\ c &= \sqrt{25} = 5. \end{aligned}$$ ◀

Practice Exercise 1

If the legs of a right triangle are 9 and 12, find the hypotenuse.

Answer: 15

EXAMPLE 2

If $c = 7$ and $a = 3$, find b. See Figure 9.3.

$$\begin{aligned} a^2 + b^2 &= c^2 & \text{Pythagorean theorem} \\ 3^2 + b^2 &= 7^2 & a = 3 \text{ and } c = 7 \\ 9 + b^2 &= 49 \\ b^2 &= 40 \end{aligned}$$

Practice Exercise 2

If $c = 9$ and $b = 3$, find a.

RADICALS

$$b = \sqrt{40}$$ Take principal square root
$$b = \sqrt{4 \cdot 10} = 2\sqrt{10}$$ ◂◂

Answer: $6\sqrt{2}$

EXAMPLE 3

If the base of a 20 foot ladder is 8 feet from the base of a wall, how far is the top of the ladder from the ground? See Figure 9.4.

$$a^2 + b^2 = c^2$$
$$a^2 + 8^2 = 20^2$$
$$a^2 + 64 = 400$$
$$a^2 = 400 - 64$$
$$a = \sqrt{336}$$
$$= \sqrt{16 \cdot 21} = 4\sqrt{21}$$

Practice Exercise 3

A guy wire from the top of a pole to the ground measures 45 feet. If the pole is 30 feet in length, how far is the wire anchored from the base of the pole?

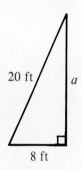

Figure 9.4

The top of the ladder is $4\sqrt{21}$ ft (approximately 18.3 ft) from the ground. ◂◂

Answer: $15\sqrt{5} \approx 33.5$ ft

We turn now to other applications that use square roots. The first is a business application.

EXAMPLE 4

A manufacturer has found that his total cost per day is given by the equation

$$c = 10\sqrt{n + 25},$$

where c is the total cost and n is the number of widgets produced.

(A) Find the cost when no widgets are produced.

$$c = 10\sqrt{n + 25} \quad \text{Cost equation}$$
$$c = 10\sqrt{0 + 25} \quad n = 0$$
$$= 10\sqrt{25}$$
$$= 10 \cdot 5 = 50$$

Thus, the cost if the manufacturer produces nothing (called overhead) is $50 per day.

(B) Find the cost when 600 widgets are produced.

Practice Exercise 4

A manufacturer of puzzles has found that his daily expenses for production can be estimated by $e = 6\sqrt{n + 49}$, where e represents total expenses and n is the number of puzzles produced.

(A) Find the total expenses when no puzzles are produced.

(B) Find the total expense of producing 95 puzzles.

$$c = 10\sqrt{n+25} \quad \text{Cost equation}$$
$$c = 10\sqrt{600+25} \quad n = 600$$
$$= 10\sqrt{625}$$
$$= 10(25) = 250 \quad (25)^2 = 625$$

The cost is $250 per day when 600 widgets are produced.

(C) How many widgets are made when the cost is $120 per day?

$$c = 10\sqrt{n+25} \quad \text{Cost equation}$$
$$120 = 10\sqrt{n+25} \quad c = 120, \text{ solve for } n$$
$$12 = \sqrt{n+25} \quad \text{Simplify first}$$
$$(12)^2 = (\sqrt{n+25})^2 \quad \text{Square both sides}$$
$$144 = n + 25$$
$$119 = n$$

Thus, 119 widgets were made on the day that the cost was $120. ◀◀

(C) How many puzzles were made on a day when the expenses were $78?

Answers: **(A)** $42 **(B)** $72 **(C)** 120

EXAMPLE 5

When an airplane is flying above the earth, an equation that gives the approximate distance to the horizon in terms of the altitude of the plane is

$$d = \sqrt{8000h},$$

where d is the distance to the horizon in miles and h is the altitude (distance above the earth) in miles.

(A) Determine the distance to the horizon when viewed from a plane at an altitude of 5 miles.

$$d = \sqrt{8000h} \quad \text{Equation relating } d \text{ and } h$$
$$d = \sqrt{(8000)(5)} \quad h = 5 \text{ miles}$$
$$d = \sqrt{40000}$$
$$= \sqrt{(200)^2} \quad (200)^2 = 40{,}000$$
$$= 200$$

Thus, the horizon is approximately 200 miles away.

(B) Determine the altitude of a plane when the horizon is 50 miles away.

$$d = \sqrt{8000h} \quad \text{Equation relating } d \text{ and } h$$
$$50 = \sqrt{8000h} \quad d = 50 \text{ miles, solve for } h$$
$$(50)^2 = (\sqrt{8000h})^2 \quad \text{Square both sides}$$
$$2500 = 8000h$$
$$\frac{2500}{8000} = h$$
$$\frac{5}{16} = h$$

Thus, the plane is about $\frac{5}{16}$ of a mile (about 1650 ft) above the earth. ◀◀

Practice Exercise 5

Use the equation $d = \sqrt{8000h}$ in the following:

(A) Determine the distance to the horizon when viewed from a balloon at an altitude of 1 mile.

(B) Determine the altitude of a balloon when the horizon is 150 miles away.

Answers: **(A)** $40\sqrt{5} \approx 89.4$ miles **(B)** Approximately 2.8 miles.

9.7 EXERCISES A

Find the unknown leg or hypotenuse.

1. $a = 12$, $b = 5$
2. $a = 7$, $c = 10$
3. $b = 5$, $c = 9$

4. $a = 6$, $b = 8$
5. $a = 1$, $c = 5$
6. $b = 10$, $c = 20$

Solve.

7. How long would a ladder need to be to reach the top of a 10 meter building if the base of the ladder is 4 meters from the building?

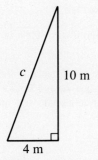

8. The diagonal of a square is 12 centimeters. Find the length of the sides. [*Hint:* $a^2 + a^2 = 12^2$.]

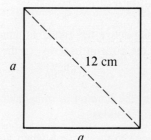

9. Find the length of the diagonal of a rectangle if the sides are 20 inches and 30 inches.

10. How long must a wire be to reach from the top of a 40 foot telephone pole to a point on the ground 30 feet from the base of the pole?

11. Three times the square root of a number is 6. Find the number.

12. Twice the square root of a number is 6 less than 12. Find the number.

13. The square root of 5 more than a number is twice the square root of the number. Find the number.

14. Twice the square root of one more than a number is the same as the square root of 13 more than the number. Find the number.

A manufacturer finds that the total cost per day of producing appliances is given by the equation $c = 100\sqrt{n + 36}$ where c is the total cost and n is the number produced. Use this information for Exercises 15–18.

15. What is c when no appliances are produced (overhead cost)?

16. What is the cost when 64 appliances are produced?

17. What is the cost when 39 appliances are produced?

18. How many appliances were made on a day when the cost was $2000?

A retailer finds that the total cost per day of selling sportcoats is given by the equation $c = 10\sqrt{n} + 100$, where c is the total cost and n is the number sold. Use this information in Exercises 19–22.

19. Find c when no sportcoats are sold (overhead cost).

20. Find the cost when 64 sportcoats are sold.

21. Find the number of coats sold on a day when the cost was $240.

22. Find the number of coats sold on a day when the cost was $50.

The distance in miles to the horizon from a point h miles above the earth is given by $d = \sqrt{8000h}$. Use this information in Exercises 23–26.

23. Find d when the altitude is 20 miles.

24. Find the distance to the horizon when the altitude is $\frac{1}{2}$ mile.

25. Find the altitude when the distance to the horizon is 120 miles.

26. Find the altitude when the distance to the horizon is 20 miles.

For Review

Solve.

27. $\sqrt{3x + 1} = 5$

28. $\sqrt{4x - 5} - \sqrt{2x + 9} = 0$

29. $2\sqrt{y^2 + 1} - 2y - 1 = 0$

30. $2\sqrt{3y + 1} - \sqrt{15y - 11} = 0$

ANSWERS: **1.** 13 **2.** $\sqrt{51}$ **3.** $2\sqrt{14}$ **4.** 10 **5.** $2\sqrt{6}$ **6.** $10\sqrt{3}$ **7.** $2\sqrt{29}$ m (approximately 10.8 m) **8.** $6\sqrt{2}$ cm (approximately 8.5 cm) **9.** $10\sqrt{13}$ in (approximately 36.1 in) **10.** 50 ft **11.** 4 **12.** 9 **13.** $\frac{5}{3}$ **14.** 3 **15.** $600 **16.** $1000 **17.** $500\sqrt{3}$ (approximately $866) **18.** 364 **19.** $100 **20.** $180 **21.** 196 **22.** no solution **23.** 400 mi **24.** $20\sqrt{10}$ mi (approximately 63.2 mi) **25.** 1.8 mi **26.** 0.05 mi (264 ft) **27.** 8 **28.** 7 **29.** $\frac{3}{4}$ **30.** 5

9.7 EXERCISES B

Find the unknown leg or hypotenuse.

1. $a = 4$, $b = 4$
2. $a = 5$, $c = 8$
3. $b = 14$, $c = 20$
4. $a = 9$, $b = 12$
5. $a = 3$, $c = 7$
6. $b = 20$, $c = 30$

Solve.

7. A ladder 8 meters long is placed against a building. If the base is 3 meters from the building, how high up the side of the building is the top of the ladder?

8. The diagonal of a square is 10 centimeters. Find the length of the sides.

9. Find the length of the diagonal of a rectangular pasture if the sides are 40 yards and 60 yards.

10. A wire 50 feet long is attached to the top of a tower and to the ground 20 feet from the base of the tower. How tall is the tower?

11. Four times the square root of a number is 20. Find the number.

12. Twice the square root of a number is 5 more than 15. Find the number.

13. The square root of 8 more than a number is 3 times the square root of the number. Find the number.

14. Twice the square root of 4 less than a number is the same as the square root of 1 less than the number. Find the number.

A manufacturer finds that the total cost per day of producing dresses is given by the equation $c = 20\sqrt{n} + 16$, where c is the total cost and n is the number produced. Use this information in Exercises 15–18.

15. What is c when no dresses are produced (overhead cost)?

16. What is the cost when 84 dresses are produced?

17. What is the cost when 24 dresses are produced?

18. How many dresses were made on a day when the cost was $360?

A retailer finds that the total cost per day of selling widgets is given by the equation $c = 50\sqrt{n} + 200$, where c is the total cost and n is the number sold. Use this information for Exercises 19–22.

19. Find c when no widgets are sold (overhead cost).

20. Find the cost when 81 widgets are sold.

21. Find the number of widgets sold on a day when the cost was $950.

22. Find the number of widgets sold on a day when the cost was $150.

The distance d in miles to the horizon from a point h miles above the earth is given by $d = 10\sqrt{80h}$. Use this information in Exercises 23–26.

23. Find d when the altitude is $\frac{1}{5}$ mile.

24. Find the distance to the horizon when the altitude is 10 miles.

25. Find the altitude when the distance to the horizon is 320 miles.

26. Find the altitude when the distance to the horizon is 8 miles.

For Review

Solve.

27. $\sqrt{2x - 3} = 3$

28. $\sqrt{3x + 2} - \sqrt{2x + 3} = 0$

29. $\sqrt{y^2 + 2} - y - 1 = 0$

30. $\sqrt{2y + 1} + \sqrt{y + 1} = 0$

9.7 EXERCISES C

Solve.

1. An airplane flew 120 kilometers north from an airport and then 200 kilometers west. How far was the plane from the airport?

2. A scientist uses the formula $a = \sqrt{1 + \frac{b}{x}}$. Solve this equation for b and find b when $a = 3$ and $x = \frac{1}{2}$. [Answer: $b = x(a^2 - 1)$; 4]

CHAPTER 9 SUMMARY

Key Words and Phrases for Review

9.1 square root
     principal square root
     radical
     radicand
9.2 rationalize the denominator
9.4 like radicals
     unlike radicals

9.5 conjugate
9.7 right triangle
     legs
     hypotenuse
     Pythagorean theorem

Key Concepts

9.1 The radical is used to indicate the principal (nonnegative) square root of a number. Thus $\sqrt{9} = 3$, not -3.

9.2 1. If $a \geq 0$ and $b \geq 0$, then
9.3 $\sqrt{ab} = \sqrt{a} \sqrt{b}$.   Multiplication rule

  2. If $a \geq 0$ and $b > 0$, then $\sqrt{\frac{a}{b}} = \frac{\sqrt{a}}{\sqrt{b}}$.   Division rule

3. When removing a perfect square from under a radical, multiply all factors; do not add. For example,

$3\sqrt{4a} = 3\sqrt{4} \sqrt{a} = 3 \cdot 2\sqrt{a} = 6\sqrt{a}$, *not* $(3 + 2)\sqrt{a}$, which is $5\sqrt{a}$.

9.4 $\sqrt{a + b} \neq \sqrt{a} + \sqrt{b}$ and $\sqrt{a - b} \neq \sqrt{a} - \sqrt{b}$.

438 RADICALS

9.5 To rationalize a binomial denominator, multiply both numerator and denominator of the fraction by the conjugate of the denominator and simplify.

9.6 1. To solve a radical equation, isolate a radical on one side of the equation and square both sides.

2. Check all solutions to a radical equation in the *original* equation.

9.7 1. Pythagorean theorem: In a right triangle with legs a and b and hypotenuse c, $a^2 + b^2 = c^2$.

2. Read applied problems carefully before attempting to solve them. Drawing a sketch may be helpful.

Review Exercises

9.1 *Determine the number of square roots of each number.*

1. 81
2. −81
3. 0

Evaluate each square root. Assume all variables and expressions are nonnegative.

4. $\sqrt{16}$
5. $\sqrt{9^2}$
6. $\sqrt{(-9)^2}$

7. $\pm\sqrt{144}$
8. $\sqrt{\dfrac{25}{16}}$
9. $\sqrt{-16}$

10. $\sqrt{0}$
11. $\sqrt{4a^2}$
12. $\sqrt{x^2 + 2x + 1}$

9.2 *Simplify each of the radicals.*

13. $\sqrt{50}$
14. $5\sqrt{8}$
15. $\sqrt{3^4}$

16. $\sqrt{16x^3}$
17. $\sqrt{8y^2}$
18. $\sqrt{50x^2y^3}$

19. $\sqrt{\dfrac{32}{81}}$
20. $\sqrt{\dfrac{50}{8}}$
21. $\sqrt{\dfrac{16y^2}{x^2}}$

22. $\sqrt{\dfrac{x^2y^3}{81}}$

23. $\dfrac{6}{\sqrt{7}}$

24. $\sqrt{\dfrac{50x^2}{y}}$

9.3 *Multiply and simplify.*

25. $\sqrt{8}\,\sqrt{18}$

26. $\sqrt{7}\,\sqrt{7}$

27. $3\sqrt{5}\,\sqrt{7}$

28. $\sqrt{12y}\,\sqrt{3y}$

29. $\sqrt{3xy}\,\sqrt{2xy^2}$

30. $\sqrt{5x}\,\sqrt{5y}\,\sqrt{5xy}$

Divide and simplify.

31. $\dfrac{\sqrt{99}}{\sqrt{11}}$

32. $\dfrac{\sqrt{50}}{\sqrt{8}}$

33. $\dfrac{3\sqrt{35}}{\sqrt{20}}$

34. $\dfrac{\sqrt{25x^2}}{\sqrt{2x^2}}$

35. $\dfrac{\sqrt{18x^2y}}{\sqrt{8y}}$

36. $\dfrac{\sqrt{25x^2}}{\sqrt{3y}}$

9.4 *Add or subtract.*

37. $5\sqrt{7} + 12\sqrt{7}$

38. $2\sqrt{5} - 7\sqrt{5}$

39. $7\sqrt{5} - 3\sqrt{3}$

40. $3\sqrt{8} + 7\sqrt{8}$

41. $\sqrt{49} - \sqrt{144}$

42. $3\sqrt{16x} + 4\sqrt{36x}$

43. $\sqrt{50} - \sqrt{98} + \sqrt{32}$

44. $2\sqrt{28} - 5\sqrt{112} + 3\sqrt{63}$

45. $3\sqrt{12x} + 2\sqrt{27x} - \sqrt{48x}$

46. $5\sqrt{48x^2} - \sqrt{27x^2} - \sqrt{3x^2}$

47. $\sqrt{5} + \dfrac{1}{\sqrt{5}}$

48. $\sqrt{\dfrac{3}{2}} - \sqrt{6}$

9.5 *Perform the indicated operations and simplify.*

49. $\sqrt{5}\sqrt{125} + \sqrt{3}\sqrt{27}$

50. $\dfrac{\sqrt{5}}{\sqrt{2}} - \dfrac{\sqrt{3}}{\sqrt{2}}$

51. $\sqrt{5}(\sqrt{3} - \sqrt{5})$

52. $\sqrt{7}\left(\sqrt{7} - \dfrac{1}{\sqrt{7}}\right)$

53. $(\sqrt{7} + 1)(\sqrt{7} + 1)$

54. $(\sqrt{7} + 1)(\sqrt{7} - 1)$

55. $(\sqrt{2} + \sqrt{3})(2\sqrt{2} - \sqrt{3})$

56. $(\sqrt{5} + 3)(\sqrt{5} - 1)$

Rationalize the denominator.

57. $\dfrac{2}{\sqrt{3} + 1}$

58. $\dfrac{-4}{\sqrt{7} - \sqrt{5}}$

59. $\dfrac{\sqrt{3}}{\sqrt{3} - \sqrt{2}}$

60. $\dfrac{\sqrt{7}}{\sqrt{7} - 1}$

9.6 *Solve.*

61. $2\sqrt{x} = \sqrt{x+2}$

62. $\sqrt{7a-1} - \sqrt{5a+3} = 0$

63. $\sqrt{x^2 - 13} - x + 1 = 0$

64. $\sqrt{y+5} + y - 1 = 0$

9.7 *Find the unknown leg or hypotenuse.*

65. $a = 6$, $b = 12$

66. $a = 5$, $c = 13$

Solve.

67. Find the length of a diagonal of a rectangular table with sides 8 feet and 12 feet.

68. Twice the square root of 4 less than a number is equal to the square root of 8 more than the number. Find the number.

A manufacturer finds that the total cost per day of producing wabbles is given by the equation $c = 500\sqrt{n+144}$, where c is the total cost and n is the number produced. Use this information in Exercises 69–70.

69. Find the cost when 25 wabbles are produced.

70. How many wabbles were produced on a day when the cost was $10,000?

The distance d in miles to the horizon from a point h miles above the earth is given by $d = 10\sqrt{80h}$. Use this information in Exercises 71–72.

71. Find the distance to the horizon when the altitude is 0.05 miles.

72. Find the altitude when the distance to the horizon is 80 miles.

ANSWERS: **1.** 2   **2.** 0   **3.** 1   **4.** 4   **5.** 9   **6.** 9   **7.** ±12 (12 and −12)   **8.** $\dfrac{5}{4}$ **9.** meaningless   **10.** 0
11. $2a$ **12.** $x+1$ **13.** $5\sqrt{2}$ **14.** $10\sqrt{2}$ **15.** 9   **16.** $4x\sqrt{x}$ **17.** $2y\sqrt{2}$ **18.** $5xy\sqrt{2y}$ **19.** $\dfrac{4\sqrt{2}}{9}$ **20.** $\dfrac{5}{2}$
21. $\dfrac{4y}{x}$ **22.** $\dfrac{xy\sqrt{y}}{9}$ **23.** $\dfrac{6\sqrt{7}}{7}$ **24.** $\dfrac{5x\sqrt{2y}}{y}$ **25.** 12   **26.** 7   **27.** $3\sqrt{35}$ **28.** $6y$ **29.** $xy\sqrt{6y}$ **30.** $5xy\sqrt{5}$
31. 3   **32.** $\dfrac{5}{2}$ **33.** $\dfrac{3\sqrt{7}}{2}$ **34.** $\dfrac{5\sqrt{2}}{2}$ **35.** $\dfrac{3x}{2}$ **36.** $\dfrac{5x\sqrt{3y}}{3y}$ **37.** $17\sqrt{7}$ **38.** $-5\sqrt{5}$ **39.** $7\sqrt{5} - 3\sqrt{3}$
40. $20\sqrt{2}$ **41.** −5   **42.** $36\sqrt{x}$ **43.** $2\sqrt{2}$ **44.** $-7\sqrt{7}$ **45.** $8\sqrt{3x}$ **46.** $16x\sqrt{3}$ **47.** $\dfrac{6\sqrt{5}}{5}$ **48.** $-\dfrac{\sqrt{6}}{2}$
49. 34   **50.** $\dfrac{\sqrt{10}-\sqrt{6}}{2}$ **51.** $\sqrt{15}-5$ **52.** 6   **53.** $8+2\sqrt{7}$ **54.** 6   **55.** $1+\sqrt{6}$ **56.** $2+2\sqrt{5}$
57. $\sqrt{3}-1$ **58.** $-2\sqrt{7}-2\sqrt{5}$ **59.** $3+\sqrt{6}$ **60.** $\dfrac{7+\sqrt{7}}{6}$ **61.** $\dfrac{2}{3}$ **62.** 2   **63.** 7   **64.** −1   **65.** $6\sqrt{5}$
66. 12   **67.** $4\sqrt{13}$ ft (approximately 14.4 ft)   **68.** 8   **69.** $6500   **70.** 256   **71.** 20 mi   **72.** 0.8 mi

CHAPTER 9 TEST

1. How many square roots does 36 have?

1. _____

Evaluate each square root. Assume all variables and expressions are nonnegative.

2. $\sqrt{(-5)^2}$

2. _____

3. $\pm\sqrt{121}$

3. _____

4. $\sqrt{\dfrac{81}{16}}$

4. _____

5. $\sqrt{16x^2}$

5. _____

6. $3\sqrt{48}$

6. _____

7. $\sqrt{\dfrac{8x}{y^2}}$

7. _____

8. $\sqrt{\dfrac{98x^2}{y}}$

8. _____

Multiply or divide and simplify.

9. $2\sqrt{6}\,\sqrt{24}$

9. _____

10. $\dfrac{\sqrt{27x^3}}{\sqrt{3x}}$

10. _____

Perform the indicated operations and simplify.

11. $\sqrt{20} + \sqrt{80}$

11. _____

CHAPTER 9 TEST Continued

12. $3\sqrt{18} - 5\sqrt{50}$

12. _____

13. $6\sqrt{12x} + 3\sqrt{3x} - \sqrt{75x}$

13. _____

14. $\sqrt{11} + \dfrac{1}{\sqrt{11}}$

14. _____

15. $\sqrt{2}\sqrt{8} + \sqrt{3}\sqrt{12}$

15. _____

16. $(\sqrt{5} + \sqrt{3})(\sqrt{5} - \sqrt{3})$

16. _____

17. $(\sqrt{2} + \sqrt{7})(2\sqrt{2} - \sqrt{7})$

17. _____

Rationalize the denominator.

18. $\dfrac{\sqrt{5}}{\sqrt{5} - 1}$

18. _____

Solve.

19. $\sqrt{x + 6} = \sqrt{3x - 1}$

19. _____

20. $\sqrt{x^2 + 7} + x - 1 = 0$

20. _____

21. Find the length of the diagonal of a rectangle with sides 6 cm and 8 cm.

21. _____

22. Given $d = 10\sqrt{80h}$.

    (A) Find d when h is $\dfrac{1}{16}$.

    22(A) _____

    (B) Find h when d is 400.

    22(B) _____

10 QUADRATIC EQUATIONS

10.1 FACTORING AND TAKING ROOTS

STUDENT GUIDEPOSTS

1 Quadratic equation
2 General form
3 Solving by factoring
4 Solving by taking roots

Most of the equations we have solved have had the variable raised only to the first power. The only exception to this occurred in Section 7.5 where we briefly introduced the notion of a quadratic equation. An equation that can be expressed in the form

$$ax^2 + bx + c = 0,$$

1 where a, b, and c are constant numbers with $a \neq 0$, is called a **quadratic equation.** (If a were 0, the term in which the variable is squared would be missing
2 and the equation would not be quadratic.) We call $ax^2 + bx + c = 0$ the **general form** of a quadratic equation.

EXAMPLE 1

Write each equation in general form and identify each constant a, b, and c.

(A) $3x^2 + 2x - 6 = 0$ $(3x^2 + 2x + (-6) = 0)$

This equation is already in general form with $a = 3$, $b = 2$ and $c = -6$. Note that $c = -6$ and not 6. In the general form of an equation, all signs are $+$ so any negative sign goes with the constant a, b, or c.

(B) $2y^2 = 3y + 1$

Collect all terms on one side of the equation to obtain

$$2y^2 - 3y - 1 = 0$$

so that $a = 2$, $b = -3$, and $c = -1$.

Practice Exercise 1

Write each equation in general form and identify each constant a, b, and c.

(A) $5x^2 - 4x + 8 = 0$

(B) $6y^2 = 1 - y$

(C) $2 + x - x^2 = 0$

Rewrite the left member in descending order to obtain
$$-x^2 + x + 2 = 0$$
so that $a = -1$, $b = 1$, and $c = 2$. Note that the coefficient of x is 1 so that $b = 1$.

(D) $y^2 = 3y + 1$

Collect all terms on the left side to obtain
$$y^2 - 3y - 1 = 0$$
so that $a = 1$, $b = -3$, and $c = -1$. If we collect all terms on the right side we obtain
$$0 = -y^2 + 3y + 1$$
so that $a = -1$, $b = 3$, and $c = 1$. Both of these are correct since the equations are equivalent. However, it is common practice to write the equation so that the coefficient of the squared term is positive.

(E) $2x + 1 = 3 - 4x + 2$

Since there is no term in which the variable in squared, this is not a quadratic equation.

(F) $2y - 5y^2 = 3y^2 + 2y + 1$

Collect all terms to obtain
$$8y^2 + 1 = 0 \quad \text{or} \quad 8y^2 + 0 \cdot y + 1 = 0$$
so that $a = 8$, $b = 0$, and $c = 1$.

(G) $4 = 6x + 12x^2$

Collect all terms on one side to obtain
$$12x^2 + 6x - 4 = 0.$$
This is a general form of the equation. However, we should simplify by factoring out 2 and multiplying both sides by $\tfrac{1}{2}$. That is,
$$2(6x^2 + 3x - 2) = 0 \quad \text{or} \quad 6x^2 + 3x - 2 = 0.$$
Having thus removed all common factors, we have $a = 6$, $b = 3$, and $c = -2$. ◂◂

(C) $7x - x^2 + 4 = 0$

(D) $-4y + 3 = -2y^2$

(E) $6 - 3x = 2 - (x - 5)$

(F) $3y + 7 + 7y^2 = 8y^2 - 3y + 2$

(G) $5 + 20x = 10x^2$

Answers: **(A)** $5x^2 - 4x + 8 = 0$; $a = 5$, $b = -4$, $c = 8$
(B) $6y^2 + y - 1 = 0$; $a = 6$, $b = 1$, $c = -1$
(C) $x^2 - 7x - 4 = 0$; $a = 1$, $b = -7$, $c = -4$
(D) $2y^2 - 4y + 3 = 0$; $a = 2$, $b = -4$, $c = 3$
(E) not a quadratic equation
(F) $y^2 - 6y - 5 = 0$; $a = 1$, $b = -6$, $c = -5$
(G) $2x^2 - 4x - 1 = 0$; $a = 2$, $b = -4$, $c = -1$

Recall that an equation of the type
$$(x - 2)(x + 1) = 0$$
is solved by using the zero-product rule, which states that if the product of two numbers is zero then one or both of the numbers must be zero. Using this rule we set each factor equal to zero and solve.

$$x - 2 = 0 \quad \text{or} \quad x + 1 = 0$$
$$x = 2 \quad \text{or} \quad x = -1$$

Since $(x - 2)(x + 1) = x^2 - x - 2$, in effect we have solved the quadratic equation
$$x^2 - x - 2 = 0.$$

QUADRATIC EQUATIONS

> **To solve a quadratic equation by factoring**
> 1. Write the equation in general form.
> 2. Divide any common numerical factor from each term.
> 3. Factor the left side into a product of two factors.
> 4. Use the zero-product rule to obtain two equations whose solutions are the solutions to the original quadratic equation.

EXAMPLE 2

Solve $2x^2 - x - 1 = 0$.

$$(2x + 1)(x - 1) = 0 \quad \text{Factor}$$
$$2x + 1 = 0 \quad \text{or} \quad x - 1 = 0 \quad \text{Zero-product rule}$$
$$2x = -1 \quad\quad\quad x = 1$$
$$x = -\frac{1}{2}$$

Check:
$$2\left(-\frac{1}{2}\right)^2 - \left(-\frac{1}{2}\right) - 1 \stackrel{?}{=} 0 \quad\quad 2(1)^2 - (1) - 1 \stackrel{?}{=} 0$$
$$2\left(\frac{1}{4}\right) + \frac{1}{2} - 1 \stackrel{?}{=} 0 \quad\quad 2 - 1 - 1 \stackrel{?}{=} 0$$
$$\frac{2}{4} + \frac{1}{2} - 1 \stackrel{?}{=} 0 \quad\quad 2 - 2 = 0$$
$$1 - 1 = 0$$

The solutions are 1 and $-\frac{1}{2}$. ◀◀

Practice Exercise 2

Solve $3x^2 + x - 2 = 0$.

Answer: $\frac{2}{3}, -1$

EXAMPLE 3

Solve $2x^2 - 2x = 12$.

$$2x^2 - 2x - 12 = 0 \quad \text{Write in general form}$$
$$2(x^2 - x - 6) = 0 \quad \text{2 is a common factor}$$
$$x^2 - x - 6 = 0 \quad \text{Divide both sides by 2}$$
$$(x - 3)(x + 2) = 0 \quad \text{Factor}$$
$$x - 3 = 0 \quad \text{or} \quad x + 2 = 0 \quad \text{Zero-product rule}$$
$$x = 3 \quad\quad\quad x = -2$$

The solutions are 3 and -2. Check. ◀◀

Practice Exercise 3

Solve $4x^2 + 18x = 10$.

Answer: $\frac{1}{2}, -5$

EXAMPLE 4

Solve $5x^2 = 3x$.

$$5x^2 - 3x = 0 \quad \text{Write in general form}$$
$$x(5x - 3) = 0 \quad \text{Factor}$$
$$x = 0 \quad \text{or} \quad 5x - 3 = 0 \quad \text{Zero-product rule}$$
$$5x = 3$$
$$x = \frac{3}{5}$$

Practice Exercise 4

Solve $7x = 14x^2$.

The solutions are 0 and $\frac{3}{5}$. Check. ◀◀

Answer: $0, \frac{1}{2}$

Whenever we have a quadratic equation in general form with $c = 0$, that is, $ax^2 + bx = 0$, one factor is always x so zero is one solution to the equation.

▶▶ **CAUTION** ◀◀ Never divide both sides of an equation by an expression involving the variable. In Example 4 if we had divided by x we would have lost the solution 0. ◀◀

EXAMPLE 5

Solve $5z(z - 6) - 8z = 2z$.

$5z^2 - 30z - 8z = 2z$

$5z^2 - 38z = 2z$

$5z^2 - 40z = 0$ Write in general form

$5(z^2 - 8z) = 0$

$z^2 - 8z = 0$ Divide out the common factor of 5

$z(z - 8) = 0$ Factor

$z = 0$ or $z - 8 = 0$

$z = 8$

The solutions are 0 and 8. Check. ◀◀

Practice Exercise 5

Solve $y(y - 3) - 7 = 2y + 7$.

Answer: $7, -2$

4 ▶▶ A quadratic equation with $b = 0$, that is, $ax^2 + c = 0$, can be solved using a different technique. Consider the equation

$$x^2 - 4 = 0$$

or

$$x^2 = 4.$$

What number squared is 4? Clearly $2^2 = 4$ and $(-2)^2 = 4$ so that x is either 2 or -2. By taking the square root of both sides of the equation and using the symbol $\pm$ to indicate both the positive (principal) square root and the negative square root, we have

$$x = \pm\sqrt{4}$$

$$x = \pm 2.$$

Remember that $x = \pm 2$ is simply a shorthand way of writing $x = +2$ or $x = -2$.

To solve a quadratic equation of the form $ax^2 + c = 0$

1. Subtract c from both sides, obtaining $ax^2 = -c$.

2. Divide both sides by a, obtaining $x^2 = -\dfrac{c}{a}$,

     where $-\dfrac{c}{a} > 0$ if a and c have different signs.

3. Take the square root of both sides, obtaining $x = \pm\sqrt{-\dfrac{c}{a}}$.

450 QUADRATIC EQUATIONS

▶▶ **CAUTION** ◀◀ This method gives real-number solutions only when the radicand $-\frac{c}{a}$ is positive. For this to happen a and c must have opposite signs. ◀◀

EXAMPLE 6

Solve $3x^2 - 75 = 0$.

$3x^2 = 75$ Add 75 to both sides

$x^2 = \frac{75}{3}$ Divide both sides by 3

$x^2 = 25$

$x = \pm\sqrt{25} = \pm 5$ Take square root of both sides

The solutions are 5 and -5. Check. ◀◀

Practice Exercise 6

Solve $64 - 4x^2 = 0$.

Answer: ± 4

EXAMPLE 7

Solve $5x^2 - 40 = 0$.

$5x^2 = 40$ Add 40

$x^2 = \frac{40}{5}$ Divide by 5

$x = \pm\sqrt{8}$ Take square root of both sides

$x = \pm\sqrt{4 \cdot 2} = \pm 2\sqrt{2}$

The solutions are $2\sqrt{2}$ and $-2\sqrt{2}$. Check. ◀◀

Practice Exercise 7

Solve $8x^2 - 96 = 0$.

Answer: $\pm 2\sqrt{3}$

EXAMPLE 8

Solve $x^2 + 5 = 0$.

$x^2 = -5$ Subtract 5

Recall from Section 9.1 that no real number squared can be negative. Hence, there are no real-number solutions to the equation $x^2 = -5$, nor to any equation that has x^2 equal to a negative number. ◀◀

Practice Exercise 8

Solve $x^2 + 17 = 0$.

Answer; no real number solutions

EXAMPLE 9

Solve $(y + 3)^2 = 16$.

Although not in the form $ax^2 + c = 0$, we can use the same basic technique of taking the square root of both sides to solve.

$y + 3 = \pm\sqrt{16}$

$y + 3 = \pm 4$

$y = -3 \pm 4$ Subtract 3 from both sides

$y = \begin{cases} -3 + 4 = 1 & \text{Use } + \text{ to obtain solution } 1 \\ -3 - 4 = -7 & \text{Use } - \text{ to obtain solution } -7 \end{cases}$

It is easy to substitute 1 and -7 into the original equation to see that they check and are the solutions. ◀◀

Practice Exercise 9

Solve $(x + 7)^2 = 49$.

Answer: $0, -14$

10.1 EXERCISES A

Write each equation in general quadratic form and identify the constants a, b, and c.

1. $2x^2 + x = 5$
2. $y - y^2 = 5$
3. $x^2 = 3x$

4. $y^2 = -7y - 12$
5. $2y^2 - y = 3 + y - y^2$
6. $2(x + x^2) = 3x - 1$

7. $x^2 + 1 = 0$
8. $\frac{1}{2}y^2 - y = 3 - y$
9. $35y - 2 = 0$

10. $2(x^2 + x) + x^3 = x^3 - 5$
11. $2y^3 + y^2 + 1 = 0$
12. $3x = 3(x + 1)$

13. Given the quadratic equation $-x^2 + 2x - 3 = 0$, we would probably conclude that $a = -1$, $b = 2$, and $c = -3$. Would $a = 1$, $b = -2$, and $c = 3$ be incorrect? Explain.

Solve.

14. $x^2 - 3x - 28 = 0$
15. $y^2 + 5y + 6 = 0$
16. $x^2 = 2x + 35$

17. $2z^2 - 5z = 3$
18. $6x^2 - 2x - 8 = 0$
19. $4z^2 = 7z$

20. $12x^2 + 6x = 0$
21. $y^2 - 10y = -25$
22. $z^2 - 6 = z$

23. $3(y^2 + 8) = 24 - 15y$ **24.** $z(8 - 2z) = 6$ **25.** $x(3x + 20) = 7$

26. $(3z - 1)(2z + 1) = 3(2z + 1)$ **27.** $(2y - 3)(y + 1) = 4(2y - 3)$ **28.** $4x^2 - 9 = 3$

29. $x^2 = 16$ **30.** $z^2 - 81 = 0$ **31.** $3x^2 - 48 = -48$

32. $2y^2 - 128 = 0$ **33.** $y^2 - \dfrac{9}{4} = 0$ **34.** $-3z^2 = -27$

35. $2x^2 - 150 = 0$ **36.** $4y^2 = 72$ **37.** $2x^2 + 50 = 0$

38. $-3y^2 = -6$ **39.** $3 - y^2 = y^2 - 3$ **40.** $z^2 + 2z = 2z - 4$

41. $(x - 2)^2 = 9$ **42.** $(y + 1)^2 = 49$ **43.** $(z - 1)^2 = 5$

44. The square of a number, decreased by 2, is equal to the negative of the number. Find the number.

45. The square of 2 more than a number is 64. Find the number.

ANSWERS: **1.** $2x^2 + x - 5 = 0$; $a = 2$, $b = 1$, $c = -5$ **2.** $-y^2 + y - 5 = 0$; $a = -1$, $b = 1$, $c = -5$ **3.** $x^2 - 3x = 0$; $a = 1$, $b = -3$, $c = 0$ **4.** $y^2 + 7y + 12 = 0$; $a = 1$, $b = 7$, $c = 12$ **5.** $3y^2 - 2y - 3 = 0$; $a = 3$, $b = -2$, $c = -3$ **6.** $2x^2 - x + 1 = 0$; $a = 2$, $b = -1$, $c = 1$ **7.** $x^2 + 1 = 0$; $a = 1$, $b = 0$, $c = 1$ **8.** $\frac{1}{2}y^2 - 3 = 0$; $a = \frac{1}{2}$, $b = 0$, $c = -3$ or $y^2 - 6 = 0$ (multiply by 2 to clear fractions and make a, b, and c integers); $a = 1$, $b = 0$, $c = -6$ **9.** not a quadratic equation **10.** $2x^2 + 2x + 5 = 0$; $a = 2$, $b = 2$, $c = 5$ **11.** not a quadratic equation **12.** not a quadratic equation **13.** No; multiplying the original equation on both sides by -1 yields a general form with the given constants. **14.** $7, -4$ **15.** $-2, -3$ **16.** $7, -5$ **17.** $-\frac{1}{2}, 3$ **18.** $\frac{4}{3}, -1$ **19.** $0, \frac{7}{4}$ **20.** $0, -\frac{1}{2}$ **21.** 5 **22.** $3, -2$ **23.** $0, -5$ **24.** $1, 3$ **25.** $\frac{1}{3}, -7$ **26.** $\frac{4}{3}, -\frac{1}{2}$ **27.** $3, \frac{3}{2}$ **28.** $\sqrt{3}, -\sqrt{3}$ **29.** ± 4 **30.** ± 9 **31.** 0 **32.** ± 8 **33.** $\pm\frac{3}{2}$ **34.** ± 3 **35.** $\pm 5\sqrt{3}$ **36.** $\pm 3\sqrt{2}$ **37.** no solution **38.** $\pm\sqrt{2}$ **39.** $\pm\sqrt{3}$ **40.** no solution **41.** $5, -1$ **42.** $6, -8$ **43.** $1 \pm \sqrt{5}$ **44.** $1, -2$ **45.** $6, -10$

10.1 EXERCISES B

Write each equation in general form and identify the constants a, b, and c.

1. $3x^2 + x = 7$

2. $y - y^2 = 4$

3. $z^2 = 9z$

4. $y^2 = -10y - 18$

5. $3z^2 - z = 5 + z - z^2$

6. $4(z + z^2) = 2z + 1$

7. $w^2 + 5 = 0$

8. $\frac{1}{3}x^2 - x = 5 - x$

9. $4x - 8 = 0$

10. $3(y^2 + y) - y^3 = 6 - y^3$

11. $5x^3 = x^2 + 7$

12. $3(w - 3) = 4w$

13. Given the quadratic equation $-y^2 + y - 7 = 0$, we would probably conclude that $a = -1$, $b = 1$, and $c = -7$. Would $a = 1$, $b = -1$, and $c = 7$ be incorrect? Explain.

Solve.

14. $x^2 + x - 6 = 0$

15. $y^2 + 3y + 2 = 0$

16. $x^2 = 2x + 24$

17. $2z^2 + 3 = 7z$

18. $3x^2 - 2x - 5 = 0$

19. $2z^2 = 3z$

20. $3x^2 - 18x = 0$

21. $y^2 - 8y + 16 = 0$

22. $z^2 - 20 = z$

23. $3(y^2 + 4y) = -6y$

24. $z(z - 3) = -2$

25. $x(2x + 9) = 5$

26. $(z - 2)(z + 1) = 3(z + 1)$

27. $5x^2 - 10 = 5$

28. $y(2y - 1) = 7(1 - 2y)$

29. $x^2 = 49$

30. $z^2 - 36 = 0$

31. $5x^2 - 7 = -7$

32. $2y^2 - 98 = 0$

33. $y^2 - \frac{25}{4} = 0$

34. $-8z^2 = -32$

35. $2x^2 - 100 = 0$

36. $3y^2 = 60$

37. $3x^2 + 75 = 0$

38. $-6y^2 = -18$

39. $7 - x^2 = x^2 - 7$

40. $z^2 + 5z = 5z - 1$

454 QUADRATIC EQUATIONS

41. $(x - 5)^2 = 4$ **42.** $(y + 2)^2 = 1$ **43.** $(z - 3)^2 = 3$

44. The square of a number, less 10 is equal to three times the number. Find the number.

45. The square of 5 more than a number is 9. Find the number.

10.1 EXERCISES C

Solve.

1. $6x^2 + 47x - 8 = 0$

2. $(2x - 1)(x + 5) = (4x + 3)(x + 5)$
[Answer: $-5, -2$]

3. $(x^2 - 5)^2 = 8$

10.2 SOLVING BY COMPLETING THE SQUARE

STUDENT GUIDEPOSTS

1 Completing the square
2 Solving by completing the square

A trinomial $x^2 + bx + c$ is a perfect square trinomial if it factors into the square of a binomial. Notice that the trinomials in the left column below are perfect square trinomials. Also notice in the right column that for each perfect square trinomial, the constant term, c, is the square of one half the coefficient of the x term, b.

$$x^2 + 6x + 9 = (x + 3)^2 \qquad 9 = \left(\frac{1}{2} \cdot 6\right)^2$$

$$x^2 - 10x + 25 = (x - 5)^2 \qquad 25 = \left(\frac{1}{2} \cdot (-10)\right)^2$$

$$x^2 - x + \frac{1}{4} = \left(x - \frac{1}{2}\right)^2 \qquad \frac{1}{4} = \left(\frac{1}{2} \cdot (-1)\right)^2$$

1 In general, whenever we are given the first two terms of a trinomial, $x^2 + bx$, we can add $\left(\frac{1}{2} \cdot b\right)^2$ to it to get a perfect square trinomial. This process, called **completing the square,** can be used only when the coefficient in the squared term is 1.

EXAMPLE 1

What must be added to complete the square?

(A) $x^2 + 12x +$ _____

Add $\left(\frac{1}{2} \cdot 12\right)^2 = 6^2 = 36$.

(B) $x^2 - 20x +$ _____

Add $\left(\frac{1}{2} \cdot (-20)\right)^2 = (-10)^2 = 100$.

(C) $x^2 + \frac{2}{3}x +$ _____

Practice Exercises 1

What must be added to complete the square?

(A) $x^2 + 14x +$ _____

(B) $x^2 - 22x +$ _____

(C) $x^2 - \frac{2}{3}x +$ _____

Add $\left(\dfrac{1}{2} \cdot \dfrac{2}{3}\right)^2 = \left(\dfrac{1}{3}\right)^2 = \dfrac{1}{9}.$ ◀◀

Answers: (A) 49 (B) 121
(C) $\dfrac{1}{9}$

The method of completing the square can be used to solve quadratic equations. For example, the equation

$$x^2 + 3x - 10 = 0,$$

which can be solved by factoring,

$$(x + 5)(x - 2) = 0$$
$$x = -5, 2,$$

can also be solved by completing the square. To do so, first isolate the constant term on the right side (leave space as indicated).

$$x^2 + 3x \phantom{+ \tfrac{9}{4}} = 10$$

Then complete the square on the left side by adding the square of half the coefficient of x.

$$\left(\dfrac{1}{2} \cdot 3\right)^2 = \left(\dfrac{3}{2}\right)^2 = \dfrac{9}{4}$$

But if $\dfrac{9}{4}$ is added on the left, to retain equality we must add $\dfrac{9}{4}$ on the right.

$$x^2 + 3x + \dfrac{9}{4} = 10 + \dfrac{9}{4}$$

$$\left(x + \dfrac{3}{2}\right)^2 = \dfrac{49}{4}$$

Always check your work by squaring the left side. Take the square root of each side, remembering to indicate both the positive and negative root.

$$x + \dfrac{3}{2} = \pm\sqrt{\dfrac{49}{4}} = \pm\dfrac{7}{2}$$

$$x = -\dfrac{3}{2} \pm \dfrac{7}{2} \qquad \text{Subtract } \dfrac{3}{2} \text{ from both sides}$$

$$x = -\dfrac{3}{2} + \dfrac{7}{2} \quad \text{or} \quad x = -\dfrac{3}{2} - \dfrac{7}{2}$$

$$x = \dfrac{4}{2} = 2 \qquad\qquad x = \dfrac{-10}{2} = -5$$

Thus, we obtain the same solutions, 2 and -5, as before.

To solve a quadratic equation by completing the square

1. Isolate the constant term on the right side of the equation.
2. If the coefficient in the squared term is 1, proceed to step 4.
3. If the coefficient of the squared term is not 1, divide each term by that coefficient.
4. Complete the square on the left side and add the same number to the right side.
5. Factor (check by multiplying the factors) and take the square root of both sides (positive and negative roots).
6. Use both the positive root and the negative root to obtain the solutions to the original equation.

456 QUADRATIC EQUATIONS

EXAMPLE 2

Solve $y^2 - 6y + 8 = 0$ by completing the square.

$$y^2 - 6y = -8 \quad \text{Isolate the constant}$$

$$y^2 - 6y + 9 = -8 + 9 \quad \text{Add 9 to both sides,}$$
$$\left(\frac{-6}{2}\right)^2 = (-3)^2 = 9$$

$$(y - 3)^2 = 1$$

$$y - 3 = \pm\sqrt{1} \quad \text{Take square root of both sides}$$

$$y - 3 = \pm 1 \quad \sqrt{1} = 1$$

$$y = 3 \pm 1$$

$$y = 3 + 1 \quad \text{or} \quad y = 3 - 1$$

$$y = 4 \quad\quad\quad\quad y = 2$$

The solutions are 2 and 4. ◀◀

Practice Exercise 2

Solve $y^2 - 8y + 12 = 0$ by completing the square.

Answer: 2, 6

As mentioned earlier, we can use the above technique to complete the square on a quadratic only when the coefficient of the squared term is 1. If this is not the case, we must divide each term of the equation by the coefficient of the squared term before we can use the method.

EXAMPLE 3

Solve $2x^2 + 2x - 3 = 0$ by completing the square.

$$2x^2 + 2x = 3 \quad \text{Isolate the constant}$$

$$x^2 + x = \frac{3}{2} \quad \text{Divide through by 2}$$

$$x^2 + x + \frac{1}{4} = \frac{3}{2} + \frac{1}{4} \quad \text{Complete the square by adding } \frac{1}{4} \text{ to both sides}$$

$$\left(x + \frac{1}{2}\right)^2 = \frac{7}{4} \quad\quad \frac{3}{2} + \frac{1}{4} = \frac{6}{4} + \frac{1}{4} = \frac{7}{4}$$

$$x + \frac{1}{2} = \pm\sqrt{\frac{7}{4}} = \pm\frac{\sqrt{7}}{\sqrt{4}} = \pm\frac{\sqrt{7}}{2}$$

$$x + \frac{1}{2} = \frac{\sqrt{7}}{2} \quad \text{or} \quad x + \frac{1}{2} = -\frac{\sqrt{7}}{2}$$

$$x = -\frac{1}{2} + \frac{\sqrt{7}}{2} \quad\quad x = -\frac{1}{2} - \frac{\sqrt{7}}{2}$$

$$x = \frac{-1 + \sqrt{7}}{2} \quad\quad x = \frac{-1 - \sqrt{7}}{2}$$

The solutions are $\frac{-1 \pm \sqrt{7}}{2}$. ◀◀

Practice Exercise 3

Solve $3x^2 - 2x - 6 = 0$ by completing the square.

Answer: $\frac{1 \pm \sqrt{19}}{3}$

10.2 EXERCISES A

What must be added to complete the square?

1. $x^2 - 8x + \underline{}$
2. $y^2 - 18y + \underline{}$
3. $z^2 + 7z + \underline{}$

4. $x^2 - x + \underline{}$
5. $y^2 + 9y + \underline{}$
6. $z^2 + \frac{1}{3}z + \underline{}$

7. $x^2 - \frac{1}{2}x + \underline{}$
8. $y^2 - \frac{4}{3}y + \underline{}$
9. $z^2 + z + \underline{}$

Solve by completing the square.

10. $x^2 + 8x + 15 = 0$
11. $y^2 + 10y + 21 = 0$
12. $z^2 - 2z - 8 = 0$

13. $x^2 - 2x - 1 = 0$
14. $y^2 + 4y = 96$
15. $3z^2 + 12z = 135$

16. $2x^2 + x - 1 = 0$
17. $2y^2 - 5y = -2$
18. $z^2 + 16z + 50 = 0$

19. $x^2 - 10x + 12 = 0$
20. $y^2 + 2y + 5 = 0$
21. $2z^2 - z - 3 = 0$

Solve by any method.

22. $x^2 - 5x = -4$
23. $5y^2 - 4y - 33 = 0$
24. $z^2 + 6z = 16$

458 QUADRATIC EQUATIONS

25. $2x^2 - 5x - 3 = 0$ **26.** $2y^2 - 3y = 1$ **27.** $z^2 + z = z + 10$

28. A free-falling object, starting from rest, will fall d feet in t seconds according to the formula $d = 16t^2$.

(A) How far will the object fall in 1 second?

(B) How far will the object fall in 5 seconds?

(C) How many seconds does it take for the object to fall 64 feet?

ANSWERS: **1.** 16 **2.** 81 **3.** $\frac{49}{4}$ **4.** $\frac{1}{4}$ **5.** $\frac{81}{4}$ **6.** $\frac{1}{36}$ **7.** $\frac{1}{16}$ **8.** $\frac{4}{9}$ **9.** $\frac{1}{4}$ **10.** $-3, -5$ **11.** $-3, -7$ **12.** $4, -2$ **13.** $1 \pm \sqrt{2}$ **14.** $-12, 8$ **15.** $5, -9$ **16.** $\frac{1}{2}, -1$ **17.** $2, \frac{1}{2}$ **18.** $-8 \pm \sqrt{14}$ **19.** $5 \pm \sqrt{3}$ **20.** no solution (square root of a negative number does not exist) **21.** $\frac{3}{2}, -1$ **22.** $4, 1$ (best to factor) **23.** $3, -\frac{11}{5}$ (factor) **24.** $2, -8$ (factor) **25.** $3, -\frac{1}{2}$ (factor) **26.** $\frac{3 \pm \sqrt{17}}{4}$ (complete the square) **27.** $\pm\sqrt{10}$ (take roots) **28.** (A) 16 feet (B) 400 feet (C) 2 seconds

10.2 EXERCISES B

What must be added to complete the square?

1. $x^2 + 24x + $ _____ **2.** $y^2 - 4y + $ _____ **3.** $z^2 + 3z + $ _____

4. $x^2 + x + $ _____ **5.** $y^2 - 11y + $ _____ **6.** $z^2 + \frac{1}{2}z + $ _____

7. $x^2 - \frac{1}{3}x + $ _____ **8.** $y^2 + \frac{2}{3}y + $ _____ **9.** $z^2 - z + $ _____

Solve by completing the square.

10. $x^2 + 6x + 5 = 0$ **11.** $y^2 + 3y - 18 = 0$ **12.** $z^2 - 3z - 10 = 0$

13. $x^2 - 4x - 1 = 0$ **14.** $y^2 = 8y + 20$ **15.** $z^2 + 5z = 24$

16. $2x^2 - 3x + 1 = 0$ **17.** $2y^2 - 5y = 3$ **18.** $z^2 - z = 3$

19. $2x^2 + 6x + 1 = 0$ **20.** $y^2 + 2y = -9$ **21.** $z^2 + 15 = 8z$

Solve by any method.

22. $x^2 + 8x - 9 = 0$ **23.** $3y^2 + 11x - 4 = 0$ **24.** $z^2 - 14z + 49 = 0$

25. $2x^2 - 3x - 2 = 0$ **26.** $3y^2 - 6 = 2y$ **27.** $z^2 - z = 8 - z$

28. A free falling object, starting at rest, will fall d feet in t seconds according to the formula $d = 16t^2$.

(A) How far will the object fall in 2 seconds?

(B) How far will the object fall in 10 seconds?

(C) How many seconds does it take for the object to fall 144 feet?

10.2 EXERCISES C

Solve by completing the square.

1. $5x^2 - 2x - 10 = 0$

2. $x^2 + bx + c = 0$
$$\left[\text{Answer: } x = \frac{-b \pm \sqrt{b^2 - 4c}}{2}\right]$$

10.3 SOLVING BY THE QUADRATIC FORMULA

STUDENT GUIDEPOSTS

1 Quadratic formula
2 Using the quadratic formula
3 General solving procedure
4 Discriminant

The method of factoring or the method of taking roots is usually faster and easier than the method of completing the square. Completing the square becomes rather tedious since we are repeating the same process time and time again. When such repetition occurs it is often better to carry through the process once in a general case and memorize the result. This is how the quadratic formula is developed. Start with a quadratic equation in general form

$$ax^2 + bx + c = 0 \quad (a \neq 0)$$

and complete the square.

$$ax^2 + bx = -c \quad \text{Isolate the constant term}$$

$$\frac{\cancel{a}x^2}{\cancel{a}} + \frac{b}{a}x = -\frac{c}{a} \quad \text{Divide by the coefficient of } x^2$$

$$x^2 + \frac{b}{a}x + \left(\frac{b}{2a}\right)^2 = -\frac{c}{a} + \left(\frac{b}{2a}\right)^2 \quad \text{Add } \left(\frac{1}{2} \cdot \frac{b}{a}\right)^2 = \left(\frac{b}{2a}\right)^2$$

$$\left(x + \frac{b}{2a}\right)^2 = -\frac{c}{a} + \frac{b^2}{4a^2} \quad \text{Factor}$$

$$\left(x + \frac{b}{2a}\right)^2 = -\frac{4ac}{4a^2} + \frac{b^2}{4a^2} \quad \text{Find a common denominator}$$

$$= \frac{b^2 - 4ac}{4a^2} \quad \text{Subtract the fractions}$$

$$x + \frac{b}{2a} = \pm\sqrt{\frac{b^2 - 4ac}{4a^2}} \quad \text{Take square roots}$$

$$= \pm\frac{\sqrt{b^2 - 4ac}}{2a}$$

460 QUADRATIC EQUATIONS

$$x = -\frac{b}{2a} \pm \frac{\sqrt{b^2 - 4ac}}{2a} \qquad \text{Subtract } \frac{b}{2a}$$

$$x = \frac{-b \pm \sqrt{b^2 - 4ac}}{2a} \qquad \text{Note that } 2a \text{ is the denominator of the entire expression}$$

1 This last formula is called the **quadratic formula** and it must be memorized. To use it to solve a quadratic equation, we identify the constants a, b, and c and substitute into the quadratic formula.

2 **To solve a quadratic equation using the quadratic formula**

1. Write the equation in general form ($ax^2 + bx + c = 0$).
2. Identify the constants a, b, and c.
3. Substitute the values for a, b, and c into the quadratic formula,
$$x = \frac{-b \pm \sqrt{b^2 - 4ac}}{2a}.$$
4. Simplify the numerical expression to obtain the solutions.

EXAMPLE 1

Solve $x^2 - 5x + 6 = 0$ using the quadratic formula.

We have $a = 1$, $b = -5$ (not 5), and $c = 6$.

$$x = \frac{-b \pm \sqrt{b^2 - 4ac}}{2a} = \frac{-(-5) \pm \sqrt{(-5)^2 - 4(1)(6)}}{2(1)} \quad \text{Substitute}$$

$$= \frac{5 \pm \sqrt{25 - 24}}{2} \quad \text{Watch all signs}$$

$$= \frac{5 \pm \sqrt{1}}{2} = \frac{5 \pm 1}{2} = \begin{cases} \dfrac{5+1}{2} = \dfrac{6}{2} = 3 \\ \dfrac{5-1}{2} = \dfrac{4}{2} = 2 \end{cases}$$

The solutions are 2 and 3. ◀◀

Practice Exercise 1

Solve $x^2 + 6x + 9 = 0$ using the quadratic formula.

Answer: -3

Answer: -3

EXAMPLE 2

Solve $3x^2 - 5 = -4x$ using the quadratic formula.

First we write the equation in general form, $3x^2 + 4x - 5 = 0$, and identify $a = 3$, $b = 4$, and $c = -5$ (not 5), then substitute.

$$x = \frac{-b \pm \sqrt{b^2 - 4ac}}{2a} = \frac{-(4) \pm \sqrt{(4)^2 - 4(3)(-5)}}{2(3)}$$

$$= \frac{-4 \pm \sqrt{16 + 60}}{6} \quad \text{Watch the signs}$$

$$= \frac{-4 \pm \sqrt{76}}{6} = \frac{-4 \pm \sqrt{4 \cdot 19}}{6}$$

Practice Exercise 2

Solve $7x - 2 = 2x^2$ using the quadratic formula.

$$= \frac{-4 \pm 2\sqrt{19}}{6} = \frac{\cancel{2}(-2 \pm \sqrt{19})}{\cancel{2} \cdot 3}$$

$$= \frac{-2 \pm \sqrt{19}}{3}$$

The solutions are $\dfrac{-2 \pm \sqrt{19}}{3}$. ◀

Answer: $\dfrac{7 \pm \sqrt{33}}{4}$

---

3 **To solve a quadratic equation**
1. First try factoring or taking roots.
2. If the first two methods are not applicable, go directly to the quadratic formula.

Notice that the quadratic equation in Example 1 could be solved much more quickly by factoring than by using the quadratic formula.

$$x^2 - 5x + 6 = 0$$
$$(x - 2)(x - 3) = 0$$
$$x - 2 = 0 \quad \text{or} \quad x - 3 = 0$$
$$x = 2 \quad \text{or} \quad x = 3$$

4 When solving a quadratic equation, $ax^2 + bx + c = 0$, the number $b^2 - 4ac$, called the **discriminant,** can be used to determine when the equation has solutions. Notice that the discriminant is the number under the radical sign in the quadratic formula $x = \dfrac{-b \pm \sqrt{b^2 - 4ac}}{2a}$. Since the square root of a negative number is not defined as a real number, a quadratic equation has no real-number solutions when the discriminant is a negative number. For example, consider

$$x^2 + x + 1 = 0.$$

Here $a = 1$, $b = 1$, and $c = 1$, so $b^2 - 4ac = (1) - 4(1)(1) = 1 - 4 = -3$. Since -3 is a negative number, $\sqrt{-3}$ is not a real number and $x^2 + x + 1 = 0$ has no real-number solutions.

Another helpful technique when solving quadratic equations involving fractional coefficients is to clear all fractions first. For example, to solve

$$\frac{1}{3}x^2 - x + \frac{1}{3} = 0,$$

it would be difficult to use $a = \frac{1}{3}$, $b = -1$, and $c = \frac{1}{3}$ in the quadratic formula. However, if we multiply through by the LCD 3, we obtain the equivalent equation

$$x^2 - 3x + 1 = 0,$$

in which $a = 1$, $b = -3$, and $c = 1$. Avoiding fractions whenever possible makes the arithmetic simpler.

In many applied problems involving quadratic equations, solutions involving radicals must be estimated by using an approximate value for the radical. This is illustrated in the next example.

EXAMPLE 3

The hypotenuse of a right triangle is 5 feet long and one leg is 4 feet longer than the other. Find the measure of each leg.

The triangle is shown in Figure 10.1. Recall that the Pythagorean theorem states that the sum of the squares of the legs of a right triangle is equal to the square of the hypotenuse (the longest side opposite the right angle).

Let $\quad x =$ the measure of one leg,

$x + 4 =$ the measure of the other leg.

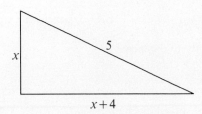

Figure 10.1

Use the Pythagorean theorem.

$$x^2 + (x + 4)^2 = 5^2$$
$$x^2 + x^2 + 8x + 16 = 25 \quad \text{Don't forget the } 8x$$
$$2x^2 + 8x - 9 = 0$$

Since we cannot factor, we use the quadratic formula.

$$x = \frac{-b \pm \sqrt{b^2 - 4ac}}{2a} = \frac{-8 \pm \sqrt{64 - 4(2)(-9)}}{2(2)} \quad a = 2, b = 8, c = -9$$

$$= \frac{-8 \pm \sqrt{64 + 72}}{4}$$

$$= \frac{-8 \pm \sqrt{136}}{4}$$

$$= \frac{-8 \pm \sqrt{4 \cdot 34}}{4}$$

$$= \frac{-8 \pm 2\sqrt{34}}{4} \quad \sqrt{4 \cdot 34} = \sqrt{4}\sqrt{34} = 2\sqrt{34}$$

$$= \frac{2[-4 \pm \sqrt{34}]}{2 \cdot 2} \quad \text{Factor out 2}$$

$$= \frac{-4 \pm \sqrt{34}}{2}$$

Since $\dfrac{-4 - \sqrt{34}}{2}$ is a negative number, we can discard it as a possible solution (a length must be positive or zero). Thus, the length of one leg is exactly $\dfrac{-4 + \sqrt{34}}{2}$ feet. However, in a practical situation, we would probably use a decimal approximation. Using a calculator or the table on page 511, we find $\sqrt{34} \approx 5.8$. Thus,

$$\frac{-4 + \sqrt{34}}{2} \approx \frac{-4 + 5.8}{2} = \frac{1.8}{2} = 0.9,$$

Practice Exercise 3

Meg and Joe need a water line the length of their garden. If the garden is a rectangle with diagonal 8 meters and the length is 2 meters longer than the width, approximately how much pipe will be needed?

and the lengths of the legs are approximately 0.9 feet and 4.9 feet ($x = 0.9$, $x + 4 = 4.9$). As a check,

$$(0.9)^2 + (4.9)^2 = 0.81 + 24.01 = 24.82 \approx 25 = 5^2.$$ ◀

Answer: approximately 6.6 meters

Unless you are working an applied problem in which an approximate value would make more sense, or unless instructed otherwise, always leave answers in radical form.

10.3 EXERCISES A

Solve using the quadratic formula.

1. $x^2 + x - 20 = 0$
2. $y^2 - 3y = 10$
3. $z^2 + 8 = 9z$

4. $x^2 - 9 = 0$
5. $0 = 7y - 2 + 15y^2$
6. $5z - 2 = 3z^2$

7. $5x^2 - x = 1$
8. $2y^2 - 4y + 1 = 0$
9. $4z^2 + 9z + 3 = 0$

10. $x^2 + 2x = 5$
11. $z^2 = z - 5$
12. $y^2 - 6y = 1$

13. $x^2 - 3x - 9 = 0$
14. $y(y + 1) + 2y = 4$
15. $(z + 1)^2 = 5 + 3z$

16. $x^2 - 2x + 5 = 0$

17. $4y^2 - 1 = 0$

18. $(2z - 1)(z - 2) - 11 = 2(z + 4) - 8$

Solve using any method.

19. $2x^2 - 7x + 3 = 0$

20. $3z^2 + 10z - 1 = 0$

21. $3y^2 - 12 = 0$

22. $3x^2 - 2 = -5x$

23. $x^2 + x - 1 = 0$

24. $-x^2 - x + 1 = 0$

25. How do the solutions to Exercises 23 and 24 compare? Explain.

26. $\dfrac{2}{3}x^2 - \dfrac{1}{3}x - 1 = 0$

27. $\dfrac{1}{2}x^2 + x - 1 = 0$

28. $\dfrac{1}{4}x^2 + \dfrac{1}{2}x - \dfrac{3}{4} = 0$

29. Write the solution to Exercise 7 using a decimal approximation.

30. Write the solution to Exercise 8 using a decimal approximation.

Solve.

31. The hypotenuse of a right triangle is 3 cm long, and one leg is 1 cm more than the other. Find the approximate measure of each leg.

32. Winnie has a square garden measuring 4 yards on each side. She wishes to place a picket fence diagonally across the garden. What is the approximate length of the fence?

For Review

33. Solve $x^2 + x = 13 + x$ by taking roots.

34. Solve $2y^2 + 2y = 3$ by completing the square.

35. If an object is thrown upward with initial velocity of 128 feet per second, its height, h, above the ground in t seconds is given by $h = 128t - 16t^2$. How long will it take for the object to reach a height of 240 feet?

ANSWERS: **1.** 4, −5 **2.** 5, −2 **3.** 1, 8 **4.** 3, −3 **5.** $\frac{1}{5}, -\frac{2}{3}$ **6.** 1, $\frac{2}{3}$ **7.** $\frac{1 \pm \sqrt{21}}{10}$ **8.** $\frac{2 \pm \sqrt{2}}{2}$ **9.** $\frac{-9 \pm \sqrt{33}}{8}$ **10.** $-1 \pm \sqrt{6}$ **11.** no solution **12.** $3 \pm \sqrt{10}$ **13.** $\frac{3 \pm 3\sqrt{5}}{2}$ **14.** 1, −4 **15.** $\frac{1 \pm \sqrt{17}}{2}$ **16.** no solution **17.** $\frac{1}{2}, -\frac{1}{2}$ **18.** $\frac{9}{2}, -1$ **19.** 3, $\frac{1}{2}$ **20.** $\frac{-5 \pm 2\sqrt{7}}{3}$ **21.** 2, −2 **22.** $\frac{1}{3}, -2$ **23.** $\frac{-1 \pm \sqrt{5}}{2}$ **24.** $\frac{1 \pm \sqrt{5}}{-2}$ **25.** Same (Multiply numerator and denominator of $\frac{-1 \pm \sqrt{5}}{2}$ by −1 to obtain $\frac{1 \pm \sqrt{5}}{-2}$). The equations are equivalent (multiply both sides of **23** by −1 to get **24**). **26.** $\frac{3}{2}, -1$ **27.** $-1 \pm \sqrt{3}$ **28.** 1, −3 **29.** 0.56, −0.36 (using $\sqrt{21} \approx 4.6$) **30.** 1.7, 0.3 (using $\sqrt{2} \approx 1.4$) **31.** 1.6 cm, 2.6 cm **32.** 5.7 yards **33.** $\pm\sqrt{13}$ **34.** $\frac{-1 \pm \sqrt{7}}{2}$ **35.** 3 seconds (on the way up) then again at 5 seconds (on the way down)

10.3 EXERCISES B

Solve using the quadratic formula.

1. $x^2 - 4x - 21 = 0$

2. $y^2 - 10y + 9 = 0$

3. $z^2 + 5 = 6z$

4. $x^2 - 49 = 0$

5. $0 = 20y^2 - y - 1$

6. $7z - 3 = 4z^2$

466 QUADRATIC EQUATIONS

7. $2x^2 - x = 2$
8. $y^2 + 5y + 5 = 0$
9. $3z^2 + 3 = 7z$
10. $x^2 - 4x = 1$
11. $3y^2 + 2y = 7$
12. $z^2 = z - 9$
13. $x^2 - 5x + 3 = 0$
14. $y(y + 3) + 2y = 6$
15. $3(z - 1)^2 = 3z$
16. $x^2 - 3x + 7 = 0$
17. $6y^2 + 7y = 3$
18. $(z + 1)^2 = 5(1 + z)$

Solve using any method.

19. $2x^2 - x - 1 = 0$
20. $7y^2 - 49 = 0$
21. $z^2 - 5z - 2 = 0$
22. $2x^2 = \frac{1}{2}(27x + 7)$
23. $x^2 + 6x - 1 = 0$
24. $-x^2 - 6x + 1 = 0$

25. How do the solutions to Exercises 23 and 24 compare? Explain.

26. $x^2 - \frac{1}{2}x - 5 = 0$
27. $\frac{1}{3}x^2 + x + \frac{1}{6} = 0$
28. $\frac{1}{11}x^2 + \frac{9}{11}x - 2 = 0$

29. Write the solution to Exercise 7 using a decimal approximation.

30. Write the solution to Exercise 8 using a decimal approximation.

31. The hypotenuse of a right triangle is 7 cm long, and one leg is 3 cm more than the other. Find the approximate measure of each leg.

32. A baseball diamond is a square 90 ft on each side. How far is it from home plate directly across to second base?

For Review

33. Solve $2y^2 - 50 = 0$ by taking roots.

34. Solve $2x^2 - 6x + 3 = 0$ by the method of completing the square.

35. If an object is thrown upward with initial velocity of 64 feet per second its height, h, above the ground in t seconds is given by $h = 64t - 16t^2$. How long will it take for the object to reach a height of 48 feet?

10.3 EXERCISES C

Solve for x using the quadratic formula.

1. $0.3x^2 - 0.1x - 1.6 = 0$ [*Hint:* Multiply by 10.]
2. $x^2 + 3xy + y^2 = 0$ $\left[\text{Answer: } \dfrac{-y(3 \pm \sqrt{5})}{2}\right]$

10.4 SOLVING FRACTIONAL AND RADICAL EQUATIONS

 **STUDENT GUIDEPOSTS**

1 Fractional equations
2 Radical equations

1 In Chapter 8 we learned to solve fractional equations by multiplying both sides by the LCD. We now consider more fractional equations that result in quadratic equations when both sides are multiplied by the LCD. Recall that we must always check our answers when solving fractional equations.

EXAMPLE 1

Solve $\dfrac{6}{x} - x = 5$.

The LCD $= x$.

$$x\left[\dfrac{6}{x} - x\right] = x \cdot 5 \quad \text{Multiply by LCD}$$

$$x \cdot \dfrac{6}{x} - x \cdot x = 5x$$

$$6 - x^2 = 5x$$

$$x^2 + 5x - 6 = 0$$

$$(x + 6)(x - 1) = 0$$

$$x + 6 = 0 \quad \text{or} \quad x - 1 = 0$$

$$x = -6 \qquad\qquad x = 1$$

Check: $\dfrac{6}{(-6)} - (-6) \stackrel{?}{=} 5 \qquad \dfrac{6}{(1)} - (1) \stackrel{?}{=} 5$

$\qquad\qquad -1 + 6 = 5 \qquad\qquad 6 - 1 = 5$

The solutions are -6 and 1. ◀◀

Practice Exercise 1

Solve $x - \dfrac{11}{x} = \dfrac{10}{x} + 4$.

Answer: $7, -3$

EXAMPLE 2

Solve $\dfrac{12}{x^2 - 4} - \dfrac{3}{x - 2} = -1$.

The LCD $= (x - 2)(x + 2)$.

$$(x - 2)(x + 2)\left[\dfrac{12}{x^2 - 4} - \dfrac{3}{x - 2}\right] = (x - 2)(x + 2)(-1) \quad \text{Multiply both sides by LCD}$$

$$(x - 2)(x + 2) \cdot \dfrac{12}{(x - 2)(x + 2)} = (x - 2)(x + 2) \cdot \dfrac{3}{(x - 2)} \quad \text{Distribute}$$

$$= (x - 2)(x + 2)(-1)$$

$$12 - 3(x + 2) = (x^2 - 4)(-1) \quad \text{Watch parentheses}$$

$$12 - 3x - 6 = -x^2 + 4 \quad \text{Watch signs}$$

$$x^2 - 3x + 2 = 0 \quad \text{Collect terms}$$

$$(x - 1)(x - 2) = 0 \quad \text{Factor}$$

$$x - 1 = 0 \quad \text{or} \quad x - 2 = 0$$

$$x = 1 \qquad\qquad x = 2$$

Check: $\dfrac{12}{(1)^2 - 4} - \dfrac{3}{(1) - 2} \stackrel{?}{=} -1 \qquad \dfrac{12}{(2)^2 - 4} - \dfrac{3}{(2) - 2} \stackrel{?}{=} -1$

$\qquad \dfrac{12}{-3} - \dfrac{3}{-1} \stackrel{?}{=} -1 \qquad\qquad \dfrac{2}{4 - 4} - \dfrac{3}{2 - 2} \stackrel{?}{=} -1$

$\qquad\qquad -4 + 3 = -1 \qquad\qquad \dfrac{12}{0} - \dfrac{3}{0} \neq -1$

$\qquad\qquad\qquad\qquad\qquad\qquad$ Cannot divide by 0

The only solution is 1. Notice that 2 makes a denominator zero in the original equation. In general, any number for which a fraction is undefined must be discarded as a solution. ◀◀

Practice Exercise 2

Solve $\dfrac{3}{x + 3} + \dfrac{4}{x^2 - 9} = 1$.

Answer: $4, -1$

EXAMPLE 3

Solve $\dfrac{y+1}{y} = \dfrac{-2}{y-2}$.

The LCD $= y(y-2)$.

$$y(y-2)\left[\dfrac{y+1}{y}\right] = y(y-2)\left[\dfrac{-2}{y-2}\right] \quad \text{Multiply by LCD}$$

$$(y-2)(y+1) = -2y$$
$$y^2 - y - 2 = -2y$$
$$y^2 + y - 2 = 0$$
$$(y+2)(y-1) = 0$$
$$y + 2 = 0 \quad \text{or} \quad y - 1 = 0$$
$$y = -2 \quad\quad\quad\quad y = 1$$

Check:
$$\dfrac{(-2)+1}{(-2)} \stackrel{?}{=} \dfrac{-2}{(-2)-2} \quad\quad \dfrac{(1)+1}{(1)} \stackrel{?}{=} \dfrac{-2}{(1)-2}$$

$$\dfrac{-1}{-2} \stackrel{?}{=} \dfrac{-2}{-4} \quad\quad\quad \dfrac{2}{1} \stackrel{?}{=} \dfrac{-2}{-1}$$

$$\dfrac{1}{2} = \dfrac{1}{2} \quad\quad\quad\quad 2 = 2$$

The solutions are -2 and 1. ◂◂

Practice Exercise 3

Solve $\dfrac{x^2}{x-10} = \dfrac{3}{x-10}$. Start by multiplying both sides by the LCD, $x-10$.

Answer: $\pm\sqrt{3}$

In a proportion equation like the one in Example 3, one step could be eliminated by forming the cross-product equation $(y+1)(y-2) = -2 \cdot y$. However, this method works only with a proportion equation.

2 ▸▸ When solving radical equations in Chapter 9, we squared both sides of the equation in order to eliminate a radical. Now we consider additional radical equations that become quadratic equations when both sides are squared. We must always check our answers when solving radical equations, since answers that do not satisfy the *original* equation may be introduced by the process of squaring.

EXAMPLE 4

Solve $\sqrt{12-x} = x$.

$$(\sqrt{12-x})^2 = x^2 \quad \text{Indicate the square of both sides}$$
$$12 - x = x^2 \quad \text{Square both sides}$$
$$x^2 + x - 12 = 0 \quad \text{Collect terms}$$
$$(x-3)(x+4) = 0 \quad \text{Factor}$$
$$x - 3 = 0 \quad \text{or} \quad x + 4 = 0$$
$$x = 3 \quad\quad\quad\quad x = -4$$

Check: $\sqrt{12-(3)} \stackrel{?}{=} 3 \quad\quad \sqrt{12-(-4)} \stackrel{?}{=} (-4)$

$\quad\quad\quad \sqrt{9} \stackrel{?}{=} 3 \quad\quad\quad\quad \sqrt{16} \neq -4 \quad (\sqrt{16} = 4,$

$\quad\quad\quad\quad 3 = 3 \quad\quad\quad\quad\quad\quad\quad\quad\quad\quad \text{not } -4.)$

The only solution is 3. ◂◂

Practice Exercise 4

Solve $x = \sqrt{8x+20}$.

Answer: 10 (-2 does not check)

EXAMPLE 5

Solve $x + 2 = \sqrt{x+8}$.

$$(x+2)^2 = (\sqrt{x+8})^2 \quad \text{Indicate the square of both sides}$$

Practice Exercise 5

Solve $\sqrt{13-6x} = x - 3$.

10.4 SOLVING FRACTIONAL AND RADICAL EQUATIONS

$x^2 + 4x + 4 = x + 8$ Square both sides, $(x + 2)^2 \neq (x^2 + 4)$

$x^2 + 3x - 4 = 0$ Collect terms

$(x + 4)(x - 1) = 0$ Factor

$x + 4 = 0$ or $x - 1 = 0$

$x = -4$ $x = 1$

Check: $(-4) + 2 \stackrel{?}{=} \sqrt{(-4) + 8}$ $(1) + 2 \stackrel{?}{=} \sqrt{(1) + 8}$

$-2 \neq \sqrt{4}$ $3 \stackrel{?}{=} \sqrt{9}$

$3 = 3$

The only solution is 1. ◂◂

Answer: no solution

EXAMPLE 6

Solve $x = 4\sqrt{x + 1} - 4$.

$x + 4 = 4\sqrt{x + 1}$ Add 4 to both sides to isolate the radical

$(x + 4)^2 = (4\sqrt{x + 1})^2$ Indicate the square of both sides

$x^2 + 8x + 16 = 16(x + 1)$ Do not forget the middle term on left side and do not forget to square 4 on the right

$x^2 + 8x + 16 = 16x + 16$ Distribute the 16

$x^2 - 8x = 0$ Collect terms

$x(x - 8) = 0$ Factor out x

$x = 0$ or $x - 8 = 0$

$x = 8$

Check: $(0) \stackrel{?}{=} 4\sqrt{(0) + 1} - 4$ $(8) \stackrel{?}{=} 4\sqrt{(8) + 1} - 4$

$0 \stackrel{?}{=} 4\sqrt{1} - 4$ $8 \stackrel{?}{=} 4\sqrt{9} - 4$

$0 \stackrel{?}{=} 4 - 4$ $8 \stackrel{?}{=} 4 \cdot 3 - 4$

$0 = 0$ $8 \stackrel{?}{=} 12 - 4$

$8 = 8$

The solutions are 0 and 8. ◂◂

Practice Exercise 6

Solve $\sqrt{x + 5} - x + 1 = 0$.

Answer: 4

10.4 EXERCISES A

Solve.

1. $\dfrac{3}{x - 4} = 1 + \dfrac{5}{x + 4}$

2. $\dfrac{y - 3}{y} = \dfrac{-4}{y + 1}$

3. $\dfrac{3}{1 + z} + \dfrac{2}{1 - z} = -1$

QUADRATIC EQUATIONS

4. $\dfrac{a^2}{a+1} - \dfrac{9}{a+1} = 0$

5. $\dfrac{1}{x+2} + \dfrac{x}{x-2} = \dfrac{1}{2}$

6. $\dfrac{60}{y+3} = \dfrac{60}{y} - 1$

7. $\dfrac{5}{z+3} - \dfrac{1}{z+3} = \dfrac{z+3}{z+2}$

8. $\dfrac{x-1}{x} = \dfrac{x}{x+1}$

9. $\dfrac{16}{a+2} = 1 + \dfrac{2}{a-4}$

10. $\dfrac{1}{x-3} + \dfrac{1}{x+3} = \dfrac{1}{x^2-9}$

11. $\dfrac{x^2}{2x+1} - \dfrac{5}{2x+1} = 0$
    [*Hint:* Multiply through by $2x+1$.]

12. $\dfrac{x+2}{2} = \dfrac{x+5}{x+2}$

13. $x = \sqrt{3x+10}$

14. $\sqrt{5y+6} - y = 0$

15. $z - 7 = \sqrt{z-5}$

16. $\sqrt{15-a} = a - 3$

17. $1 + 2\sqrt{x-1} = x$

18. $y = 1 + 6\sqrt{y-9}$

19. $\sqrt{1-2a} = a - 1$

20. $\sqrt{2x+7} - 4 = x$

21. $\sqrt{z^2+2} = z + 1$

22. $3\sqrt{y+1} - y - 1 = 0$ **23.** $\sqrt{5-4x} = 2 - x$ **24.** $\sqrt{2x^2 - 5} = x$

25. A number increased by its reciprocal is the same as $\frac{5}{2}$. Find the number.

26. The principal square root of 2 more than a number is equal to the number itself. Find the number.

For Review

Solve.

27. $z^2 - 8z + 8 = 0$ **28.** $x^2 - \frac{2}{3}x - 1 = 0$

29. $2y^2 + 3y - 2 = 0$ **30.** $5z^2 + 2z + 1 = 0$

31. It can be shown that a polygon (many sided geometric figure) with x sides has a total of n diagonals where $n = \frac{1}{2}x^2 - \frac{3}{2}x$. A hexagon has 6 sides; how many diagonals does it have? Draw a hexagon and verify your answer by actual count.

32. Use the formula in Exercise 31 to find the number of sides that a polygon has if it is known to have 20 diagonals. Draw a polygon with the number of sides that you determine and count its diagonals.

ANSWERS: **1.** 6, −8 **2.** 1, −3 **3.** −3, 2 **4.** 3, −3 **5.** 0, −6 **6.** 12, −15 **7.** −1 **8.** no solution **9.** 6, 10 **10.** $\frac{1}{2}$ **11.** $\pm\sqrt{5}$ **12.** $-1 \pm \sqrt{7}$ **13.** 5 **14.** 6 **15.** 9 **16.** 6 **17.** 5, 1 **18.** 25, 13 **19.** no solution **20.** −3 **21.** $\frac{1}{2}$ **22.** −1, 8 **23.** 1, −1 **24.** $\sqrt{5}$ **25.** 2, $\frac{1}{2}$ **26.** 2 **27.** $4 \pm 2\sqrt{2}$ **28.** $\frac{1 \pm \sqrt{10}}{3}$ **29.** $\frac{1}{2}$, −2 **30.** no solution **31.** 9 **32.** 8

10.4 EXERCISES B

Solve.

1. $\dfrac{3}{x-1} = 1 + \dfrac{2}{x+1}$ **2.** $\dfrac{y}{1+y} = \dfrac{3-y}{4}$ **3.** $\dfrac{1}{x+2} + \dfrac{1}{x-2} = \dfrac{3}{8}$

472 QUADRATIC EQUATIONS

4. $\dfrac{a^2}{a+5} - \dfrac{4}{a+5} = 0$
5. $\dfrac{6}{x+3} + \dfrac{x}{x-3} = 1$
6. $\dfrac{32}{y+5} = \dfrac{6}{y} + 2$

7. $\dfrac{4}{z+1} - \dfrac{1}{z+1} = \dfrac{z-1}{z+3}$
8. $x + \dfrac{2}{x-2} = \dfrac{1}{x-2}$
9. $\dfrac{z+2}{z} = \dfrac{1}{z+2}$

10. $\dfrac{3}{a-1} + 1 = \dfrac{6}{a^2-1}$
11. $\dfrac{x^2}{1+3x} - \dfrac{11}{1+3x} = 0$
12. $\dfrac{x+3}{x+1} = \dfrac{x+1}{3}$

13. $x = \sqrt{2x+24}$
14. $\sqrt{4y+5} - y = 0$
15. $z + 3 = \sqrt{12z+9}$

16. $\sqrt{5y+21} = y + 3$
17. $\sqrt{2x-1} = x - 2$
18. $z = \sqrt{z+7} - 1$

19. $\sqrt{7-3a} = 1 + a$
20. $2\sqrt{3-x} - 5 = x$
21. $\sqrt{y^2+8} = y - 2$

22. $2\sqrt{y+15} + y - 9 = 0$
23. $\sqrt{4x+13} = x + 2$
24. $\sqrt{5x^2-11} = 2x$

25. A number decreased by its reciprocal is the same as $\dfrac{15}{4}$. Find the number.

26. The principal square root of 8 more than a number is three times the number. Find the number.

For Review

Solve.

27. $x^2 - 6x + 6 = 0$
28. $y^2 - \dfrac{1}{3}y - 1 = 0$

29. $5z^2 + 14z - 3 = 0$
30. $4x^2 + x + 2 = 0$

31. It can be shown that a polygon with x sides has a total of n diagonals where $n = \dfrac{1}{2}x^2 - \dfrac{3}{2}x$. An octagon has 8 sides; how many diagonals does it have? Draw an octagon and verify your answer by actual count.

32. Use the formula in Exercise 31 to find the number of sides that a polygon has if it is known to have 5 diagonals. Draw a polygon with the number of sides that you determine and count its diagonals.

10.4 EXERCISES C

Solve.

1. $\dfrac{3}{x^2-5x+6} - \dfrac{2}{x^2+3x-10} = \dfrac{x+2}{x^2+2x-15}$
   $\left[\text{Answer: } \dfrac{1 \pm \sqrt{101}}{2}\right]$

2. $\sqrt{2x-2} - \sqrt{x+6} = -1$

10.5 APPLICATIONS OF QUADRATIC EQUATIONS

STUDENT GUIDEPOSTS

1. Age and number problems
2. Geometry problems
3. Work problems

1 Many applied problems can be solved using quadratic equations. We illustrated this in the preceding sections using several types of problems, and now we consider a wider variety of these applications.

10.5 APPLICATIONS OF QUADRATIC EQUATIONS

EXAMPLE 1

Murphy's age in 3 years will be four times the square of his age now. How old is Murphy?

Let x = Murphy's present age,

$x + 3$ = Murphy's age in 3 years,

$4x^2$ = four times the square of his present age.

The last two expressions are equal, so we solve the following equation.

$$4x^2 = x + 3$$
$$4x^2 - x - 3 = 0$$
$$(4x + 3)(x - 1) = 0$$
$$4x + 3 = 0 \quad \text{or} \quad x - 1 = 0$$
$$4x = -3 \qquad\qquad x = 1$$
$$x = -\frac{3}{4}$$

Since $-\frac{3}{4}$ could not be a person's age, $x = 1$ is the only possible solution.

Check: $4(1)^2 \stackrel{?}{=} 1 + 3$
$\qquad\quad 4 = 4$

Murphy is 1 year old. ◀◀

Practice Exercise 1

The square of Roberta's age plus the square of her age in 5 years is 97. How old is Roberta?

Answer: 4 years old

EXAMPLE 2

Twice the product of two consecutive positive even integers is 160. Find the integers.

Let x = the first positive even integer,

$x + 2$ = the next consecutive even integer.

Twice the product of these two integers is 160, so we should solve the following equation.

$$2x(x + 2) = 160$$
$$x(x + 2) = 80 \quad \text{Multiply both sides by } \frac{1}{2}$$
$$x^2 + 2x = 80$$
$$x^2 + 2x - 80 = 0$$
$$(x - 8)(x + 10) = 0$$
$$x - 8 = 0 \quad \text{or} \quad x + 10 = 0$$
$$x = 8 \qquad\qquad x = -10$$

Since x must be positive, 8 is the only possible solution.

Check: $2 \cdot 8(8 + 2) \stackrel{?}{=} 160$
$\qquad\quad 2 \cdot 8(10) \stackrel{?}{=} 160$
$\qquad\qquad\quad 160 = 160$

The numbers are 8 and 10. ($x + 2 = 10$.) ◀◀

Practice Exercise 2

Four times the product of two consecutive negative odd integers is 780. Find the integers.

Answer: $-15, -13$

---

2 When working a geometry problem remember to sketch a figure as shown in the next two examples.

474 QUADRATIC EQUATIONS

EXAMPLE 3

Find the length and width of a rectangle if the length is 3 cm more than the width and the area is 180 cm^2.

Make a sketch as in Figure 10.2.

Let x = width of the rectangle,

$x + 3$ = length of the rectangle.

Use the formula for the area of a rectangle, $A = l \cdot w$, to get the following equation.

$$x(x + 3) = 180$$
$$x^2 + 3x = 180$$
$$x^2 + 3x - 180 = 0$$
$$(x + 15)(x - 12) = 0$$
$$x + 15 = 0 \quad \text{or} \quad x - 12 = 0$$
$$x = -15 \quad\quad\quad x = 12$$
$$(x + 3 = 15)$$

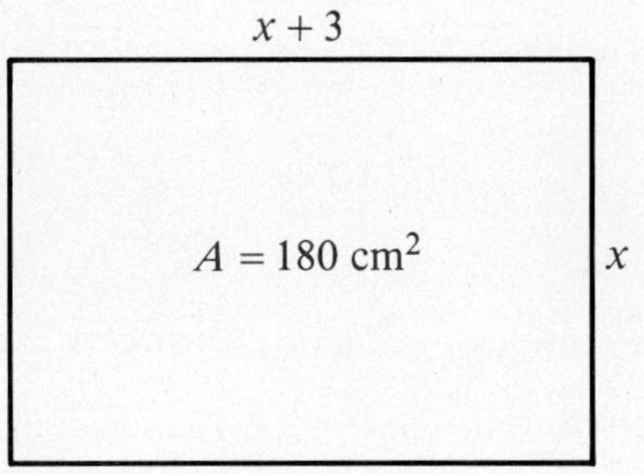

$x + 3$

$A = 180$ cm^2 x

Figure 10.2

Since -15 could not be the width of a rectangle, 12 is the only possible solution.

Check: $12(12 + 3) \stackrel{?}{=} 180$

$12 \cdot 15 \stackrel{?}{=} 180$

$180 = 180$

The width is 12 cm and the length is 15 cm. ◀◀

Practice Exercise 3

Find the base and height of a triangle if the base is 2 cm more than the height and the area is 24 cm^2.

Answer: 8 cm, 6 cm

EXAMPLE 4

The number of square inches in the area of a square is 12 more than the number of inches in its perimeter. Find the length of a side.

A sketch of the square is shown in Figure 10.3.

Let x = length of a side (all sides are the same length, x),

x^2 = area of the square ($A = l \cdot w = x \cdot x$),

$4x$ = perimeter of the square ($P = 2 \cdot l + 2 \cdot w = 2 \cdot x + 2 \cdot x = 4x$).

Since the area is 12 more than the perimeter, adding 12 to the perimeter, $4x$, will give us the area, x^2.

$$x^2 = 4x + 12$$

Practice Exercise 4

A box is 12 inches high and the volume is 420 in^3. Find the length and width if the length is 2 inches more than the width.

Let x = width of box

$x + 2$ = length of box

$12x(x + 2) =$ _____

$$x^2 - 4x - 12 = 0$$
$$(x + 2)(x - 6) = 0$$
$$x + 2 = 0 \quad \text{or} \quad x - 6 = 0$$
$$x = -2 \qquad\qquad x = 6$$

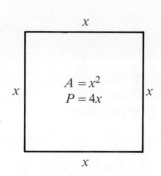

Figure 10.3

Since the length of a side cannot be negative, 6 is the only possible solution. Each side is 6 in. ◂◂

Answer: 7 inches, 5 inches

 Another variation of the work problems introduced in Section 8.5 results in a quadratic equation.

EXAMPLE 5

When each works alone, Ralph can do a job in 3 hours less time than Bert. When they work together, it takes 2 hours. How long does it take each to do the job by himself?

$2 = $ number of hrs to do the job together

$\dfrac{1}{2} = $ amount done together in 1 hr

Let $x = $ number of hrs for Bert to do the job,

$\dfrac{1}{x} = $ amount done by Bert in 1 hr.

Then $x - 3 = $ number of hrs for Ralph to do the job,

$\dfrac{1}{(x-3)} = $ amount done by Ralph in 1 hr.

(Note: $\dfrac{1}{(x-3)} \neq \dfrac{1}{x} - \dfrac{1}{3}$. To see this substitute $x = 2$.)

We must solve the following equation.

$\dfrac{1}{x} + \dfrac{1}{(x-3)} = \dfrac{1}{2}$ \qquad (amount by Bert) + (amount by Ralph) = (amount together)

$2x(x-3)\left(\dfrac{1}{x} + \dfrac{1}{(x-3)}\right) = 2x(x-3) \cdot \dfrac{1}{2}$ \qquad The LCD is $2x(x-3)$

$2\cancel{x}(x-3)\dfrac{1}{\cancel{x}} + 2x\cancel{(x-3)}\dfrac{1}{\cancel{(x-3)}} = x(x-3)$

$2(x - 3) + 2x = x^2 - 3x$

$2x - 6 + 2x = x^2 - 3x$

Practice Exercise 5

Gloria can do a job in 2 hours less time than Evelyn. How long would it take each working alone, if together they can do the job in 5 hours?

476 QUADRATIC EQUATIONS

$$0 = x^2 - 7x + 6$$
$$0 = (x - 6)(x - 1)$$

$x - 6 = 0$ or $x - 1 = 0$
$x = 6$ $x = 1$

If Bert did the job in 1 hr, Ralph would do it in $1 - 3 = -2$ hr, which makes no sense. Thus, Bert would take 6 hr and Ralph 3 hr. ◀◀

Answer: Gloria approximately 9.1 hr; Evelyn approximately 11.1 hr

---

It is important when working any word problem to be neat and complete. Do not attempt to take shortcuts, especially when writing down the pertinent information. Writing complete descriptions of the variables can eliminate time-consuming errors.

10.5 EXERCISES A

Solve. (Some problems have been started.)

1. If five is added to the square of a number the result is 41. Find the number.

   Let x = the desired number,
        $x^2 + 5$ = five added to the square of the number.

2. If the square of Mary's age is decreased by 44, the result is 100. How old is Mary?

   Let x = Mary's age,
        $x^2 - 44$ = the square of Mary's age decreased by 44.

3. If the square of a number is decreased by 10 the result is three times the number. Find the number.

   Let x = the number

4. Twice the square of Ernie's age, less 9, is the same as seventeen times his age. How old is Ernie?

   Let x = Ernie's age

5. The product of 1 more than a number and 1 less than the number is 99. Find the number.

   Let x = the desired number,
        $x + 1$ = 1 more than the number,
        $x - 1$ = 1 less than the number.

6. Sam's present age times his age in five years is 84. How old is Sam?

7. The product of two positive consecutive even integers is 120. Find the integers.

   Let x = the first even integer,
   $x + 2$ = the next even integer.

8. Three times the product of two positive consecutive odd integers is 297. Find the integers.

9. Adding 4 to the square of Marvin's age is the same as subtracting 3 from eight times his age. How old is Marvin?

10. The sum of the squares of two consecutive even positive integers is 100. Find the integers.

11. One number is 6 larger than another. The square of the larger is 96 more than the square of the smaller. Find the numbers.

12. Find the length and width of a rectangle if the length is 4 cm longer than the width and the area is 140 cm².

    Let x = width of rectangle,
    $x + 4$ = length of rectangle.

    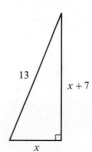

13. The number of square inches in the area of a square is 21 more than the number of inches in its perimeter. Find the length of a side.

    Let x = length of sides of square

    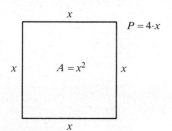

14. If the hypotenuse of a right triangle is 13 feet long and one leg is 7 feet longer than the other, find the measure of each leg.

QUADRATIC EQUATIONS

15. If the sides of a square are lengthened by 2 cm, its area becomes 169 cm². Find the length of a side.

Let x = length of sides of square,
  $x + 2$ = length of sides increased by 2 cm.

16. The area of a triangle is 24 ft² and the base is 1/3 as long as the altitude. Find the base and altitude.

17. A box is 6 inches high. The length is 8 inches longer than the width and the volume is 1080 in³. Find the width and length.

18. The area of a circle is 1256 cm². Find the radius. (Use $\pi \approx 3.14$.)

19. The area of a parallelogram is 55 ft². If the base is 1 ft greater than twice the altitude, find the base and altitude.

20. The perimeter of a rectangle is 62 inches and the length is 3 inches more than the width. Find the dimensions.

21. It takes one pipe 6 hours longer to fill a tank than it takes a second pipe. Working together they fill the tank in 4 hours. How long would it take each working alone to fill the tank?

22. It takes Joe 9 hours longer to grade a set of tests than Dan. If they work together they can grade them in 20 hours. How long would it take each to grade the tests if they worked alone?

23. A picture frame is 24 inches by 18 inches as shown in the sketch below. If the area of the picture itself is 216 in^2, what is the width of the frame?

Let x = the width of the frame.
$(24 - 2x)(18 - 2x) = 216$

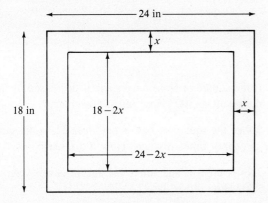

24. The pressure p, in pounds per square foot, of a wind blowing v miles per hour can be approximated by the equation $p = 0.003v^2$. What is the approximate velocity of the wind when it creates a pressure of 1.875 pounds per square foot against the side of a building?

25. The amount of money A that will result if a principal P is invested at r percent interest compounded annually for 2 years is given by $A = P(1 + r)^2$. If $1000 grows to $1210 in 2 years using this formula, what is the interest rate?

26. A boat travels 48 miles upstream and then returns in a total time of 10 hours. If the speed of the stream is 2 mph, what is the speed of the boat in still water?

27. Mary Conners can walk 8 miles up a mountain and then return in a total time of 6 hours. Her speed downhill is 2 mph faster than her speed uphill. What is her speed uphill?

For Review

Solve.

28. $\dfrac{x^2}{x - 3} = \dfrac{9}{x - 3} + 10$

29. $x = \sqrt{7x + 18} - 4$

30. The square root of Neil's age in 6 years is the same as one fifth of his present age. How old is Neil?

ANSWERS: **1.** 6, −6 **2.** 12 years old **3.** 5, −2 **4.** 9 years old **5.** 10, −10 **6.** 7 years old **7.** 10, 12 **8.** 9, 11 **9.** 1 year old or 7 years old **10.** 6, 8 **11.** 5, 11 **12.** 14 cm, 10 cm **13.** 7 in **14.** 5 ft, 12 ft **15.** 11 cm **16.** 12 ft, 4 ft **17.** 10 in, 18 in **18.** 20 cm **19.** 11 ft, 5 ft **20.** 17 in, 14 in **21.** 6 hr, 12 hr **22.** 36 hr, 45 hr **23.** 3 in **24.** 25 mph **25.** 10% **26.** 10 mph **27.** 2 mph **28.** 7 **29.** 1, −2 **30.** 30 years old

10.5 EXERCISES B

Solve.

1. If 4 is added to the square of a number, the result is 85. Find the number.

2. If the square of Rosemary's age is decreased by 24, the result is 300. How old is Rosemary?

3. If the square of a number is decreased by 32, the result is four times the number. Find the number.

4. Twice the square of Arlo's age, less 22, is the same as twenty times his age. How old is Arlo?

5. The product of 1 more than a number and 1 less than the number is 224. Find the number.

6. Troy's present age times his age in seven years is 260. How old is Troy?

7. The product of two positive consecutive even integers is 168. Find the integers.

8. Twice the product of two positive consecutive odd integers is 126. Find the integers.

9. Subtracting 100 from the square of Raul's age is the same as adding 60 to twelve times his age. How old is Raul?

10. The sum of the squares of two consecutive even positive integers is 164. Find the integers.

11. One number is 5 more than another. The square of the larger exceeds the square of the smaller by 95. Find the number.

12. Find the length and width of a rectangle if the length is 7 meters more than the width and the area is 198 m².

13. The number of square inches in the area of a square is 5 more than the number of inches in its perimeter. Find the length of a side.

14. If the hypotenuse of a right triangle is 26 cm long and one leg is 14 cm longer than the other, find the measure of each leg.

15. If the sides of a square are lengthened by 2 feet, its area becomes 100 ft². Find the length of a side.

16. The area of a triangle is 18 m² and the base is $\frac{1}{4}$ as long as the altitude. Find the base and altitude.

17. A box is 8 yards high. The length is 4 yards longer than the width and the volume is 360 yd³. Find the width and length.

18. The area of a circle is 200.96 cm². Find the radius. (Use $\pi \approx 3.14$.)

19. The area of a parallelogram is 175 ft². If the base is 4 feet more than three times the altitude, find the base and altitude.

20. The perimeter of a rectangular garden is 28 meters and the length is 4 meters more than the width. Find the dimensions of the garden.

21. It takes Jeff 16 hours longer to repair his car than it takes his father, who is a mechanic, to do the same job. If they could do the job together in 6 hours, how long would it take each if they worked alone?

22. Graydon, the registered cheese cutter at Perko's Delicatessen, is training a new assistant, Burford. It takes Burford 24 hours longer to process the Tillamook cheese shipment than it takes Graydon. If together they could process the cheese in 5 hours, how long would it take each, working alone, to cut and display the cheese?

23. A rectangular back yard is to have a sidewalk placed entirely around its perimeter in such a way that 875 ft² of lawn area are enclosed inside the walk. If the dimensions of the yard are 40 ft by 30 ft, what is the width of the sidewalk?

24. The pressure p, in pounds per square foot, of a wind blowing v miles per hour can be approximated by the equation $p = 0.003v^2$. What is the approximate wind velocity when a pressure of 2.7 pounds per square foot is exerted against the side of a camper?

25. The amount of money A that will result if a principal P is invested at r percent interest compounded annually for 2 years is given by $A = P(1 + r)^2$. If $2000 grows to $2645 in 2 years using this formula, what is the interest rate?

26. Pat Marx swims 4 miles downstream and then returns in a total time of 3 hours. If the speed of the stream is 1 mph, what is Pat's speed in still water?

27. Chuck Little rode a bicycle with the wind for 18 miles. He then returned against the wind and the total time of his trip was 5 hours. If his speed with the wind was 3 mph faster than against the wind, what was his speed against the wind?

For Review

Solve.

28. $\dfrac{z^2}{z - 2} = \dfrac{4}{z - 2} + 5$

29. $\sqrt{8y - 7} = y$

30. The square root of 6 more than a number is the same as 6 less than the number. Find the number.

10.5 EXERCISES C

Solve.

1. Two boats leave an island with one heading south and the other west. After 4 hours they are 100 miles apart. What is the speed of each boat if one travels 5 mph faster than the other? [Answer: 15 mph, 20 mph]

2. A boat requires one hour longer to go 80 miles upstream than to make the return trip downstream. What is the speed of the boat in still water if the speed of the stream is 2 mph?

10.6 SOLVING FORMULAS

STUDENT GUIDEPOSTS

1 Linear equations
2 Quadratic and radical equations

Many times, equations involve a variable and constants represented by letters instead of numerical constants. Formulas are excellent examples of these types of equations. For instance, we have used such formulas as

$$A = lw \qquad P = 2l + 2w \qquad c^2 = a^2 + b^2 \qquad c = \pi d.$$

To solve a given formula for a particular letter, remember that the letters play the same role as numbers and all of our equation-solving rules apply. If you have trouble solving for a letter in a formula, it may be helpful to make up a similar equation with numbers instead of letters, and pattern your solution steps after the procedure involved in solving the new equation. We will demonstrate this technique in the examples.

 Recall that solving a formula for a particular variable is a process of isolating the variable on one side of the equation.

482 QUADRATIC EQUATIONS

EXAMPLE 1

Solve $a + x = b$ for x.

Formula	Similar numerical equation
$a + x = b$	$3 + x = 7$
$a - a + x = b - a$ Subtract a	$3 - 3 + x = 7 - 3$ Subtract 3
$0 + x = b - a$	$0 + x = 7 - 3$
$x = b - a$	$x = 4$

The solution is $b - a$. ◂◂

Practice Exercise 1

Solve $2y + z = x$ for z.

Answer: $z = x - 2y$

EXAMPLE 2

Solve $cx = d$ for x.

Formula	Similar numerical equation
$cx = d$	$5x = 8$
$\frac{1}{c} \cdot cx = \frac{1}{c} \cdot d$ Multiply by $\frac{1}{c}$	$\frac{1}{5} \cdot 5x = \frac{1}{5} \cdot 8$ Multiply by $\frac{1}{5}$
$1 \cdot x = \frac{d}{c}$	$1 \cdot x = \frac{8}{5}$
$x = \frac{d}{c}$	$x = \frac{8}{5}$

The solution is $\frac{d}{c}$. ◂◂

Practice Exercise 2

Solve $5uv = w$ for u.

Answer: $u = \frac{w}{5v}$

EXAMPLE 3

Solve $ax + b = c$ for x.

Formula	Similar numerical equation
$ax + b = c$	$3x + 5 = 20$
$ax + b - b = c - b$ Subtract b	$3x + 5 - 5 = 20 - 5$ Subtract 5
$ax = c - b$	$3x = 15$
$\frac{1}{a} \cdot ax = \frac{1}{a}(c - b)$ Multiply by $\frac{1}{a}$	$\frac{1}{3} \cdot 3x = \frac{1}{3} \cdot 15$ Multiply by $\frac{1}{3}$
$x = \frac{c - b}{a}$	$x = \frac{15}{3} = 5$

The solution is $\frac{c - b}{a}$. ◂◂

Practice Exercise 3

Solve $mn - k = 6$ for n.

Answer: $n = \frac{k + 6}{m}$

EXAMPLE 4

Solve $A = lw$ for l.

Formula	Similar numerical equation
$a = lw$	$12 = l \cdot 4$
$A \cdot \frac{1}{w} = lw \cdot \frac{1}{w}$ Multiply by $\frac{1}{w}$	$12 \cdot \frac{1}{4} = l \cdot 4 \cdot \frac{1}{4}$ Multiply by $\frac{1}{4}$

Practice Exercise 4

Solve $PV = RT$ for R.

$$\frac{A}{w} = l \cdot 1$$

$$\frac{A}{w} = l$$

The solution is $\frac{A}{w}$. ◀◀

$$\frac{12}{4} = l \cdot 1$$

$$3 = l$$

Answer: $R = \dfrac{PV}{T}$

EXAMPLE 5

Solve $a = bcx$ for x.

$$a = bcx$$

$$\frac{1}{bc} \cdot a = \frac{1}{bc} \cdot bcx \quad \text{Multiply by } \frac{1}{bc}, \text{ the reciprocal of the coefficient of } x$$

$$\frac{a}{bc} = 1 \cdot x$$

$$\frac{a}{bc} = x$$

The solution is $\dfrac{a}{bc}$. ◀◀

Practice Exercise 5

Solve $3xyz = uv$ for x.

Answer: $x = \dfrac{uv}{3yz}$

---

2 ▶ Some formulas involve quadratic equations or radical equations. We shall assume that the variables and constants are chosen so that division by zero and negative numbers under a radical sign are avoided.

EXAMPLE 6

Solve $a = bx^2$ for x.

Formula

$$a = bx^2$$

$$\frac{1}{b} \cdot a = \frac{1}{b} \cdot bx^2 \quad \text{Multiply by } \frac{1}{b}$$

$$\frac{a}{b} = x^2$$

$$\pm\sqrt{\frac{a}{b}} = x \quad \text{Take square root of both sides}$$

The solutions are $\pm\sqrt{\dfrac{a}{b}}$. ◀◀

Similar numerical equation

$$50 = 2x^2$$

$$\frac{1}{2} \cdot 50 = \frac{1}{2} \cdot 2x^2 \quad \text{Multiply by } \frac{1}{2}$$

$$25 = x^2$$

$$\pm 5 = \pm\sqrt{25} = x$$

Practice Exercise 6

Solve $uy^2 - v = 0$ for y.

Answer: $y = \pm\sqrt{\dfrac{v}{u}}$

Answer: $y = \pm\sqrt{\dfrac{v}{u}}$

EXAMPLE 7

Solve $a = \sqrt{\dfrac{x}{b}}$ for x.

Formula

$$a = \sqrt{\frac{x}{b}}$$

$$(a)^2 = \left(\sqrt{\frac{x}{b}}\right)^2 \quad \text{Square both sides}$$

Similar numerical equation

$$5 = \sqrt{\frac{x}{2}}$$

$$(5)^2 = \left(\sqrt{\frac{x}{2}}\right)^2$$

Practice Exercise 7

Solve $m = \sqrt{\dfrac{2k}{n}}$ for n.

$a^2 = \dfrac{x}{b}$

$b \cdot a^2 = b \cdot \dfrac{x}{b}$ Multiply by b

$ba^2 = x$

The solution is ba^2. ◂◂

$25 = \dfrac{x}{2}$

$2 \cdot 25 = 2 \cdot \dfrac{x}{2}$

$50 = x$

Answer: $n = \dfrac{2k}{m^2}$

10.6 EXERCISES A

1. Solve $g = x + h$ for x.

2. Solve $t - x = k$ for x.

3. Solve $bx + d = e$ for x.

4. Solve $g - ax = m$ for x.

5. Solve $d = rt$ for t.

6. Solve $d = rt$ for r.

7. Solve $E = IR$ for I.

8. Solve $I = prt$ for t.

9. Solve $I = prt$ for r.

10. Solve $I = prt$ for p.

11. Solve $A = \dfrac{1}{2}bh$ for b.

12. Solve $A = \dfrac{1}{2}bh$ for h.

13. Solve $r = \dfrac{d}{t}$ for d.

14. Solve $r = \dfrac{d}{t}$ for t.

15. Solve $A = \pi r^2$ for r.

16. Solve $S = \dfrac{1}{2}gt^2$ for t.

17. Solve $E = mc^2$ for m.

18. Solve $E = mc^2$ for c.

19. Solve $A = \dfrac{4}{3}\pi r^2$ for r.

20. Solve $U = \sqrt{\dfrac{a}{c}}$ for c.

21. Solve $U = \sqrt{\dfrac{a}{c}}$ for a.

22. Solve $\sqrt{x+a} = b$ for x.

For Review

Solve.

23. The product of 2 less than a number and 3 more than the number is 176. Find the number.

24. The hypotenuse of a right triangle is 5 cm and one leg is 1 cm longer than the other. Find the measure of each leg.

25. The area of a triangle is 24 ft², and the base is 13 as long as the altitude. Find the base and altitude.

26. Laura takes 2 days to type a manuscript and Holly takes 8 days. How long would it take if they worked together?

ANSWERS: **1.** $g-h$ **2.** $t-k$ **3.** $\dfrac{e-d}{b}$ **4.** $\dfrac{g-m}{a}$ **5.** $\dfrac{d}{r}$ **6.** $\dfrac{d}{t}$ **7.** $\dfrac{E}{R}$ **8.** $\dfrac{I}{pr}$ **9.** $\dfrac{I}{pt}$ **10.** $\dfrac{I}{rt}$
11. $\dfrac{2A}{h}$ **12.** $\dfrac{2A}{b}$ **13.** rt **14.** $\dfrac{d}{r}$ **15.** $\pm\sqrt{\dfrac{A}{\pi}}$ **16.** $\pm\sqrt{\dfrac{2S}{g}}$ **17.** $\dfrac{E}{c^2}$ **18.** $\pm\sqrt{\dfrac{E}{m}}$ **19.** $\pm\sqrt{\dfrac{3A}{4\pi}}$ **20.** $\dfrac{a}{U^2}$
21. cU^2 **22.** $b^2 - a$ **23.** $-14, 13$ **24.** 3 cm, 4 cm **25.** 12 ft, 4 ft **26.** $\dfrac{8}{5}$ days

---

10.6 EXERCISES B

1. Solve $x + k = w$ for x.

2. Solve $t = m - x$ for x.

3. Solve $ax + w = m$ for x.

4. Solve $p - bx = w$ for x.

5. Solve $A = bh$ for h.

6. Solve $A = bh$ for b.

7. Solve $T = cn$ for n.

8. Solve $V = lwh$ for h.

9. Solve $V = lwh$ for w.

10. Solve $V = lwh$ for l.

11. Solve $W = \dfrac{1}{3}pq$ for p.

12. Solve $W = \dfrac{1}{3}pq$ for q.

13. Solve $a = \dfrac{b}{m}$ for b.

14. Solve $a = \dfrac{b}{m}$ for m.

15. Solve $V = \pi r^2 h$ for h.

16. Solve $V = \pi r^2 h$ for r.

17. Solve $S = a\pi r^2$ for r.

18. Solve $S = a\pi r^2$ for a.

19. Solve $W = \dfrac{5}{4} a m^2$ for m.

20. Solve $T = \sqrt{\dfrac{a}{w}}$ for a.

21. Solve $T = \sqrt{\dfrac{a}{w}}$ for w.

22. Solve $\sqrt{x + t} = p$ for x.

For Review

Solve.

23. Of two positive numbers, one is 4 larger than the other and the sum of the squares of the two is 136. Find the numbers.

24. The altitude of a triangle is 8 inches more than the base. If the area is 90 in^2, find the altitude and base.

25. The length of a rectangle is 4 cm more than the width and the area is 96 cm^2. Find the dimensions.

26. The amount of money A that will result if a principal P is invested at r percent interest compounded annually for 2 years is given by $A = P(1 + r)^2$. If $5000 grows to $6498 in 2 years using this formula, what is the interest rate?

10.6 EXERCISES C

1. Solve $\dfrac{u}{1 + v^2} = 2$ for positive v.

2. Solve $\dfrac{x}{\sqrt{2x + y}} = 1$ for x.
   [Answer: $x = 1 \pm \sqrt{1 + y}$]

10.7 GRAPHING QUADRATIC EQUATIONS

STUDENT GUIDEPOSTS

1. Parabola
2. Graphs for $a > 0$ and $a < 0$
3. Vertex

In Chapter 4 we learned how to graph a linear equation of the form

$$ax + by + c = 0 \quad \text{General form}$$

or

$$y = mx + b \quad \text{Slope-intercept form}$$

by plotting the intercepts (or the intercept and one additional point if the intercepts are both (0, 0)) and drawing the straight line through them. The graph of every **quadratic equation in two variables**,

$$y = ax^2 + bx + c, \quad (a \neq 0)$$

1 is a **parabola**, a U-shaped curve similar to the one shown in Figure 10.4.

In order to graph a quadratic equation, we choose several values for x, calculate the corresponding y-values, and plot the resulting points. A table of values helps us keep a record of the points.

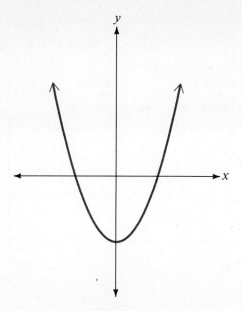

Figure 10.4

EXAMPLE 1

Graph $y = x^2$.

Find y-values for the following values of x.

If $x = 0$, $y = x^2 = 0^2 = 0$.   If $x = -1$, $y = x^2 = (-1)^2 = 1$.
If $x = 1$, $y = x^2 = 1^2 = 1$.   If $x = -2$, $y = x^2 = (-2)^2 = 4$.
If $x = 2$, $y = x^2 = 2^2 = 4$.

The completed table is shown below. Plotting the ordered pairs, we obtain the parabola in Figure 10.5. ◀◀

x	y
0	0
1	1
2	4
−1	1
−2	4

Practice Exercise 1

Graph $y = x^2 - 1$.

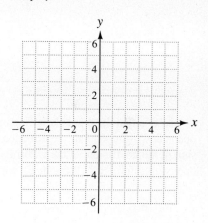

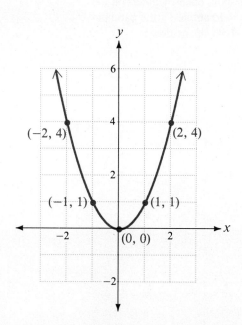

Figure 10.5

Answer: Same as Figure 10.5 except all points down one unit.

488 QUADRATIC EQUATIONS

EXAMPLE 2

Graph $y = -x^2$.

Be careful when finding values for y in this case. For example, if $x = 1$, $y = -x^2 = -(1)^2 = -1$.

The completed table appears beside the graph given in Figure 10.6.

x	y
0	0
1	−1
2	−4
−1	−1
−2	−4

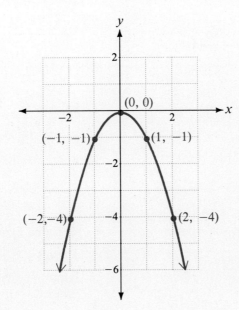

Figure 10.6

Practice Exercise 2

Graph $y = -x^2 + 2$.

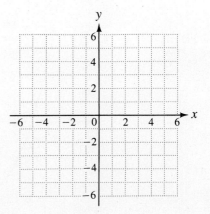

Answer: Same as Figure 10.6 except all points up two units.

Notice that the parabola in Example 1 opens up while the one in Example 2 opens down. In general, the direction in which a parabola opens depends on the sign of the coefficient of the x^2 term, a. If $a > 0$ (as in Example 1), the parabola opens up, and if $a < 0$ (as in Example 2), the parabola opens down.

 The graph of $y = ax^2 + bx + c$ ($a \neq 0$) is a parabola that

1. opens up when the coefficient of x^2 is positive

2. opens down when the coefficient of x^2 is negative.

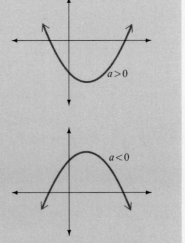

EXAMPLE 3

Graph $y = x^2 - 2x - 3$.

We know that the parabola opens up since the coefficient of x^2 is positive. Using the values computed in the table, the graph is plotted in Figure 10.7. ◀◀

x	y
0	-3
1	-4
2	-3
3	0
-1	0

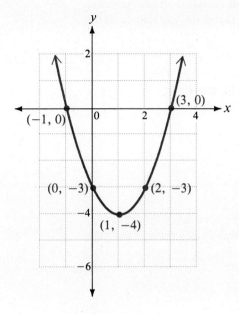

Figure 10.7

Practice Exercise 3

Graph $y = -x^2 + 2x + 3$.

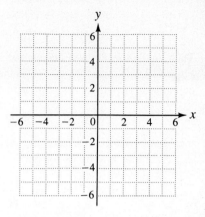

Answer: Like Figure 10.7 except opening down.

EXAMPLE 4

Graph $y = -x^2 + 2x$.

Be sure to use care when substituting for x. The graph is given in Figure 10.8. ◀◀

x	y
0	0
1	1
2	0
3	-3
-1	-3

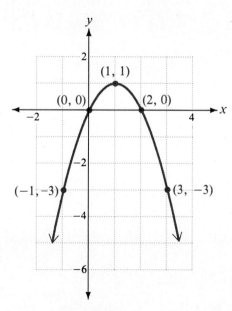

Figure 10.8

Practice Exercise 4

Graph $y = x^2 - 2x$.

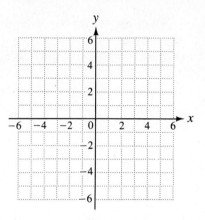

Answer: Like Figure 10.8 except opening up.

490 QUADRATIC EQUATIONS

Consider the following equation.
$$y = (x - 1)^2 - 4$$
$$= x^2 - 2x + 1 - 4$$
$$= x^2 - 2x - 3$$

Thus, $y = (x - 1)^2 - 4$ and $y = x^2 - 2x - 3$ have the same graph, given in Figure 10.7. Notice that $(1, -4)$ is the low point or **vertex** of the parabola. If the equation of a parabola is written in the form

$$y = (x - \boxed{h})^2 + \boxed{k},$$

then *(h, k)* is the vertex and the graph is a parabola opening up.

A similar rule holds for parabolas opening down. Consider
$$y = -(x - 1)^2 + 1$$
$$= -(x^2 - 2x + 1) + 1$$
$$= -x^2 + 2x - 1 + 1$$
$$= -x^2 + 2x.$$

Thus, $y = -(x - 1)^2 + 1$ and $y = -x^2 + 2x$ have the same graph, given in Figure 10.8, with vertex $(1, 1)$. If the equation of a parabola is written in the form

$$y = -(x - \boxed{h})^2 + \boxed{k},$$

then *(h, k)* is the vertex, and the graph is a parabola opening down.

EXAMPLE 5

Determine the vertex of each parabola.

(A) $y = (x + 3)^2 - 5$

We must have the form $y = (x - h)^2 + k$.
$$y = (x + 3)^2 - 5$$
$$= [x - (-3)]^2 + (-5)$$

Thus, $h = -3$ and $k = -5$. The vertex is $(-3, -5)$.

(B) $y = -(x + 1)^2 + 7$

Write in the form $y = -(x - h)^2 + k$.
$$y = -(x + 1)^2 + 7$$
$$= -[x - (-1)]^2 + 7$$

Thus, $h = -1$ and $k = 7$. The vertex is $(-1, 7)$.

(C) $y = (x + 5)^2$

Write in the form $y = (x - h)^2 + k$
$$y = (x + 5)^2$$
$$= [x - (-5)]^2 + 0$$

Thus, $h = -5$ and $k = 0$. The vertex is $(-5, 0)$.

Practice Exercise 5

Determine the vertex of each parabola.

(A) $y = (x - 4)^2 - 5$

(B) $y = -(x + 2)^2 + 4$

(C) $y = -(x + 8)^2$

Answers: **(A)** $(4, -5)$ **(B)** $(-2, 4)$ **(C)** $(-8, 0)$

10.7 EXERCISES A

Before graphing, determine whether the parabola opens up or down. Then graph the equation.

1. $y = x^2 + 1$

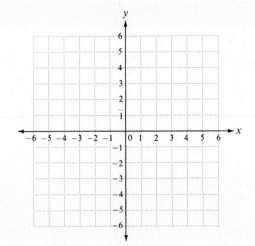

2. $y = -x^2 + 1$

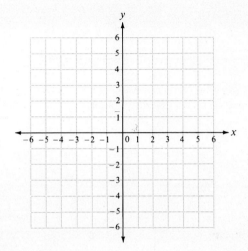

3. $y = x^2 - 2$

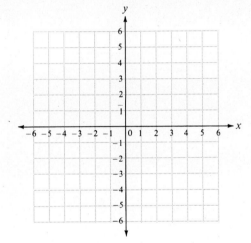

4. $y = -x^2 + 2$

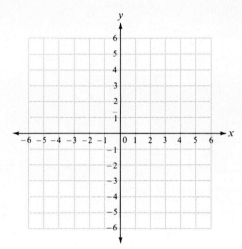

5. $y = x^2 + 2x + 1$

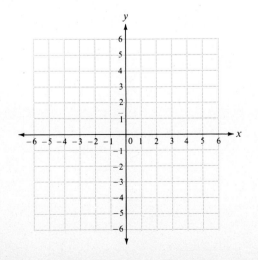

6. $y = -x^2 - 2x - 1$

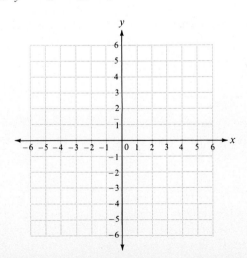

492 QUADRATIC EQUATIONS

7. $y = x^2 + 2x$

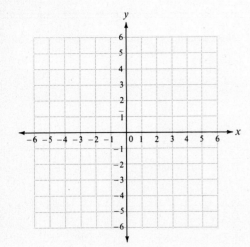

8. $y = -x^2 - 2x$

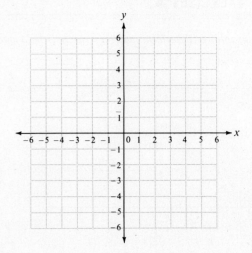

9. $y = x^2 - 3x + 2$

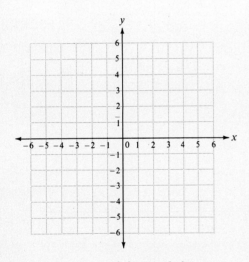

10. $y = -x^2 + 3x - 2$

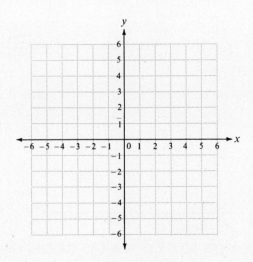

Determine the vertex of the parabola.

11. $y = (x - 4)^2 + 2$

12. $y = -(x - 1)^2 + 5$

13. $y = (x + 3)^2 - 2$

14. $y = -(x + 10)^2 - 9$

15. $y = x^2 - 3$

16. $y = -x^2 - 3$

For Review

17. Solve $p = ax + t$ for x.

18. Solve $M = \dfrac{1}{3}wt^2$ for t.

19. Solve $A = \dfrac{v}{2w}$ for v.

20. Solve $A = \dfrac{v}{2w}$ for w.
21. Solve $S = \sqrt{\dfrac{m}{r}}$ for m.
22. Solve $\sqrt{3x + u} = a$ for x.

ANSWERS:

1. 2. 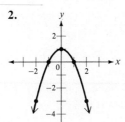 3. 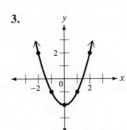 4. 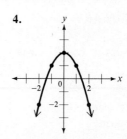 5.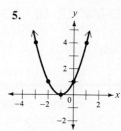

6. 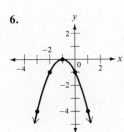 7. 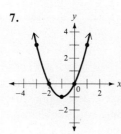 8. 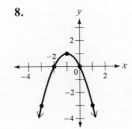 9. 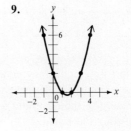 10.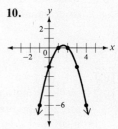

11. $(4, 2)$ 12. $(1, 5)$ 13. $(-3, -2)$ 14. $(-10, -9)$ 15. $(0, -3)$ 16. $(0, -3)$ 17. $\dfrac{p - t}{a}$ 18. $\pm\sqrt{\dfrac{3M}{w}}$

19. $2Aw$ 20. $\dfrac{v}{2A}$ 21. $S^2 r$ 22. $\dfrac{a^2 - u}{3}$

10.7 EXERCISES B

Before graphing, determine whether the parabola opens up or down. Then graph the equation.

1. $y = x^2 + 2$
2. $y = -x^2 - 1$
3. $y = x^2 - 1$
4. $y = -x^2 + 3$
5. $y = x^2 - 2x + 1$
6. $y = -x^2 + 2x - 1$
7. $y = x^2 + 4x$
8. $y = -x^2 - 4x$
9. $y = x^2 + 3x + 2$
10. $y = -x^2 - 3x - 2$

Determine the vertex of the parabola.

11. $y = (x - 7)^2 + 1$
12. $y = -(x - 4)^2 + 8$
13. $y = (x + 12)^2 - 8$
14. $y = -(x + 9)^2 - 14$
15. $y = (x + 1)^2$
16. $y = -x^2 + 8$

For Review

17. Solve $q = bx - w$ for x.
18. Solve $V = \dfrac{4}{3} am^2$ for m.
19. Solve $R = \dfrac{I}{2E}$ for I.
20. Solve $R = \dfrac{I}{2E}$ for E.
21. Solve $W = \sqrt{\dfrac{a}{n}}$ for n.
22. Solve $\sqrt{a + 2x} = z$ for x.

10.7 EXERCISES C

Find the vertex of each parabola by completing the square.

1. $y = x^2 + 10x + 20$ [Answer: $(-5, -5)$]

2. $y = -x^2 + 6x - 4$

CHAPTER 10 SUMMARY

Key Words and Phrases for Review

- **10.1** quadratic equation
  solving by factoring
  solving by taking roots
- **10.2** completing the square
- **10.3** quadratic formula
  discriminant
- **10.6** formula
- **10.7** quadratic equation in two variables
  parabola
  vertex

Key Concepts

10.1
10.2 The best ways to solve a quadratic equation are by factoring and taking roots. If these methods are not appropriate, use the quadratic formula instead of completing the square.

10.3 The quadratic formula for solving $ax^2 + bx + c = 0$ is
$$x = \frac{-b \pm \sqrt{b^2 - 4ac}}{2a}.$$

Make sure that an equation is put into the above general form before trying to identify the constants a, b, and c. Also, the entire numerator $-b \pm \sqrt{b^2 - 4ac}$, is divided by $2a$, *not* just the radical term.

10.4
1. To solve a fractional equation, multiply both sides by the LCD of all fractions. Be sure to check all possible solutions in the original equations and omit those that make any denominator zero.

2. To solve a radical equation, isolate a radical and square both sides of the equation. If a binomial occurs on one side, don't forget the middle term when squaring. For example,
$$x + 4 = 3\sqrt{x + 1}$$
becomes
$$x^2 + 8x + 16 = 9(x + 1).$$

Be sure to square 3 also. Finally, check all possible answers in the *original* equation and remember that the radical only represents the positive (principal) square root.

10.5 Be precise and write out all details (including complete descriptions of the variable) when solving word problems.

10.6 Solving a similar numerical equation can often help when solving a formula for a particular variable.

10.7 The graph of a quadratic equation $y = ax^2 + bx + c (a \neq 0)$ is a parabola that opens up if $a > 0$ and down if $a < 0$.

Review Exercises

10.1 *Solve the following quadratic equations.*

1. $2x^2 - 32 = 0$
2. $-3z^2 = -15$
3. $x^2 + 6x - 72 = 0$

4. $2y^2 + y = 21$
5. $(z - 5)^2 = 25$
6. $21x^2 - 4x - 32 = 0$

7. The square of a number, less 7, is equal to 9. Find the number.

10.2 *What must be added to complete the square?*

8. $x^2 + x +$ _____
9. $y^2 + \frac{1}{4}y +$ _____

10. Solve $x^2 - 3x + 1 = 0$ by completing the square.

10.3 *Solve.*

11. $2y^2 + y = 5$
12. $2z^2 + 11z = -(10 + z)$

13. $x^2 + 2x - 5 = 0$
14. $2y^2 + y + 5 = 0$

15. Lennie has a garden in the shape of a square 12 yards on a side. He wishes to lay a water pipe from one corner diagonally across to the other. What is the approximate length of this pipe?

10.4 *Solve.*

16. $\dfrac{z+2}{-z} = \dfrac{1}{z+2}$

17. $\dfrac{3}{1+x} = -1 - \dfrac{2}{1-x}$

18. $x = 1 + \sqrt{4-4x}$

19. $4 = 4\sqrt{x+1} - x$

20. Twice the square root of 2 more than a number is the same as 1 less than the number. Find the number.

10.5

21. The product of two positive consecutive integers is 420. Find the integers.

22. Mike's present age times his age in 7 years is 30. Find his present age.

23. The number of square inches in a square is 3 less than the number of inches in its perimeter. Find the length of its sides.

24. Find the number whose square is 18 more than three times the number.

25. The pressure p, in pounds per square foot, of a wind blowing v miles per hour can be approximated by the equation $p = 0.003v^2$. What is the approximate wind velocity when a pressure of 3.675 pounds per square foot is exerted against the side of a skyscraper?

26. Use $A = P(1+r)^2$ to find the interest rate if $3000 grows to $3630 in 2 years.

27. If it takes Bill 12 hours longer to paint a room than Peter and together they can complete the job in 8 hours, how long would it take each to do it alone?

10.6 **28.** Solve $c = \frac{1}{3}dh$ for h.   **29.** Solve $a - x = 2b$ for x.   **30.** Solve $D = au^2$ for u.

31. Solve $D = au^2$ for a.   **32.** Solve $g = \sqrt{\frac{a}{d}}$ for a.   **33.** Solve $g = \sqrt{\frac{a}{d}}$ for d.

10.7 **34.** Determine whether the parabola with the given equation opens up or down.

(A) $y = 3x^2 + x - 1$ (B) $y = -2x^2 + 5$

Graph the given equations.

35. $y = x^2 - 3$

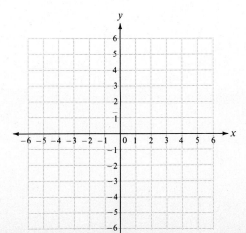

36. $y = x^2 - 6x + 8$

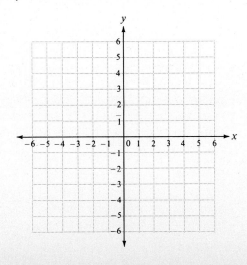

498 QUADRATIC EQUATIONS

Determine the vertex of the parabola.

37. $y = -(x + 4)^2 - 3$

38. $y = x^2 + 6$

ANSWERS: **1.** ±4 **2.** ±$\sqrt{5}$ **3.** 6, −12 **4.** 3, $-\frac{7}{2}$ **5.** 0, 10 **6.** $\frac{4}{3}$, $-\frac{8}{7}$ **7.** 4, −4 **8.** $\frac{1}{4}$ **9.** $\frac{1}{64}$
10. $\frac{3 \pm \sqrt{5}}{2}$ **11.** $\frac{-1 \pm \sqrt{41}}{4}$ **12.** −1, −5 **13.** $-1 \pm \sqrt{6}$ **14.** no solution **15.** 16.8 yd (using $\sqrt{2} \approx 1.4$)
16. −1, −4 **17.** 2, −3 **18.** 1 **19.** 0, 8 **20.** 7 **21.** 20, 21 **22.** 3 years old **23.** 1 in or 3 in **24.** 6, −3
25. 35 mph **26.** 10% **27.** Peter: 12 hr, Bill: 24 hr **28.** $\frac{3c}{d}$ **29.** $a - 2b$ **20.** $\pm\sqrt{\frac{D}{a}}$ **31.** $\frac{D}{u^2}$ **32.** dg^2
33. $\frac{a}{g^2}$ **34.** (A) up (B) down
35. 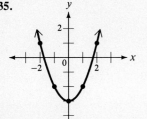 **36.** 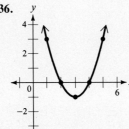 **37.** (−4, −3) **38.** (0, 6)

CHAPTER 10 TEST

Solve the following quadratic equations.

1. $x^2 + 13x + 36 = 0$

    1. _____

2. $3z^2 - z = 27 - z$

    2. _____

3. $y(y - 2) = 3y$

    3. _____

4. $x^2 - 4x + 1 = 0$

    4. _____

5. What must be added to complete the square?

    $y^2 + \dfrac{1}{2}y +$ _____

    5. _____

Solve.

6. $\dfrac{2}{x - 1} + 2 = \dfrac{3}{x + 2}$

    6. _____

CHAPTER 10 TEST Continued

7. $4 = x + \sqrt{2 + x}$

7. _____

Solve.

8. The product of two positive consecutive even integers is 120. Find the integers.

8. _____

9. It takes 8 hours less time to fill a pond using a large pipe than it takes using a smaller pipe. If when used together they fill the pond in 3 hours, how long would it take each to fill it alone?

9. _____

10. A boat travels 70 miles upstream and then returns in a total time of 12 hours. If the speed of the stream is 2 mph, what is the speed of the boat in still water?

10. _____

11. Solve $w = \sqrt{\dfrac{b}{m}}$ for m.

11. _____

12. Given the equation $y = x^2 + 2$.

    (A) Does the graph of this parabola open up or down?   12(A) _____

    (B) What is the vertex of the parabola?   12(B) _____

    (C) Graph the equation.

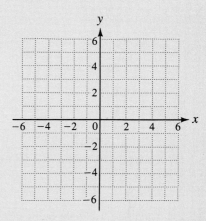

FINAL REVIEW EXERCISES

Chapter 1

1. Reduce $\frac{242}{220}$ to lowest terms.

2. Bill received 28 of the 42 votes cast. What fractional part of the votes did he receive? Reduce the fraction.

Perform the indicated operations.

3. $\frac{5}{11} \cdot \frac{33}{35}$

4. $\frac{20}{39} \div \frac{6}{13}$

5. $\frac{2}{15} + \frac{8}{3}$

6. $\frac{11}{14} - \frac{3}{4}$

7. $29.705 + 3.01$

8. $17.6 - 9.72$

9. $(0.0049)(52.8)$

10. $2.86 \div 0.052$

11. Change $\frac{43}{18}$ to a mixed number.

12. Change $8\frac{3}{13}$ to an improper fraction.

13. There are 475 light bulbs in a shipment. If $\frac{1}{25}$ of them are defective, how many are defective?

14. Harry bought $2\frac{3}{8}$ pounds of one candy and $3\frac{11}{16}$ pounds of another. What total weight of candy did he buy?

15. Convert $\frac{3}{8}$ to a percent.

16. Convert 2.72 to a fraction.

17. The tax rate in Upstate, NY is 6%. What is the tax on a $38.50 purchase?

18. Write $yyyyyy$ in exponential notation.

19. Evaluate $x^2 + 4(x - y)$ for $x = 5$ and $y = 3$.

20. If $A = P(1 + r)^t$, find A when P is $2000, the interest is 8% compounded annually, and t is 2 years.

Chapter 2

21. Find the reciprocal of $-\frac{3}{4}$.

22. Find $|-8|$.

Place the appropriate symbol =, <, or > between the pairs of numbers.

23. $-8 \quad -10$

24. $\frac{8}{31} \quad \frac{4}{17}$

Perform the indicated operations.

25. $-4 - (-16)$ **26.** $14 + (-24)$ **27.** $\left(-\dfrac{5}{9}\right) \cdot \left(-\dfrac{3}{20}\right)$ **28.** $(7.5) \div (-2.5)$

Evaluate.

29. $2 \cdot 4 + (8 - 12) - 2(3 - 1)$ **30.** $-8 - [-(-8)]$

Factor.

31. $2y + 12$ **32.** $-14x - 21$

Collect like terms.

33. $-8a + 3b - 4a - 5 - b$ **34.** $7y - (2y - 5) + 3$

35. Evaluate $2u^2 - v^2$ when $u = -2$ and $v = 4$.   **36.** Write $-5x^4$ without exponents.

Simplify and write without negative exponents.

37. $3x^2 x^3$ **38.** $\dfrac{y^6}{y^2}$ **39.** $(-3a^3)^2$ **40.** a^{-2}

41. Write 36,200,000 in scientific notation.   **42.** Write 7.2×10^{-4} without using scientific notation.

43. Evaluate. $-\sqrt{144}$ **44.** Evaluate. $\sqrt{\dfrac{12}{75}}$

Chapter 3

Solve.

45. $x + 10 = 19$ **46.** $16x = 4$ **47.** $x - \dfrac{5}{3} = \dfrac{7}{3}$

48. $-5.5x = 33$

49. $\dfrac{2}{3}x = 26$

50. $\dfrac{x}{\dfrac{3}{4}} = 12$

51. $4x + 8 = 7x - 4$

52. $-3x - 2 = -5x + 9$

53. $5(x + 2) - 2(x - 10) = 0$

54. $9x - (2x + 3) = 4$

55. The sum of two consecutive even integers is 134. What are the integers?

56. A suit is put on sale at a 20% discount. If the original price was $160, what is the sale price?

57. The area of a triangle is 85 in^2 and the base is 10 inches. What is the altitude?

58. Two trains leave Portland, one traveling north and the other south. If one is traveling 30 mph faster than the other, how fast is each traveling if they are 360 miles apart after 3 hours?

Solve the inequality.

59. $4x + 9 \geq 7x - 3$

60. $2x - (3x - 2) < 9x - 8$

Chapter 4

Graph on the number line.

61. $3x + 8 = 5$

62. $4(x - 2) \geq 4 - (x - 3)$

63. The point with coordinates $(-3, -5)$ is located in which quadrant?

64. Complete the ordered pair $(5,)$ so that it is a solution to $2x + 3y = 10$.

65. Cost, c, is related to number of items produced, n, by $c = 100n + 50$. Find c when n is 18.

66. True or False: $5 + 3y = 2x$ is a linear equation.

67. Find the slope of the line through the points (7, −1) and (2, −5). Determine the general form of the equation of the line.

68. Write $3x - 4y = 12$ in slope-intercept form and give the slope and the y-intercept.

Graph in the given Cartesian coordinate system.

69. $y = -3x + 6$

70. $2x - 3y + 6 \leq 0$

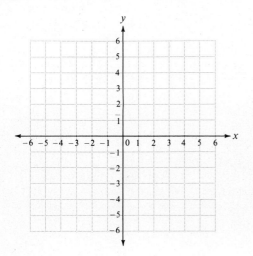
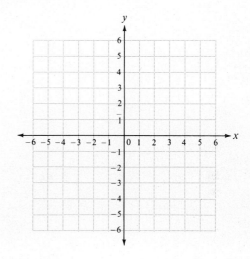

Chapter 5

71. Without graphing, determine the nature of the graphs (intersecting, parallel, or coinciding).
$3x + 4y = 7$
$9x + 12y = 5$

72. If the lines in a system of equations are intersecting, the system has _____ solution(s).

Solve the system.

73. $3x - y = -2$
$-5x + 2y = 5$

74. $2x + 3y = 8$
$4x + 7y = 16$

75. Twice one number plus a second is 13. The first minus the second is 5. Find the numbers.

76. A candy shop has one candy that sells for 95¢ per pound and a second that sells for 65¢ per pound. How much of each should be used to make 60 pounds of candy selling for 75¢ per pound?

Chapter 6

77. Is $-8x^3y$ a monomial, binomial, or trinomial?

78. Give the degree of $3x^4 - 6x^5 + 5x - 2$.

79. Write in descending order.
$3x - 4x^3 - 6x^5 + 4x^2 + x^7$

80. Profit in dollars when x coats are sold is given by $2x^2 - 5$. What is the profit when 6 coats are sold?

Add.

81. $(-5a + 2) + (6a - 1)$

82. $(2x^2y^2 - 3xy + 2) + (-6x^2y^2 - xy - 5)$

Subtract.

83. $(9y - 3) - (-2y + 5)$

84. $(3x^2 - 2x + 5) - (4x^2 + 9x - 6)$

Multiply.

85. $-3a^2(4a - 2)$

86. $(3x + 4y)(9x - 2y)$

87. $(a - 2)(3a^2 + 2a - 5)$

88. $(5x + 7)^2$

89. $(2x + 5y)(2x - 5y)$

90. $(x^2 - 2)^2$

Divide.

91. $\dfrac{6y^9 - 2y^6 + 10y^3}{2y^2}$

92. $(2x^3 - 27x + 20) \div (x + 4)$

Chapter 7

Factor by removing the greatest common factor.

93. $24x - 16$

94. $6x^3y^2 - 3x^2y^3 + 9x^2y^2$

Factor by grouping.

95. $3x^3 + 2x^2 + 6x + 4$

96. $a^2b^2 - a^2 + 3b^2 - 3$

Factor.

97. $x^2 + 13x + 40$

98. $y^2 - 18y + 81$

99. $6x^2 - 24y^2$

100. $x^2 + xy - 20y^2$

101. $2x^2 - 9x - 5$

102. $9x^2 + 12xy + 4y^2$

Solve.

103. $(2x - 1)(x + 8) = 0$

104. $x^2 - 8x - 20 = 0$

105. The length of a rectangle is four times the width. If the area is 144 cm², find the dimensions.

106. Profit is given by $P = n^2 - 2n + 3$, where n is the number of items sold. How many items were sold when the profit was $38?

Chapter 8

107. What values of the variable must be omitted in $\dfrac{3x}{(x + 1)(x - 8)}$?

108. Reduce $\dfrac{a^2 - b^2}{(a - b)^2}$ to lowest terms.

Perform the indicated operations.

109. $\dfrac{x^2 - x - 2}{x^2 - 2x - 3} \cdot \dfrac{x^2 - 3x}{x + 2}$

110. $\dfrac{y^2 - 1}{y^2 + 2y} \div \dfrac{y - 1}{y + 2}$

111. $\dfrac{5}{y^2 - 4y + 3} + \dfrac{7}{y^2 + y - 2}$

112. $\dfrac{x + 6}{x^2 - 4} - \dfrac{x}{x - 2}$

113. Simplify. $\dfrac{x - \dfrac{1}{x}}{1 - \dfrac{1}{x}}$

114. Solve. $\dfrac{1}{y + 2} + \dfrac{1}{y - 2} = \dfrac{1}{y^2 - 4}$

115. A car goes 205 miles on 15 gallons of gasoline. How many miles can the car go on 75 gallons of gasoline?

116. The denominator of a fraction is 5 more than the numerator. If 2 is added to both the numerator and the denominator, the result is $\frac{1}{2}$. Find the fraction.

Chapter 9

Evaluate. Assume all variables and expressions are nonnegative.

117. $\sqrt{(-16)^2}$

118. $\pm\sqrt{169}$

119. $\sqrt{81x^2}$

120. $5\sqrt{8x^2y^2}$

121. $\sqrt{\dfrac{27x^2}{4y^2}}$

122. $\sqrt{\dfrac{49y^2}{x}}$

Perform the indicated operations and simplify.

123. $5\sqrt{15}\,\sqrt{45}$

124. $\dfrac{\sqrt{24x^3y}}{\sqrt{3xy}}$

125. $\sqrt{98}+\sqrt{50}$

126. $3\sqrt{45}-2\sqrt{80}+\sqrt{125}$

127. $\sqrt{28x}+3\sqrt{7x}-2\sqrt{63x}$

128. $\sqrt{3}-\dfrac{2}{\sqrt{3}}$

129. $(\sqrt{7}-2\sqrt{2})(\sqrt{7}+\sqrt{2})$

130. $\dfrac{\sqrt{6}}{\sqrt{3}-\sqrt{2}}$

Solve.

131. $\sqrt{2x-1}-\sqrt{3x-7}=0$

132. $\sqrt{x^2+3}+1-x=0$

133. One side of a rectangle is 12 m long and the diagonal is 13 m. Find the length of the other side.

134. Three times the square root of 5 more than a number is equal to 15. Find the number.

Chapter 10

Solve.

135. $x^2 - 12x + 35 = 0$

136. $x^2 + 14x + 49 = 0$

137. $2x^2 - 11x + 5 = 0$

138. $x^2 - 3x - 1 = 0$

139. $3y^2 - 5 = y(y + 3)$

140. $(z - 2)(z + 2) = 5z$

141. $\sqrt{7 - 2x} = x - 2$

142. $\dfrac{x + 2}{x} = \dfrac{2}{x - 1}$

143. The number of square inches in a square is 3 less than the perimeter of the square. Find the length of a side.

144. Al can do a job in 3 hours less time than Joe. Together they can do the job in 2 hours. How long would it take each to do the job?

145. Solve $a = \sqrt{\dfrac{2b}{c}}$ for b.

146. Graph $y = -x^2 + 1$.

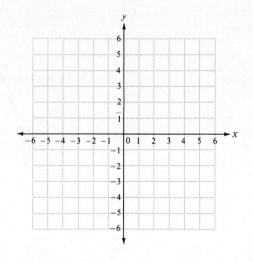

ANSWERS: **1.** $\dfrac{11}{10}$ **2.** $\dfrac{2}{3}$ **3.** $\dfrac{3}{7}$ **4.** $\dfrac{10}{9}$ **5.** $\dfrac{14}{5}$ **6.** $\dfrac{1}{28}$ **7.** 32.715 **8.** 7.88 **9.** 0.25872 **10.** 55 **11.** $2\dfrac{7}{18}$ **12.** $\dfrac{107}{13}$ **13.** 19 **14.** $6\dfrac{1}{16}$ **15.** 37.5% **16.** $2\dfrac{18}{25}$ **17.** \$2.31 **18.** y^6 **19.** 33 **20.** \$2332.80 **21.** $-\dfrac{4}{3}$ **22.** 8 **23.** > **24.** > **25.** 12 **26.** -10 **27.** $\dfrac{1}{12}$ **28.** -3 **29.** 0 **30.** -16 **31.** $2(y + 6)$ **32.** $-7(2x + 3)$ **33.** $-12a + 2b - 5$ **34.** $5y + 8$ **35.** -8 **36.** $-5xxxx$ **37.** $3x^5$ **38.** y^4 **39.** $9a^6$ **40.** $\dfrac{1}{a^2}$ **41.** 3.62×10^7 **42.** 0.00072 **43.** -12 **44.** $\dfrac{2}{5}$ **45.** 9 **46.** $\dfrac{1}{4}$ **47.** 4 **48.** -6 **49.** 39 **50.** 9 **51.** 4 **52.** $\dfrac{11}{2}$ **53.** -10 **54.** 1 **55.** 66, 68 **56.** \$128 **57.** 17 in **58.** 45 mph, 75 mph **59.** $x \leq 4$ **60.** $x > 1$

61. [number line with solid dot at −1]   **62.** [number line with ray from 3 rightward]   **63.** third
64. (5, 0)   **65.** $1850   **66.** true   **67.** $\frac{4}{5}$, $4x - 5y - 33 = 0$ **68.** $y = \frac{3}{4}x - 3$, $\frac{3}{4}$, $(0, -3)$

69. [graph with points (0, 6) and (2, 0)]   **70.**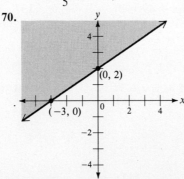
71. parallel
72. exactly one   **73.** (1, 5)
74. (4, 0)   **75.** (6, 1)
76. 20 lbs of 95¢ candy and 40 lbs of 65¢ candy   **77.** monomial   **78.** 5
79. $x^7 - 6x^5 - 4x^3 + 4x^2 + 3x$
80. $67   **81.** $a + 1$
82. $-4x^2y^2 - 4xy - 3$

83. $11y - 8$ **84.** $-x^2 - 11x + 11$ **85.** $-12a^3 + 6a^2$ **86.** $27x^2 + 30xy - 8y^2$ **87.** $3a^3 - 4a^2 - 9a + 10$ **88.** $25x^2 + 70x + 49$ **89.** $4x^2 - 25y^2$ **90.** $x^4 - 4x^2 + 4$ **91.** $3y^7 - y^4 + 5y$ **92.** $2x^2 - 8x + 5$ **93.** $8(3x - 2)$
94. $3x^2y^2(2x - y + 3)$ **95.** $(x^2 + 2)(3x + 2)$ **96.** $(a^2 + 3)(b^2 - 1)$ **97.** $(x + 8)(x + 5)$ **98.** $(y - 9)^2$
99. $6(x + 2y)(x - 2y)$ **100.** $(x - 4y)(x + 5y)$ **101.** $(2x + 1)(x - 5)$ **102.** $(3x + 2y)^2$ **103.** $\frac{1}{2}, -8$ **104.** $-2, 10$

105. 6 cm, 24 cm   **106.** 7   **107.** $-1, 8$ **108.** $\frac{a+b}{a-b}$ **109.** $\frac{x(x-2)}{x+2}$ **110.** $\frac{y+1}{y}$ **111.** $\frac{12y - 11}{(y+2)(y-3)(y-1)}$
112. $\frac{-x-3}{x+2}$ **113.** $x + 1$ **114.** $\frac{1}{2}$ **115.** 1025 mi   **116.** $\frac{3}{8}$ **117.** 16   **118.** ±13   **119.** $9x$ **120.** $10xy\sqrt{2}$
121. $\frac{3x\sqrt{3}}{2y}$ **122.** $\frac{7y\sqrt{x}}{x}$ **123.** $75\sqrt{3}$ **124.** $2x\sqrt{2}$ **125.** $12\sqrt{2}$ **126.** $6\sqrt{5}$ **127.** $-\sqrt{7x}$ **128.** $\frac{\sqrt{3}}{3}$
129. $3 - \sqrt{14}$ **130.** $3\sqrt{2} + 2\sqrt{3}$ **131.** 6   **132.** no solution   **133.** 5 m   **134.** 20   **135.** 5, 7   **136.** -7
137. $\frac{1}{2}, 5$ **138.** $\frac{3 \pm \sqrt{13}}{2}$ **139.** $\frac{5}{2}, -1$ **140.** $\frac{5 \pm \sqrt{41}}{2}$ **141.** 3   **142.** 2, -1 **143.** 3 in, 1 in   **144.** Al needs 3 hr and Joe 6 hr   **145.** $b = \frac{a^2c}{2}$ **146.**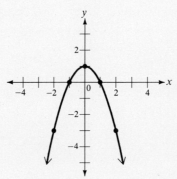

TABLE OF SQUARES AND SQUARE ROOTS

n	n^2	$\sqrt{n}$	$\sqrt{10n}$	n	n^2	$\sqrt{n}$	$\sqrt{10n}$
1	1	1.000	3.162	51	2601	7.141	22.583
2	4	1.414	4.472	52	2704	7.211	22.804
3	9	1.732	5.477	53	2809	7.280	23.022
4	16	2.000	6.325	54	2916	7.348	23.238
5	25	2.236	7.071	55	3025	7.416	23.452
6	36	2.449	7.746	56	3136	7.483	23.664
7	49	2.646	8.367	57	3249	7.550	23.875
8	64	2.828	8.944	58	3364	7.616	24.083
9	81	3.000	9.487	59	3481	7.681	24.290
10	100	3.162	10.000	60	3600	7.746	24.495
11	121	3.317	10.488	61	3721	7.810	24.698
12	144	3.464	10.954	62	3844	7.874	24.900
13	169	3.606	11.402	63	3969	7.937	25.100
14	196	3.742	11.832	64	4096	8.000	25.298
15	225	3.873	12.247	65	4225	8.062	25.495
16	256	4.000	12.649	66	4356	8.124	25.690
17	289	4.123	13.038	67	4489	8.185	25.884
18	324	4.243	13.416	68	4624	8.246	26.077
19	361	4.359	13.784	69	4761	8.307	26.268
20	400	4.472	14.142	70	4900	8.367	26.458
21	441	4.583	14.491	71	5041	8.426	26.646
22	484	4.690	14.832	72	5184	8.485	26.833
23	529	4.796	15.166	73	5329	8.544	27.019
24	576	4.899	15.492	74	5476	8.602	27.203
25	625	5.000	15.811	75	5625	8.660	27.386
26	676	5.099	16.125	76	5776	8.718	27.568
27	729	5.196	16.432	77	5929	8.775	27.749
28	784	5.292	16.733	78	6084	8.832	27.928
29	841	5.385	17.029	79	6241	8.888	28.107
30	900	5.477	17.321	80	6400	8.944	28.284
31	961	5.568	17.607	81	6561	9.000	28.460
32	1024	5.657	17.889	82	6724	9.055	28.636
33	1089	5.745	18.166	83	6889	9.110	28.810
34	1156	5.831	18.439	84	7056	9.165	28.983
35	1225	5.916	18.708	85	7225	9.220	29.155
36	1296	6.000	18.974	86	7396	9.274	29.326
37	1369	6.083	19.235	87	7569	9.327	29.496
38	1444	6.164	19.494	88	7744	9.381	29.665
39	1521	6.245	19.748	89	7921	9.434	29.833
40	1600	6.325	20.000	90	8100	9.487	30.000
41	1681	6.403	20.248	91	8281	9.539	30.166
42	1764	6.481	20.494	92	8464	9.592	30.332
43	1849	6.557	20.736	93	8649	9.644	30.496
44	1936	6.633	20.976	94	8836	9.695	30.659
45	2025	6.708	21.213	95	9025	9.747	30.822
46	2116	6.782	21.448	96	9216	9.798	30.984
47	2209	6.856	21.679	97	9409	9.849	31.145
48	2304	6.928	21.909	98	9604	9.899	31.305
49	2401	7.000	22.136	99	9801	9.950	31.464
50	2500	7.071	22.361	100	10000	10.000	31.623

APPENDIX: UNIT CONVERSION AND THE METRIC SYSTEM

STUDENT GUIDEPOSTS

1 Converting in the English system
2 Converting in the metric system
3 Converting between systems

The units of measure with which we are most familiar in the United States are inches, feet, yards, miles, ounces, pounds, tons, pints, quarts, and gallons. These make up the **English system** of measurement. However, in most of the rest of the world, the **metric system** is used. Here we will consider conversions in the English system, then conversions in the metric system, and finally conversions from one system to the other.

The following table summarizes the conversion identities for the English system:

Identities	To convert	Multiply by	To convert	Multiply by
1 ft = 12 in	ft to in	12	in to ft	1/12
1 yd = 3 ft	yd to ft	3	ft to yd	1/3
1 mi = 5280 ft	mi to ft	5280	ft to mi	1/5280
1 ft^2 = 144 in^2	sq ft to sq in	144	sq in to sq ft	1/144
1 yd^2 = 9 ft^2	sq yd to sq ft	9	sq ft to sq yd	1/9
1 ton = 2000 lb	tons to lb	2000	lb to tons	1/2000
1 lb = 16 oz	lb to oz	16	oz to lb	1/16
1 hr = 60 min	hr to min	60	min to hr	1/60
1 min = 60 sec	min to sec	60	sec to min	1/60
1 yd^3 = 27 ft^3	cu yd to cu ft	27	cu ft to cu yd	1/27
1 ft^3 = 1728 in^3	cu ft to cu in	1728	cu in to cu ft	1/1728
1 gal = 4 qt	gal to qt	4	qt to gal	1/4
1 qt = 2 pt	qt to pt	2	pt to qt	1/2

EXAMPLE 1

(A) Convert 13 yards to feet.

$$13 \text{ yd} = 13 \cdot 3 \text{ ft} = 39 \text{ ft}$$

(B) Convert 432 square inches to square feet.

$$432 \text{ in}^2 = 432 \cdot \frac{1}{144} \text{ ft}^2 = 3 \text{ ft}^2$$

(C) Convert 7 tons to pounds.

$$7 \text{ ton} = 7 \cdot 2000 \text{ lb} = 14{,}000 \text{ lb}$$

(D) Convert 5 square yards to square inches.

$$5 \text{ yd}^2 = 5 \cdot 9 \text{ ft}^2 = 45 \text{ ft}^2 = 45 \cdot 144 \text{ in}^2 = 6480 \text{ in}^2$$

(E) Convert 54 cubic feet to cubic yards.

$$54 \text{ ft}^3 = 54 \cdot \frac{1}{27} \text{ yd}^3 = 2 \text{ yd}^3$$

Practice Exercise 1

(A) Convert 15 feet to inches.

(B) Convert 162 square feet to square yards.

(C) Convert 6 hours to minutes.

(D) Convert 9072 square inches to square yards.

(E) Convert 31 cubic yards to cubic feet.

Answers: **(A)** 180 in **(B)** 18 yd^2 **(C)** 360 min **(D)** 7 yd^2 **(E)** 837 ft^3

When converting from one metric unit to another, the conversion factors are always powers of ten or one tenth. A meter (approximately 39 inches) is the standard unit of length, a gram (approximately the weight of a paper clip) is the standard unit of weight, and a liter (slightly larger than a quart) is the standard unit of capacity. The following prefixes may be attached to all three of these units.

$$\text{kilo-} = 1000 \qquad \text{deci-} = \frac{1}{10} = 0.1$$

$$\text{hecto-} = 100 \qquad \text{centi-} = \frac{1}{100} = 0.01$$

$$\text{deka-} = 10 \qquad \text{milli-} = \frac{1}{1000} = 0.001$$

Thus, we have the following equivalences.

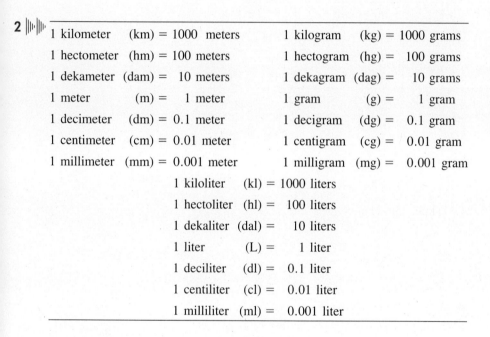

1 kilometer	(km) = 1000 meters	1 kilogram	(kg) = 1000 grams
1 hectometer	(hm) = 100 meters	1 hectogram	(hg) = 100 grams
1 dekameter	(dam) = 10 meters	1 dekagram	(dag) = 10 grams
1 meter	(m) = 1 meter	1 gram	(g) = 1 gram
1 decimeter	(dm) = 0.1 meter	1 decigram	(dg) = 0.1 gram
1 centimeter	(cm) = 0.01 meter	1 centigram	(cg) = 0.01 gram
1 millimeter	(mm) = 0.001 meter	1 milligram	(mg) = 0.001 gram
		1 kiloliter	(kl) = 1000 liters
		1 hectoliter	(hl) = 100 liters
		1 dekaliter	(dal) = 10 liters
		1 liter	(L) = 1 liter
		1 deciliter	(dl) = 0.1 liter
		1 centiliter	(cl) = 0.01 liter
		1 milliliter	(ml) = 0.001 liter

Since conversion from one unit to another is accomplished by multiplying by a power of ten, conversions can be made by simply moving the decimal point left or right. The following table will be useful.

1000	100	10	1	$\frac{1}{10}$	$\frac{1}{100}$	$\frac{1}{1000}$
kilo-unit	hecto-unit	deka-unit	unit	deci-unit	centi-unit	milli-unit

To convert hecto-units to deci-units we multiply by 1000. This is the same as moving the decimal point three units to the right, which is the same number of positions that deci-unit is to the right of hecto-unit in our table. Since deka-unit is four positions to the left of milli-unit in the table, to convert 13.3 ml to dal we move the decimal point four positions to the left (multiply by 0.0001) obtaining 0.00133 dal.

EXAMPLE 2

(A) To convert 23.4 hm to dm we move the decimal point three units (multiply by 1000) to the right, obtaining 23,400 dm.

(B) To convert 121 cg to kg we move the decimal point five units to the left (multiply by 0.00001), obtaining 0.00121 kg.

(C) To convert 22.41 kl to dal we move the decimal point two units to the right (multiply by 100), obtaining 2241 dal.

Practice Exercise 2

(A) Convert 581.2 dag to kg.

(B) Convert 0.381 hm to mm.

(C) Convert 43,200 dl to dal.

Answers: **(A)** 5.812 kg
(B) 38,100 mm   **(C)** 432 dal

If the United States ever changes completely to the metric system, it will not be necessary to convert from one system to the other. But while both systems are in use, it is important to know the relationship between the two. The following table summarizes the conversion identities.

Identities	To convert	Multiply by	To convert	Multiply by
1 km = 0.621 mi	km to mi	0.621	mi to km	1.61
1 m = 39.37 in	m to in	39.37	in to m	0.0254
1 L = 1.06 qt	L to qt	1.06	qt to L	0.946
1 kg = 2.20 lb	kg to lb	2.20	lb to kg	0.454

EXAMPLE 3

(A) Convert 20.0 km to mi.

$$20.0 \text{ km} = 20.0(0.621) \text{ mi} \approx 12.4 \text{ mi}$$

(B) Convert 35.0 mi to km.

$$35.0 \text{ mi} = 35.0(1.61) \text{ km} \approx 56.4 \text{ km}$$

(C) Convert 5.27 m to in.

$$5.27 \text{ m} = 5.27(39.37) \text{ in} \approx 207.5 \text{ in}$$

(D) Convert 18.75 in to cm.

$$18.75 \text{ in} = 18.75(0.0254) \text{ m} \approx 0.476 \text{ m} = 47.6 \text{ cm}$$

(E) Convert 520 ml to qt.

$$520 \text{ ml} = 0.52 \text{ L} = 0.52(1.06) \text{ qt} \approx 0.55 \text{ qt}$$

(F) Convert 0.035 lb to g.

$$0.035 \text{ lb} = 0.035(0.454) \text{ kg} \approx 0.0159 \text{ kg} = 15.9 \text{ g}$$

Practice Exercise 3

(A) Convert 32.5 kg to lb.

(B) Convert 600 lb to kg.

(C) Convert 425 in to m.

(D) Convert 72.15 cm to in.

(E) Convert 2.65 qt to ml.

(F) Convert 0.45 km to ft.

Answers: **(A)** 71.5 lb   **(B)** 272.4 kg
(C) 10.8 m   **(D)** 28.4 in   **(E)** 2507 m
(F) 1475 ft

EXERCISES A

Complete by filling in the blanks.

1. 1 ft = _____ in
2. 1 yd^3 = _____ ft^3
3. 1 hr = _____ min
4. 1 ton = _____ lb
5. 1 ft = _____ yd
6. 1 in^3 = _____ ft^3
7. 1 sec = _____ min
8. 1 oz = _____ lb
9. 1 min = _____ hr
10. 1 lb = _____ ton
11. 1 in^2 = _____ ft^2
12. 8 yd = _____ ft
13. 36 in = _____ ft
14. 2 mi = _____ ft
15. 44 yd^2 = _____ ft^2

16. 1440 in² = _____ ft²
17. 288 in² = _____ yd²
18. 70 min = _____ sec
19. 6 hr = _____ sec
20. 6912 in³ = _____ ft³
21. 1 ml = _____ L
22. 1 cm = _____ m
23. 1 L = _____ cl
24. 1 dg = _____ g
25. 1 hm = _____ m
26. 1 cg = _____ g
27. 0.43 cm = _____ km
28. 13.35 mg = _____ dag
29. 1.03 kl = _____ cl
30. 41.15 mm = _____ dam
31. 1 km = _____ mi
32. 1 m = _____ in
33. 1 kg = _____ lb
34. 1 L = _____ qt
35. 16 km = _____ mi
36. 2.6 m = _____ in
37. 3.75 m = _____ in
38. 3.2 ft = _____ cm
39. 20 qt = _____ L
40. 17.5 pt = _____ L
41. 5 lb = _____ g
42. 625 lb = _____ kg

43. Terry Larsen bought a foreign car having a gas tank with capacity 48 L. How many gallons of gas will the tank hold?

ANSWERS: **1.** 12 **2.** 27 **3.** 60 **4.** 2000 **5.** $\frac{1}{3}$ **6.** $\frac{1}{1728}$ **7.** $\frac{1}{60}$ **8.** $\frac{1}{16}$ **9.** $\frac{1}{60}$ **10.** $\frac{1}{2000}$ **11.** $\frac{1}{144}$
12. 24 **13.** 3 **14.** 10,560 **15.** 396 **16.** 10 **17.** $\frac{2}{9}$ **18.** 4200 **19.** 21,600 **20.** 4 **21.** 0.001 **22.** 0.01
23. 100 **24.** 0.1 **25.** 100 **26.** 0.01 **27.** 0.0000043 **28.** 0.001335 **29.** 103,000 **30.** 0.004115 **31.** 0.621
32. 39.37 **33.** 2.20 **34.** 1.06 **35.** 9.9 **36.** 102.4 **37.** 147.6 **38.** 97.5 **39.** 18.9 **40.** 8.3 **41.** 2270
42. 284 **43.** 12.7 gal

EXERCISES B

Complete by filling in the blanks.

1. 1 ft² = _____ yd²
2. 1 ft = _____ mi
3. 1 in = _____ ft
4. 1 ft³ = _____ yd³
5. 1 lb = _____ oz
6. 1 yd² = _____ ft²
7. 1 mi = _____ ft
8. 1 ft² = _____ in²
9. 1 yd = _____ ft
10. 1 ft³ = _____ in³
11. 1 min = _____ sec
12. 12 yd = _____ in
13. 18 ft = _____ yd
14. 7 mi = _____ yd
15. 40 ft² = _____ in²
16. 36 ft² = _____ yd²
17. 5 hr = _____ min
18. 5 sec = _____ min
19. 81 ft³ = _____ yd³
20. 1 km = _____ m
21. 1 g = _____ kg
22. 1 dag = _____ g
23. 1 hl = _____ L
24. 1 m = _____ mm
25. 1 dal = _____ kl
26. 10.5 hm = _____ cm
27. 0.024 hg = _____ cg
28. 3.42 cl = _____ hl
29. 35 kg = _____ dg
30. 425 ml = _____ dl
31. 1 mi = _____ km
32. 1 in = _____ m
33. 1 lb = _____ kg
34. 1 qt = _____ L
35. 128 mi = _____ km
36. 72 in = _____ cm
37. 3 km = _____ ft
38. 8 L = _____ qt
39. 40 L = _____ gal
40. 50 kg = _____ lb
41. 0.25 kg = _____ oz
42. 12.6 gal = _____ L

43. While driving in Mexico, Inez Schwanke read a sign showing the speed limit to be 100 km/hr. What is this in miles per hour?

EXERCISES C

Complete by filling in the blanks.

1. $30 \dfrac{\text{mi}}{\text{hr}} = \underline{} \dfrac{\text{ft}}{\text{sec}}$
2. $500 \dfrac{\text{cm}}{\text{sec}} = \underline{} \dfrac{\text{km}}{\text{hr}}$
3. $20 \dfrac{\text{mi}}{\text{hr}} = \underline{} \dfrac{\text{m}}{\text{sec}}$
   [Answer: Approximately 9]

Answers to Chapter Tests

CHAPTER 1

1. $\dfrac{23}{25}$ 2. $\dfrac{8}{52}$ 3. 42 4. $\dfrac{3}{4}$ 5. $\dfrac{3}{11}$ 6. $\dfrac{9}{4}$ 7. $\dfrac{29}{18}$ 8. $\dfrac{1}{24}$ 9. $3\dfrac{4}{9}$ 10. $\dfrac{158}{27}$ 11. 157 12. $4\dfrac{7}{10}$ liters
13. 17.117 14. 18.859 15. 2.1252 16. 44 17. 20% 18. $\dfrac{37}{20}$ 19. \$1.08 20. a^3 21. 1 22. 75 23. 225
24. 125 25. 35 26. 5 27. \$594.05

CHAPTER 2

1. false 2. 13 3. true 4. < 5. 141° 6. −22 7. 8 8. $-\dfrac{1}{8}$ 9. 3 10. 5 11. −18 12. −19 13. −6
14. $-3(y-5)$ 15. $5-b-2y$ 16. 3 17. 15 18. $-12y+6$ 19. 5^2a^3 20. $2y^7$ 21. a^4 22. $-8x^6$ 23. $\dfrac{1}{z^3}$
24. $\dfrac{y^7}{x^8}$ 25. $-\dfrac{1}{64}$ 26. 3×10^{-2} 27. 11 28. $-\dfrac{5}{4}$

CHAPTER 3

1. 15 2. 8 3. $\dfrac{1}{2}$ 4. −2 5. 36 6. 4 7. −3 8. 13 9. 3 10. 83, 85 11. \$62.10 12. 40 cm
13. 48 mph, 57 mph 14. $x \geq -3$ 15. $x < \dfrac{1}{2}$

CHAPTER 4

1.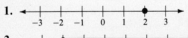
2.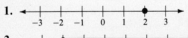
3. IV 4. $\left(-3, \dfrac{2}{5}\right)$ 5. 400 items
6. false 7. $\dfrac{1}{3}$ 8. $y = -\dfrac{2}{5}x + 4$; $-\dfrac{2}{5}$; $(0, 4)$
9. $6x + y - 19 = 0$

10.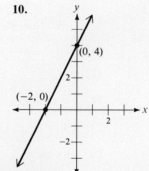
11.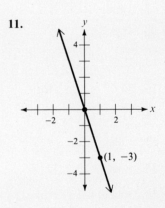
12.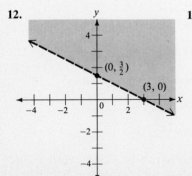
13.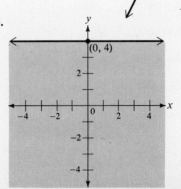

CHAPTER 5

1. intersecting lines **2.** no **3.** $(3, -1)$ **4.** no solution **5.** infinitely many **6.** 3, 15 **7.** 24°, 156° **8.** 30 nickels, 22 dimes **9.** 450 mph

CHAPTER 6

1. binomial **2.** 4 **3.** $-6y^5 + 4y^3 - 6y + 6$ **4.** $-4x^2y^2 + 5x^2y - 4x^2$ **5.** $285 **6.** $-2x + 10$ **7.** $3x^4 + 10x^3 - 3x^2 + 7$ **8.** $7a^2b^2 + 6ab + 2$ **9.** $-7x + 7$ **10.** $-4x^4 - x^3 + 6x^2 + 1$ **11.** $4a^2b - 4ab^2 - ab$ **12.** $-12x^4 + 16x^3 - 4x^2$ **13.** $x^2 - 3x - 40$ **14.** $8a^2 + 22a - 21$ **15.** $3y^3 + 22y^2 - 15y + 8$ **16.** $4x^2 - y^2$ **17.** $9x^2 - 24xy + 16y^2$ **18.** $4x^2 + 20x + 25$ **19.** $6x^3 - 5x^2 + 6x + 8$ **20.** $x^4 - 9$ **21.** $5x^2 - 2x - 3$ **22.** $3x^2 - 2x + 4$

CHAPTER 7

1. $4(5x + 3)$ **2.** $7(5y - 1)$ **3.** $6x^2(4x^2 - 2x + 3)$ **4.** $5x^3y^2(y + 8)$ **5.** $(y^2 + 2)(7x^2 - 1)$ **6.** $(x + 7)(x + 8)$ **7.** $(x - 8)^2$ **8.** $3(y + 5)(y - 5)$ **9.** $(x + 2)(x - 10)$ **10.** $(2x - 3)(x + 8)$ **11.** $(3x + y)(x + 5y)$ **12.** $\frac{5}{4}, -\frac{3}{2}$ **13.** $6, -5$ **14.** 1 or -4 **15.** 4 cm, 6 cm **16.** **(A)** $332 **(B)** 12

CHAPTER 8

1. $0, -5$ **2.** yes **3.** $\frac{1}{a}$ **4.** $\frac{5(x + 4)}{x}$ **5.** $\frac{1}{y - 3}$ **6.** $\frac{1}{a - b}$ **7.** $\frac{2x}{x + 2}$ **8.** $\frac{y - 5}{(y - 1)(y + 1)(y - 2)}$ **9.** no solution **10.** $\frac{10}{13}$ **11.** 37 gallons **12.** $13.80 **13.** $\frac{15}{8}$ days **14.** $\frac{1}{1 + y}$

CHAPTER 9

1. 2 **2.** 5 **3.** ± 11 **4.** $\frac{9}{4}$ **5.** $4x$ **6.** $12\sqrt{3}$ **7.** $\frac{2\sqrt{2x}}{y}$ **8.** $\frac{7x\sqrt{2y}}{y}$ **9.** 24 **10.** $3x$ **11.** $6\sqrt{5}$ **12.** $-16\sqrt{2}$ **13.** $10\sqrt{3x}$ **14.** $\frac{12\sqrt{11}}{11}$ **15.** 10 **16.** 2 **17.** $-3 + \sqrt{14}$ **18.** $\frac{5 + \sqrt{5}}{4}$ **19.** $\frac{7}{2}$ **20.** -3 **21.** 10 cm **22.** **(A)** $10\sqrt{5}$ **(B)** 20

CHAPTER 10

1. $-4, -9$ **2.** $3, -3$ **3.** $0, 5$ **4.** $2 \pm \sqrt{3}$ **5.** $\frac{1}{16}$ **6.** $3, \frac{1}{2}$ **7.** 2 **8.** 10, 12 **9.** large pipe: 4 hours, small pipe: 12 hours **10.** 12 mph **11.** $m = \frac{b}{w^2}$ **12.** **(A)** up **(B)** $(0, 2)$ **(C)**

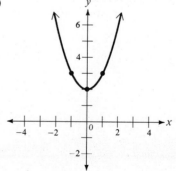

Solutions to Selected Exercises

All exercises are selected from the A set of exercises.

1.2

8. $4\dfrac{2}{5} \cdot \dfrac{3}{11} = \dfrac{22}{5} \cdot \dfrac{3}{11} = \dfrac{2 \cdot 11}{5} \cdot \dfrac{3}{11}$

   $= \dfrac{2 \cdot \cancel{11} \cdot 3}{5 \cdot \cancel{11}} = \dfrac{6}{5}$

14. $3\dfrac{1}{5} \div \dfrac{2}{5} = \dfrac{16}{5} \div \dfrac{2}{5} = \dfrac{16}{5} \cdot \dfrac{5}{2} =$

    $\dfrac{\cancel{2} \cdot 2 \cdot 2 \cdot 2 \cdot \cancel{5}}{\cancel{5} \cdot \cancel{2}} =$

    $\dfrac{2 \cdot 2 \cdot 2}{1} = \dfrac{8}{1} = 8$

43. $\dfrac{5}{6}$ of 72 means $\dfrac{5}{6} \cdot 72$.

    $\dfrac{5}{6} \cdot 72 = \dfrac{5}{6} \cdot \dfrac{72}{1} = \dfrac{5}{2 \cdot 3} \cdot \dfrac{2 \cdot 2 \cdot 2 \cdot 3 \cdot 3}{1} =$

    $\dfrac{5 \cdot \cancel{2} \cdot 2 \cdot 2 \cdot \cancel{3} \cdot 3}{\cancel{2} \cdot \cancel{3}} = \dfrac{5 \cdot 2 \cdot 2 \cdot 3}{1} = 60$ girls

    $72 - 60 = 12$ boys

46. Divide $12\dfrac{1}{2}$ by $\dfrac{3}{4}$.

    $12\dfrac{1}{2} \div \dfrac{3}{4} = \dfrac{25}{2} \div \dfrac{3}{4} = \dfrac{25}{2} \cdot \dfrac{4}{3} = \dfrac{5 \cdot 5}{2} \cdot \dfrac{2 \cdot 2}{3} =$

    $\dfrac{5 \cdot 5 \cdot \cancel{2} \cdot 2}{\cancel{2} \cdot 3} = \dfrac{50}{3} = 16\dfrac{2}{3}$ gallons

1.3

41. $7\dfrac{3}{4} + 9\dfrac{1}{8} = \dfrac{31}{4} + \dfrac{73}{8} = \dfrac{31 \cdot 2}{2 \cdot 2 \cdot 2} + \dfrac{73}{2 \cdot 2 \cdot 2}$

    $= \dfrac{62 + 73}{2 \cdot 2 \cdot 2} = \dfrac{135}{8}$ Total distance

    $\dfrac{135}{8} \cdot \dfrac{1}{2} = \dfrac{135}{16} = 8\dfrac{7}{16}$ mi   One half distance

1.4

37. $\dfrac{15}{20} = \dfrac{3 \cdot \cancel{5}}{4 \cdot \cancel{5}} = \dfrac{3}{4}$ Fraction of shots

    $\dfrac{3}{4} = 0.75 = 75\%$ Percent of shots

39. Find 55% of 5000.
    $(0.55)(5000) = 2750$ votes

1.5

47. $x - [a(b + 1) - c] = 12 - [2(3 + 1) - 5] =$
    $12 - [2 \cdot 4 - 5] = 12 - [8 - 5] = 12 - 3 = 9$

56. $A = 2\pi rh + 2\pi r^2 = 2(3.14)(2)(10) + 2(3.14)(2)^2 =$
    $125.60 + 25.12 = 150.72$ cm^2

2.1

26. Given $\dfrac{17}{43}$ and $\dfrac{28}{79}$. The first cross product is $17 \cdot 79 = 1343$, and the second cross product is $43 \cdot 28 = 1204$. Since $1343 > 1204$, we have $\dfrac{17}{43} > \dfrac{28}{79}$.

2.2

51. The difference in temperature was from $+35°$ to $-13°$. That is, we must find $35 - (-13) = 35 + 13 = 48°$. Thus, the total change in temperature was a decrease of $48°$.

2.3

44. Since 20 penalty points must be deducted from the total 3 times, we have that $3(-20) = -60$ represents the deduction. Thus, $423 + (-60) = 363$ points is her total after losing the penalty points.

50. $3 - \dfrac{2 + (-4)}{1 - 5} = 3 - \dfrac{-2}{-4}$

    $= 3 - \dfrac{1}{2}$

    $= \dfrac{6}{2} - \dfrac{1}{2} = \dfrac{5}{2}$

63. When $a = -1$, $b = 2$, and $c = -4$,
    $2[3(a + 1) + 2(4 + c)]$
    $= 2[3((-1) + 1) + 2(4 + (-4))]$
    $= 2[3(0) + 2(0)] = 2[0 + 0] = 2[0] = 0.$

75. -3000 represents the amount owed, $+1200$ represents the amount repaid, and -650 is the amount again borrowed. We must compute $(-3000) + (1200) + (-650) = -2450$. Thus, the Brisebois family owes $2450, so that -2450 represents the status of their account.

2.4

37. $\dfrac{1}{2}x - \dfrac{2}{3}y - \dfrac{5}{2}x - \dfrac{1}{3}y$

    $= \dfrac{1}{2}x - \dfrac{5}{2}x - \dfrac{2}{3}y - \dfrac{1}{3}y$

    $= \left(\dfrac{1}{2} - \dfrac{5}{2}\right)x - \left(\dfrac{2}{3} + \dfrac{1}{3}\right)y$

    $= -2x - y$

43. $2x - (-x + 1) + 3$
    $= 2x + x - 1 + 3$
    $= 3x + 2$

47. $a - [3a - (1 - 2a)] = a - [3a - 1 + 2a] =$
    $a - [5a - 1] = a - 5a + 1 = -4a + 1$

55. When $x = -3$, $-2x^2 = -2(-3)^2 = -2 \cdot 9 = -18$.
    Note that we square -3 before multiplying the result by -2.

56. When $x = -3$, $(-2x)^2 = ((-2)(-3))^2 = (6)^2 = 36$.
    Note that we first multiply -2 by -3, obtaining 6, then square the result. Compare Exercise 55 with Exercise 56.

68. When $x = -3$ and $y = -1$, $|x - y| = |(-3) - (-1)|$
    $= |(-3) + 1| = |-2| = 2$. Notice that we first simplify inside the absolute value bars.

2.5

27. $\left(\dfrac{2y}{x^3}\right)^{-2} = \dfrac{(2y)^{-2}}{(x^3)^{-2}}$ $\left(\dfrac{a}{b}\right)^n = \dfrac{a^n}{b^n}$

    $= \dfrac{2^{-2}y^{-2}}{(x^3)^{-2}}$ $(ab)^n = a^n b^n$

    $= \dfrac{2^{-2}y^{-2}}{x^{-6}}$ $(a^m)^n = a^{mn}$

    $= \dfrac{x^6}{2^2 y^2}$

    $= \dfrac{x^6}{4y^2}$

30. $\dfrac{a^{-2}b^2}{a^4 b^{-3}} = a^{-2-4} b^{2-(-3)}$ $\dfrac{a^m}{a^n} = a^{m-n}$

    $= a^{-6} b^5$

    $= \dfrac{1}{a^6} \cdot b^5$ $a^{-n} = \dfrac{1}{a^n}$

    $= \dfrac{b^5}{a^6}$

48. When $a = -2$, $a^{-2} = (-2)^{-2} = \dfrac{1}{(-2)^2} = \dfrac{1}{4}$.

49. When $a = -2$, $-2a = -2(-2) = 4$.

50. When $a = -2$, $-a^2 = -(-2)^2 = -(4) = -4$.
    Note the difference between Exercises 48, 49, and 50.

51. $a^{-2} + b^{-2} = (-2)^{-2} + (3)^{-2}$ Substitute -2 for a and 3 for b

    $= \dfrac{1}{(-2)^2} + \dfrac{1}{3^2}$ $a^{-n} = \dfrac{1}{a^n}$

    $= \dfrac{1}{4} + \dfrac{1}{9}$

    $= \dfrac{13}{36}$

52. $(a + b)^{-2} = ((-2) + 3)^{-2}$ Substitute -2 for a and 3 for b

    $= (1)^{-2}$

    $= \dfrac{1}{1^2} = 1$

    Note the difference between 51 and 52.

56. $a^{-1} b^{-1} = (-2)^{-1}(3)^{-1}$

    $= \left(\dfrac{1}{-2}\right)\left(\dfrac{1}{3}\right)$ $a^{-1} = \dfrac{1}{a^1} = \dfrac{1}{a}$

    $= -\dfrac{1}{6}$

57. $(ab)^{-1} = ((-2)(3))^{-1}$

    $= (-6)^{-1}$

    $= \dfrac{1}{(-6)^1}$

    $= -\dfrac{1}{6}$

    Note the difference between 51 and 52.

2.6

19. $(0.0000022)(300) = (2.2 \times 10^{-6})(3.0 \times 10^2) =$
    $(2.2)(3.0) \times (10^{-6})(10^2) = 6.6 \times 10^{-4}$

2.7

17. Always reduce fractions when they appear under a radical. Thus,

    $\sqrt{\dfrac{8}{18}} = \sqrt{\dfrac{2 \cdot 4}{2 \cdot 9}} = \sqrt{\dfrac{4}{9}} = \dfrac{2}{3}$.

29. Since

    $3^2 = 9 < 15 < 16 = 4^2$,

    we have

    $\sqrt{3^2} = 3 < \sqrt{15} < 4 = \sqrt{4^2}$.

    Thus, $\sqrt{15}$ is between 3 and 4.

3.1

28. Change $2\dfrac{1}{2}$ to $\dfrac{5}{2}$ and subtract $\dfrac{5}{2}$ from both sides of the equation.

    $x + 2\dfrac{1}{2} = 5$

    $x + \dfrac{5}{2} = 5$

    $x + \dfrac{5}{2} - \dfrac{5}{2} = 5 - \dfrac{5}{2}$

    $x + 0 = \dfrac{10}{2} - \dfrac{5}{2}$

    $x = \dfrac{5}{2} = 2\dfrac{1}{2}$

3.2

15. Since $-\dfrac{1}{8}$ is the divisor of z we must multiply both sides by $-\dfrac{1}{8}$.

$$\dfrac{z}{-\dfrac{1}{8}} = 16$$

$$\left(-\dfrac{1}{8}\right)\dfrac{z}{-\dfrac{1}{8}} = 16\left(-\dfrac{1}{8}\right)$$

$$\dfrac{\left(-\dfrac{1}{8}\right)}{\left(-\dfrac{1}{8}\right)}z = -\dfrac{16}{8} \qquad \dfrac{-\dfrac{1}{8}}{-\dfrac{1}{8}} = 1$$

$$1 \cdot z = -2$$
$$z = -2$$

16. First multiply both sides by 3.

$$\dfrac{2x}{3} = 4$$

$$(3)\dfrac{2x}{3} = 4(3)$$

$$\dfrac{3}{3} \cdot 2x = 12$$

$$2x = 12$$

$$\dfrac{2x}{2} = \dfrac{12}{2} \qquad \text{Now divide by 2}$$

$$x = 6$$

22. First change $2\dfrac{1}{2}$ to $\dfrac{5}{2}$.

$$2\dfrac{1}{2}x = 10$$

$$\dfrac{5}{2}x = 10$$

$$\dfrac{2}{5} \cdot \dfrac{5}{2}x = 10 \cdot \dfrac{2}{5} \qquad \text{The product } \dfrac{2}{5} \cdot \dfrac{5}{2} \text{ is 1}$$

$$1 \cdot x = \dfrac{10}{1} \cdot \dfrac{2}{5}$$

$$x = 4 \qquad 10 \div 5 = 2 \text{ and } 2 \cdot 2 = 4$$

3.3

11. Collect like terms on the left side.

$$-\dfrac{1}{7}y + y = 18$$

$$\left(-\dfrac{1}{7} + 1\right)y = 18$$

$$\left(-\dfrac{1}{7} + \dfrac{7}{7}\right)y = 18$$

$$\dfrac{6}{7}y = 18$$

$$\dfrac{7}{6} \cdot \dfrac{6}{7}y = 18 \cdot \dfrac{7}{6} \qquad \dfrac{7}{6} \text{ is the reciprocal of } \dfrac{6}{7}$$

$$1 \cdot y = \dfrac{18}{1} \cdot \dfrac{7}{6}$$

$$y = 21 \qquad 18 \div 6 = 3$$
$$ \text{and } 3 \cdot 7 = 21$$

14. $-7 + 21y + 23 = 16$

$\qquad 21y + 16 = 16 \qquad$ Collect like terms

$\qquad 21y = 0 \qquad$ Subtract 16 from both sides

$\qquad y = 0 \qquad$ Divide both sides by 21

17. $6y - 4y + 1 = 12 + 2y - 11$

$\qquad 2y + 1 = 1 + 2y \qquad$ Collect like terms

Since this is an identity, every real number is a solution.

19. Clear fractions by multiplying through by 2 and collect like terms on both sides.

$$3y + \dfrac{5}{2}y + \dfrac{3}{2} = \dfrac{1}{2}y + \dfrac{5}{2}y$$

$$2\left(3y + \dfrac{5}{2}y + \dfrac{3}{2}\right) = 2\left(\dfrac{1}{2}y + \dfrac{5}{2}y\right)$$

$\qquad 6y + 5y + 3 = y + 5y \qquad$ Multiply each term by 2

$\qquad 11y + 3 = 6y \qquad$ Collect like terms

$\qquad 11y = 6y - 3 \qquad$ Subtract 3 from both sides

$\qquad 11y - 6y = -3 \qquad$ Subtract $6y$ from both sides

$\qquad 5y = -3$

$\qquad y = -\dfrac{3}{5} \qquad$ Divide both sides by 5

29. $5(2 - 3x) = 15 - (x + 7)$

$\qquad 10 - 15x = 15 - x - 7 \qquad$ Clear parentheses

$\qquad 10 - 15x = 8 - x \qquad$ Collect like terms

$\qquad 10 = 8 + 14x \qquad$ Add $15x$ to both sides

$\qquad 2 = 14x \qquad$ Subtract 8 from both sides

$\qquad \dfrac{2}{14} = x \qquad$ Divide both sides by 14

$\qquad \dfrac{1}{7} = x$

36. The fractions could be cleared by multiplying both sides by the LCD of 2 and 3, which is 6. However, in this exercise the fractions are cleared when we use the distributive law.

$$\dfrac{1}{3}(6x - 9) = \dfrac{1}{2}(8x - 4)$$

$$\frac{1}{3} \cdot 6x - \frac{1}{3} \cdot 9 = \frac{1}{2} \cdot 8x - \frac{1}{2} \cdot 4$$

$$2x - 3 = 4x - 2$$

$$2x = 4x - 2 + 3$$

$$2x - 4x = 1$$

$$-2x = 1$$

$$\frac{-2x}{-2} = \frac{1}{-2}$$

$$x = -\frac{1}{2}$$

37. $5(x + 1) - 4x = x - 5$

$5x + 5 - 4x = x - 5$	Clear parentheses
$x + 5 = x - 5$	Collect like terms
$5 = -5$	Subtract x from both sides

Since we obtain a contradiction, the equation has no solution.

38. Clear all parentheses first.

$$3(2 - 4x) = 4(2x - 1) - 2(1 + x)$$

$$6 - 12x = 8x - 4 - 2 - 2x$$

$$6 - 12x = 8x - 2x - 4 - 2 \quad \text{Collect like terms}$$

$$-12x = 6x - 6 - 6$$

$$-12x - 6x = -12$$

$$-18x = -12$$

$$\frac{-18x}{-18} = \frac{-12}{-18}$$

$$x = \frac{2}{3}$$

41. $-4(-x + 1) + 3(2x + 3) - 7x = -10$

$4x - 4 + 6x + 9 - 7x = -10$	Clear parentheses
$3x + 5 = -10$	Collect like terms
$3x = -15$	Subtract 5 from both sides
$x = -5$	Divide both sides by 3

3.5

17. Write out complete details when solving a word problem.

Let x = length of one piece of rope,

$x + 7$ = length of other piece of rope (it is 7 feet longer than the first).

$x + (x + 7) = 17$	The sum of the lengths must be 17

$$2x + 7 = 17$$
$$2x = 10$$
$$x = 5$$
$$x + 7 = 12$$

Thus, one length is 5 ft and the other is 12 ft.

18. Let x = Tony's age.

$\frac{1}{4}x$ = Angela's age (If Tony is 4 times as old as Angela, then Angela is $\frac{1}{4}$ as old as Tony.)

$2x$ = Theresa's age (If Tony is $\frac{1}{2}$ as old as Theresa, then Theresa is 2 times as old as Tony.)

$x + \frac{1}{4}x + 2x = 39$	The sum of their ages is 39
$4\left(x + \frac{1}{4}x + 2x\right) = 4(39)$	Clear fractions

$$4x + x + 8x = 156$$
$$13x = 156$$
$$\frac{13}{13}x = \frac{156}{13}$$

$x = 12$	Tony's age
$\frac{1}{4}x = 3$	Angela's age
$2x = 24$	Theresa's age

32. Write out complete details. Let x = amount that Percy had invested. We are given that the amount of interest was $31.50 and the percent interest was 6%.

(% interest)·(amount invested) = (amount of interest)

$0.06 \quad \cdot \quad x \quad = \quad 31.50$

$$0.06x = 31.50$$

$$\frac{0.06}{0.06}x = \frac{31.50}{0.06}$$

$$x = 525$$

The amount invested was $525.

34. Let x = the original price of the dress.

$0.30x$ = amount of discount.

(original price) − (amount of discount) = (new price)

$x \quad - \quad 0.30x \quad = \quad 51.45$

$$(1 - 0.30)x = 51.45$$

$$0.7x = 51.45$$

$$x = \frac{51.45}{0.7} = 73.50$$

The original price of the dress was $73.50.

38. x = sum invested. $0.045x$ = amount of interest.

(sum invested) + (amount of interest) = (amt. in one year)
$$x + 0.045x = 1254$$
$$(1 + 0.045)x = 1254$$
$$1.045x = 1254$$
$$x = \frac{1254}{1.045} = 1200$$

The sum invested was $1200.

3.6

14. The volume of a rectangular solid is $V = lwh$. Since we have $V = 4725$ m^3, $w = 15$ m, and $h = 15$ m, substitute and solve for l.

$$4725 = l(15)(15)$$
$$4725 = l(225)$$
$$\frac{4725}{225} = l$$
$$21 = l$$

Thus, the length is 21 m.

17. Let x = number of feet that the water rises. The volume of the cube is $10 \cdot 10 \cdot 10 = 1000$ ft^3. Thus, 1000 ft^3 of water will be the increase in the level of the tank. But this volume is represented by $40 \cdot 50 \cdot x$. Thus,

$$(40)(50)x = 1000$$
$$2000x = 1000$$
$$x = \frac{1000}{2000} = \frac{1}{2} = 0.5.$$

The level of the tank will increase 0.5 ft.

19. Let x = number of meters that the level rises. From Exercise 18 the volume increase is 33.5 m^3 (the volume of the sphere). Thus,

$$(10)(8)x = 33.5$$
$$80x = 33.5$$
$$x = \frac{33.5}{80} \approx 0.4.$$

Thus, the level increased about 0.4 m.

20. Let x = cost in dollars to enclose the pasture. The perimeter of the pasture is $P = 2l + 2w = 2(3) + 2(2) = 6 + 4 = 10$ mi. The length in feet is $10(5280) = 52,800$. Now multiply the number of feet (52,800 ft) by the cost per foot in dollars, $0.35. $x = (52,800)(0.35) = 18,480$.
The total cost is $18,480.

25. The distance one travels in 8 hours is $(8)(22) = 176$ nautical miles. The distance the second travels in 8 hours is $(8)(17) = 136$ nautical miles. Thus, in 8 hours they will be $176 - 136 = 40$ nautical miles apart.

Supplementary Applied Problems

5. Let x = the length of the pole. We can obtain an equation to use by setting the length of the pole equal to the sum of the lengths.

$$(\text{total length}) = \binom{\text{length in}}{\text{sand}} + \binom{\text{length in}}{\text{water}} + \binom{\text{length in}}{\text{air}}$$

$$x = \frac{1}{4}x + 9 + \frac{3}{8}x$$

$$x = \frac{1}{4}x + 9 + \frac{3}{8}x$$

$$8x = 2x + 72 + 3x \quad \text{Multiply by the LCD 8}$$
$$8x = 5x + 72$$
$$3x = 72$$
$$x = \frac{72}{3} = 24$$

The pole is 24 ft long.

10. Let x = measure of first angle of the triangle.

$x + 30$ = measure of second angle of the triangle.

$4(x + 30)$ = measure of third angle of the triangle (If the second is $\frac{1}{4}$ the third, then the third is 4 times the second).

$$x + (x + 30) + 4(x + 30) = 180 \quad \text{Sum of the angles is 180}$$
$$x + x + 30 + 4x + 120 = 180$$
$$6x + 150 = 180$$
$$6x = 30$$
$$x = 5$$
$$x + 30 = 35$$
$$4(x + 30) = 140$$

The first angle is 5°, the second is 35°, and the third is 140°.

13. Let x = width of rectangle.

$2x$ = length of rectangle.

$x + 8$ = width increased by 8.

$2x - 8$ = length decreased by 8.

For a square the length and width are equal.

$$x + 8 = 2x - 8$$
$$x = 2x - 16$$
$$-x = -16$$
$$x = 16$$
$$2x = 32$$

Thus, the width is 16 ft and the length is 32 ft.

19. x = population in 1970.

$0.20x$ = amount of decrease.

(population in 1970) − (amt. of decrease) = (pop. in 1980)
$$x \quad - \quad 0.20x \quad = \quad 740$$
$$(1 - 0.20)x = 740$$
$$0.8x = 740$$
$$x = \frac{740}{0.8} = 925$$

There were 925 people in Deserted in 1970.

20. $x =$ amount invested at 14% interest.

$10,000 - x =$ amount invested at 11% interest (If x dollars are invested at 14%, the amount left to invest at 11% is $10,000 - x$).

$0.14x =$ amount of interest earned on 14% investment.

$0.11(10,000 - x) =$ amount of interest earned on 11% investment.

$$0.14x + 0.11(10,000 - x) = 1310$$
$$14x + 11(10,000 - x) = 131,000 \quad \text{Multiply by 100 to clear decimals}$$
$$14x + 110,000 - 11x = 131,000$$
$$3x = 21,000$$
$$x = 7000$$
$$10,000 - x = 3000$$

Thus, $7000 was invested at 14% and $3000 at 11%.

24. $x =$ score Becky must make to have 90 average. The average of her 4 scores must be 90.

$$\frac{96 + 78 + 91 + x}{4} = 90$$
$$\frac{265 + x}{4} = 90$$
$$4\left(\frac{265 + x}{4}\right) = 90(4)$$
$$265 + x = 360$$
$$x = 95$$

Becky must make at least 95 on her fourth test to get an A.

3.7

7. $2a + 1 > a - 3$
$2a > a - 4$
$a > -4$

13. Multiply both sides by -4 to make the coefficient of y 1. When multiplying by a negative number we must reverse the inequality.

$$-\frac{1}{4}y \geq 2$$

$$-4\left(-\frac{1}{4}\right)y \leq 2(-4)$$
$$y \leq -8$$

25. $3(2x + 3) - (3x + 2) < 12$
$6x + 9 - 3x - 2 < 12$
$3x + 7 < 12$
$3x < 5$
$$x < \frac{5}{3}$$

30. Multiply by the LCD, 8, to clear the fractions. Since 8 is positive the inequality stays the same.

$$\frac{3}{4}z - \frac{3}{8} < \frac{3}{2} + \frac{1}{8}z$$
$$8\left(\frac{3}{4}z - \frac{3}{8}\right) < 8\left(\frac{3}{2} + \frac{1}{8}z\right)$$
$$6z - 3 < 12 + z$$
$$5z < 15$$
$$z < \frac{15}{5} \quad \text{5 is positive}$$
$$z < 3$$

37. $x =$ the number.
$3x =$ product of the number and 3.
$x - 8 =$ the number less 8.
$3x \geq x - 8$
$2x \geq -8$
$$x \geq \frac{-8}{2}$$
$$x \geq -4$$

40. Let $x =$ number of cars he must sell in August. The average number of cars in the three month period is given by $\frac{47 + 62 + x}{3}$.

To win the trip

$$\frac{47 + 62 + x}{3} \geq 50.$$
$$47 + 62 + x \geq 150 \quad \text{Multiply both sides by 3}$$
$$109 + x \geq 150$$
$$x \geq 41$$

Thus, Darrell must sell at least 41 cars in August.

4.2

17. (A) Substitute 0 for x in $x - 5 = 0$.
$0 - 5 = 0$
$-5 = 0$

This is a contradiction so x cannot be 0. In fact, x must always be 5.

(B) Substitute 0 for y in $x - 5 = 0$. But y does not occur in the equation. Thus, if y is 0, x is still 5, so that (5, 0) is the completed ordered pair.

(C) Substitute 5 for x in $x - 5 = 0$

$$5 - 5 = 0$$
$$0 = 0$$

Thus, x is 5 and y can be any number. That is, (5, any number) is a solution to the equation.

(D) As in part (b), y can be -10 but x is still 5. Thus, $(5, -10)$ is the completed ordered pair.

22. Using the equation $y = 55x$, we must complete each pair.

(A) For (1,): Substitute 1 for x to obtain $y = 55 \cdot 1 = 55$. Thus, (1, 55) is the completed pair and the car travels 55 miles in 1 hour.

(B) For (5,): Substitute 5 for x to obtain $y = 55 \cdot 5 = 275$. Thus, (5, 275) is the completed pair and the car travels 275 miles in 5 hours.

(C) For (10,): Substitute 10 for x to obtain $y = 55 \cdot 10 = 550$. Thus, the completed pair is (10, 550) and the car travels 550 miles in 10 hours.

4.4

13. First choose two points on $5x + 1 = 0$. Since all points on this line have x-coordinate $-\frac{1}{5}$ ($5x + 1 = 0$ means $5x = -1$ or $x = -\frac{1}{5}$), two such points are $\left(-\frac{1}{5}, 1\right)$ and $\left(-\frac{1}{5}, 2\right)$. Substitute into the slope formula.

$$m = \frac{y_2 - y_1}{x_2 - x_1} = \frac{2 - 1}{\left(-\frac{1}{5}\right) - \left(-\frac{1}{5}\right)} = \frac{1}{0}$$

Since m is undefined, the line has no slope (it is parallel to the y-axis).

29. Substitute $-\frac{1}{2}$ for m and $(3, -2)$ for (x_1, y_1) in

$$y - y_1 = m(x - x_1).$$

$$y - (-2) = -\frac{1}{2}(x - 3)$$

$$2(y + 2) = -(x - 3) \quad \text{Clear fractions}$$
$$2y + 4 = -x + 3$$
$$x + 2y + 1 = 0 \quad \text{General form}$$

32. First find the slope.

$$m = \frac{y_2 - y_1}{x_2 - x_1} = \frac{-2 - (-1)}{3 - (-1)}$$

$$= \frac{-2 + 1}{3 + 1}$$

$$= \frac{-1}{4}$$

Then substitute into the slope-intercept form using $(3, -2)$ for (x_1, y_1) and $-\frac{1}{4}$ for m.

$$y - y_1 = m(x - x_1)$$

$$y - (-2) = -\frac{1}{4}(x - 3)$$

$$4(y + 2) = -(x - 3) \quad \text{Multiply both sides by 4}$$
$$4y + 8 = -x + 3$$
$$x + 4y + 5 = 0 \quad \text{General form}$$

35. The three points will all lie on the same straight line if the slopes of the lines through (2, 3) and (0, 2) and through (0, 2) and $(-2, 1)$ are equal. The slope of the line through (2, 3) and (0, 2) is

$$\frac{3 - 2}{2 - 0} = \frac{1}{2},$$

and the slope of the line through (0, 2) and $(-2, 1)$ is

$$\frac{2 - 1}{0 - (-2)} = \frac{1}{2}.$$

Thus, the three points all lie on the same straight line.

5.1

7. Solve for y.

$$2x - 7y + 1 = 0 \qquad -6x + 21y + 3 = 0$$
$$-7y = -2x - 1 \qquad 21y = 6x - 3$$
$$y = \frac{2}{7}x + \frac{1}{7} \qquad y = \frac{6}{21}x - \frac{3}{21}$$
$$\qquad\qquad\qquad\qquad y = \frac{2}{7}x - \frac{1}{7}$$

Since the coefficients of x are both $\frac{2}{7}$, the lines are either parallel or coinciding (they have the same slope). Since the constants are unequal, the y-intercepts are different so the lines do not coincide. Thus, the lines are parallel.

10. Since the line $2y = 8$ is the same as $y = 4$, its graph is a horizontal line with y-intercept (0, 4). Also, the line $2x = 8$ is the same as $x = 4$, and its graph is a vertical line with x-intercept (4, 0). Since a horizontal line and a vertical line are perpendicular, they are intersecting.

14. To determine whether $(-3, 2)$ is a solution to the system, substitute into both equations

$$2x + 3y = 0 \qquad\qquad x + 8y = 19$$
$$2(-3) + 3(2) \stackrel{?}{=} 0 \qquad (-3) + 8(2) \stackrel{?}{=} 19$$
$$-6 + 6 \stackrel{?}{=} 0 \qquad\qquad -3 + 16 \stackrel{?}{=} 19$$
$$0 = 0 \qquad\qquad\qquad 13 \neq 19$$

Thus, although $(-3, 2)$ is a solution to the first equation, it is not a solution to the system since it does not solve the second equation.

5.2

5. First determine the number of solutions by writing each equation in slope-intercept form.

$$x + 3y = 4 \qquad -2x - 6y = -8$$
$$3y = -x + 4 \qquad -6y = 2x - 8$$
$$y = -\frac{1}{3}x + \frac{4}{3} \qquad y = -\frac{2}{6}x + \frac{8}{6}$$
$$= -\frac{1}{3}x + \frac{4}{3}$$

Since the equations are the same, the lines are coinciding so there will be infinitely many solutions. We might also have observed sooner that if the second equation is multiplied by $-\frac{1}{2}$ on both sides the result is the first equation. Thus, obviously the two equations are the same.

8. The system will have exactly one solution since the two lines have slope -1 and $-\frac{1}{3}$. Graph each equation by finding its intercepts.

$x + y = 3 \qquad x + 3y = -1$

x	y
0	3
3	0

x	y
0	$-\frac{1}{3}$
-1	0

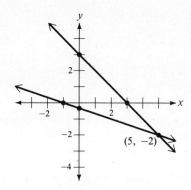

The point of intersection appears to have coordinates $(5, -2)$. Check by substitution.

$$x + y = 3 \qquad x + 3y = -1$$
$$5 + (-2) \stackrel{?}{=} 3 \qquad 5 + 3(-2) \stackrel{?}{=} -1$$
$$3 = 3 \qquad -1 = -1$$

Thus, the solution is the ordered pair $(5, -2)$.

13. Writing in slope-intercept form, we have:

$$x + 4y = 1 \qquad -x - 4y = 1$$
$$4y = -x + 1 \qquad -4y = x + 1$$
$$y = -\frac{1}{4}x + \frac{1}{4} \qquad y = -\frac{1}{4}x - \frac{1}{4}$$

Although the slopes are the same $\left(\text{both are } -\frac{1}{4}\right)$, the y-intercepts are different. Thus, the lines are parallel, and the system has no solution.

5.3

7. $3x - 5y = 19$
$2x - 4y = 16$

Simplify by multiplying the second equation by $\frac{1}{2}$.

$x - 2y = 8$

Solve for x.

$x = 2y + 8$

Substitute into the first equation.

$$3(2y + 8) - 5y = 19$$
$$6y + 24 - 5y = 19$$
$$y + 24 = 19$$
$$y = -5$$

Substitute -5 for y in $x - 2y = 8$.

$$x - 2(-5) = 8$$
$$x + 10 = 8$$
$$x = -2$$

Thus, the solution is the ordered pair $(-2, -5)$.

10. $3x - 3y = 1$
$x - y = -1$

Solve the second equation for x, $x = y - 1$, and substitute into the first.

$$3(y - 1) - 3y = 1$$
$$3y - 3 - 3y = 1$$
$$-3 = 1$$

Since a contradiction is obtained, the system has no solution.

11. $2x + 2y = -6$
$-x - y = 3$

Solve the second equation for y.

$$-y = x + 3$$
$$y = -x - 3$$

Substitute into the first equation.

$$2x + 2(-x - 3) = -6$$
$$2x - 2x - 6 = -6$$
$$-6 = -6$$

Since an identity is obtained, the system has infinitely many solutions.

14. $3x + 5y = 30$
$5x + 3y = 34$

Solving for either variable in either equation will result in fractions. Perhaps it is best to solve for x in the first.

$3x = -5y + 30$

$x = -\dfrac{5}{3}y + 10$

Substitute into the second.

$5\left(-\dfrac{5}{3}y + 10\right) + 3y = 34$

$-\dfrac{25}{3}y + 50 + 3y = 34$

$-25y + 150 + 9y = 102$ Multiply through by 3 to clear the fraction

$-16y = -48$

$y = 3$

Substitute 3 for y in the second equation.

$5x + 3(3) = 34$

$5x = 25$

$x = 5$

Thus, the solution pair is (5, 3).

5.4

5. $6x + 5y = 11$
$3x - 7y = -4$

Multiply the second equation by -2 and add.

$6x + 5y = 11$
$-6x + 14y = 8$
$19y = 19$
$y = 1$

Substitute 1 for y in the first equation and solve for x.

$6x + 5(1) = 11$

$6x + 5 = 11$

$6x = 6$

$x = 1$

Thus, the solution is the pair (1, 1).

10. $2x + 3y = -4$
$1 + 2y = -3x$

First rewrite the system.

$2x + 3y = -4$
$3x + 2y = -1$

Multiply the first equation by 3 and the second equation by -2 and add.

$6x + 9y = -12$
$-6x - 4y = 2$
$5y = -10$
$y = -2$

Now substitute -2 for y in the first equation.

$2x + 3(-2) = -4$

$2x - 6 = -4$

$2x = 2$

$x = 1$

Thus, the solution is (1, -2).

11. $3x + 7y = -21$
$7x + 3y = -9$

If we multiply the first equation by -7 and the second by 3, the coefficients of x will be -21 and 21. Then by adding, we eliminate x and obtain an equation in the one variable y.

$-21x - 49y = 147$
$21x + 9y = -27$
${-40y} = 120$
$y = -3$

Substitute -3 for y in the first equation.

$3x + 7(-3) = -21$

$3x = 0$

$x = 0$

Thus, the solution is the pair (0, -3).

16. $3x + 3y = 3$
$4x + 4y = -3$

Multiply the first equation by -4, the second by 3, and add.

$-12x - 12y = -12$
$12x + 12y = -9$
$0 = -21$

Since a contradiction is obtained, the system has no solution.

17. $2x + 3y = -4$
$5x + 7y = -10$

Multiply the first equation by -5, the second by 2, and add to eliminate x.

$-10x - 15y = 20$
$10x + 14y = -20$
${-y} = 0$
$y = 0$

Substitute 0 for y in the first equation.

$2x + 3(0) = -4$

$2x = -4$

$x = -2$

Thus, the solution is $(-2, 0)$.

5.5

11. Let x = number of pounds of 90¢ candy.

y = number of pounds of \$1.50 candy.

Since there are 30 pounds in the mixture,

$$x + y = 30.$$

The value of the mixture is

$$90x + 150y = 110(30).$$

Notice that the units were changed to cents, and don't forget to multiply 110¢ by 30 pounds on the right side. Solve the first equation for x and substitute.

$$x = 30 - y$$

$90(30 - y) + 150y = 110(30)$

$9(30 - y) + 15y = 11(30)$ Divide through by 10

$270 - 9y + 15y = 330$

$6y = 60$

$y = 10$

Then $x = 30 - y = 30 - 10 = 20$. Thus, there are 20 lb of the 90¢ candy and 10 lb of the \$1.50 candy.

17. Let x = Terry's average speed riding in still air.

y = average wind speed.

$x + y$ = Terry's average speed riding with the wind.

$x - y$ = Terry's average speed riding against the wind.

Use the distance formula, $d = rt$, twice noting that both distances are 40 miles.

$40 = (x - y)4$ Distance against the wind

$40 = (x + y)2$ Distance with the wind

Thus, dividing the first equation by 4 and the second by 2, we have the system:

$x - y = 10$
$x + y = 20$
$\overline{2x = 30}$
$x = 15$

Substitute 15 for x in the first equation.

$15 - y = 10$

$-y = -5$

$y = 5$

Thus, Terry's rate riding is 15 mph, and the wind speed is 5 mph.

19. Let x = amount to be invested at 10% interest.

y = amount to be invested at 12% interest.

Then,

$$x + y = 10,000. \quad \text{The total to invest is \$10,000}$$

The total interest is the sum of the interests earned on each part.

$$(0.10)x + (0.12)y = 1160$$

Multiply through by 100 to clear all decimals,

$$10x + 12y = 116,000,$$

then divide through by 2 to decrease the coefficients.

$$5x + 6y = 58,000$$

Solve the first equation for x and substitute.

$5(10,000 - y) + 6y = 58,000$

$50,000 - 5y + 6y = 58,000$

$y = 8000$

Then $x = 10,000 - y = 10,000 - 8000 = 2000$. Thus, \$2000 is invested at 10% and \$8000 is invested at 12%.

21. Let x = daily rate charged.

y = mileage rate charged.

The system of equations to solve is:

$3x + 400y = 82$

$5x + 500y = 120$

Divide the second equation by 5 and solve for x.

$x + 100y = 24$

$x = 24 - 100y$

Substitute into the first equation.

$3(24 - 100y) + 400y = 82$

$72 - 300y + 400y = 82$

$100y = 10$

$y = 0.10$

Then $x = 24 - 100y = 24 - 100(0.10) = 24 - 10 = 14$. Thus, the charges are \$14 per day and 10¢ per mile.

6.1

28. $-8x^{10} + x^5 - 2x^{10} - 7x^5 + 1 - x^{10} + 3$

$= (-8x^{10} - 2x^{10} - x^{10}) + (x^5 - 7x^5) + (1 + 3)$

$= (-8 - 2 - 1)x^{10} + (1 - 7)x^5 + (1 + 3)$

$= -11x^{10} - 6x^5 + 4$

35. $-4x^2y + 2xy^2 + x^2 - 3x^2y$

$= (-4x^2y - 3x^2y) + 2xy^2 + x^2$

$= (-4 - 3)x^2y + 2xy^2 + x^2$

$= -7x^2y + 2xy^2 + x^2$

528 SOLUTIONS TO SELECTED EXERCISES

38. $7y^2 - 2y - 5 = 7(-2)^2 - 2(-2) - 5$
$ = 7 \cdot 4 + 4 - 5$
$ = 28 + 4 - 5 = 27$

41. We must evaluate $0.05x + 15.5$ when $x = 440$.

$0.05(440) + 15.5$
$= 22 + 15.5$
$= 37.5$

Thus, the cost is $37.50 to manufacture 440 bolts.

6.2

10. $(-4a^3 + 7a^4 + 3a + 2) + (5 - 3a + 7a^3) +$
$(17a^4 - 5 + 12a^3)$
$= (7a^4 - 4a^3 + 3a + 2) + (7a^3 - 3a + 5) +$
$ (17a^4 + 12a^3 - 5)$
$= 7a^4 + 17a^4 - 4a^3 + 7a^3 + 12a^3 + 3a - 3a +$
$ 2 + 5 - 5$
$= (7 + 17)a^4 + (-4 + 7 + 12)a^3 + (3 - 3)a +$
$ (2 + 5 - 5)$
$= 24a^4 + 15a^3 + 0 \cdot a + 2$
$= 24a^4 + 15a^3 + 2$

16. $\begin{array}{r} 0.03y^3 - 0.75y^2 - 3y + 2 \\ -0.15y^3 + 5y - 0.3 \\ \underline{0.21y^3 - 0.13y^2 + 0.6} \\ 0.09y^3 - 0.88y^2 + 2y + 2.3 \end{array}$

Simply add the coefficients of like terms (down the columns) to obtain the coefficients in the sum.

23. $(56x - 93x^3 + 21x^4 + 32x^5) - (3x^4 - 7x^5 + 15x - 32)$
$= (32x^5 + 21x^4 - 93x^3 + 56x) -$
$ (-7x^5 + 3x^4 + 15x - 32)$
$= 32x^5 + 21x^4 - 93x^3 + 56x + 7x^5 - 3x^4 -$
$ 15x + 32$
$= 32x^5 + 7x^5 + 21x^4 - 3x^4 - 93x^3 + 56x -$
$ 15x + 32$
$= (32 + 7)x^5 + (21 - 3)x^4 - 93x^3 +$
$ (56 - 15)x + 32$
$= 39x^5 + 18x^4 - 93x^3 + 41x + 32$

28. $(8y^{10} - y^8) - (3y^{12} + 2y^{10} - y^8)$
$= 8y^{10} - y^8 - 3y^{12} - 2y^{10} + y^8$ Clear parentheses
$= -3y^{12} + 8y^{10} - 2y^{10} - y^8 + y^8$
$= -3y^{12} + (8 - 2)y^{10} + (-1 + 1)y^8$ Distributive laws
$= -3y^{12} + 6y^{10}$

31. $(9x^4 + 3x^3 + 8x) + (3x^4 + x^3 - 7x^2) -$
$(12x^4 - 3x^2 + x)$

$= 9x^4 + 3x^3 + 8x + 3x^4 + x^3 - 7x^2 - 12x^4 +$
$ 3x^2 - x$
$= (9 + 3 - 12)x^4 + (3 + 1)x^3 + (-7 + 3)x^2 +$
$ (8 - 1)x$
$= 0 \cdot x^4 + 4x^3 - 4x^2 + 7x$
$= 4x^3 - 4x^2 + 7x$

38. $(-2a^2b + ab - 4ab^2) - (6a^2b + 4ab^2)$
$= -2a^2b + ab - 4ab^2 - 6a^2b - 4ab^2$
$= (-2 - 6)a^2b + ab + (-4 - 4)ab^2$
$= -8a^2b + ab - 8ab^2$

40. $(6x^2y - xy) - (3x^2y - 7xy^2) - (4xy - 5xy^2)$
$= 6x^2y - xy - 3x^2y + 7xy^2 - 4xy + 5xy^2$
                                       Clear parentheses
$= 6x^2y - 3x^2y - xy - 4xy + 7xy^2 + 5xy^2$
$= (6 - 3)x^2y + (-1 - 4)xy + (7 + 5)xy^2$
                                       Distributive laws
$= 3x^2y - 5xy + 12xy^2$

6.3

16. $-10x^2(x^5 - 6x^3 + 7x^2)$
$= (-10x^2)(x^5) - (-10x^2)(6x^3) +$ Distributive
$ (-10x^2)(7x^2)$ laws
$= -10x^2x^5 + 60x^2x^3 - 70x^2x^2$ Multiply
                                       coefficients
$= -10x^7 + 60x^5 - 70x^4$ $a^m a^n = a^{m+n}$

23. $(x^2 + 4x - 2)(2x^2 - x + 3)$
$= (x^2 + 4x - 2)(2x^2) + (x^2 + 4x - 2)(-x) +$
$ (x^2 + 4x - 2)(3)$
$= (x^2)(2x^2) + (4x)(2x^2) + (-2)(2x^2) + (x^2)(-x) +$
$ (4x)(-x) + (-2)(-x) + (x^2)(3) + (4x)(3) +$
$ (-2)(3)$
$= 2x^4 + 8x^3 - 4x^2 - x^3 - 4x^2 + 2x + 3x^2 +$
$ 12x - 6$
$= 2x^4 + (8 - 1)x^3 + (-4 + 3)x^2 + (2 + 12)x - 6$
$= 2x^4 + 7x^3 - x^2 + 14x - 6$

37. $(5x - 2)(3x + 4)$
$= (5x)(3x) + (5x)(4) + (-2)(3x) + (-2)(4)$
$= 15x^2 + 20x - 6x - 8$
$= 15x^2 + 14x - 8$

41. $(2z^2 + 1)(z^2 - 2)$
$= (2z^2)(z^2) + (2z^2)(-2) + (1)(z^2) + (1)(-2)$
$= 2z^4 - 4z^2 + z^2 - 2$
$= 2z^4 - 3z^2 - 2$ Collect like terms

52. $\begin{array}{r} 0.3x^2 + 0.2 \\ \underline{0.5x - 0.7} \\ 0.15x^3 + 0.10x \\ \underline{ - 0.21x^2 - 0.14} \\ 0.15x^3 - 0.21x^2 + 0.10x - 0.14 \end{array}$

55. $\begin{array}{r} 3y^2 + 5y - 6 \\ \underline{y^2 - 3y + 2} \\ 3y^4 + 5y^3 - 6y^2 \\ -9y^3 - 15y^2 + 18y \\ \underline{ 6y^2 + 10y - 12} \\ 3y^4 - 4y^3 - 15y^2 + 28y - 12 \end{array}$

59. $(2a - 3b)(5a + b)$
   $= (2a)(5a) + (2a)(b) + (-3b)(5a) + (-3b)(b)$
   $= 10a^2 + 2ab - 15ab - 3b^2$
   $= 10a^2 - 13ab - 3b^2$

65. To find the product $(x - 1)(x + 1)(x + 2)$, we multiply the first two factors then the result times the third.

   $(x - 1)(x + 1) = x^2 + x - x - 1$ FOIL
   $ = x^2 - 1$

   Now multiply by $(x + 2)$.

   $(x^2 - 1)(x + 2) = x^3 + 2x^2 - x - 2$ FOIL

   Thus,

   $(x - 1)(x + 1)(x + 2) = x^3 + 2x^2 - x - 2.$

67. Use the formula $A = lw$ and substitute $(2x + 1)$ for l and $(3x - 2)$ for w.

   $A = (2x + 1)(3x - 2)$
   $ = (2x)(3x) + (2x)(-2) + (1)(3x) + (1)(-2)$ FOIL
   $ = 6x^2 - 4x + 3x - 2$
   $ = 6x^2 - x - 2$ Collect like terms

   Thus, the area is $(6x^2 - x - 2)$ ft².

6.4

13. $(4y - 9)(4y + 9)$
   $= (4y)^2 - (9)^2 \qquad (a - b)(a + b) = a^2 - b^2$
   $= 4^2 y^2 - 81$
   $= 16y^2 - 81$

31. $(0.7y - 3)^2 = (0.7y)^2 - 2(0.7y)(3) + (3)^2$
   $= (0.7)^2 y^2 - 6(0.7)y + 9 = 0.49y^2 - 4.2y + 9$

37. $(2x^2 - y)^2 = (2x^2)^2 - 2(2x^2)(y) + (y)^2$
   $= 2^2(x^2)^2 - 4x^2 y + y^2 = 4x^4 - 4x^2 y + y^2$

39. $(a^2 + 2b)^2$
   $= (a^2)^2 + 2(a^2)(2b) + (2b)^2$ Use a^2 for a and $2b$ for b in $(a + b)^2 = a^2 + 2ab + b^2$
   $= a^4 + 4a^2 b + 4b^2$

40. $(x + 1)^2 - (x - 1)^2$
   $= (x^2 + (2)(x)(1) + 1^2) - (x^2 - (2)(x)(1) + 1^2)$
   Use both perfect square formulas and be sure to enclose the products in parentheses.
   $= (x^2 + 2x + 1) - (x^2 - 2x + 1)$
   $= x^2 + 2x + 1 - x^2 + 2x - 1$ Clear parentheses
   $= 4x$
   Remember to use parentheses when subtracting in order to avoid a common sign error.

44. Substitute $(2a + 7)$ for l and $(a - 1)$ for w in $A = lw$.

   $A = (2a + 7)(a - 1)$
   $ = 2a^2 - 2a + 7a - 7$ FOIL
   $ = 2a^2 + 5a - 7$ Collect like terms

   Thus, the area is $(2a^2 + 5a - 7)$ mi².

6.5

7. $(3a^{12} - 9a^6 + 27a^5 + 81a^4) \div 9a^2$
   $= \dfrac{3a^{12} - 9a^6 + 27a^5 + 81a^4}{9a^2}$
   $= \dfrac{3a^{12}}{9a^2} - \dfrac{9a^6}{9a^2} + \dfrac{27a^5}{9a^2} + \dfrac{81a^4}{9a^2}$
   $= \dfrac{1}{3}a^{12-2} - 1 \cdot a^{6-2} + 3a^{5-2} + 9a^{4-2}$
   $= \dfrac{1}{3}a^{10} - a^4 + 3a^3 + 9a^2$

10. $\dfrac{-8x^3 + 6x^2 - 4x}{0.2x} = \dfrac{-8x^3}{0.2x} + \dfrac{6x^2}{0.2x} - \dfrac{4x}{0.2x}$
   $= -40x^{3-1} + 30x^{2-1} - 20x^{1-1} \qquad \dfrac{8}{0.2} = \dfrac{80}{2} = 40$
   $= -40x^2 + 30x - 20x^0$
   $= -40x^2 + 30x - 20 \qquad x^0 = 1$

13. $\dfrac{-5a^2 b + 3ab - 2a}{-a} = \dfrac{-5a^2 b}{-a} + \dfrac{3ab}{-a} - \dfrac{2a}{-a} =$
   $5a^{2-1}b - 3a^{1-1}b + 2a^{1-1} = 5ab - 3a^0 b + 2a^0 =$
   $5ab - 3b + 2$

22. $(6 + 8y - y^2) \div (4 - y)$
   $= \dfrac{6 + 8y - y^2}{4 - y}$
   $= \dfrac{(-1)(6 + 8y - y^2)}{(-1)(4 - y)}$ To make y in denominator positive
   $= \dfrac{-6 - 8y + y^2}{-4 + y}$
   $= \dfrac{y^2 - 8y - 6}{y - 4}$ Descending order

$$y - 4 \overline{\smash{\big)}\, y^2 - 8y - 6}$$
$$\underline{y^2 - 4y}$$
$$-4y - 6$$
$$\underline{-4y + 16}$$
$$-22 \quad \text{The answer is } y - 4 - \frac{22}{y-4}.$$

23. $a - 5 \overline{\smash{\big)}\, 3a^2 - 23a + 40}$ with quotient $3a - 8$
$$\underline{3a^2 - 15a}$$
$$-8a + 40 \quad \text{Remember to subtract at this step}$$
$$\underline{-8a + 40} \quad \text{then bring down the 40}$$
$$0 \quad \text{Remember again to subtract}$$

Thus, the quotient is $3a - 8$.

28. $y + 2 \overline{\smash{\big)}\, y^5 + 32}$ with quotient $y^4 - 2y^3 + 4y^2 - 8y + 16$
$$\underline{y^5 + 2y^4}$$
$$-2y^4$$
$$\underline{-2y^4 - 4y^3}$$
$$4y^3$$
$$\underline{4y^3 + 8y^2}$$
$$-8y^2$$
$$\underline{-8y^2 - 16y}$$
$$16y + 32$$
$$\underline{16y + 32}$$

30. $x^2 - 1 \overline{\smash{\big)}\, x^4 - 1}$ with quotient $x^2 + 1$
$$\underline{x^4 - x^2}$$
$$+x^2 - 1 \quad \text{Subtract, don't add}$$
$$\underline{x^2 - 1}$$
$$0$$

7.1

14. $36y^4 = 2^2 \cdot 3^2 \cdot y^4$
$6y^3 = 2 \cdot 3 \cdot y^3$
$42y^5 = 2 \cdot 3 \cdot 7 \cdot y^5$

The factors 2, 3, and y are in all three monomials. The lowest power of 2 and 3 is 1 and the lowest power of y is 3. Thus, the GCF is $2 \cdot 3 \cdot y^3 = 6y^3$.

27. The only common factor of $3a(a + 2)$ and $5(a + 2)$ is $(a + 2)$. Thus, the GCF is $(a + 2)$.

29. Since 5 is the GCF of 5, -15, and 10, and since ab^2 is the GCF of a^3b^3, a^2b^2, and ab^2, the GCF of $5a^3b^3 - 15a^2b^2 + 10ab^2$ is $5ab^2$.

41. $-6y^{10} - 8y^8 - 4y^5$
$= (-2y^5) \cdot 3y^5 + (-2y^5) \cdot 4y^3 + (-2y^5) \cdot 2$
$= -2y^5(3y^5 + 4y^3 + 2)$

47. $a^2(a + 2) + 3(a + 2)$
$= (a + 2) \cdot a^2 + (a + 2) \cdot 3 \quad$ GCF is $(a + 2)$
$= (a + 2)(a^2 + 3)$

50. $x^2a + x^2b + y^2a + y^2b = (x^2a + x^2b) + (y^2a + y^2b)$
$ = x^2(a + b) + y^2(a + b)$
$ = (x^2 + y^2)(a + b)$

56. $-x^2y - x^2 - 3y - 3$
$= (-1)(x^2y + x^2 + 3y + 3) \quad$ First factor -1 from all terms
$= (-1)[x^2(y + 1) + 3(y + 1)] \quad$ Factor x^2 from first two and 3 from second two
$= (-1)[(x^2 + 3)(y + 1)] \quad (y + 1)$ is common to the two terms in brackets
$= -(x^2 + 3)(y + 1)$

7.2

10. $y^2 + 5y - 24 = (y + \underline{})(y + \underline{})$
$b = 5$ and $c = -24$

Factors of $c = -24$	Sum of factors
12, -2	10
-2, 12	-10
8, -3	5

With c negative, we know that the factors must have opposite signs. Since we have found factors of c whose sum is $b = 5$, we can stop the table.
$y^2 + 5y - 24 = (y + 8)(y - 3)$

14. $x^2 - 2x - 63 = (x + \underline{})(x + \underline{})$
$b = -2$ and $c = -63$

Factors of $c = -63$	Sum of factors
21, -3	18
-21, 3	-18
9, -7	2
-9, 7	-2

With c negative, we only try factors with opposite signs. Since $b = -2$ the factors are -9 and 7.
$x^2 - 2x - 63 = (x - 9)(x + 7)$

20. $u^2 - 9uv + 20v^2 = (u + \underline{}v)(u + \underline{}v)$
$b = -9$ and $c = 20$

Factors of $c = 20$	Sum of factors
-10, -2	-12
-5, -4	-9

With c positive and b negative, we only try factors that are both negative. Thus, $u^2 - 9uv + 20v^2 = (u - 5v)(u - 4v)$.

35. $u^2 - 22uv + 121v^2 = (u + \underline{}v)(u + \underline{}v)$
$b = -22$ and $c = 121$

Factors of $c = 121$	Sum of factors
-121, -1	-122
-11, -11	-22

With c positive and b negative, we only try factors that are both negative. Thus, $u^2 - 22uv + 121v^2 = (u - 11v)(u - 11v)$.

7.3

10. For $6z^2 - 13z - 28$ we have $a = 6$, $b = -13$, and $c = -28$. The factors of $c = -28$ will have opposite signs.

Factors of $a = 6$ Factors of $c = -28$
   6, 1   28, -1 and -28, 1
   3, 2   14, -2 and -14, 2
                       7, -4 and -7, 4
$6z^2 - 13z - 28 = (\_\_z + \_\_)(\_\_z + \_\_)$
$\stackrel{?}{=} (6z - 1)(z + 28)$ Does not work
$\stackrel{?}{=} (3z - 2)(2z + 14)$ There would have to be a common factor of 2 for the 14, 2 factors to work
$\stackrel{?}{=} (3z - 4)(2z + 7)$ Does not work
$\stackrel{?}{=} (3z + 4)(2z - 7)$ This works
$6z^2 - 13z - 28 = (3z + 4)(2z - 7)$

17. $-45x^2 + 150x - 125$
$= (-5)9x^2 + (-5)(-30x) + (-5)(25)$
$= -5(9x^2 - 30x + 25)$

Now factor $9x^2 - 30x + 25$.
Factors of $a = 9$ Factors of $c = 25$
   9, 1 $-25, -1$
   3, 3 $-5, -5$
$9x^2 - 30x + 25 = (\_\_x + \_\_)(\_\_x + \_\_)$
$\stackrel{?}{=} (9x - 1)(x - 25)$ Does not work
$\stackrel{?}{=} (3x - 5)(3x - 5)$ This works
$9x^2 - 30x + 25 = (3x - 5)(3x - 5)$

We must include the common factor in the final answer.
$-45x^2 + 150x - 125 = -5(3x - 5)(3x - 5)$

22. $2x^2 + 7xy + 5y^2$ has $a = 2$, $b = 7$, and $c = 5$.
Factors of $a = 2$ Factors of $c = 5$
   2, 1   5, 1
$2x^2 + 7xy + 5y^2 = (\_\_x + \_\_y)(\_\_x + \_\_y)$
$\stackrel{?}{=} (2x + y)(x + 5y)$ Does not work
$\stackrel{?}{=} (2x + 5y)(x + y)$ This works
$2x^2 + 7xy + 5y^2 = (2x + 5y)(x + y)$

31. $4u^2 - v^2$ has $a = 4$, $b = 0$, and $c = -1$.
Factors of $a = 4$ Factors of $c = -1$
   4, 1   1, -1
   2, 2
$4u^2 - v^2 = (\_\_u + \_\_v)(\_\_u + \_\_v)$
$\stackrel{?}{=} (4u + v)(u - v)$ Does not work
$\stackrel{?}{=} (2u + v)(2u - v)$ This works
$4u^2 - v^2 = (2u + v)(2u - v)$

7.4

11. $9u^2 + 6u + 1$
$= (3u)^2 + 6u + 1^2$ $(3u)^2$ and 1^2 are perfect squares
$= (3u)^2 + 2 \cdot 3u \cdot 1 + 1^2$ $3u = a$ and $1 = b$
$= (3u + 1)^2$ $a^2 + 2ab + b^2 = (a + b)^2$

19. $-12y^2 + 60y - 75$
$= (-3)4y^2 + (-3)(-20y) + (-3)(25)$ Common factor first
$= -3(4y^2 - 20y + 25)$
$= -3[(2y)^2 - 2 \cdot 2y \cdot 5 + (5)^2]$ $2y = a$ and $5 = b$
$= -3(2y - 5)^2$

28. $25x^2 - 10xy + y^2 = (5x)^2 - 2 \cdot 5x \cdot y + (y)^2$
$= (5x - y)^2$

30. $u^4 - v^4 = (u^2)^2 - (v^2)^2$
$= (u^2 + v^2)(u^2 - v^2)$
$= (u^2 + v^2)(u + v)(u - v)$

41. In $4y^2 - 16y + 15$ the constants are $a = 4$, $b = -16$, and $c = 15$.
Factors of $a = 4$ Factors of $c = 15$
   4, 1 $-15, -1$
   2, 2 $-5, -3$
Since $b = -16$, the most likely combinations are 4, 1 with $-5, -3$ or 2, 2 with $-5, -3$.
$4y^2 - 16y + 15 = (\_\_y + \_\_)(\_\_y + \_\_)$
$\stackrel{?}{=} (4y - 5)(y - 3)$ Does not work
$\stackrel{?}{=} (2y - 5)(2y - 3)$ This works
$4y^2 - 16y + 15 = (2y - 5)(2y - 3)$

43. $3x^9 - 147x^3 = (3x^3)x^6 - (3x^3)49$ Common factor is $3x^3$
$= 3x^3(x^6 - 49)$
$= 3x^3[(x^3)^2 - (7)^2]$ $x^3 = a$ and $7 = b$
$= 3x^3(x^3 + 7)(x^3 - 7)$

49. $100x^5 - 60x^4y + 9x^3y^2$
$= (x^3)100x^2 - (x^3)60xy + x^3(9y^2)$
$= x^3(100x^2 - 60xy + 9y^2)$
$= x^3[(10x)^2 - 2(10x)(3y) + (3y)^2]$
$= x^3(10x - 3y)^2$

7.5

13. $x(3x + 7) = 0$
$x = 0$ or $3x + 7 = 0$ Zero-product rule; do not divide both sides by x
               $3x = -7$
$x = 0$ or $x = -\dfrac{7}{3}$

Thus, the solutions are 0 and $-\dfrac{7}{3}$. Remember that whenever one factor is x, 0 will always be one solution.

532 SOLUTIONS TO SELECTED EXERCISES

23.
$$4u^2 - 8u = 0$$
$$(4u) \cdot u - (4u) \cdot 2 = 0$$
$$4u(u - 2) = 0$$
$u = 0$ or $u - 2 = 0$
$$u = 2$$

The solutions are 0 and 2. Do not forget the zero solution.

26. $(3x - 5) - (x + 7) = 0$
Do not try to use the zero-product rule on a problem like this. It involves a zero-difference, not a zero-product. Clear parentheses and solve.

$3x - 5 - x - 7 = 0$ Watch the signs
$2x - 12 = 0$ Collect like terms
$2x = 12$
$x = 6$

Thus, the solution is 6.

27.
$$y(y + 3) = 10$$
$$y^2 + 3y = 10$$
$$y^2 + 3y - 10 = 0$$
$$(y - 2)(y + 5) = 0$$
$y - 2 = 0$ or $y + 5 = 0$
$y = 2$ $y = -5$

The solutions are 2 and -5.

7.6

5. Let n = first integer,
$n + 1$ = next consecutive integer,
$n(n + 1)$ = product of consecutive integers.
$$n(n + 1) = 240$$
$$n^2 + n = 240$$
$$n^2 + n - 240 = 0$$
$$(n - 15)(n + 16) = 0$$
$n - 15 = 0$ or $n + 16 = 0$
$n = 15$ $n = -16$
$n + 1 = 16$ $n + 1 = -15$

Thus, 15 and 16 form one solution and -16 and -15 the other.

8. Let n = the first even positive integer,
$n + 2$ = the next consecutive even positive integer.
The equation to solve is
$$n^2 + (n + 2)^2 = 100$$
$$n^2 + n^2 + 4n + 4 = 100$$
$$2n^2 + 4n - 96 = 0$$
$$n^2 + 2n - 48 = 0 \quad \text{Divide out common factor 2}$$

$(n + 8)(n - 6) = 0$ Factor
$n + 8 = 0$ or $n - 6 = 0$ Zero-product rule
$n = -8$ $n = 6$

Since n must be positive, we discard -8. Thus, the integers are 6 and 8 (since $n + 2 = 8$).

10. Let x = length of one side.
x^2 = area of square.
$4x$ = perimeter of square.

If we add 4 to the area, the value is the same as the perimeter.
$$x^2 + 4 = 4x$$
$$x^2 - 4x + 4 = 0$$
$$(x - 2)^2 = 0$$
$x - 2 = 0$ or $x - 2 = 0$
$x = 2$ $x = 2$

The side is 2 cm.

13. We must find P when n is 30.
$P = n^2 - 3n - 60$
$= (30)^2 - 3(30) - 60$ Substitute 30 for n
$= 900 - 90 - 60$
$= 750$

Thus, the profit is $750 when 30 appliances are sold.

15. We must find n when P is 120.
$P = n^2 - 3n - 60$
$120 = n^2 - 3n - 60$ Substitute 120 for P
$0 = n^2 - 3n - 180$
$0 = (n + 12)(n - 15)$
$n + 12 = 0$ or $n - 15 = 0$ Zero-product rule
$n = -12$ $n = 15$

Since the number of appliances cannot be negative, we discard -12. Thus, 15 were sold when the profit was $120.

17. We must find N when n is 7.
$$N = \frac{1}{2}n(n - 1)$$
$$= \frac{1}{2}(7)(7 - 1) \quad \text{Substitute 7 for } n$$
$$= \frac{1}{2}(7)(6) = 21$$

Thus, 21 committees of two can be formed.

19. We must find n when N is 10.
$$N = \frac{1}{2}n(n - 1)$$

$$10 = \frac{1}{2}n(n-1)$$

$20 = n(n-1)$ Clear fraction

$20 = n^2 - n$

$0 = n^2 - n - 20$

$0 = (n+4)(n-5)$

$n + 4 = 0$ or $n - 5 = 0$

$n = -4 \qquad\qquad n = 5$

Since n cannot be negative, there are 5 members of the organization.

8.1

7. $\dfrac{2x+7}{x^2+2x+1}$ We must solve $x^2 + 2x + 1 = 0$. First factor, then use the zero-product rule.

$$x^2 + 2x + 1 = 0$$
$$(x+1)(x+1) = 0$$
$x + 1 = 0$ or $x + 1 = 0$
$x = -1$ or $x = -1$

Thus, -1 is the only number to be omitted.

16. $\dfrac{2}{x^2}$ and $\dfrac{2+x}{x^2+x}$ are not equivalent since one cannot be obtained from the other by multiplying both numerator and denominator by the same expression. Notice that the second fraction can be formed from the first by adding x to both numerator and denominator, but this process does not make the fraction equivalent. To see this, substitute 1 for x in both fractions.

17. $\dfrac{x-1}{3x-1}$ and $\dfrac{x}{3x}$ are not equivalent.

To obtain one from the other we would have to add or subtract the same number in both the numerator and denominator, and this process (unlike multiplying or dividing) does not yield equivalent fractions.

29. $\dfrac{x^2 - 3x - 10}{x^2 - 6x + 5} = \dfrac{(x-5)(x+2)}{(x-5)(x-1)} = \dfrac{x+2}{x-1}$

32. $\dfrac{x^2 - 9}{3 - x} = \dfrac{(x-3)(x+3)}{3-x} = \dfrac{x-3}{3-x} \cdot (x+3) =$

$(-1)(x+3) = -(x+3)$

36. $\dfrac{x^2 - 7xy + 6y^2}{x^2 - 4xy - 12y^2}$

$= \dfrac{(x-y)(x-6y)}{(x+2y)(x-6y)}$ Factor numerator and denominator

$= \dfrac{x-y}{x+2y}$ Cancel factors $(x-6y)$

Do not cancel the x's since they are terms, not factors, of the numerator and denominator.

8.2

10. $\dfrac{z^2 - z - 20}{z^2 + 7z + 12} \cdot \dfrac{z+3}{z^2 - 25}$

$= \dfrac{(z-5)(z+4)}{(z+3)(z+4)} \cdot \dfrac{(z+3)}{(z-5)(z+5)}$

$= \dfrac{(z-5)(z+4)(z+3)}{(z+3)(z+4)(z-5)(z+5)} = \dfrac{1}{z+5}$

14. $\dfrac{2x^2 - 5xy + 3y^2}{x^2 - y^2} \cdot (x^2 + 2xy + y^2)$

$= \dfrac{(2x - 3y)(x - y)}{(x - y)(x + y)} \cdot \dfrac{(x+y)(x+y)}{1}$

$= \dfrac{(2x-3y)(x-y)(x+y)(x+y)}{(x-y)(x+y)}$

$= (2x - 3y)(x + y)$

22. The reciprocal of $z + 2$ is $\dfrac{1}{z+2}$ and *not* $\dfrac{1}{z} + \dfrac{1}{2}$; a common error.

29. $(x + 6) \div \dfrac{x^2 - 36}{x^2 - 6x}$

$= \dfrac{(x+6)}{1} \cdot \dfrac{x^2 - 6x}{x^2 - 36}$ Multiply by the reciprocal

$= \dfrac{(x+6)}{1} \cdot \dfrac{x(x-6)}{(x+6)(x-6)}$ Factor

$= \dfrac{(x+6)x(x-6)}{(x+6)(x-6)}$ Cancel common factors

$= x$

34. $\dfrac{x^2 + 16x + 64}{2x^2 - 128} \div \dfrac{3x^2 + 30x + 48}{x^2 - 6x - 16}$

$= \dfrac{x^2 + 16x + 64}{2x^2 - 128} \cdot \dfrac{x^2 - 6x - 16}{3x^2 + 30x + 48}$

$= \dfrac{(x+8)(x+8)}{2(x-8)(x+8)} \cdot \dfrac{(x-8)(x+2)}{3(x+8)(x+2)}$

$= \dfrac{(x+8)(x+8)(x-8)(x+2)}{2(x-8)(x+8) \cdot 3 \cdot (x+8)(x+2)}$

$= \dfrac{1}{2 \cdot 3} = \dfrac{1}{6}$

35. $\dfrac{x^2 - 5xy + 6y^2}{x^2 - 4y^2} \div (x^2 - 2xy - 3y^2)$

$= \dfrac{x^2 - 5xy + 6y^2}{x^2 - 4y^2} \cdot \dfrac{1}{x^2 - 2xy - 3y^2}$

$= \dfrac{(x-2y)(x-3y)}{(x-2y)(x+2y)} \cdot \dfrac{1}{(x-3y)(x+y)}$

$= \dfrac{1}{(x+2y)(x+y)}$

38. $\dfrac{a^2 - y^2}{a^2 - ay} \cdot \dfrac{2a^2 + ay}{a^2 - 4y^2} \div \dfrac{a+y}{a+2y}$

$$= \frac{(a-y)(a+y)}{a(a-y)} \cdot \frac{a(2a+y)}{(a-2y)(a+2y)} \cdot \frac{(a+2y)}{a+y}$$

$$= \frac{(a-y)(a+y)a(2a+y)(a+2y)}{a(a-y)(a-2y)(a+2y)(a+y)}$$

$$= \frac{2a+y}{a-2y}$$

8.3

5. $\dfrac{2y}{y+2} - \dfrac{y+1}{y+2} = \dfrac{2y-(y+1)}{y+2}$ Use parentheses

$$= \frac{2y-y-1}{y+2} = \frac{y-1}{y+2}$$

10. $\dfrac{z}{2} + \dfrac{3z-1}{-2}$

$$= \frac{z}{2} + \frac{(-1)(3z-1)}{(-1)(-2)}$$

$$= \frac{z}{2} + \frac{1-3z}{2}$$

$$= \frac{z+(1-3z)}{2}$$

$$= \frac{z+1-3z}{2} = \frac{-2z+1}{2}$$

14. $\dfrac{2a}{a-1} + \dfrac{3a}{1-a} = \dfrac{2a}{a-1} + \dfrac{3a}{(-1)(a-1)} =$

$$\frac{2a}{a-1} + \frac{-3a}{a-1} = \frac{2a-3a}{a-1} = \frac{-a}{a-1}$$

20. $\dfrac{2x}{x^2+x-6} + \dfrac{x-3}{6-x-x^2}$

$$= \frac{2x}{x^2+x-6} + \frac{x-3}{(-1)(-6+x+x^2)}$$

$$= \frac{2x}{x^2+x-6} + \frac{(-1)(x-3)}{x^2+x-6}$$

$$= \frac{2x+(-1)(x-3)}{x^2+x-6}$$

$$= \frac{2x-x+3}{x^2+x-6} = \frac{(x+3)}{(x+3)(x-2)} = \frac{1}{x-2}$$

24. $\dfrac{x+y}{x-y} - \dfrac{x+y}{y-x}$

$$= \frac{x+y}{x-y} - \frac{(-1)(x+y)}{(-1)(y-x)}$$

$$= \frac{x+y}{x-y} + \frac{x+y}{x-y}$$

$$= \frac{(x+y)+(x+y)}{x-y}$$

$$= \frac{2x+2y}{x-y} = \frac{2(x+y)}{x-y}$$

25. $\dfrac{5x}{x+1} + \dfrac{2x-1}{x+1} - \dfrac{3x}{x+1} = \dfrac{5x+2x-1-3x}{x+1}$

$$= \frac{4x-1}{x+1}$$

8.4

10. $\dfrac{1}{x^2+2x+1}$ and $\dfrac{2}{x^2-1}$

$x^2+2x+1 = (x+1)(x+1)$ The LCD consists of one $(x-1)$ and two $(x+1)$'s

$x^2-1 = (x+1)(x-1)$

Thus, LCD $= (x-1)(x+1)^2$.

14. $\dfrac{5x}{x+y}, \dfrac{7y}{2x+2y}$, and $\dfrac{2xy}{3x+3y}$

The factored denominators are $x+y$, $2(x+y)$, and $3(x+y)$. The LCD must consist of one $(x+y)$, one 2, and one 3. Thus, LCD $= 2 \cdot 3 \cdot (x+y) = 6(x+y)$.

23. $\dfrac{4y}{y^2-36} - \dfrac{4}{y+6}$

$$= \frac{4y}{(y-6)(y+6)} - \frac{4(y-6)}{(y+6)(y-6)}$$

$$= \frac{4y-4(y-6)}{(y-6)(y+6)}$$

$$= \frac{4y-4y+24}{(y-6)(y+6)} = \frac{24}{(y-6)(y+6)}$$

31. $\dfrac{3}{a^2-a-12} - \dfrac{2}{a^2-9}$

$$= \frac{3}{(a-4)(a+3)} - \frac{2}{(a-3)(a+3)}$$

$$= \frac{3(a-3)}{(a-4)(a+3)(a-3)} - \frac{2(a-4)}{(a-3)(a+3)(a-4)}$$

$$= \frac{3(a-3)-2(a-4)}{(a-4)(a+3)(a-3)}$$

$$= \frac{3a-9-2a+8}{(a-4)(a+3)(a-3)} = \frac{a-1}{(a-4)(a+3)(a-3)}$$

35. $\dfrac{3}{x+1} + \dfrac{5}{x-1} - \dfrac{10}{x^2-1}$

$$= \frac{3}{x+1} + \frac{5}{x-1} - \frac{10}{(x-1)(x+1)}$$

Factor denominators, the LCD $= (x-1)(x+1)$

$$= \frac{3(x-1)}{(x-1)(x+1)} + \frac{5(x+1)}{(x-1)(x+1)} - \frac{10}{(x-1)(x+1)}$$

Supply missing factors

$$= \frac{3(x-1)+5(x+1)-10}{(x-1)(x+1)}$$

Add and subtract numerators over LCD

$$= \frac{3x - 3 + 5x + 5 - 10}{(x - 1)(x + 1)}$$

$$= \frac{8x - 8}{(x - 1)(x + 1)} \quad \text{Collect like terms}$$

$$= \frac{8(x - 1)}{(x - 1)(x + 1)} = \frac{8}{x + 1} \quad \text{Cancel common factor}$$

37. $\dfrac{2y}{y + 5} - \dfrac{3y}{y - 2} - \dfrac{2y^2}{y^2 + 3y - 10}$

$$= \frac{2y}{y + 5} - \frac{3y}{y - 2} - \frac{2y^2}{(y + 5)(y - 2)} =$$

$$\frac{2y(y - 2)}{(y + 5)(y - 2)} - \frac{3y(y + 5)}{(y - 2)(y + 5)} - \frac{2y^2}{(y + 5)(y - 2)}$$

$$= \frac{2y(y - 2) - 3y(y + 5) - 2y^2}{(y + 5)(y - 2)}$$

$$= \frac{2y^2 - 4y - 3y^2 - 15y - 2y^2}{(y + 5)(y - 2)}$$

$$= \frac{-3y^2 - 19y}{(y + 5)(y - 2)} = \frac{-y(3y + 19)}{(y + 5)(y - 2)}$$

8.5

4. $\dfrac{1}{z - 3} + \dfrac{3}{z - 5} = 0$

$\dfrac{1}{z - 3} = -\dfrac{3}{z - 5}$ Subtract $\dfrac{3}{z - 5}$ from both sides

$(1)(z - 5) = (-3)(z - 3)$ Cross-product equation

$z - 5 = -3z + 9$ Clear parentheses

$4z = 14$

$z = \dfrac{14}{4} = \dfrac{7}{2}$

Since $\dfrac{7}{2}$ will check in the original equation, it is the solution.

8. $\dfrac{z}{z + 1} - 1 = \dfrac{1}{z}$

Multiply through by the LCD, $z(z + 1)$.

$z(z + 1)\dfrac{z}{z + 1} - z(z + 1)(1) = z(z + 1)\dfrac{1}{z}$

$z^2 - (z^2 + z) = z + 1$ Use parentheses in this step to avoid a common sign error

$z^2 - z^2 - z = z + 1$

$-2z = 1$

$z = -\dfrac{1}{2}$

The solution is $-\dfrac{1}{2}$, which does indeed check in the original equation.

11. $\dfrac{4}{z - 3} + \dfrac{2z}{z^2 - 9} = \dfrac{1}{z + 3}$

The LCD $= (z - 3)(z + 3)$; so multiply both sides by it.

$$\left[\frac{4}{z - 3} + \frac{2z}{(z - 3)(z + 3)}\right](z - 3)(z + 3)$$

$$= \frac{1}{z + 3}(z - 3)(z + 3)$$

$$\frac{4}{(z - 3)}(z - 3)(z + 3) + \frac{2z}{(z - 3)(z + 3)}(z - 3)(z + 3)$$

$$= \frac{1}{(z + 3)}(z - 3)(z + 3)$$

$4(z + 3) + 2z = z - 3$

$4z + 12 + 2z = z - 3$

$6z + 12 = z - 3$

$5z = -15$

$z = -3$

However, if -3 is substituted into the original equation, we obtain zero in two denominators. That is, -3 must be omitted as a replacement for z. Thus, the equation has no solution.

16. $\dfrac{3}{x - 3} + \dfrac{2}{x + 3} = -1$

The LCD is $(x - 3)(x + 3)$. Multiply both sides by it, and do not forget to multiply on the right side

$$\left[\frac{3}{x - 3} + \frac{2}{x + 3}\right](x - 3)(x + 3)$$

$$= (-1)(x - 3)(x + 3)$$

$$\frac{3}{(x - 3)}(x - 3)(x + 3) + \frac{2}{(x + 3)}(x - 3)(x + 3) =$$

$(-1)(x^2 - 9)$

$3(x + 3) + 2(x - 3) = -x^2 + 9$

$3x + 9 + 2x - 6 = -x^2 + 9$

$5x + 3 = -x^2 + 9$

$x^2 + 5x - 6 = 0$

$(x + 6)(x - 1) = 0$

$x + 6 = 0 \quad \text{or} \quad x - 1 = 0 \quad$ Zero-product rule

$x = -6 \quad \text{or} \quad x = 1$

Both -6 and 1 are solutions since both check in the original equation.

23. Let $x =$ the desired number.
The equation to solve is
$\dfrac{5 + x}{7 + x} = \dfrac{4}{5}$.

$5(5 + x) = 4(7 + x)$ Cross-product equation
$25 + 5x = 28 + 4x$ Clear parentheses
$x = 3$

25. Let x = the numerator of the fraction,
$x + 4$ = the denominator of the fraction.

Then the fraction is $\dfrac{x}{x+4}$.

Now we increase *both* the numerator and the denominator by 1, and set the result equal to $\dfrac{2}{3}$.

$$\frac{x+1}{x+4+1} = \frac{2}{3}$$

$$\frac{x+1}{x+5} = \frac{2}{3} \quad \text{Multiply both sides by the LCD} = 3(x+5)$$

$$3(x+5) \cdot \frac{(x+1)}{(x+5)} = 3(x+5) \cdot \frac{2}{3}$$

$$3(x+1) = (x+5)2$$
$$3x + 3 = 2x + 10$$
$$x = 7$$

Thus, the fraction (which was what we were asked to find) is $\dfrac{7}{11}$.

28. Let n = the desired number,
$n - 4$ = 4 less than the number,
$\dfrac{1}{n-4}$ = the reciprocal of 4 less than the number,
$\dfrac{1}{n}$ = the reciprocal of the desired number.

The equation we must solve is

$$\frac{1}{n-4} = 3 \cdot \frac{1}{n}$$

$$\frac{1}{n-4} = \frac{3}{n}$$

Multiplying both sides by $n(n-4)$ results in the equation

$$n = 3(n-4)$$
$$n = 3n - 12$$
$$-2n = -12$$
$$n = 6.$$

Thus, the desired number is 6, and 6 does satisfy the description of the problem.

30. Let t = time required for Bob's father to paint house,
12 = time required for Bob to paint house,
9 = time required for Bob and his father to paint the house working together.

Then,

$$\frac{1}{12} + \frac{1}{t} = \frac{1}{9}$$

That is, the amount Bob does in 1 day plus the amount his father does in 1 day equals the amount they do together in 1 day.
Multiply through by $36t$, the LCD.

$$36t \cdot \frac{1}{12} + 36t \cdot \frac{1}{t} = 36t \cdot \frac{1}{9}$$

$$3t + 36 = 4t$$
$$36 = t$$

Thus, it would take Bob's father 36 days to paint the house.

32. Let x = time for Max to repair the car working alone,
$4x$ = time for Irv to repair the car working alone,
$\dfrac{1}{x}$ = amount done by Max in 1 hr,
$\dfrac{1}{4x}$ = amount done by Irv in 1 hr,
$\dfrac{1}{8}$ = amount done together in 1 hr.

The equation to solve is

$$\frac{1}{x} + \frac{1}{4x} = \frac{1}{8}.$$

Multiply through by the LCD, $8x$.

$$8x\left(\frac{1}{x}\right) + 8x\left(\frac{1}{4x}\right) = 8x\left(\frac{1}{8}\right)$$

$$8 + 2 = x$$
$$10 = x$$

Thus, it will take Max 10 hr and Irv 40 hr to repair the car working alone.

8.6

11. $\dfrac{y-3}{y+2} = \dfrac{y-2}{y+3}$ Form the cross-product equation and solve

$$(y-3)(y+3) = (y+2)(y-2)$$
$$y^2 - 9 = y^2 - 4$$
$$-9 = -4$$

Since we obtain a contradiction, the proportion has no solutions.

17. Let x = the number of defective tires expected in a sample of 1288 tires. Then

$$\frac{6}{184} = \frac{x}{1288}.$$

$$6 \cdot 1288 = 184x$$
$$7728 = 184x$$
$$42 = x$$

Thus, you would expect 42 tires to be defective.

26. Let $x =$ approximate number of fish in the lake. The ratio of total fish to the number tagged should be about the same as the ratio of the number of fish caught in the sample to the number of fish in the sample that were tagged. That is, we must solve

$$\frac{x}{100} = \frac{70}{14}$$

$$14x = (70)(100) \quad \text{Cross-product equation}$$
$$x = 500$$

Thus, there are about 500 fish in the lake.

30.
$c = kw$	This is the direct variation equation
$480 = k \cdot 12$	Substitute values for c and w, and let c be in cents to avoid decimals
$40 = k$	The value of the constant k is 40
$c = 40w$	This is the variation equation using the known value for the constant
$c = 40 \cdot 66 = 2640$	Substitute the new value of w

Thus, the cost of 66 lb of peaches would be $26.40.

34. The variation equation is

$$a = kt,$$

where a is the amount spent on recreation and t is total income. Then

$$1920 = k(16{,}000)$$
$$\frac{1920}{16{,}000} = k$$
$$0.12 = k$$

Thus, the variation equation is

$a = 0.12t$.

Next we find a when his total income is increased by $5000; that is, when $t = 21{,}000$.

$$a = 0.12(21{,}000)$$
$$a = 2520$$

Hence, the man would spend about $2520 on recreation under these conditions.

8.7

11. $\dfrac{\dfrac{1}{2x} - 1}{\dfrac{1}{x} - 2}$ The LCD of all fractions is $2x$. Multiply both numerator and denominator by $2x$.

$$\frac{\left[\dfrac{1}{2x} - 1\right]2x}{\left[\dfrac{1}{x} - 2\right]2x} = \frac{\dfrac{1}{2x} \cdot 2x - 1 \cdot 2x}{\dfrac{1}{x} \cdot 2x - 2 \cdot 2x} = \frac{1 - 2x}{2 - 4x} =$$

$$\frac{(1 - 2x)}{2(1 - 2x)} = \frac{1}{2}$$

15. $\dfrac{z - 2 - \dfrac{3}{z}}{1 + \dfrac{1}{z}} = \dfrac{\left[z - 2 - \dfrac{3}{z}\right]z}{\left[1 + \dfrac{1}{z}\right]z} = \dfrac{z^2 - 2z - 3}{z + 1}$

$$= \frac{(z - 3)(z + 1)}{(z + 1)} = z - 3$$

17. $\dfrac{x - 3 + \dfrac{2}{x}}{x - 4 + \dfrac{3}{x}} = \dfrac{x\left(x - 3 + \dfrac{2}{x}\right)}{x\left(x - 4 + \dfrac{3}{x}\right)}$ The LCD is x

$$= \frac{x^2 - 3x + 2}{x^2 - 4x + 3}$$

$$= \frac{(x - 1)(x - 2)}{(x - 1)(x - 3)}$$

$$= \frac{x - 2}{x - 3}$$

9.1

14. $\sqrt{\dfrac{1000}{10}} = \sqrt{100}$ Reduce the fraction

$$= 10$$

19. $\sqrt{49a^2} = \sqrt{7^2 a^2}$ Factor 49

$$= \sqrt{(7a)^2} \quad a^n b^n = (ab)^n$$
$$= 7a$$

23. $\sqrt{x^2 + 10x + 25} = \sqrt{(x + 5)^2}$ Factor the radicand
$$= x + 5$$

Do not make the mistake of trying to take the square roots term-by-term.

25. We are given that $p = \sqrt{w}$ and that w is 25. Thus,

$$p = \sqrt{w} = \sqrt{25} = 5.$$

The productivity is 5.

538 SOLUTIONS TO SELECTED EXERCISES

28. With $p = \sqrt{w}$ and $p = 4$ we have
$$p = \sqrt{w}$$
$$4 = \sqrt{w}$$
$$(4)^2 = (\sqrt{w})^2$$
$$16 = w$$

9.2

28. $\sqrt{\dfrac{x^3y^3}{49}} = \dfrac{\sqrt{x^3y^3}}{\sqrt{49}} = \dfrac{\sqrt{x^2 \cdot x \cdot y^2 \cdot y}}{7} = \dfrac{\sqrt{x^2}\sqrt{y^2}\sqrt{xy}}{7}$

$= \dfrac{xy\sqrt{xy}}{7}$

31. $\sqrt{\dfrac{9}{5}} = \dfrac{\sqrt{9}}{\sqrt{5}} = \dfrac{3}{\sqrt{5}}$

$= \dfrac{3\sqrt{5}}{\sqrt{5}\sqrt{5}}$ Rationalize the denominator

$= \dfrac{3\sqrt{5}}{5}$

34. $\sqrt{\dfrac{36y^2}{x}} = \dfrac{\sqrt{36y^2}}{\sqrt{x}} = \dfrac{\sqrt{36}\sqrt{y^2}}{\sqrt{x}} = \dfrac{6y}{\sqrt{x}} =$

$\dfrac{6y\sqrt{x}}{\sqrt{x}\sqrt{x}} = \dfrac{6y\sqrt{x}}{x}$

35. $\sqrt{\dfrac{25x^3}{y^3z^4}} = \sqrt{\dfrac{25x^2x}{y^2y(z^2)^2}}$ Identify perfect squares

$= \dfrac{5x}{yz^2}\sqrt{\dfrac{x}{y}}$ Remove perfect squares

$= \dfrac{5x}{yz^2}\sqrt{\dfrac{xy}{y^2}}$ Multiply numerator and denominator of radicand by y to obtain a perfect square denominator

$= \dfrac{5x}{yz^2} \cdot \dfrac{\sqrt{xy}}{y}$

$= \dfrac{5x\sqrt{xy}}{y^2z^2}$

9.3

14. $\sqrt{3a^2}\sqrt{12a} = \sqrt{3a^2 \cdot 12a} = \sqrt{3^2 \cdot 2^2 \cdot a^2 \cdot a} =$
$\sqrt{3^2}\sqrt{2^2}\sqrt{a^2}\sqrt{a} = 3 \cdot 2 \cdot a\sqrt{a} = 6a\sqrt{a}$

17. $\sqrt{6x^2y}\sqrt{2xy^2} = \sqrt{6x^2y \cdot 2xy^2} =$
$\sqrt{2^2 \cdot 3 \cdot x^2 \cdot y \cdot x \cdot y^2} = \sqrt{2^2x^2y^2 \cdot 3xy} =$
$\sqrt{2^2}\sqrt{x^2}\sqrt{y^2}\sqrt{3xy} = 2xy\sqrt{3xy}$

19. $\sqrt{5x}\sqrt{15xy}\sqrt{3y} = \sqrt{5x \cdot 15xy \cdot 3y} = \sqrt{5^2 \cdot 3^2 \cdot x^2y^2}$
$= \sqrt{5^2}\sqrt{3^2} \cdot \sqrt{x^2} \cdot \sqrt{y^2} = 5 \cdot 3 \cdot xy = 15xy$

23. $\dfrac{\sqrt{98}}{\sqrt{18}} = \sqrt{\dfrac{98}{18}}$ $\dfrac{\sqrt{a}}{\sqrt{b}} = \sqrt{\dfrac{a}{b}}$

$= \sqrt{\dfrac{\cancel{2} \cdot 49}{\cancel{2} \cdot 9}}$ Factor, look for perfect squares, cancel the 2's

$= \dfrac{\sqrt{49}}{\sqrt{9}} = \dfrac{7}{3}$

26. $\dfrac{5\sqrt{15}}{\sqrt{75}} = 5\sqrt{\dfrac{15}{75}}$

$= 5\sqrt{\dfrac{\cancel{3} \cdot 5}{\cancel{3} \cdot 25}}$

$= 5\sqrt{\dfrac{5}{25}}$ Do not cancel 5's since 25 is a perfect square

$= 5\dfrac{\sqrt{5}}{\sqrt{25}} = \dfrac{\cancel{5}\sqrt{5}}{\cancel{5}} = \sqrt{5}$

31. $\dfrac{\sqrt{3y}}{\sqrt{4y}} = \sqrt{\dfrac{3\cancel{y}}{4\cancel{y}}} = \sqrt{\dfrac{3}{4}} = \dfrac{\sqrt{3}}{\sqrt{4}} = \dfrac{\sqrt{3}}{2}$

32. $\dfrac{\sqrt{27a^3}}{\sqrt{3}} = \sqrt{\dfrac{27a^3}{3}}$ $\dfrac{\sqrt{a}}{\sqrt{b}} = \sqrt{\dfrac{a}{b}}$

$= \sqrt{9a^2a}$

$= 3a\sqrt{a}$

37. $\dfrac{5\sqrt{2xy}}{\sqrt{25x}} = 5\sqrt{\dfrac{2xy}{25x}} = 5\sqrt{\dfrac{2y}{25}} = 5\dfrac{\sqrt{2y}}{\sqrt{25}} =$

$\dfrac{\cancel{5}\sqrt{2y}}{\cancel{5}} = \sqrt{2y}$

38. $\dfrac{\sqrt{18x}}{\sqrt{2y}} = \sqrt{\dfrac{18x}{2y}} = \sqrt{\dfrac{9x}{y}} = \dfrac{\sqrt{9}\sqrt{x}}{\sqrt{y}} =$

$\dfrac{3\sqrt{x}\sqrt{y}}{\sqrt{y}\sqrt{y}} = \dfrac{3\sqrt{xy}}{y}$

9.4

5. $5\sqrt{12} - 3\sqrt{12} = (5 - 3)\sqrt{12}$ Distributive law

$= 2\sqrt{12}$

$= 2\sqrt{4 \cdot 3}$ Look for perfect squares

$= 2 \cdot 2\sqrt{3}$

$= 4\sqrt{3}$

10. $4\sqrt{50} + 7\sqrt{18} = 4\sqrt{25 \cdot 2} + 7\sqrt{9 \cdot 2} =$
$4\sqrt{25}\sqrt{2} + 7\sqrt{9}\sqrt{2} = 4 \cdot 5\sqrt{2} + 7 \cdot 3\sqrt{2} =$
$20\sqrt{2} + 21\sqrt{2} = (20 + 21)\sqrt{2} = 41\sqrt{2}$

14. $\sqrt{147} - 2\sqrt{75} = \sqrt{49 \cdot 3} - 2\sqrt{25 \cdot 3}$

$= 7\sqrt{3} - 2 \cdot 5\sqrt{3}$

$= 7\sqrt{3} - 10\sqrt{3}$

$= (7 - 10)\sqrt{3}$

$= -3\sqrt{3}$

SOLUTIONS TO SELECTED EXERCISES **539**

18. $2\sqrt{4x} + 7\sqrt{9x} = 2\sqrt{4}\sqrt{x} + 7\sqrt{9}\sqrt{x} =$
    $2 \cdot 2\sqrt{x} + 7 \cdot 3\sqrt{x} = 4\sqrt{x} + 21\sqrt{x} =$
    $(4 + 21)\sqrt{x} = 25\sqrt{x}$

21. $x\sqrt{y^3} + y\sqrt{x^2y} = x\sqrt{y^2 \cdot y} + y\sqrt{x^2y} =$
    $x\sqrt{y^2}\sqrt{y} + y\sqrt{x^2}\sqrt{y} = xy\sqrt{y} + yx\sqrt{y} =$
    $(xy + xy)\sqrt{y} = 2xy\sqrt{y}$

22. $\sqrt{18} + \sqrt{50} + \sqrt{72} =$
    $\sqrt{9 \cdot 2} + \sqrt{25 \cdot 2} + \sqrt{36 \cdot 2} =$
    $\sqrt{9}\sqrt{2} + \sqrt{25}\sqrt{2} + \sqrt{36}\sqrt{2} =$
    $3\sqrt{2} + 5\sqrt{2} + 6\sqrt{2} = (3 + 5 + 6)\sqrt{2} =$
    $14\sqrt{2}$

25. $3\sqrt{75} + 2\sqrt{12} - 5\sqrt{48} =$
    $3\sqrt{25 \cdot 3} + 2\sqrt{4 \cdot 3} - 5\sqrt{16 \cdot 3} =$
    $3\sqrt{25}\sqrt{3} + 2\sqrt{4}\sqrt{3} - 5\sqrt{16}\sqrt{3} =$
    $3 \cdot 5\sqrt{3} + 2 \cdot 2\sqrt{3} - 5 \cdot 4\sqrt{3} =$
    $15\sqrt{3} + 4\sqrt{3} - 20\sqrt{3} = -\sqrt{3}$

29. $4\sqrt{300} - 2\sqrt{500} + 4\sqrt{125}$
    $= 4\sqrt{100 \cdot 3} - 2\sqrt{100 \cdot 5} + 4\sqrt{25 \cdot 5}$ Look for perfect squares
    $= 4 \cdot 10\sqrt{3} - 2 \cdot 10\sqrt{5} + 4 \cdot 5\sqrt{5}$
    $= 40\sqrt{3} - 20\sqrt{5} + 20\sqrt{5}$
    $= 40\sqrt{3}$

30. $\sqrt{4x} + \sqrt{16x} - \sqrt{25x} =$
    $\sqrt{4}\sqrt{x} + \sqrt{16}\sqrt{x} - \sqrt{25}\sqrt{x} =$
    $2\sqrt{x} + 4\sqrt{x} - 5\sqrt{x} = \sqrt{x}$

32. $9\sqrt{100a^2b^2} - 4\sqrt{36a^2b^2} - 10\sqrt{a^2b^2}$
    $= 9\sqrt{100}\sqrt{a^2}\sqrt{b^2} - 4\sqrt{36}\sqrt{a^2}\sqrt{b^2} -$
    $10\sqrt{a^2}\sqrt{b^2}$
    $= 9 \cdot 10 \cdot ab - 4 \cdot 6 \cdot ab - 10 \cdot ab$
    $= 90ab - 24ab - 10ab = 56ab$

34. $\sqrt{3} + \dfrac{1}{\sqrt{3}} = \sqrt{3} + \dfrac{\sqrt{3}}{\sqrt{3}\sqrt{3}} =$
    $\sqrt{3} + \dfrac{\sqrt{3}}{3} = \left(1 + \dfrac{1}{3}\right)\sqrt{3} = \left(\dfrac{3}{3} + \dfrac{1}{3}\right)\sqrt{3} =$
    $\dfrac{4}{3}\sqrt{3} = \dfrac{4\sqrt{3}}{3}$

38. $3\sqrt{7} - \sqrt{\dfrac{1}{7}} = 3\sqrt{7} - \dfrac{1}{\sqrt{7}}$
    $= 3\sqrt{7} - \dfrac{1 \cdot \sqrt{7}}{\sqrt{7}\sqrt{7}}$ Rationalize
    $= 3\sqrt{7} - \dfrac{\sqrt{7}}{7}$

$\quad = \left(3 - \dfrac{1}{7}\right)\sqrt{7}$ Distributive law

$\quad = \left(\dfrac{21}{7} - \dfrac{1}{7}\right)\sqrt{7}$

$\quad = \dfrac{20\sqrt{7}}{7}$

39. $-3\sqrt{5} - \dfrac{9}{\sqrt{5}} = -3\sqrt{5} - \dfrac{9\sqrt{5}}{\sqrt{5}\sqrt{5}} =$
    $-3\sqrt{5} - \dfrac{9\sqrt{5}}{5} = \left(-3 - \dfrac{9}{5}\right)\sqrt{5} =$
    $\left(-\dfrac{15}{5} - \dfrac{9}{5}\right)\sqrt{5} = -\dfrac{24\sqrt{5}}{5}$

41. $\dfrac{\sqrt{2}}{\sqrt{10}} - \sqrt{5} = \sqrt{\dfrac{2}{10}} - \sqrt{5} = \sqrt{\dfrac{1}{5}} - \sqrt{5} =$
    $\dfrac{1}{\sqrt{5}} - \sqrt{5} = \dfrac{\sqrt{5}}{\sqrt{5}\sqrt{5}} - \sqrt{5} = \dfrac{\sqrt{5}}{5} - \sqrt{5} =$
    $\left(\dfrac{1}{5} - 1\right)\sqrt{5} = -\dfrac{4\sqrt{5}}{5}$

9.5

5. $\dfrac{\sqrt{25x^2}}{\sqrt{3y}} = \dfrac{\sqrt{25}\sqrt{x^2}}{\sqrt{3y}} = \dfrac{5x}{\sqrt{3y}} \cdot \dfrac{\sqrt{3y}}{\sqrt{3y}} = \dfrac{5x\sqrt{3y}}{3y}$

10. $\dfrac{\sqrt{2}}{\sqrt{50}} - \dfrac{\sqrt{3}}{\sqrt{75}} = \sqrt{\dfrac{2}{50}} - \sqrt{\dfrac{3}{75}} =$
    $\sqrt{\dfrac{\cancel{2}}{25 \cdot \cancel{2}}} - \sqrt{\dfrac{\cancel{3}}{25 \cdot \cancel{3}}} = \sqrt{\dfrac{1}{25}} - \sqrt{\dfrac{1}{25}} = 0$

14. $\sqrt{x}\sqrt{y} + \dfrac{\sqrt{xy^2}}{\sqrt{y}} = \sqrt{xy} + \sqrt{\dfrac{xy^2}{y}} =$
    $\sqrt{xy} + \sqrt{xy} = 2\sqrt{xy}$

17. $\sqrt{x^3}\sqrt{y} - 3x\sqrt{x}\sqrt{y} + \dfrac{4x\sqrt{xy^2}}{\sqrt{y}}$
    $= \sqrt{x^3y} - 3x\sqrt{xy} + 4x\sqrt{\dfrac{xy^2}{y}}$
    $= \sqrt{x^2xy} - 3x\sqrt{xy} + 4x\sqrt{xy}$
    $= x\sqrt{xy} - 3x\sqrt{xy} + 4x\sqrt{xy}$
    $= (x - 3x + 4x)\sqrt{xy} = 2x\sqrt{xy}$

20. $\sqrt{2}(\sqrt{27} - \sqrt{12}) = \sqrt{2}\sqrt{27} - \sqrt{2}\sqrt{12}$
    $= \sqrt{2 \cdot 27} - \sqrt{2 \cdot 12}$
    $= \sqrt{9 \cdot 6} - \sqrt{4 \cdot 6}$
    $= 3\sqrt{6} - 2\sqrt{6}$
    $= \sqrt{6}$

26. $\sqrt{15}\left(\dfrac{\sqrt{3}}{\sqrt{5}} + \dfrac{1}{\sqrt{15}}\right) = \dfrac{\sqrt{15}\sqrt{3}}{\sqrt{5}} + \dfrac{\sqrt{15}}{\sqrt{15}} =$
    $\dfrac{\sqrt{5 \cdot 3 \cdot 3}}{\sqrt{5}} + 1 = \dfrac{\sqrt{\cancel{5}}\sqrt{9}}{\sqrt{\cancel{5}}} + 1 = 3 + 1 = 4$

540 SOLUTIONS TO SELECTED EXERCISES

28. $\sqrt{2}\left(\sqrt{6} + \dfrac{\sqrt{2}}{\sqrt{3}}\right) = \sqrt{2}\sqrt{6} + \dfrac{\sqrt{2}\sqrt{2}}{\sqrt{3}} =$

$\sqrt{2 \cdot 2 \cdot 3} + \dfrac{2}{\sqrt{3}} = \sqrt{4}\sqrt{3} + \dfrac{2}{\sqrt{3}} =$

$2\sqrt{3} + \dfrac{2}{\sqrt{3}} \cdot \dfrac{\sqrt{3}}{\sqrt{3}} = 2\sqrt{3} + \dfrac{2\sqrt{3}}{3} =$

$\left(2 + \dfrac{2}{3}\right)\sqrt{3} = \dfrac{8\sqrt{3}}{3}$

35. $(\sqrt{2} - 2)(\sqrt{2} + 1)$
$= \sqrt{2} \cdot \sqrt{2} + \sqrt{2} \cdot 1 + (-2)\sqrt{2} + (-2) \cdot 1$
$= 2 + \sqrt{2} - 2\sqrt{2} - 2$
$= 2 - 2 + \sqrt{2} - 2\sqrt{2} = -\sqrt{2}$

37. $(3\sqrt{3} - \sqrt{2})(2\sqrt{3} + \sqrt{2})$
$= (3\sqrt{3})(2\sqrt{3}) + (3\sqrt{3})(\sqrt{2}) + (-\sqrt{2})(2\sqrt{3}) +$
$ (-\sqrt{2})(\sqrt{2}) = 6 \cdot 3 + 3\sqrt{6} - 2\sqrt{6} - 2$
$= 18 - 2 + 3\sqrt{6} - 2\sqrt{6} = 16 + \sqrt{6}$

43. $\dfrac{-8}{\sqrt{3} - \sqrt{7}} = \dfrac{-8(\sqrt{3} + \sqrt{7})}{(\sqrt{3} - \sqrt{7})(\sqrt{3} + \sqrt{7})} =$

$\dfrac{-8(\sqrt{3} + \sqrt{7})}{(\sqrt{3})^2 - (\sqrt{7})^2} = \dfrac{-8(\sqrt{3} + \sqrt{7})}{3 - 7} =$

$\dfrac{-8(\sqrt{3} + \sqrt{7})}{-4} = 2(\sqrt{3} + \sqrt{7}) = 2\sqrt{3} + 2\sqrt{7}$

44. $\dfrac{2\sqrt{3}}{\sqrt{3} - 1} = \dfrac{2\sqrt{3}(\sqrt{3} + 1)}{(\sqrt{3} - 1)(\sqrt{3} + 1)}$ Rationalize

$= \dfrac{2\sqrt{3}(\sqrt{3} + 1)}{3 - 1}$

$= \dfrac{\cancel{2}\sqrt{3}(\sqrt{3} + 1)}{\cancel{2}}$

$= 3 + \sqrt{3}$

47. $\dfrac{x - 1}{\sqrt{x} - 1} = \dfrac{(x - 1)(\sqrt{x} + 1)}{(\sqrt{x} - 1)(\sqrt{x} + 1)} =$

$\dfrac{(x - 1)(\sqrt{x} + 1)}{(\sqrt{x})^2 - (1)^2} =$

$\dfrac{(\cancel{x - 1})(\sqrt{x} + 1)}{\cancel{x - 1}} = \sqrt{x} + 1$

9.6

2. $3\sqrt{2x} - 9 = 0$
$3\sqrt{2x} = 9$ Isolate the radical
$\sqrt{2x} = 3$ Simplify
$(\sqrt{2x})^2 = (3)^2$ Square both sides
$2x = 9$
$x = \dfrac{9}{2}$

Check: $3\sqrt{2 \cdot \dfrac{9}{2}} - 9 = 3\sqrt{9} - 9 = 3 \cdot 3 - 9 = 0$

6. $3\sqrt{a - 3} + 6 = 0$
$3\sqrt{a - 3} = -6$ Isolate the radical
$\sqrt{a - 3} = -2$ Divide by 3
$(\sqrt{a - 3})^2 = (-2)^2$ Square both sides
$a - 3 = 4$
$a = 7$

Check: $3\sqrt{a - 3} + 6 = 3\sqrt{7 - 3} + 6$
$ = 3\sqrt{4} + 6 = 3 \cdot 2 + 6 \neq 0$

Since 7 does not check, there is no solution.

13. $3\sqrt{x - 1} - \sqrt{x + 31} = 0$
$3\sqrt{x - 1} = \sqrt{x + 31}$ Isolate the radicals
$(3\sqrt{x - 1})^2 = (\sqrt{x + 31})^2$ Square both sides
$9(x - 1) = x + 31$ Square the 3 also
$9x - 9 = x + 31$
$8x = 40$
$x = 5$

Check: $3\sqrt{5 - 1} - \sqrt{5 + 31} = 3\sqrt{4} - \sqrt{36} =$
$ 3 \cdot 2 - 6 = 0$

17. $\sqrt{a^2 - 5} + a - 5 = 0$
$\sqrt{a^2 - 5} = 5 - a$ Subtract a and add 5
$(\sqrt{a^2 - 5})^2 = (5 - a)^2$
$a^2 - 5 = 25 - 10a + a^2$ Do not forget the $-10a$
$-5 = 25 - 10a$ Subtract a^2
$10a = 30$
$a = 3$

Check: $\sqrt{3^2 - 5} + 3 - 5 = \sqrt{9 - 5} - 2 =$
$ \sqrt{4} - 2 = 2 - 2 = 0$

20. $\sqrt{x^2 + 3} + x = 6$
$\sqrt{x^2 + 3} = 6 - x$
$(\sqrt{x^2 + 3})^2 = (6 - x)^2$
$x^2 + 3 = 36 - 12x + x^2$
$3 = 36 - 12x$
$12x = 33$
$x = \dfrac{33}{12} = \dfrac{11}{4}$

Check: $\sqrt{\left(\dfrac{11}{4}\right)^2 + 3} + \dfrac{11}{4} = \sqrt{\dfrac{121}{16} + \dfrac{48}{16}} +$
$\dfrac{11}{4} = \sqrt{\dfrac{169}{16}} + \dfrac{11}{4} = \dfrac{13}{4} + \dfrac{11}{4} = \dfrac{24}{4} = 6$

25. $\sqrt{x - 3} - x + 5 = 0$
$\sqrt{x - 3} = x - 5$

$$(\sqrt{x-3})^2 = (x-5)^2$$
$$x - 3 = x^2 - 10x + 25$$
$$0 = x^2 - 11x + 28$$
$$0 = (x-7)(x-4)$$
$$x - 7 = 0 \quad \text{or} \quad x - 4 = 0$$
$$x = 7 \quad \text{or} \quad x = 4$$

Check: $\sqrt{7-3} - 7 + 5 = \sqrt{4} - 2 =$
$2 - 2 = 0$
$\sqrt{4-3} - 4 + 5 = \sqrt{1} + 1 =$
$1 + 1 \neq 0$

Since 7 checks and 4 does not, the only solution is 7.

28. $\sqrt{x+10} + x - 2 = 0$

$\sqrt{x+10} = 2 - x$ Isolate radical

$(\sqrt{x+10})^2 = (2-x)^2$ Square both sides

$x + 10 = 4 - 4x + x^2$ Don't forget the middle term $-4x$

$0 = x^2 - 5x - 6$
$0 = (x-6)(x+1)$
$x - 6 = 0 \quad \text{or} \quad x + 1 = 0$
$x = 6 \quad\quad\quad\quad x = -1$

Check: $x = 6$ 　　　　Check: $x = -1$
$\sqrt{6+10} + 6 - 2 \stackrel{?}{=} 0$ 　 $\sqrt{-1+10} + (-1) - 2 \stackrel{?}{=} 0$
$\sqrt{16} + 6 - 2 \stackrel{?}{=} 0$ 　　　 $\sqrt{9} - 1 - 2 \stackrel{?}{=} 0$
$4 + 6 - 2 \neq 0$ 　　　　 $3 - 1 - 2 = 0$

Since 6 does not check, the only solution is -1.

9.7

10. Let $x =$ length of wire needed to reach top. Use the Pythagorean theorem.

$x^2 = (40)^2 + (30)^2$
$x^2 = 1600 + 900$
$x^2 = 2500$
$x = \sqrt{2500} = 50$

The wire must be 50 ft long.

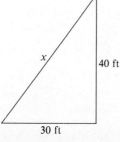

13. 　 Let $x =$ the number,
$x + 5 =$ five more than the number,
$\sqrt{x+5} =$ square root of 5 more than the number,
$2\sqrt{x} =$ twice the square root of the number.
$\sqrt{x+5} = 2\sqrt{x}$
$(\sqrt{x+5})^2 = (2\sqrt{x})^2$

$x + 5 = 4x$
$5 = 3x$
$\dfrac{5}{3} = x$

Check: $\sqrt{\dfrac{5}{3} + 5} = \sqrt{\dfrac{5}{3} + \dfrac{15}{3}} = \sqrt{\dfrac{20}{3}} = \sqrt{\dfrac{4 \cdot 5}{3}} =$
$\sqrt{4}\sqrt{\dfrac{5}{3}} = 2\sqrt{\dfrac{5}{3}}$

16. We are given that $n = 64$ and asked to find c.

$c = 100\sqrt{n + 36} = 100\sqrt{64 + 36} = 100\sqrt{100} =$
$100(10) = 1000$

The cost is \$1000 when 64 appliances are produced.

22. We must find n when c is 50.

$c = 10\sqrt{n} + 100$
$50 = 10\sqrt{n} + 100$ Substitute 50 for c
$-50 = 10\sqrt{n}$
$-5 = \sqrt{n}$

At this point we can stop if we realize that $\sqrt{n}$ must be nonnegative and hence cannot be -5. Thus, there is no value of n for $c = 50$.

10.1

19.　　　$4z^2 = 7z$
$4z^2 - 7z = 0$
$z(4z - 7) = 0$
$z = 0 \quad \text{or} \quad 4z - 7 = 0$
$\quad\quad\quad\quad\quad\quad\quad 4z = 7$
$\quad\quad\quad\quad\quad\quad\quad z = \dfrac{7}{4}$

Thus, the solutions are 0 and $\dfrac{7}{4}$.

26. $(3z - 1)(2z + 1) = 3(2z + 1)$
$6z^2 + z - 1 = 6z + 3$
$6z^2 - 5z - 4 = 0$
$(3z - 4)(2z + 1) = 0$
$3z - 4 = 0 \quad \text{or} \quad 2z + 1 = 0$
$3z = 4 \quad\quad\quad\quad 2z = -1$
$z = \dfrac{4}{3} \quad\quad\quad\quad z = -\dfrac{1}{2}$

Thus, the solutions are $\dfrac{4}{3}$ and $-\dfrac{1}{2}$.

28. $4x^2 - 9 = 3$
$\quad\;\; 4x^2 = 12$

542 SOLUTIONS TO SELECTED EXERCISES

$$x^2 = \frac{12}{4}$$
$$x^2 = 3$$
$$x = \pm\sqrt{3}$$

Thus, the solutions are $\sqrt{3}$ and $-\sqrt{3}$.

31. $3x^2 - 48 = -48$

$3x^2 = 0$	Add 48 to both sides
$x^2 = 0$	Divide both sides by 3
$x = \pm\sqrt{0}$	Take square root of both sides
$x = \pm 0 = 0$	$+0$ and -0 are both 0

Thus, 0 is the only solution.

35. $2x^2 - 150 = 0$
$$2x^2 = 150$$
$$x^2 = 75$$
$$x = \pm\sqrt{75} = \pm\sqrt{25 \cdot 3} = \pm 5\sqrt{3}$$

Thus, the two solutions are $5\sqrt{3}$ and $-5\sqrt{3}$.

37. $2x^2 + 50 = 0$
$$2x^2 = -50$$
$$x^2 = -25$$

Since $\sqrt{-25}$ is not a real number, there are no real solutions to this equation.

41.
$(x - 2)^2 = 9$	
$(x - 2) = \pm\sqrt{9}$	Take square root of both sides
$x - 2 = \pm 3$	
$x = 2 \pm 3$	Add 2 to both sides
$x = 2 + 3 = 5$ or $x = 2 - 3 = -1$	

Thus, the solutions are 5 and -1.

10.2

6. $z^2 + \frac{1}{3}z + \underline{}$

$$\left[\frac{1}{2} \text{ of } \frac{1}{3}\right]^2 = \left[\frac{1}{2} \cdot \frac{1}{3}\right]^2 = \left[\frac{1}{6}\right]^2 = \frac{1}{36}$$

Thus, we must add $\frac{1}{36}$ to complete the square.

13. $x^2 - 2x - 1 = 0$
$$x^2 - 2x = 1$$
$$x^2 - 2x + 1 = 1 + 1 \qquad 1 = \left[\frac{1}{2} \text{ coefficient of } x\right]^2$$
$$(x - 1)^2 = 2$$
$$x - 1 = \pm\sqrt{2} \qquad \text{Take square root of both sides}$$
$$x = 1 \pm \sqrt{2} \qquad \text{Add 1 to both sides}$$

15. $3z^2 + 12z = 135$
$$3z^2 + 12z - 135 = 0$$
$$z^2 + 4z - 45 = 0 \qquad \text{Divide through by 3}$$
$$z^2 + 4z = 45$$
$$z^2 + 4z + 4 = 45 + 4 \qquad \left[\frac{1}{2} \text{ of } 4\right]^2 =$$
$$[2]^2 = 4$$
$$(z + 2)^2 = 49$$
$$z + 2 = \pm\sqrt{49} = \pm 7$$
$$z = -2 \pm 7$$
$$z = -2 + 7 = 5 \quad \text{or} \quad z = -2 - 7 = -9$$

Thus, the solutions are 5 and -9.

16. $2x^2 + x - 1 = 0$
$$x^2 + \frac{1}{2}x - \frac{1}{2} = 0 \qquad \text{Divide through by 2 to make coefficient of } x^2\text{-term equal to 1}$$
$$x^2 + \frac{1}{2}x + \frac{1}{16} = \frac{1}{2} + \frac{1}{16} \qquad \left[\frac{1}{2} \text{ of } \frac{1}{2}\right]^2 =$$
$$\left[\frac{1}{4}\right]^2 = \frac{1}{16}$$
$$\left(x + \frac{1}{4}\right)^2 = \frac{9}{16}$$
$$x + \frac{1}{4} = \pm\sqrt{\frac{9}{16}} = \pm\frac{3}{4}$$
$$x = -\frac{1}{4} \pm \frac{3}{4}$$
$$x = -\frac{1}{4} + \frac{3}{4} = \frac{2}{4} = \frac{1}{2} \quad \text{or}$$
$$x = -\frac{1}{4} - \frac{3}{4} = -1$$

Thus, the solutions are $\frac{1}{2}$ and -1.

20. $y^2 + 2y + 5 = 0$
$$y^2 + 2y = -5$$
$$y^2 + 2y + 1 = -5 + 1$$
$$(y + 1)^2 = -4$$

But since $\sqrt{-4}$ is not a real number, there are no real solutions.

23. $5y^2 - 4y - 33 = 0$

We first try to factor.

$$(5y + 11)(y - 3) = 0$$
$$5y + 11 = 0 \qquad \text{or} \qquad y - 3 = 0 \qquad \text{Zero-product rule}$$
$$5y = -11 \qquad \qquad y = 3$$
$$y = -\frac{11}{5}$$

Thus, the solutions are $-\dfrac{11}{5}$ and 3.

10.3

5. $0 = 7y - 2 + 15y^2$

$15y^2 + 7y - 2 = 0$ Write in general form

$y = \dfrac{-7 \pm \sqrt{49 - 4(15)(-2)}}{2 \cdot 15}$ $a = 15,\ b = 7,$ $c = -2$

$= \dfrac{-7 \pm \sqrt{49 + 120}}{30} = \dfrac{-7 \pm \sqrt{169}}{30} = \dfrac{-7 \pm 13}{30}$

$y = \dfrac{-7 + 13}{30} = \dfrac{6}{30} = \dfrac{1}{5}$ or

$y = \dfrac{-7 - 13}{30} = \dfrac{-20}{30} = -\dfrac{2}{3}$

Thus, the solutions are $\dfrac{1}{5}$ and $-\dfrac{2}{3}$.

7. $5x^2 - x = 1$

$5x^2 - x - 1 = 0$ General form, $a = 5,\ b = -1,$ $c = -1$

$x = \dfrac{-b \pm \sqrt{b^2 - 4ac}}{2a}$

$= \dfrac{-(-1) \pm \sqrt{(-1)^2 - 4(5)(-1)}}{2(5)}$

$= \dfrac{1 \pm \sqrt{1 - (-20)}}{10}$

$= \dfrac{1 \pm \sqrt{21}}{10}$

11. $z^2 = z - 5$

$z^2 - z + 5 = 0$ General form, $a = 1,\ b = -1,$ $c = 5$

$z = \dfrac{-b \pm \sqrt{b^2 - 4ac}}{2a}$

$= \dfrac{-(-1) \pm \sqrt{(-1)^2 - 4(1)(5)}}{2(1)}$

$= \dfrac{1 \pm \sqrt{1 - 20}}{2}$

$= \dfrac{1 \pm \sqrt{-19}}{2}$

Since $\sqrt{-19}$ is not a real number, the equation has no real solution.

14. $y(y + 1) + 2y = 4$

$y^2 + y + 2y = 4$

$y^2 + 3y - 4 = 0$

$y = \dfrac{-3 \pm \sqrt{9 - 4(1)(-4)}}{2 \cdot 1}$ $a = 1,\ b = 3,$ $c = -4$

$= \dfrac{-3 \pm \sqrt{9 + 16}}{2} = \dfrac{-3 \pm \sqrt{25}}{2} = \dfrac{-3 \pm 5}{2}$

$y = \dfrac{-3 + 5}{2} = \dfrac{2}{2} = 1$ or

$y = \dfrac{-3 - 5}{2} = \dfrac{-8}{2} = -4$

Thus, the solutions are 1 and -4.

18. $(2z - 1)(z - 2) - 11 = 2(z + 4) - 8$

$2z^2 - 5z + 2 - 11 = 2z + 8 - 8$

$2z^2 - 7z - 9 = 0$

$z = \dfrac{7 \pm \sqrt{49 - 4(2)(-9)}}{2 \cdot 2}$ $a = 2,\ b = -7,$ $c = -9$

$= \dfrac{7 \pm \sqrt{49 + 72}}{4} = \dfrac{7 \pm \sqrt{121}}{4} = \dfrac{7 \pm 11}{4}$

$z = \dfrac{7 + 11}{4} = \dfrac{18}{4} = \dfrac{9}{2}$ or

$z = \dfrac{7 - 11}{4} = \dfrac{-4}{4} = -1$

Thus, the solutions are $\dfrac{9}{2}$ and -1.

20. $3z^2 + 10z - 1 = 0$

$z = \dfrac{-10 \pm \sqrt{100 - 4(3)(-1)}}{2 \cdot 3}$ $a = 3,\ b = 10,$ $c = -1$

$= \dfrac{-10 \pm \sqrt{100 + 12}}{6} = \dfrac{-10 \pm \sqrt{112}}{6}$

$= \dfrac{-10 \pm \sqrt{16 \cdot 7}}{6}$

$= \dfrac{-10 \pm 4\sqrt{7}}{6} = \dfrac{\cancel{2}(-5 \pm 2\sqrt{7})}{\cancel{2} \cdot 3}$

$= \dfrac{-5 \pm 2\sqrt{7}}{3}$

Thus, the solutions are $\dfrac{-5 + 2\sqrt{7}}{3}$ and $\dfrac{-5 - 2\sqrt{7}}{3}$.

26. $\dfrac{2}{3}x^2 - \dfrac{1}{3}x - 1 = 0$

$2x^2 - x - 3 = 0$ Multiply by 3 to clear fractions

$(2x - 3)(x + 1) = 0$

$2x - 3 = 0$ or $x + 1 = 0$

$2x = 3$ $x = -1$

$x = \dfrac{3}{2}$

Thus, the solutions are $\dfrac{3}{2}$ and -1.

32. Let $x =$ length of the fence.
By the Pythagorean theorem,

$x^2 = 4^2 + 4^2$

$x^2 = 16 + 16$

$x^2 = 16 \cdot 2$

$x = \pm\sqrt{16 \cdot 2} = \pm 4\sqrt{2}$.

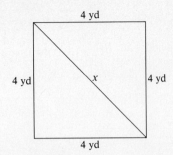

Since we are interested in the length of the fence, we discard the negative solution, $-4\sqrt{2}$. Thus, the length of the fence is $4\sqrt{2}$ yd, which is approximately 5.7 yards (using $\sqrt{2} \approx 1.414$).

10.4

7. $\dfrac{5}{z+3} - \dfrac{1}{z+3} = \dfrac{z+3}{z+2}$ LCD $= (z+3)(z+2)$

$\left[\dfrac{5}{z+3} - \dfrac{1}{z+3}\right](z+3)(z+2)$

$= \dfrac{z+3}{z+2}(z+3)(z+2)$ Multiply by LCD

$\dfrac{5}{z+3}(z+3)(z+2) - \dfrac{1}{z+3}(z+3)(z+2)$

$= (z+3)(z+3)$

$5(z+2) - (z+2) = z^2 + 6z + 9$

$5z + 10 - z - 2 = z^2 + 6z + 9$

$0 = z^2 + 2z + 1$

$0 = (z+1)(z+1)$

Thus, $z + 1 = 0$ or $z = -1$, which does indeed check in the original equation.

8. $\dfrac{x-1}{x} = \dfrac{x}{x+1}$ LCD $= x(x+1)$

$\dfrac{x-1}{x} x(x+1) = \dfrac{x}{x+1} x(x+1)$ Multiply by LCD

$x^2 - 1 = x^2$

$-1 = 0$

Since we obtain a contradiction, there is no solution.

12. $\dfrac{x+2}{2} = \dfrac{x+5}{x+2}$

$(x+2)(x+2) = 2(x+5)$ Cross-product equation

$x^2 + 4x + 4 = 2x + 10$

$x^2 + 2x - 6 = 0$

Since this will not factor, we use the quadratic formula with $a = 1$, $b = 2$, and $c = -6$.

$x = \dfrac{-b \pm \sqrt{b^2 - 4ac}}{2a}$

$= \dfrac{-2 \pm \sqrt{2^2 - 4(1)(-6)}}{2(1)}$

$= \dfrac{-2 \pm \sqrt{4 + 24}}{2}$

$= \dfrac{-2 \pm \sqrt{28}}{2}$

$= \dfrac{-2 \pm 2\sqrt{7}}{2}$ $\sqrt{28} = \sqrt{4 \cdot 7} = 2\sqrt{7}$

$= \dfrac{2(-1 \pm \sqrt{7})}{2}$

$= -1 \pm \sqrt{7}$

Thus, the solutions are $-1 \pm \sqrt{7}$.

17. $1 + 2\sqrt{x-1} = x$

$2\sqrt{x-1} = x - 1$ Isolate the radical

$(2\sqrt{x-1})^2 = (x-1)^2$ Square both sides

$4(x-1) = x^2 - 2x + 1$ Remember to square 2 on the left

$4x - 4 = x^2 - 2x + 1$

$0 = x^2 - 6x + 5$

$0 = (x-1)(x-5)$

$x - 1 = 0$ or $x - 5 = 0$

$x = 1$ or $x = 5$

Thus, since both 1 and 5 check in the original equation, they are the solutions.

22. $3\sqrt{y+1} - y - 1 = 0$

$3\sqrt{y+1} = y + 1$ Isolate the radical

$(3\sqrt{y+1})^2 = (y+1)^2$ Square both sides

$9(y+1) = y^2 + 2y + 1$

$9y + 9 = y^2 + 2y + 1$

$0 = y^2 - 7y - 8$

$0 = (y-8)(y+1)$

$y - 8 = 0$ or $y + 1 = 0$

$y = 8$ $y = -1$

Thus, the solutions are 8 and -1, which do check.

24. $\sqrt{2x^2 - 5} = x$

$(\sqrt{2x^2 - 5})^2 = x^2$ Square both sides

$2x^2 - 5 = x^2$

$x^2 = 5$

$x = \pm\sqrt{5}$

But if we check $-\sqrt{5}$, we have

$\sqrt{2(-\sqrt{5})^2 - 5} \stackrel{?}{=} -\sqrt{5}$

$\sqrt{5} \neq -\sqrt{5}$

Since $-\sqrt{5}$ does not check, it must be discarded. Thus, $\sqrt{5}$ is the only solution that checks.

25. Let n = the desired number,

$\dfrac{1}{n}$ = reciprocal of the number,

$n + \dfrac{1}{n}$ = the number increased by its reciprocal.

$n + \dfrac{1}{n} = \dfrac{5}{2}$ LCD = $2n$

$2n\left[n + \dfrac{1}{n}\right] = 2n \cdot \dfrac{5}{2}$

$2n^2 + 2 = 5n$

$2n^2 - 5n + 2 = 0$

$(2n - 1)(n - 2) = 0$

$2n - 1 = 0$ or $n - 2 = 0$

$2n = 1$ or $n = 2$

$n = \dfrac{1}{2}$

Thus, the solutions are $\dfrac{1}{2}$ and 2.

10.5

8. Let x = a positive odd integer,

$x + 2$ = the next consecutive positive odd integer.

$3x(x + 2) = 297$

$x(x + 2) = 99$ Divide both sides by 3

$x^2 + 2x - 99 = 0$

$(x - 9)(x + 11) = 0$

$x - 9 = 0$ or $x + 11 = 0$

$x = 9$ or $x = -11$

Since x must be positive, we discard -11. Thus, the two integers are 9 and 11.

16. Let x = altitude of a triangle,

$\dfrac{1}{3}x$ = base of the triangle.

Since the formula for the area of a triangle is
$A = \dfrac{1}{2}bh = \dfrac{1}{2}$ (base)(altitude), we must solve:

$24 = \dfrac{1}{2}\left(\dfrac{1}{3}x\right)(x)$

$24 = \dfrac{1}{6}x^2$

$144 = x^2$ Multiply both sides by 6

$\pm\sqrt{144} = x$

$\pm 12 = x$

But since the altitude of a triangle must be positive we discard -12. Thus, the altitude is 12 ft and the base is 4 ft.

19. The area of a parallelogram is given by $A = bh$.

Let x = altitude of parallelogram,

$2x + 1$ = base of parallelogram.

The equation to solve is

$55 = x(2x + 1)$.

$55 = 2x^2 + x$

$0 = 2x^2 + x - 55$

$0 = (2x + 11)(x - 5)$

$2x + 11 = 0$ or $x - 5 = 0$

$x = -\dfrac{11}{2}$ $x = 5$

Since the altitude cannot be negative, we discard $-\dfrac{11}{2}$. Thus, the altitude is 5 ft and the base is 11 ft (since $2x + 1 = 2(5) + 1 = 11$).

22. Let x = number of hrs required for Dan to grade tests,

$x + 9$ = number of hrs required for Joe to grade tests.

$\dfrac{1}{x} + \dfrac{1}{x + 9} = \dfrac{1}{20}$ Number each grades in 1 hr is $\dfrac{1}{20}$ of the total

$20\cancel{x}(x + 9)\dfrac{1}{\cancel{x}} + 20x\cancel{(x + 9)}\dfrac{1}{\cancel{(x + 9)}} = \cancel{20}x(x + 9)\dfrac{1}{\cancel{20}}$

$20x + 180 + 20x = x^2 + 9x$

$0 = x^2 - 31x - 180$

$0 = (x + 5)(x - 36)$

$x + 5 = 0$ or $x - 36 = 0$

$x = -5$ or $x = 36$

Discard -5. Thus, Dan takes 36 hrs and Joe takes 45 hrs to grade the tests.

25. $A = P(1 + r)^2$

$1210 = 1000(1 + r)^2$

$\dfrac{1210}{1000} = (1 + r)^2$

$\pm\sqrt{1.21} = 1 + r$

$\pm 1.1 = 1 + r$

Thus, $r = -1 \pm 1.1$, or $r = -2.1$ and $r = 0.1$. Since the negative value for r must be discarded, $r = 0.1$ or $r = 10\%$.

26. Let x = speed of boat in still water,

$x + 2$ = speed of boat downstream,

$x - 2$ = speed of boat upstream.

We use the distance formula $d = rt$, solved for t, to obtain:

$\dfrac{48}{x - 2}$ = time of travel upstream

$$\frac{48}{x+2} = \text{time of travel downstream}$$

Since the total time is 10 hr, we must solve

$$\frac{48}{x-2} + \frac{48}{x+2} = 10.$$

Multiply both sides by the LCD, $(x-2)(x+2)$ to obtain

$$48(x+2) + 48(x-2) = 10(x-2)(x+2).$$
$$48x + 96 + 48x - 96 = 10x^2 - 40$$
$$0 = 10x^2 - 96x - 40$$
$$0 = 5x^2 - 48x - 20$$
$$0 = (5x+2)(x-10)$$

$5x + 2 = 0$ or $x - 10 = 0$

$$x = -\frac{2}{5} \qquad\qquad x = 10$$

We can discard $-\frac{2}{5}$, so the speed of the boat in still water is 10 mph.

10.6

4. $g - ax = m$

$\quad -ax = m - g$ Subtract g from both sides

$\quad x = \dfrac{m-g}{-a}$ Divide both sides by $-a$

$\quad x = \dfrac{g-m}{a}$ Multiplying numerator and denominator by -1

11. Solve $A = \dfrac{1}{2}bh$ for b

$\quad 2A = bh$ Multiply both sides by 2

$\quad \dfrac{2A}{h} = b$ Divide both sides by h

14. $r = \dfrac{d}{t}$

$\quad rt = d$ Multiply both sides by t

$\quad t = \dfrac{d}{r}$ Divide both sides by r

16. Solve $S = \dfrac{1}{2}gt^2$ for t.

$\quad 2S = gt^2$ Multiply by 2

$\quad \dfrac{2S}{g} = t^2$ Divide by g

$\quad \pm\sqrt{\dfrac{2S}{g}} = t$ Take square root of both sides

20. Solve $U = \sqrt{\dfrac{a}{c}}$ for c.

$\quad U^2 = \dfrac{a}{c}$ Square both sides

$\quad U^2 c = a$ Multiply both sides by c

$\quad c = \dfrac{a}{U^2}$ Divide both sides by U^2

10.7

15. To determine the vertex of the parabola with equation

$$y = x^2 - 3$$

we must observe that the equation can be written in the form

$$y = (x - 0)^2 - 3.$$

Then it is clear that $h = 0$ and $k = -3$, so the vertex is $(0, -3)$.

INDEX

Absolute value, 54
Addition
  associative law of, 60
  and the column method, 255
  commutative law of, 60
  of decimals, 21
  of fractions, 10–14
  of like algebraic fractions, 347, 352–56
  on the number line, 57, 58
  of polynomials, 254
  of radicals, 410–12
  of rational numbers, 57–61
  of unlike algebraic fractions, 352–56
Addition-subtraction method
  solving systems of equations by, 229–32
Addition-subtraction rule, 109, 147
Additive identity, 60
Additive inverse, 59
Algebraic expressions, 38, 248
  evaluation of, 38
Algebraic fractions, 332
  addition of, 347, 352–56
  division of, 341
  equivalent, 334
  least common denominator of, 352
  like, 347
  multiplication of, 339, 340
  rational, 332
  reduced to lowest terms, 334
  subtraction of, 347
  unlike, 347
  values to be omitted, 332
Angles
  complementary, 236
  supplementary, 236
Applied problems
  business, 251, 322, 373, 375, 432
  geometry, 136–38, 322, 431, 462, 474
  language of, 123
  mixture, 236, 237
  motion, 138, 139, 238
  number and age, 127–29, 235, 236, 321, 363, 473
  percent, 32, 34, 129–31
  proportion, 373–75
  and quadratic equations, 321–23, 473–76
  ratio, 372
  rules of solving, 127
  and square roots, 431–33
  using systems of equations, 235–39

Applied problems
  variation, 374
  work, 364–66, 475
Approximately equal ($\approx$), 138
Ascending order, 249
Associative law
  of addition, 60
  of multiplication, 67
Axes, 168

Base, 35
Binomial(s), 248

Canceling factors, 3, 4, 334
Cartesian coordinate system, 167
  See also Graphing.
Clearing fractions, 360
Clearing parentheses, 78
Coefficient
  numerical, of terms, 76, 248
Coinciding lines, 213, 218
Collecting like terms, 77, 78, 250
Column method
  of addition, 255
  of multiplication, 264
  of subtraction, 256
Commission, 130
  rate of, 130
Common factor(s), 289
Commutative law
  of addition, 60
  of multiplication, 67
Complementary angles, 236
Completing the square, 454–56
Complex fraction(s), 380
  simplifying, 380
Conditional equation, 109
Conjugates, 419
Consecutive integers, 128
Constants of variation, 374
Contradiction, 109, 226
Converting
  a decimal to a percent, 30
  a fraction to a decimal, 27
  a fraction to a percent, 30
  a percent to a decimal, 29
  a percent to a fraction, 29
  a terminating decimal to a fraction, 28
  units of measure, 512–14
Coordinate(s), 168
Coordinate plane. *See* Cartesian coordinate system; Graphing
Counting numbers, 1

Cross product(s), 53, 361
Cube, 35

Decimal(s), 20–22
  addition of, 21
  conversion, 26–31
  division of, 22
  multiplication of, 21, 22
  subtraction of, 21
Decimal notation, 20
Decimal point, 20
Degree
  of a polynomial, 249
  of a term, 249
Denominator, 2
  least common denominator, 10, 352
  rationalizing the, 399, 419
Descending order, 249
Difference of two cubes, 316
Difference of two squares, 270, 312, 313
Direct variation, 374
Digit(s), 20
Discriminant, 461
Distributive law, 75
Dividend, 9
Division
  of algebraic fractions, 341
  of decimals, 22
  of exponents, 84, 85
  of fractions, 9
  of polynomials, 275–78
  of radicals, 404
  of rational numbers, 68, 69
  by zero, 8
Divisor, 9, 340
Double negative property, 70
Double sign properties, 70

Element(s), 1
English system of measures, 512
Equal to (=), 52
  approximately equal ($\approx$), 138
Equation(s), 108
  addition-subtraction rule, 109
  conditional, 109
  contradiction, 109
  cross-product, 53, 361, 373
  equivalent, 112
  first-degree, *see* Linear equations.
  fractional, 360, 467
  graphing, 163, 164, 174–81, 486–89
  identity, 109

Equation(s)
  left side, 108
  linear, 109, 178
  literal, 481
  multiplication-division rule, 112
  quadratic, 318, 446
  radical, 425, 468
  right side, 108
  solving, 108–19, 316–18, 447–63
    by factoring, 316, 448
  systems of, 212–15, 223–26, 229–32
    in two variables, 169, 170, 178
  of variation, 374
Equations of a line
  general form, 178
  point-slope form, 191
  slope-intercept form, 190
Equivalent equations, 112
Equivalent fractions, 2
  algebraic, 334
Evaluate expression, 36–39
Expanded numeral, 20
Exponents, 35
  division of, 84, 85
  integer, 82–89
  multiplication of, 83–85
  negative, 87
  power rule, 85
  product rule, 83
  quotient rule, 84
  rules for, 88
  zero as, 86
Exponential notation, 35

Factor(s), 35, 289
  canceling, 3, 4, 334
  common, 289
  prime, 3, 289
Factoring, 76
  applications, 316–18, 447–49
  a difference of two cubes, 316
  a difference of two squares, 312, 313
  by grouping, 292, 293, 306, 307
  perfect square trinomials, 311, 312
  into primes, 3, 289
  by removing the greatest common factor, 291
  to solve equations, 316–18, 447–49
  summary of techniques, 314
  a sum of two cubes, 316
  trinomials, 296–314
First cross product, 53
First-degree equations, 109, 178
FOIL method
  and binomial radical expressions, 417, 418
  for multiplying binomials, 263, 296
Formula(s), 39
  solving, 481–84
Fraction(s), 2
  addition of, 10–14, 347, 348, 352–56
  algebraic, *see* Algebraic fractions.
  build up of, 2

Fraction(s)
  clearing, 360
  complex, 380
  conversions, 27–30
  denominator, 2
  division of, 9
  equivalent, 2
  fundamental principle of, 2
  improper, 2
  least common denominator of, 10, 352
  multiplication of, 8, 339, 340
  numerator, 2
  as a percent, 29
  proper, 2
  reciprocal of, 9
  reduced to lowest terms, 2, 3, 334
  simplified, 380
  subtraction of, 10, 14, 347, 348
  terminating decimals, 28
Fractional equation(s), 360, 467
  solving, 360, 467, 468

General form
  of linear equations, 178
  of quadratic equations, 446
Graphing
  linear equations, 163, 164, 174–81
  linear inequalities, 164, 165
    with one variable, 164
    with two variables, 197–201
  on the number line, 162–65
  plotting points, 52, 163
  quadratic equations, 486–89
  of systems of equations, 213
Greater than (>), 51, 52
Greater than or equal to (≥), 52
Greatest common factor (GCF), 289
Grouping
  factoring by, 292, 293
  symbols, 36, 37

Horizontal axis, 168
Hypotenuse, 430

Identity, 109, 226
  additive, 60
  multiplicative, 66
Improper fraction(s), 2
Inequalities
  addition-subtraction rule, 147
  combination of rules, 149
  linear, 147
    graphing, 164, 165, 196–201
    solving, 147–51
  multiplication-division rule, 148
Integer(s), 52
  consecutive, 128
  as exponents, 82–89
  negative, 52
  nonnegative, 52
  positive, 52
Intercepts, 178
Interest, 131
  rate of, 131
  simple, 131
Intersecting lines, 213, 218

Inverse(s)
  additive, 59
  multiplicative, 66
Inverse variation, 380
Irrational numbers, 28, 96

Least common denominator (LCD), 10, 352
Legs of a right triangle, 430
Less than (<), 51
Less than or equal to (≤), 52
Like fractions, 347
Like radicals, 409
Like terms, 76, 77, 250
Line(s)
  coinciding, 213, 218
  intersecting, 213, 218
  number, 50, 57, 58, 162
  parallel, 189, 213, 218
  slope of a, 186–88
Linear equations, 109
  addition-subtraction rule, 109
  general form of, 178
  graphing, 163, 164, 174–81
  multiplication-division rule, 112
  point-slope form, 191
  slope-intercept form, 190
  solving, 108–19
  system of two, 212–15, 223–26
  in two variables, 169, 170, 178
Linear inequalities, 147
  addition-subtraction rule, 147
  combination of rules, 149
  graphing, 164, 165, 196–201
    in two variables, 197–201
  multiplication-division rule, 148
  solving, 147–51
Literal equations, 481
Lowest terms, 2, 3

Method of taking roots, 449, 450
Metric system of measures, 512
Mixed number, 12, 13
Mixture problems, 236, 237
Monomial(s), 248
Motion problems, 138, 139, 238
Multiplication
  of algebraic fractions, 339, 340
  the associative law of, 67
  and the column method, 264
  the commutative law of, 67
  of decimals, 21, 22
  and difference of two squares, 270
  distributive law of, 75
  of exponents, 83
  of fractions, 8
  perfect square trinomial in, 270, 271
  of polynomials, 262–65
  of radicals, 403, 418
  of rational numbers, 65–68
Multiplication-division rule, 112, 148
Multiplicative identity, 66
Multiplicative inverse, 66

Natural number(s), 1
Negative exponents, 87

Negative integers, 52
Negative numbers, 53
Notation
  decimal, 20
  exponential, 35
  percent, 28
  scientific, 92–94
Number(s), 1
  counting, 1
  decimal, 20–22
  irrational, 28, 96
  mixed, 12, 13
  natural, 1
  negative, 53
  prime, 3
  rational, 28, 53, 54, 57–61, 65–69, 96
  real, 96
  whole, 1
Number line, 50, 162
  addition on, 57, 58
  graphing on, 162–65
  plotting on, 52, 163
Numeral(s), 1, 20
Numerator, 2
Numerical coefficient, 76, 248

Order of operations, 36, 37
Ordered pair, 168
Origin, 162, 168
Overhead, 193, 196

Parabola, 486
  vertex of, 490
Parallel lines, 189, 213, 218
Percent, 28–31
  conversions, 29–31
  word problems with, 32, 34, 129–31
Perfect square(s), 96, 97
  fractional, 97
Perfect square trinomials, 270, 271, 311, 312
Place value system, 20
Plotting, 52
  points, 163, 169
Point-slope form, 191
Polynomials, 248
  addition of, 254
  ascending order of, 249
  binomials, 248
  column method, 255, 264
  degree of, 249
  descending order of, 249
  and the difference of two squares, 270, 312, 313
  division of, 275–78
  evaluation of, 254
  factoring, 289–314
  FOIL method and, 263
  like terms, 250
  monomials, 248
  multiplication of, 262–65
  in one variable, 249
  and the perfect square trinomial, 270–72, 311, 312

Polynomials
  prime, 291
  in several variables, 249
  subtraction of, 255
  trinomials, 248
Positive integers, 52
Power(s)
  in exponential notation, 35
  See also Exponents.
Power rule for exponents, 85
Prime factors, 3, 289
Prime numbers, 3
Prime polynomials, 291
Principal, 131
Principal square root, 96, 393
Product(s)
  first cross, 53
  second cross, 53
Product rule for exponents, 83–85
Proper fraction(s), 2
Proportion, 373
Pythagorean theorem, 431

Quadrants, 169
Quadratic equation(s), 318, 446
  applications using, 321–23, 473–76
  general form of, 446
  graphing, 486–89
  solving by completing the square, 454–56
  solving by taking roots, 449, 450
  solving by factoring, 316–18, 447–49
  solving by the quadratic formula, 459–63
  in two variables, 486–90
Quadratic formula, 459, 460
Quotient, 9, 340
Quotient rule for exponents, 84, 85

Radical(s), 96, 393
  addition of, 410–12
  division of, 404
  like, 409
  multiplication of, 403, 418
  simplifying, 396–400, 416–19
  solving equations with, 425–27, 468, 469
  subtraction of, 410–12
  unlike, 410
Radicand, 96
Ratio, 95, 372
Rational expression, 332
Rationalizing the denominator, 399, 419
Rational numbers, 28, 53, 96
  addition of, 57–61
  absolute value of, 54
  division of, 68, 69
  multiplication of, 65–68
  negative, 53
  positive, 53
  subtraction of, 61, 62
Real numbers, 96
Reciprocal(s), 9, 66, 340
  of a fraction, 9

Rectangular coordinate system, see Cartesian coordinate system.
Remainder, 276
Removing parentheses, 78
Right triangle, 430
  hypotenuse, 430
  legs, 430
  Pythagorean theorem, 431
Root(s), 96, 393
  method of taking, 449, 450
  square, 96, 97, 393
Rule of squaring, 425

Scientific notation, 92–94
Second cross product, 53
Set, 1
Similar terms, 76, 77
Simplifying an expression, 77, 78
Simplifying radicals, 396–400, 416–19
Slope, 186–88
Slope-intercept form, 190
Solution(s), 108, 169, 197
  to a system of equations, 212
Solving equations, 108, 316, 447
  formulas, 481–84
  fractional, 360, 467
  quadratic, 316–18, 447–63
  by the quadratic formula, 459–63
  radical, 425–27, 468
Square, 35
  table of, 97, 511
Square root(s), 96, 97, 393
  negative, 96
  positive, 96
  principal, 96, 393
  table of, 97, 511
Substitution method
  solving systems of equations by, 223–26
Subtraction
  and the column method, 256
  of decimals, 21
  of fractions, 10, 14
  of like algebraic fractions, 347
  of polynomials, 255
  of radicals, 410–12
  of rational numbers, 61, 62
  of unlike algebraic fractions, 347
Sum of two cubes, 316
Supplementary angles, 236
Symbols
  approximately equal to, 138
  equal to, 52
  greater than, 51, 52
  greater than or equal to, 52
  of grouping, 36, 37
  less than, 51, 52
  less than or equal to, 52
  unequal to, 52
Systems of equations, 212
  solutions to, 212
  solving by addition-subtraction, 229–32
  solving by graphing, 217, 218
  solving by substitution, 223–26
  work problems using, 235–39

Taking roots, 449, 450
Term(s), 76, 248
  coefficient of, 76, 248
  collecting like, 77, 78
  degree of, 249
  factors of, 76
  lowest, 2, 3
  similar, 76, 77, 250
Test point, 198
Total value, 236
Trinomials, 248
  factoring, 296–314
  perfect square, 270, 271, 311, 312
    and factoring, 311, 312

Unequal to ($\neq$), 52
Units of measure, 512
  English system, 512
  metric system, 512
Unlike fractions, 347
Unlike radicals, 410
Unlike terms, 251

Variable(s), 34, 108, 248
Variation, 374
  constants of, 374
Varies directly, 374
Varies inversely, 380
Vertex of parabola, 490
Vertical axis, 168

Whole numbers, 1
Word problems
  business, 251, 322, 373, 375, 432
  geometry, 136–38, 322, 431, 462, 474
  language of, 123
  mixture, 236, 237
  motion, 138, 139, 238
  number and age, 127–29, 235, 236, 321, 363, 473
  percent, 32, 34, 129–31
  proportion, 373–75
  and quadratic equations, 321–23, 473–76

Word problems
  ratio, 372
  rules for solving, 127
  and square roots, 431–33
  using systems of equations, 235–39
  variation, 374
  work, 364–66, 475

x-axis, 168
x-coordinate, 168
x-intercept, 178

y-axis, 168
y-coordinate, 168
y-intercept, 178

Zero
  division by, 8
  as exponent, 86
Zero-product rule, 316
  and quadratic equations, 316–18, 447–49